Nano-Electronic Devices

Dragica Vasileska · Stephen M. Goodnick
Editors

Nano-Electronic Devices

Semiclassical and Quantum Transport Modeling

Springer

Editors
Dragica Vasileska
School of Electrical, Computer
and Energy Engineering
Arizona State University
Tempe, Arizona
USA
vasileska@asu.edu

Stephen M. Goodnick
School of Electrical, Computer
and Energy Engineering
Arizona State University
Tempe, Arizona
USA
stephen.goodnick@asu.edu

ISBN 978-1-4419-8839-3 e-ISBN 978-1-4419-8840-9
DOI 10.1007/978-1-4419-8840-9
Springer New York Dordrecht Heidelberg London

Library of Congress Control Number: 2011928232

Printed on acid-free paper

Springer is part of Springer Science+Business Media (www.springer.com)

Preface

Within this volume, we have attempted to present a comprehensive picture of the state of the art in transport modeling relevant for the simulation of nanoscale semiconductor devices. At the time of the publication of this book, advances in conventional planar semiconductor device scaling have resulted in production devices with gate lengths approaching 22 nanometers (at the time of writing this preface), while research devices with gate lengths of just a few nanometers have been demonstrated. The semiconductor industry has been dominated by Si based Metal Oxide Semiconductor (MOS) transistors for over 40 years. However, at present, there is an increasing drive to integrate a diversity of materials such as III–V compound channel materials and high insulator dielectrics, and the introduction of radically new materials such as graphene. At the same time, there have been extraordinary advances in new types of self-assembled materials such as carbon nanotubes, and semiconductor nanowires, which offer the potential for new families of fully three-dimensional devices that will allow scaling to continue to atomic dimensions. As characteristic length scales decrease, the physics of transport changes dramatically. For large dimensions compared to the mean free path for scattering (and the related phase coherence length), the semi-classical diffusive picture of charge transport holds, governed by the Boltzmann transport equation (BTE). On the other hand, for very short length scales, much less than the scattering mean free path, transport is coherent, and described in a purely quantum mechanical framework in terms of current associated with probability flux, usually from some idealized reservoir of carriers, i.e. contacts. The actual situation in current nanoscale devices is somewhere in between these two pictures, which in the past has been referred to as a *mesoscopic* system (somewhere between microscopic and macroscopic). This regime perhaps the most interesting in terms of phenomena, but the most difficult to theoretically describe, in which both quantum mechanical phase coherent phenomena co-exist with phase randomizing, dissipative scattering processes, which requires a general theoretical approach capable of dealing with both on an equal footing. In this book, we compile different approaches to the problem of transport in mesoscopic semiconductor systems, ranging from semi-classical to fully quantum mechanical, in order to understand the advantages and limitations of each, as well as elucidating the complex and interesting phenomena encountered in ultra-small devices.

In Chap. 1, we begin with an introduction to semi-classical device modeling, starting from the BTE, and deriving the associated moment equations leading to the widely used drift-diffusion and energy transport models, with different approaches for extraction of the transport parameters, and applications of this approach in some new novel energy conversion and sensing technologies. Chapter 2 considers the inclusion of quantum mechanical effects such as tunneling and quantum confinement within the popular ensemble Monte Carlo (EMC) method for the solution of the semi-classical BTE, as well as the treatment of many body interactions between particles as well as between particles and impurities within a molecular dynamics framework. Chapter 3 introduces the full-band EMC method, in which the complete electronic bandstructure is used in the description of the electron and hole dynamics as well as scattering processes semi-classically. A formalism based on the Pauli Master Equation is then introduced which allows for simulation of quantum transport within a similar framework to the BTE, and which is applied to some specific nanoscale structures where quantum effects are important such as resonant tunneling diodes (RTDs). Chapter 4 provides the general theoretical framework for quantum transport starting with the Liouiville-von Neumann equation, and then the various approximation schemes which lead to various forms of Master equations, including the Pauli and Boltzmann formalisms. Chapter 5 gives an overview of quantum transport based on the Wigner Function method, which utilizes a quantum mechanical distribution function in place of the semi-classical distribution function appearing in the BTE to obtain the Wigner–Boltzmann equation. Numerical approaches for the solution of the Wigner–Boltzmann equation are discussed, and the application to quantum devices such as RTDs and nanoscale transistors presented. Chapter 6 provides a description of quantum transport from a scattering matrix, wavefunction approach, based on the so-called Usuki method. Applications to transport through various prototype nanostructures such as quantum dots, nanowires and molecular systems are presented, including spin dependent phenomena which can be described within the same framework. The inclusion of scattering in real space within the Usuki method is then described, and its application to nanoscale MOSFETs presented. Chapter 7 details an atomistic approach to transport appropriate for nanoscale systems, based on the empirical tight binding method for large systems of atoms such as quantum dots and nanoscale transistors.

We deeply acknowledge the valuable contributions that each of the authors made in writing these excellent chapters that this book consists of.

Tempe Arizona, USA
2011

Dragica Vasileska
Stephen M. Goodnick

Contents

Contributors

Shaik Shahid Ahamed Department of Electrical and Computer Engineering, Southern Illinois University at Carbondale, 1230 Lincoln Drive, Carbondale, IL 62901, USA, ahmed@siu.edu

Richard Akis Department of Electrical, Computer, and Energy Engineering, Arizona State University, Tempe, AZ, USA, richard.akis@asu.edu

Aron Cummings Sandia National Laboratories, Livermore, CA, USA, aron.cummings@gmail.com

P. Dollfus Institute of Fundamental Electronics, CNRS, Univ. Paris-sud, Orsay, France, philippe.dollfus@u-psud.fr

David Ferry Department of Electrical, Computer, and Energy Engineering, Arizona State University, Tempe, AZ, USA, dkferry@asu.edu

Massimo V. Fischetti Department of Materials Science and Engineering, University of Texas at Dallas, 800 W. Campbell Rd., Richardson, TX 75080, USA, max.fischetti@utdallas.edu

Bo Fu Department of Materials Science and Engineering, University of Texas at Dallas, 800 W. Campbell Rd., Richardson, TX 75080, USA, bo.fu@utdallas.edu

Vamsi Gaddipathi Department of Electrical and Computer Engineering, Southern Illinois University at Carbondale, 1230 Lincoln Drive, Carbondale, IL 62901, USA

Matthew Gilbert Department of Electrical and Computer Engineering, University of Illinois, Urbana, IL, USA, matthewg@illinois.edu

Ramya Hindupur Department of Electrical and Computer Engineering, Southern Illinois University at Carbondale, 1230 Lincoln Drive, Carbondale, IL 62901, USA

Sharnali Islam Department of Electrical and Computer Engineering, Southern Illinois University at Carbondale, 1230 Lincoln Drive, Carbondale, IL 62901, USA

Dylan John Department of Electrical and Computer Engineering, Southern Illinois University at Carbondale, 1230 Lincoln Drive, Carbondale, IL 62901, USA

Gokula Kannan Department of ECEE, Arizona State University, Tempe, AZ, USA, gokul@asu.edu

Hasanur Rahman Khan Intel Corp., Hillsboro, OR, USA, hasanur.khan@intel.com

Jiseok Kim Department of Electrical and Computer Engineering, University of Massachusetts, 100 Natural Resources Rd., Amherst, MA 01003, USA, jikim@ecs.umass.edu

Irena Knezevic University of Wisconsin-Madison, 3442 Engineering Hall, 1415 Engineering Drive, Madison, WI 53706-1691, USA, knezevic@engr.wisc.edu

H. Kosina Institute of Microelectronics, TU Vienna, Vienna, Austria, kosina@iue.tuwien.ac.at

Shareef Mohammed Department of Electrical and Computer Engineering, Southern Illinois University at Carbondale, 1230 Lincoln Drive, Carbondale, IL 62901, USA

Abdussamad Muntahi Department of Electrical and Computer Engineering, Southern Illinois University at Carbondale, 1230 Lincoln Drive, Carbondale, IL 62901, USA

Sudarshan Narayanan Department of Materials Science and Engineering, University of Texas at Dallas, 800 W. Campbell Rd., Richardson, TX 75080, USA, sudarshan.narayanan@utdallas.edu

M. Nedjalkov Institute of Microelectronics, TU Vienna, Vienna, Austria, mixi@iue.tuwien.ac.at

Bozidar Novakovic University of Wisconsin-Madison, Madison, WI 53706, USA, novakovic@wisc.edu

Joshua Ogden Department of Electrical and Computer Engineering, Southern Illinois University at Carbondale, 1230 Lincoln Drive, Carbondale, IL 62901, USA

D. Querlioz Institute of Fundamental Electronics, CNRS, Univ. Paris-sud, Orsay, France, damien.querlioz@gmail.com

Christian Ringhofer Department of Mathematics, Arizona State University, Tempe, AZ, USA, ringhofer@asu.edu

Siegfried Selberherr Institute for Microelectronics, Gußhausstraße 27–29/E360, 1040 Vienna, Austria, Selberherr@iue.tuwien.ac.at

Gil Speyer High Performance Computing Initiative, Arizona State University, Tempe, AZ, USA, speyer@asu.edu

Sasi Sundaresan Department of Electrical and Computer Engineering, Southern Illinois University at Carbondale, 1230 Lincoln Drive, Carbondale, IL 62901, USA

Viktor Sverdlov Institute for Microelectronics, Gußhausstraße 27–29/E360, 1040 Vienna, Austria, Sverdlov@iue.tuwien.ac.at

Dragica Vasileska School of Electrical, Computer and Energy Engineering, Arizona State University, Tempe, AZ, USA, vasileska@asu.edu

Thomas Windbacher Institute for Microelectronics, Gußhausstraße 27–29/E360, 1040 Vienna, Austria, Windbacher@iue.tuwien.ac.at

Krishnakumari Yalavarthi Department of Electrical and Computer Engineering, Southern Illinois University at Carbondale, 1230 Lincoln Drive, Carbondale, IL 62901, USA

Chapter 1
Classical Device Modeling

Thomas Windbacher, Viktor Sverdlov, and Siegfried Selberherr

Abstract In this chapter an overview of classical device modeling will be given. The first section is dedicated to the derivation of the Drift–Diffusion Transport model guided by physical reasoning. How to incorporate Fourier's law to add a dependence on temperature gradients into the description, is presented. Quantum mechanical effects relevant for small devices are approximately covered by quantum correction models. After a discussion of the Boltzmann Transport equation and the systematic derivation of the Drift–Diffusion Transport model, the Hydrodynamic Transport model, the Energy Transport model, and the Six-Moments Transport model via a moments based method out of the Boltzmann Transport Equation, which is the essential topic of classical transport modeling, are highlighted. The parameters required for the different transport models are addressed by an own section in conjunction with a comparison between the Six-Moments Transport model and the more rigorous Spherical Harmonics Expansion model, benchmarking the accuracy of the moments based approach. Some applications of classical transport models are presented, namely, analyses of solar cells, biologically sensitive field-effect transistors, and thermovoltaic elements. Each example is addressed with an introduction to the application and a description of its peculiarities.

Keywords Classical device modeling · Drift–Diffusion · Six moments · Hydrodynamic transport · Energy transport · Solar cells · BioFET · Biologically sensitive field-effect transistor · Boltzmann transport · Thermoelectric · Figure of merit · Electrothermal transport · Spherical harmonics expansion

1 Heuristic Derivation of the Drift–Diffusion Transport Model

Even though the method of moments, which will be presented in Sect. 5, is quite sophisticated and offers the possibility to extend a transport model to an arbitrary large and accurate set of equations, physically understanding of the model is not

T. Windbacher (✉)
Institute for Microelectronics, Gußhausstraße 27–29/E360, 1040 Vienna, Austria
e-mail: Windbacher@iue.tuwien.ac.at

D. Vasileska and S.M. Goodnick (eds.), *Nano-Electronic Devices: Semiclassical and Quantum Transport Modeling*, DOI 10.1007/978-1-4419-8840-9_1,

as instructive as a derivation via a heuristic approach. Therefore, in this section a derivation of the Drift–Diffusion Transport model with the aid of physical reasoning will be given.

One of the most general ways to treat electromagnetic phenomena is via the Maxwell equations. So we will start with a few simplifying assumptions and reduce the required equation set to the absolute minimum necessary to describe micro-electronic devices. Then we will introduce a few additional equations covering the physical behavior of semiconducting materials.

1.1 Poisson Equation

The first simplifying assumption is the quasi-static approximation. This assumption restricts one to devices exhibiting a characteristic length which is noticeably smaller than the shortest electromagnetic wavelength existent in the considered system. For instance, assuming an upper limit of 100 GHz for the frequency of the electromagnetic field yields a wavelength of $\lambda = c/f = 877\,\mu\text{m}$. Thus characteristic device dimensions in the micrometer regime and below are quite reasonable. Due to the quasi-static approximation the displacement current $\partial_t \mathbf{D}$ and the induction $\partial_t \mathbf{B}$ can be neglected. This leads to a decoupling of the former coupled system of partial differential equations for the electric field and the magnetic field. The only remaining connection between the electric field $\mathbf{E}$ and the magnetic field $\mathbf{H}$ is given by the relation between the electric field $\mathbf{E}$ and the current density $\mathbf{j}$ which raises a magnetic field $\mathbf{H}$. In order to further simplify the equation system the magnetic part is completely neglected. Due to the now vanishing right hand side of $\operatorname{curl}\mathbf{E} = -\partial_t \mathbf{B}$ it is possible to define a scalar potential $\mathbf{E} = -\nabla\varphi$. The relation between the electric displacement field and the electric field is assumed to be linear and anisotropic for an inhomogeneous material $\mathbf{D} = \varepsilon\mathbf{E}$ dependent on the spatial coordinates. Embracing all assumptions with Gauß's law yields:

$$\nabla\cdot(\varepsilon\nabla\varphi) = -\rho. \tag{1.1}$$

The space charge density ρ has to reflect the charge contributions in the semiconductor. This is accomplished by three components: the electron concentration n, the hole concentration p and the concentration of fixed ionized charges C:

$$\rho = q\,(p - n + C). \tag{1.2}$$

Assembling all derived terms and further restricting to a scalar and spatial independent permittivity we obtain the well known Poisson equation:

$$\varepsilon\,\Delta\,\varphi = q\,(n - p - C). \tag{1.3}$$

1.2 Continuity Equation

The second ingredient for the Drift–Diffusion Transport model is derived from the continuity equation which takes care of mass conservation:

$$q\frac{\partial \rho}{\partial t} + \nabla \cdot \mathbf{j} = 0. \tag{1.4}$$

Like before we decompose the contributions of the current $\mathbf{j} = \mathbf{j}_n + \mathbf{j}_p$ and the space charge density $\partial\rho/\partial t = q\partial/\partial t\,(p-n)$ (assuming all immobile charges as fixed $\partial C/\partial t = 0$) into an electron and a hole related part:

$$\nabla \cdot (\mathbf{j}_n + \mathbf{j}_p) + q\frac{\partial}{\partial t}(p-n) = 0. \tag{1.5}$$

This steps enables to separate the electron and hole related contributions into two independent equations:

$$\nabla \cdot \mathbf{j}_n - q\frac{\partial}{\partial t}n = \ qR, \tag{1.6}$$

$$\nabla \cdot \mathbf{j}_p + q\frac{\partial}{\partial t}p = -qR. \tag{1.7}$$

The new term on the right hand side of (1.6) and (1.7) denotes the so-called net generation-recombination rate R. Since electrons and holes can not just vanish or appear, every additional electron generates an additional hole and vice versa. Due to their opposing charges the quantity R enters with opposite signs into the equations for electrons and holes. The net generation-recombination rate is usually modeled by the net generation rate of electron–hole pairs minus the net recombination rate of electron–hole pairs. In equilibrium R is equal zero but also out of equilibrium R is often neglected.

1.3 Charge Transport: Drift–Diffusion Assumption

Summarizing our equations, we have the Poisson equation and two continuity equations involve five unknown quantities ($\varphi, n, p, \mathbf{j}_n$ and $\mathbf{j}_p$). Therefore, we need two more conditions to make the equation system complete. These material equations can be deduced by examination of the forces acting upon the charged carriers (n, p) on a microscopic level. The simplest model at hand is based on the so-called Drift–Diffusion assumption. The model distincts between two charge carrier transport mechanisms: the *drift* of charge carriers due to an external electric field caused by a gradient in the electric potential and the *diffusion* of the charge carriers due to a spatial gradient in the charge carrier concentration.

The drift contribution is caused by the force of an externally applied electric field **E** on the charge carriers. Since the movement of charge carriers due to the electric field **E** constitutes an electric current, the drift current density is related to the applied electric field by the charge carrier concentration times mobility times electric field strength:

$$\mathbf{j}_n^{\mathrm{Drift}} = q n \mu_n \mathbf{E} \text{ and} \tag{1.8}$$

$$\mathbf{j}_p^{\mathrm{Drift}} = q p \mu_p \mathbf{E}. \tag{1.9}$$

The carrier mobility $\mu_{n,p}$ is a material dependent parameter and relates the electric field **E** to the drift current density $\mathbf{j}_{n,p}^{\mathrm{Drift}}$. Equations (1.8) and (1.9) are related to Ohm's law by the conductivities $\sigma_n = q n \mu_n$ for electrons and $\sigma_p = q p \mu_p$ for holes:

$$\mathbf{j}_n^{\mathrm{Drift}} = \sigma_n \mathbf{E} \text{ and } \mathbf{j}_p^{\mathrm{Drift}} = \sigma_p \mathbf{E}. \tag{1.10}$$

The second transport phenomenon is given by the particle flux density **F** and due to the gradient of the particle concentration. The proportionality factor is called diffusion coefficient $D_{n,p}$ and, further distinguishing between electron and hole diffusion, one obtains:

$$\mathbf{F}_n = -D_n \nabla n, \ \mathbf{F}_p = -D_p \nabla p. \tag{1.11}$$

The diffusion related current densities are defined by their flux density multiplied with the individual charge of the charge carrier:

$$\mathbf{j}_n^{\mathrm{Diffusion}} = -q\mathbf{F}_n = q D_n \nabla n, \ \mathbf{j}_p^{\mathrm{Diffusion}} = q\mathbf{F}_p = -q D_p \nabla p. \tag{1.12}$$

Close to the equilibrium the diffusion coefficient can be related to the carrier mobility via the Einstein relation:

$$D_{n,p} = \frac{k_{\mathrm{B}} T}{q} \mu_{n,p} = V_{\mathrm{T}} \mu_{n,p}. \tag{1.13}$$

k_{B} denotes the Boltzmann constant and T the temperature in K. The quantity V_{T} denotes the thermal voltage and is around $\approx$26 mV at room temperature. The Einstein relation is only approximately valid for the non-equilibrium case and often used as a good starting guess for a numerical iterative solving algorithm.

Once more assembling all derived expressions yields a set of equations which is identical to the Drift–Diffusion Transport model derived by the method of moments:

$$\varepsilon \Delta \varphi = q(n - p - C), \tag{1.14}$$

$$qR = \nabla \cdot \mathbf{j}_n - q\frac{\partial n}{\partial t}, \tag{1.15}$$

$$-qR = \nabla \cdot \mathbf{j}_p + q\frac{\partial p}{\partial t}, \tag{1.16}$$

$$\mathbf{j}_n = -q\mu_n \left(n\nabla\varphi - V_{\mathrm{T}}\nabla n\right), \tag{1.17}$$

$$\mathbf{j}_p = -q\mu_p \left(p\nabla\varphi + V_{\mathrm{T}}\nabla p\right). \tag{1.18}$$

Even though the set of equations is now complete, it can not be solved without further description of the material parameters for the mobilities $\mu_{n,p}$ and the generation-recombination rate R. This will be taken care of in Sect. 7.

1.4 Quasi-Fermi Levels

The thermal equilibrium does not demand a position independent potential. For instance:

$$\mathscr{E}_c = \mathscr{E}_{c,0}(\mathbf{r}) - q\varphi(\mathbf{r}), \tag{1.19}$$

$$\mathscr{E}_v = \mathscr{E}_{v,0}(\mathbf{r}) - q\varphi(\mathbf{r}), \tag{1.20}$$

$$\mathscr{E}_i = \mathscr{E}_{i,0}(\mathbf{r}) - q\varphi(\mathbf{r}), \tag{1.21}$$

denoting the conduction band edge $\mathscr{E}_c$, the valence band edge $\mathscr{E}_v$ and the intrinsic Fermi level $\mathscr{E}_i$, respectively.

Treating the situation away from thermal equilibrium complicates the matter. Taking (1.17) and reformulating it:

$$\begin{aligned}
\mathbf{j}_n &= q\mu_n V_{\mathrm{T}}\nabla n - q\mu_n n\nabla\varphi \\
&= q\mu_n n\left(V_{\mathrm{T}}\frac{1}{n}\nabla n - \nabla\varphi\right) \\
&= q\mu_n n\left(V_{\mathrm{T}}\frac{n_i}{n}\nabla\frac{n}{n_i} - \nabla\varphi\right) \\
&= q\mu_n n\left(V_{\mathrm{T}}\nabla\ln\left(\frac{n}{n_i}\right) - \nabla\varphi\right) \\
&= q\mu_n n\nabla\underbrace{\left(V_{\mathrm{T}}\ln\left(\frac{n}{n_i}\right) - \varphi\right)}_{=-\phi_n},
\end{aligned}$$

with n_i as intrinsic concentration, shows that the drift and the diffusive contribution can be merged into one quantity. This quantity can be related to the quasi-Fermi level as follows [184]:

$$-q\phi_n = \mathscr{E}_{Fn} - \mathscr{E}_{i,0}. \tag{1.22}$$

Therefore, in the most general case, the current depends on the gradient of the quasi-Fermi levels and not solely on the gradient of the potential[1]:

$$\mathbf{j}_n = n\mu_n \nabla \mathscr{E}_{Fn}, \tag{1.23}$$

$$\mathbf{j}_p = p\mu_p \nabla \mathscr{E}_{Fp}. \tag{1.24}$$

The quasi-Fermi levels $\mathscr{E}_{Fn}$ and $\mathscr{E}_{Fp}$ introduced in (1.22)–(1.24) can be gained from (1.17) and (1.22) for electrons and in an analog way from (1.18) for holes, under the assumption that the solution of the equation system (1.14)–(1.18) is available:

$$\mathscr{E}_{Fn} = \mathscr{E}_{i,0} - q\varphi + qV_{\mathrm{T}} \ln\left(\frac{n}{n_i}\right), \tag{1.25}$$

$$\mathscr{E}_{Fp} = \mathscr{E}_{i,0} - q\varphi - qV_{\mathrm{T}} \ln\left(\frac{p}{n_i}\right). \tag{1.26}$$

2 Heuristic Inclusion of Heat Transport in the Drift–Diffusion Transport Model

The Drift–Diffusion Transport model assumes equality between the lattice temperature T_{L} and the charge carriers' temperature T_{n}. Furthermore, it states negligible temperature gradients in the device. However, there is an intrinsic temperature dependence in basically all microscopic phenomena in solids, which is mirrored in the basic semiconductor equations directly by the thermal voltage V_{T} and indirectly via the temperature dependence of the mobilities μ_n and μ_p and the recombination rate R. Generalizing the Drift–Diffusion Transport model by introducing a local temperature, in order to cover a more detailed view of temperature dependent phenomena, one has to employ an extra equation. Heat energy is also a conserved quantity, where the heat flux is governed by an expression similar to the continuity equation for charge:

$$\rho\, c\, \frac{\partial T_{\mathrm{L}}}{\partial t} - \nabla \cdot (\kappa \nabla T_{\mathrm{L}}) = H. \tag{1.27}$$

ρ denotes the mass density of the material and c describes the specific heat of the material, while κ expresses the thermal conductivity. Due to the phonon dominated heat transport in semiconductors the lattice temperature T_{L} is the quantity of interest. The first term on the left hand side of (1.27) characterizes the initial transient time dependent behavior of changes due to the heat sources H, while the second term takes care of the stationary temperature distribution. The heat generation term H

[1] The intrinsic energy $\mathscr{E}_{i,0}$ is globally constant.

establishes the link between the heat-flow and the current and can be approximated by a first-order Joule-term $\mathbf{j} \cdot \mathbf{E}$ and an expression for the carrier recombination. Every generation or recombination of an electron–hole pair withdraws or releases an energy amount of at least the band gap energy $\mathscr{E}_g$ from the crystal lattice. Therefore, the heat source term can be formulated as [3]:

$$H = \nabla \cdot \left(\frac{\mathscr{E}_c}{q} \mathbf{j}_n + \frac{\mathscr{E}_v}{q} \mathbf{j}_p \right), \tag{1.28}$$

with $\mathscr{E}_c$ and $\mathscr{E}_v$ denoting the conduction and valence band edge energy, respectively. Considering non-degenerate materials only [184], one can further simplify (1.28) to:

$$H = (\mathbf{j}_n + \mathbf{j}_p) \cdot \mathbf{E} + R\mathscr{E}_g. \tag{1.29}$$

Accompanying with spatial gradients in the local temperature a new driving force occurs. This additional driving force causes an extra current flow, which has to be incorporated by supplementary terms in the current density relations in (1.17) and (1.18):

$$\mathbf{j}_{n,\mathrm{th}} = qD_{n,\mathrm{th}} \nabla T_\mathrm{L} \ \text{ and } \ \mathbf{j}_{p,\mathrm{th}} = -qD_{p,\mathrm{th}} \nabla T_\mathrm{L}, \tag{1.30}$$

with thermal diffusion coefficients $D_{n,\mathrm{th}}$ and $D_{p,\mathrm{th}}$ approximately related to the diffusion coefficients D_n and D_p by [209]:

$$D_{n,p.\mathrm{th}} \backsimeq \frac{D_{n,p}}{2T}. \tag{1.31}$$

These current density contributions are essential for the description of thermoelectric effects, like the Seebeck effect or the Peltier effect.

During the derivation of the model above it was demonstrated that one can deduce a higher order transport model via physically sound reasoning and not only by the mathematically sophisticated method of moments. Van Roosbroeck [173] was the first to present a model pretty close to the description given here already in 1950.

One has to note that for higher order transport models the description of the heat source term H becomes much more challenging (see Sect. 2.4 in [124]).

3 Incorporating Quantum Mechanical Effects via Quantum Correction Models

The density of states (DOS) of a system is given by the number of states at each energy level, which are available for occupation (q.v. [14, 122]). Since quantum mechanical effects affects the DOS by causing a two-dimensional electron gas, the carrier concentration near the gate oxide decreases. This influences several device characteristics like the current–voltage or the capacitance–voltage characteristics and therefore has to be taken into account either by a rigorous self-consistent solution of the Schrödinger equation and the Poisson equation, which is computationally expensive, or via a supplemental quantum correction model in classical device simulations. Various quantum correction models stemming from different approaches have been proposed [47, 82, 112, 117, 148, 156, 225], some of them are based on empirical fits via many parameters [112, 148], some models exhibit a degraded convergence depending on the electric field [47] or demand a recalibration for each particular device [82].

The modified local density approximation (MLDA) by Paasch [156] proposes a local correction of the effective DOS N_c near the gate oxide defined by:

$$N_c = N_{c,0}\left(1-\exp\left[-\frac{(z+z_0)^2}{\chi^2\lambda_{\text{thermal}}^2}\right]\right) \quad \text{with} \quad \lambda_{\text{thermal}} = \frac{\hbar}{\sqrt{2mk_\text{B}T}}. \tag{1.32}$$

$N_{c,0}$ denotes the classical effective DOS modified by the fitting parameter χ. z describes the distance from the interface, z_0 is the tunneling distance, and λ_{thermal} constitutes the thermal wavelength. Equation (1.32) can be gained from the quantum mechanical expression governing the particle density [82]. The benefit of the MLDA procedure lies in the fact that no solution variable is needed in the correction term. Hence, this model can be employed as a preprocessing step with only minimal significance for the overall CPU time required for the solution of the entire set of the transport equations [225]. On the other hand, the drawback of the MLDA is its founding on the field-free Schrödinger equation and in conjunction the loss of validity for high fields.

An improved MLDA (IMLDA) technique has been suggested by [112, 148], introducing a heuristic wavelength parameter:

$$\lambda'_{\text{thermal}}(z, N_{\text{eff}}, T) = \chi(z, N_{\text{eff}}, T)\,\lambda_{\text{thermal}}(T), \tag{1.33}$$

where N_{eff} denotes the net doping with $\chi(z, N_{\text{eff}}, T)$ as a fit factor. Due to this adaption, the IMLDA is able to cover the important case of high-fields perpendicular to the interface [112]. The fit parameters have been extracted from results gained by a self-consistent Schrödinger Poisson solver and are calibrated for bulk MOSFET structures. However, the MLDA method is only valid for devices with one gate oxide and thus a description of double-gate SOI MOSFETs (DG SOI MOSFETs) is not possible.

A quantum correction technique capable of treating DG SOI MOSFETs is shown in [117]. The basic concept of this approach is that due to the strong quantization perpendicular to the interface, the potential in the SOI is well approximated by an infinite square well potential. The eigenstates in the quantization region can be calculated with an analytic approach and related to a quantum correction potential which adjusts the band edge in such a way that the quantum mechanical carrier concentration is reproduced.

Van Dorts approach [47] improves the modeling of the conduction band edge:

$$\mathcal{E}_c = \mathcal{E}_{\text{class}} + \frac{13}{9}\text{F}(z)\,\Delta\mathcal{E}_\text{g} \quad \text{with} \quad \Delta\mathcal{E}_\text{g} \approx \beta\left(\frac{\kappa_{\text{Si}}}{4qk_\text{B}T}\right)^{1/3}|\mathbf{E}_\perp|^{2/3}. \tag{1.34}$$

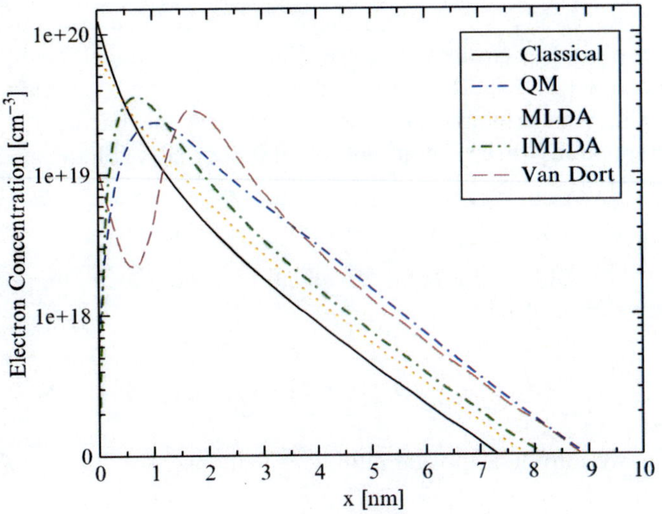

Fig. 1.1 Electron concentration of a single-gate SOI MOSFET for different modeling approaches. Illustrating the classically, quantum-mechanically, in conjunction with the quantum correction models MLDA, IMLDA, and Van Dort calculated electron concentration as a function of the distance to the interface [220]

$\mathscr{E}_{\text{class}}$ denotes the classical band energy and the correction function F depends on the distance z to the interface, while $\mathbf{E}_{\perp}$ stands for the electric field perpendicular to the interface. The proportionality factor β is gained from the shift of the long-channel threshold voltage as explained in [47].

Figure 1.1 compares the different quantum correction models against the classical model and the quantum mechanical model [116] for a single-gate SOI MOSFET. It shows the electron concentration as a function of the distance to the interface for the classical, the exact quantum mechanical, the quantum correction model MLDA, the IMLDA [112], and the model after Van Dort [47] for a gate voltage of 1 V. As can be seen the IMLDA model reproduces quite well the quantum mechanical concentration and hence is sufficient to cover quantum mechanical effects in classical device simulations [220].

4 Boltzmann Transport Equation

There are two fundamental equations for semi-classical device simulation, the Poisson equation and the Boltzmann equation. While the Poisson equation takes care of the electrostatical description of the system, the Boltzmann equation describes the propagation of the distribution function in the device. The distribution function $f(\mathbf{r},\mathbf{k},t)$ is a function describing the number of particles contained in a unit volume in phase space and depends on three values for the position $\mathbf{r} = x\mathbf{x} + y\mathbf{y} + z\mathbf{z}$, three values of the wave vector $\mathbf{k} = k_x\mathbf{k}_x + k_y\mathbf{k}_y + k_z\mathbf{k}_z$, and time t.

These two equations in conjunction have to be solved in a self-consistent manner and can be exploited as a reference for any higher-order models (see Sect. 5).

The Boltzmann Transport equation is gained from the Liouville theorem [110, 140], a fundamental principle of classical statistical mechanics. It states that the distribution function $f(\mathbf{r},\mathbf{k})$ is constant for all times t along phase-space trajectories Γ_i ((1.35), [151]):

$$f(\mathbf{r}+\mathrm{d}\mathbf{r},\mathbf{k}+\mathrm{d}\mathbf{k},t+\mathrm{d}t) = f(\mathbf{r},\mathbf{k},t), \tag{1.35}$$

which leads to the Boltzmann Transport equation without scattering, after taking the total derivative of (1.35):

$$\partial_t f(\mathbf{r},\mathbf{k},t) + \frac{\mathrm{d}\mathbf{r}}{\mathrm{d}t}\cdot\partial_{\mathbf{r}} f(\mathbf{r},\mathbf{k},t) + \frac{\mathrm{d}\mathbf{k}}{\mathrm{d}t}\cdot\partial_{\mathbf{k}} f(\mathbf{r},\mathbf{k},t) = 0. \tag{1.36}$$

Furthermore, we introduce the Hamiltonian equations:

$$\frac{\mathrm{d}\mathbf{r}}{\mathrm{d}t} = \nabla_{\mathbf{p}}\mathscr{H} \quad \text{and} \quad \frac{\mathrm{d}\mathbf{p}}{\mathrm{d}t} = -\nabla_{\mathbf{r}}\mathscr{H}, \tag{1.37}$$

with $\mathbf{p} = \hbar\mathbf{k}$ denoting the momentum, and $\mathbf{r}$ the position of a particle in phase-space, while $\mathscr{H}$ describes the Hamiltonian of the system, which will be incorporated later.

Inaugurating the scattering operator Q_{coll}, the balance equation for the distribution function must obey the conservation equation:

$$\frac{\mathrm{d}f(\mathbf{r},\mathbf{k},t)}{\mathrm{d}t} = Q_{\mathrm{coll}}\left(f(\mathbf{r},\mathbf{k},t)\right). \tag{1.38}$$

Hence, the scattering operator opens up the possibility for particles to jump from one phase-space trajectory to another. Joining the full derivative of the distribution function and (1.37), the commonly used expression for the Boltzmann Transport equation can be written as:

$$\partial_t f + \nabla_{\mathbf{p}}\mathscr{H}\nabla_{\mathbf{r}} f - \nabla_{\mathbf{r}}\mathscr{H}\nabla_{\mathbf{p}} f = Q_{\mathrm{coll}}\left(f\right). \tag{1.39}$$

Neglecting inter-band processes and by this the generation and recombination of free carriers in the semiconductor, the collision operator $Q_{\mathrm{coll}}\left(f\right)$ can be written as [138]:

$$Q_{\mathrm{eff}}(f) = \sum_{\mathbf{p}'} f(\mathbf{p}')\,(1-f(\mathbf{p}))\,S(\mathbf{p}',\mathbf{p}) - \sum_{\mathbf{p}'} f(\mathbf{p})\,\left(1-f(\mathbf{p}')\right)\,S(\mathbf{p},\mathbf{p}'). \tag{1.40}$$

The collision term accounts for in-scattering from $\mathbf{p}'$ to $\mathbf{p}$ as well as out-scattering from $\mathbf{p}$ to $\mathbf{p}'$. $f(\mathbf{p}')$ represents the probability for the state $\mathbf{p}'$ to be occupied and $1-f(\mathbf{p})$ the probability for the state $\mathbf{p}$ to be accessible for in-scattering. $S(\mathbf{p}',\mathbf{p})$ describes the transition rate from $\mathbf{p}'$ to $\mathbf{p}$. The sum governs all states accessible for scattering from and to $\mathbf{p}$. From a physical point of view, the collision term covers the interaction of the carriers with the lattice (e.g. phonon scattering), the influence of ionized impurities, as well as additional scattering due to inhomogeneities in the grid in material alloys: it can be modeled as outlined in [106, 190].

Equation (1.40) represents a seven-dimensional integro-differential semi-classical equation. While the left hand side of the equation represents Newton mechanics, the right side denotes a quantum mechanical scattering operator. In order to develop solution strategies for this equation one has to understand the incorporated assumptions and limitations:

- The initial Liouville formulation stated a many particle problem. Introducing the Hartree–Fock approximation [137] allows to reduce the problem to a particle system with a proper potential. The contribution of the surrounding electrons is approximated by a charge density. Therefore, the short-range electron–electron interaction is excluded. Nevertheless, the potential of the surrounding carriers is treated self-consistently.
- The use of a distribution function $f(\mathbf{r}, \mathbf{k}, t)$ is a classical concept. Therefore, the Heisenberg uncertainty principle is not considered, and position and momentum are always treated at the same time.
- Because of Heisenberg's principle, the Boltzmann Transport equation is only valid, if the mean free path of particles is longer than the De Broglie wavelength.
- Particles abide Newton's law, due to the semi-classical treatment of particles.
- It is assumed that collisions between particles are binary and instantaneous in time and local in space. This approximation holds true for long free flight times compared to the collision times

During the derivation of the transport models from the Boltzmann Transport equation it is important to take these limitations and implications into account. However, models based on the Boltzmann Transport equation give good results in the scattering dominated regime [19, 97, 105, 159].

5 Derivation of Transport Models from the Boltzmann Transport Equation via a Moments Based Method

Solving the Boltzmann Transport equation yields excellent results [19, 97, 105, 159], but is much more demanding than other transport models (e.g. Drift–Diffusion Transport model or Energy Transport model) due to its high dimensionality. For instance, assuming a discrete mesh with 100 ticks in each spatial coordinate and time, will result in 10^{14} points. If we assert further 7×8 bytes (8 bytes for each coordinate), the memory consumption will be already 5.600 Terrabytes for just storing the points. Therefore, one is interested in numerically cheaper, but at the same time valid transport models, for the regime of interest.

From an engineering viewpoint, the method of moments is a very efficient way to derive transport models with a reduced complexity compared to the Boltzmann

Transport equation. By multiplying the Boltzmann Transport equation with a set of weight functions and integrating over **k**-space one can deduce a set of balance and flux equations coupled with the Poisson equation.

Via this formalism an arbitrary number of equations can be generated. Each equation contains information from the next higher moment, thus exhibiting more moments than equations. Therefore, one has to truncate the equation system at a certain point and complete the system by an additional condition. This condition, relating the highest moment with the lower moments, is called closure relation. The closure relation appraises the information of the higher moments and in conjunction with it determines the error introduced in the system. For example, the Drift–Diffusion Transport model can be gained by assuming thermal equilibrium between the charge carriers and the lattice ($T_n = T_{\mathrm{L}}$) [138]. There are various theoretical approaches to tackle the closure problem [133] for an arbitrary moment (e.g. maximum entropy principle [11, 12, 146]).

The basic concept of the maximum entropy principle is that a large set of collisions is needed to relax the carrier energies to their equilibrium, while at the same time momentum, heat flow, and anisotropic stress relax within shorter time. Hence, the charge carriers are in an intermediate state. This state can be noted as partial thermal equilibrium. Only the carrier temperature T_{n} is non-zero, while all other parameters vanish. Furthermore it is assumed that the entropy density and the entropy flux are independent on the relative electron gas velocity. The Hydrodynamic Transport model is obtained by assuming a heated Maxwellian distribution for closure, while the introduction of the kurtosis leads to the Six-Moments Transport model. A more detailed explanation will be given later.

In order to obtain physically reasonable equations, it is beneficial to choose weight functions as the power of increasing orders of momentum. The moments in one, two, and three dimensions can be defined as:

$$x_{\mathrm{j,d}}(\mathbf{r}_{\mathrm{d}}) = \frac{2}{(2\pi)^{\mathrm{d}}} \int_{-\infty}^{\infty} \mathbf{X}_{\mathrm{j,d}}(\mathbf{r}_{\mathrm{d}}, \mathbf{k}_{\mathrm{d}}) f_{\mathrm{d}}(\mathbf{r}_{\mathrm{d}}, \mathbf{k}_{\mathrm{d}}, t) \mathrm{d}^{\mathrm{d}}k = n\langle \mathbf{X}_{\mathrm{j,d}}(\mathbf{k}_{\mathrm{d}})\rangle = \langle\langle \mathbf{X}_{\mathrm{j,d}}(\mathbf{k}_{\mathrm{d}})\rangle\rangle . \tag{1.41}$$

$x_{\mathrm{j}}(\mathbf{r})$ are the macroscopic values with their microscopic counterpart $X_{\mathrm{j}}(\mathbf{k})$, and $f_{\mathrm{d}}(\mathbf{r}_d, \mathbf{k}_d, t)$ denotes the time dependent distribution function spanning over the six-dimensional phase space. The letter $\mathrm{d} = 1, 2, 3$ symbolizes the one-, two-, and three-dimensional system, respectively, while n describes the carrier concentration. The notations $\langle\ \rangle$ and $\langle\langle\ \rangle\rangle$ denote the normalized statistic average and the statistic average, respectively.

During the derivation of the macroscopic transport models, the dimension indices are skipped to ease readability. Multiplying the Boltzmann transport equation with the even scalar-valued weights $X = X(\mathbf{r}, \mathbf{k})$ and integrating over **k**-space:

$$\int X\,\partial_t f\,\mathrm{d}^3\mathbf{k} + \int X\,\mathbf{v}\nabla_{\mathbf{r}} f\,\mathrm{d}^3\mathbf{k} + \int X\,\mathbf{F}\nabla_{\mathbf{p}} f\,\mathrm{d}^3\mathbf{k} = \langle\langle \partial_t X\rangle\rangle_{\mathrm{coll}}\,, \tag{1.42}$$

results in the general conservation laws. Furthermore, in the following derivation, the distribution function $f(\mathbf{r},\mathbf{k},t)$, the group velocity $\mathbf{v}(\mathbf{r},\mathbf{k})$, and the generalized force $\mathbf{F}(\mathbf{r},\mathbf{k})$ are written as f, $\mathbf{v}$, and $\mathbf{F}$, respectively. The first term on the left side of (1.42) can be simplified to:

$$\int X\,\partial_t f\,\mathrm{d}^3\mathbf{k} = \partial_t \int X\,f\,\mathrm{d}^3\mathbf{k} = \partial_t \langle\!\langle X \rangle\!\rangle, \tag{1.43}$$

while the second term can be reformulated to:

$$\int X\,\mathbf{v}\nabla_{\mathbf{r}} f\,\mathrm{d}^3\mathbf{k} = \int \nabla_{\mathbf{r}}(X\,\mathbf{v}\,f)\,\mathrm{d}^3\mathbf{k} - \int X\,f\,\nabla_{\mathbf{r}}\mathbf{v}\,\mathrm{d}^3\mathbf{k} - \int \mathbf{v}\,f\,\nabla_{\mathbf{r}} X\,\mathrm{d}^3\mathbf{k}, \tag{1.44}$$

and the third term can be written as:

$$\int X\,\mathbf{F}\,\nabla_{\mathbf{p}} f\,\mathrm{d}^3\mathbf{k} = \int \nabla_{\mathbf{p}}(X\,\mathbf{F}\,f)\,\mathrm{d}^3\mathbf{k} - \int X\nabla_{\mathbf{p}}\mathbf{F}\,f\,\mathrm{d}^3\mathbf{k} - \int \mathbf{F}\,\nabla_{\mathbf{p}} X\,f\,\mathrm{d}^3\mathbf{k}, \tag{1.45}$$

Exploiting Gauß's law in conjunction with the assumption that all surface integrals over the first Brioullin-zone vanish [147], the first term on the right side of (1.45) becomes zero. Substituting $\mathbf{F} = -\nabla_{\mathbf{r}}\mathscr{H}$ and $\mathbf{v} = \nabla_{\mathbf{p}}\mathscr{H}$ in combination with the Hamiltonian function $\mathscr{H}$ defined as:

$$\mathscr{H} = \pm\mathscr{E}_{\mathrm{c,v}}(\mathbf{r}) + \mathrm{s}_\alpha q\,\varphi + \mathscr{E}(\mathbf{r},\mathbf{k}) = \mathscr{E}(\mathbf{r},\mathbf{k}) + \mathrm{s}_\alpha q\,\tilde{\varphi}, \tag{1.46}$$

with $\mathrm{s}_\alpha = \mp 1$ for electrons and holes, respectively, into (1.44) and (1.45) results into the Boltzmann Transport equation expressed via its averages of the even scalar-valued moment:

$$\partial_t \langle\!\langle X \rangle\!\rangle + \nabla_{\mathbf{r}} \langle\!\langle \mathbf{v}X \rangle\!\rangle - \langle\!\langle \mathbf{v}\,\nabla_{\mathbf{r}} X \rangle\!\rangle - \langle\!\langle \mathbf{F}\,\nabla_{\mathbf{p}} X \rangle\!\rangle = \langle\!\langle \partial_t X \rangle\!\rangle_{\mathrm{coll}}, \tag{1.47}$$

or, after some additional calculation steps:

$$\partial_t \langle\!\langle X \rangle\!\rangle + \nabla_{\mathbf{r}} \langle\!\langle \mathbf{v}X \rangle\!\rangle - \langle\!\langle \mathbf{v}\,\nabla_{\mathbf{r}} X \rangle\!\rangle + \langle\!\langle \nabla_{\mathbf{r}}\mathscr{E}\nabla_{\mathbf{p}} X \rangle\!\rangle + \mathrm{s}_\alpha q\,\langle\!\langle \nabla_{\mathbf{p}} X \rangle\!\rangle\,\nabla_{\mathbf{r}}\tilde{\varphi} = \langle\!\langle \partial_t X \rangle\!\rangle_{\mathrm{coll}}. \tag{1.48}$$

Analog to the derivation for the even scalar-valued moments, the odd vector-valued moment's equations can be deduced:

$$\begin{aligned}\partial_t \langle\!\langle \mathbf{X} \rangle\!\rangle + \nabla_{\mathbf{r}} \langle\!\langle \mathbf{v}\otimes\mathbf{X} \rangle\!\rangle - \langle\!\langle \mathbf{v}\,\nabla_{\mathbf{r}}\otimes\mathbf{X} \rangle\!\rangle + \langle\!\langle \nabla_{\mathbf{r}}\mathscr{E}\nabla_{\mathbf{p}}\otimes\mathbf{X} \rangle\!\rangle + \mathrm{s}_\alpha q\,\langle\!\langle \nabla_{\mathbf{p}}\otimes\mathbf{X} \rangle\!\rangle\,\nabla_{\mathbf{r}}\tilde{\varphi} \\ = \langle\!\langle \partial_t \mathbf{X} \rangle\!\rangle_{\mathrm{coll}}.\end{aligned} \tag{1.49}$$

From (1.48) and (1.49) the conservation equations and fluxes of the different macroscopic transport models will be derived in the sequel.

5.1 Modeling of the Scattering Operator

Several approaches to describe the scattering operator analytically have been proposed [31,208]. Here, the emphasis will be put on the relaxation time approximation of Bløtekjær [28]:

$$\langle\!\langle \partial_t X \rangle\!\rangle_{\text{coll}} = -\frac{\langle\!\langle X \rangle\!\rangle - \langle\!\langle X_0 \rangle\!\rangle}{\tau_X(f)}. \tag{1.50}$$

Here, $\tau_X(f)$ denotes the macroscopic relaxation time concerning the weight function X. $\langle\!\langle X_0 \rangle\!\rangle$ describes the average weight function in equilibrium. Due to the dependence of the relaxation time $\tau_X(f)$ on the distribution function, (1.50) is no approximation [67]. Setting:

$$\tau_X \neq \tau_X(f) \tag{1.51}$$

Equation (1.51) assumes a solely dependence of the relaxation time τ on the moments of the distribution function and is also known as the macroscopic relaxation time approximation. This way, the relaxation times depend only on the moments of the distribution function. Therefore, the odd moments can be formulated as:

$$\langle\!\langle \partial_t \mathbf{X} \rangle\!\rangle_{\text{coll}} \approx -\frac{\langle\!\langle \mathbf{X} \rangle\!\rangle - \langle\!\langle \mathbf{X}_0 \rangle\!\rangle}{\tau_{\text{odd}}} = -\frac{\mathbf{x}}{\tau_{\text{odd}}}, \tag{1.52}$$

and the even moments can be written as:

$$\langle\!\langle \partial_t X \rangle\!\rangle_{\text{coll}} \approx -\frac{\langle\!\langle X \rangle\!\rangle - \langle\!\langle X_0 \rangle\!\rangle}{\tau_{\text{even}}} = -\frac{x - x_0}{\tau_{\text{even}}}. \tag{1.53}$$

The subscript *even* and *odd* is connected to the corresponding even and odd moments.

5.2 Macroscopic Transport Models

From (1.48) and (1.49) the hierarchy of macroscopic transport models can be deduced by means of the moments based method described before [76]. The first three even scalar valued moments are given by powers of the energy $\mathscr{E}(\mathbf{r},\mathbf{k})$:

$$X^{\text{even}} = \left(\mathscr{E}^0, \mathscr{E}^1, \mathscr{E}^2\right), \tag{1.54}$$

while the first three odd vector valued moments are formulated as:

$$X^{\text{odd}} = \left(\mathbf{p}\mathscr{E}^0, \mathbf{p}\mathscr{E}^1, \mathbf{p}\mathscr{E}^2\right). \tag{1.55}$$

Inserting the zeroth moment $\mathscr{E}^0$ and the first moment $\mathbf{p}\mathscr{E}^0$ into (1.48) and (1.49) delivers the particle balance equation and the current equation, respectively. While in the particle balance equation the particle current constitutes an unknown variable, the particle current equation contains the average kinetic energy. Postulating the diffusion approximation (neglecting the kinetic energy of the particles) and assuming the shape of a heated Maxwell distribution[2] the powers of the average energy can be expressed by the carrier temperature T_n, under the constraint of a parabolic band structure[3], as:

$$\langle\langle \mathscr{E}^i \rangle\rangle^{1\mathrm{D}} = \frac{(2i-1)!!}{2^i}(k_\mathrm{B}T_\mathrm{n})^i,\ \langle\langle \mathscr{E}^i \rangle\rangle^{2\mathrm{D}} = i!\,(k_\mathrm{B}T_\mathrm{n})^i, \quad \text{and}$$
$$\langle\langle \mathscr{E}^i \rangle\rangle^{3\mathrm{D}} = \frac{(2i+1)!!}{2^i}(k_\mathrm{B}T_\mathrm{n})^i \ \text{ for } i \geq 1, \tag{1.56}$$

for a one-, two- and three-dimensional electron gas. For example, the average energy $(i = 1)$ for the three-dimensional case is given by:

$$\langle\langle \mathscr{E} \rangle\rangle = \frac{3}{2}k_\mathrm{B}T_\mathrm{n}. \tag{1.57}$$

5.3 *Drift–Diffusion Transport Model*

The Drift–Diffusion Transport model can be derived, by closing the equation system with the assumption of a local thermal equilibrium. This is realized by setting the carrier temperature T_n equal to the lattice temperature T_L. Starting with the substitution of the zeroth moment in (1.48), the particle balance equation is obtained:

$$\underbrace{\partial_t \langle\langle \mathscr{E}^0 \rangle\rangle}_{(1)} + \underbrace{\nabla_\mathbf{r} \langle\langle \mathbf{v}\mathscr{E}^0 \rangle\rangle}_{(2)} - \underbrace{\langle\langle \mathbf{v}\nabla_\mathbf{r}\mathscr{E}^0 \rangle\rangle}_{(3)} + \underbrace{\langle\langle \nabla_\mathbf{r}\mathscr{E}\nabla_\mathbf{p}\mathscr{E}^0 \rangle\rangle}_{(4)} + \underbrace{\mathrm{s}_\alpha q \langle\langle \nabla_\mathbf{p}\mathscr{E}^0 \rangle\rangle \nabla_\mathbf{p}\tilde{\varphi}}_{(5)} = -R. \tag{1.58}$$

Due to the lacking dependence of $\mathscr{E}^0$ on $\mathbf{r}$ and $\mathbf{k}$, the third, fourth and fifth term of the left side of (1.58) vanish and one obtains:

$$\partial_t (n w_0) + \nabla_\mathbf{r}(n\mathbf{V}_0) = -R. \tag{1.59}$$

In order to simplify the mathematical expressions, the averages of the microscopic quantities defined as $w_i = \langle \mathscr{E}^i \rangle$ and $\mathbf{V}_i = \langle \mathbf{v}\mathscr{E}^i \rangle$ will be successively inserted.

By inserting the first moment $\mathbf{p}\mathscr{E}^0$ into (1.49) the particle flux is deduced. Since the relaxation time is in the order of picoseconds, the terms containing the time

---

[2] For non-degenerate semiconductors the Fermi–Dirac distribution can be approximated by the Maxwell-Boltzmann distribution ($\mathscr{E}_c - \mathscr{E}_F \gg k_\mathrm{B}T_\mathrm{L}$).

[3] Close to the band edges, the relation between the wave vector $\mathbf{k}$ and the energy, also known as dispersion relation, can be approximated by an isotropic and parabolic relation $\mathscr{E}(\mathbf{k}) = \frac{\hbar^2 k^2}{2m^*}$, which corresponds to a free electron without any potential.

derivative can be omitted and still quasi-stationary behavior even for today's fastest devices is ensured [71, 224]:

$$\underbrace{\nabla_{\mathbf{r}} \left\langle\!\left\langle \mathbf{v} \otimes \mathbf{p}\mathscr{E}^0 \right\rangle\!\right\rangle}_{(1)} - \underbrace{\left\langle\!\left\langle \mathbf{v} \nabla_{\mathbf{r}} \otimes \mathbf{p}\mathscr{E}^0 \right\rangle\!\right\rangle}_{(2)} + \underbrace{\left\langle\!\left\langle \nabla_{\mathbf{r}} \mathscr{E} \nabla_{\mathbf{p}} \otimes \mathbf{p}\mathscr{E}^0 \right\rangle\!\right\rangle}_{(3)} + \underbrace{s_\alpha q \left\langle\!\left\langle \nabla_{\mathbf{p}} \otimes \mathbf{p}\mathscr{E}^0 \right\rangle\!\right\rangle \nabla_{\mathbf{r}} \tilde{\varphi}}_{(4)}$$

$$= -\frac{\left\langle\!\left\langle \mathbf{p}\mathscr{E}^0 \right\rangle\!\right\rangle}{\tau_0}, \tag{1.60}$$

where τ_0 denotes the momentum relaxation time. Assuming an isotropic band structure and the diffusive limit, the non-diagonal elements of the tensors in (1.60) are zero. Therefore, the tensor of the first part of (1.60) can be approximated by its trace appropriately divided by the dimensionality of the system. Now multiplying the first term with the non-parabolicity factor H_i leads to:

$$\nabla_{\mathbf{r}} \left\langle\!\left\langle \mathbf{v} \otimes \mathbf{p}\mathscr{E}^0 \right\rangle\!\right\rangle \approx \frac{1}{\mathrm{d}} \nabla_{\mathbf{r}} \left\langle\!\left\langle \mathrm{Tr}\left(\mathbf{v} \otimes \mathbf{p}\right) \mathbb{1} \right\rangle\!\right\rangle = A\mathsf{H}_1 \nabla_{\mathbf{r}} \left(n w_1\right), \tag{1.61}$$

with A representing a dimension factor. A can be determined by taking the dimension of the system, the prefactors of the average energy for a parabolic bandstructure, and a Maxwell distribution into account. For example, considering a three-dimensional electron gas the value of A will take the form:

$$\left\langle\!\left\langle \mathbf{v} \otimes \mathbf{p}\mathscr{E}^0 \right\rangle\!\right\rangle \approx \frac{1}{3} \nabla_{\mathbf{r}} \left\langle\!\left\langle \mathrm{Tr}\left(\mathbf{v} \otimes \mathbf{p}\right) \mathbb{1} \right\rangle\!\right\rangle = \frac{2}{3} \mathsf{H}_1 \frac{3}{2} n k_{\mathrm{B}} T_{\mathrm{n}}. \tag{1.62}$$

Here, A exhibits the value $2/3$, while the average energy has been chosen according to (1.57). In the case of a one- and two-dimensional electron gas, A is equal to 2 and 1, respectively. Founding on the validity of the premise, that the kinetic energy can be described by a product ansatz:

$$\mathscr{E} = \nu \kappa(\mathbf{k}), \tag{1.63}$$

the second and third term on the left side of (1.60) vanish. The remaining fourth term can be approximated via:

$$s_\alpha q \left\langle\!\left\langle \nabla_{\mathbf{p}} \otimes \mathbf{p} \right\rangle\!\right\rangle \nabla_{\mathbf{r}} \tilde{\varphi} \approx s_\alpha q n w_0 \nabla_{\mathbf{r}} \tilde{\varphi}. \tag{1.64}$$

Now, assembling all derived expressions, the particle flux equation takes the following form:

$$n\mathbf{V}_0 = -\frac{\mu_0}{q} \mathsf{H}_1 A \nabla_{\mathbf{r}} \left(n w_1\right) - s_\alpha n \mu_0 w_0 \nabla_{\mathbf{r}} \tilde{\varphi}. \tag{1.65}$$

The carrier mobility is given by $\mu_0 = q\tau_0/m^*_{n,p}$, where $m^*_{n,p}$ denote the effective masses for electrons and holes respectively. In combination with the Poisson equation the Drift–Diffusion Transport model can now be expressed as:

$$\partial_t \left(n w_0\right) + \nabla_{\mathbf{r}} \left(n\mathbf{V}_0\right) = -R \;\text{ with:} \tag{1.66}$$

$$n\mathbf{V}_0 = -\frac{\mu_0}{q} \mathsf{H}_1 A \nabla_{\mathbf{r}} \left(n w_1\right) - s_\alpha n \mu_0 w_0 \nabla_{\mathbf{r}} \tilde{\varphi}. \tag{1.67}$$

If one additionally assumes a cold Maxwell distribution function, the highest moment w_1 can be written as:

$$w_1^{1D} = \frac{1}{2} k_B T_L, \ w_1^{2D} = k_B T_L \ , \quad \text{and } w_1^{3D} = \frac{3}{2} k_B T_L \tag{1.68}$$

The average carrier energy of the drift term is neglected, which is also known as the diffusion approximation. Equations (1.66)–(1.67) and the Drift–Diffusion Transport model equations (1.15)–(1.18) from Sect. 1.3 are identical under the assumption of parabolic bands, $H_1 = 1$, and for a three-dimensional electron gas.

The Drift–Diffusion Transport model is the simplest widely employed macroscopic transport model in industrial Technology Computer Aided Design (TCAD) solutions. It allows to discretize its partial differential equations on an unstructured mesh and offers a stable and robust iterative solution. There is also the possibility to generalize its mobility description in order to account for an anisotropic mobility. Furthermore, due to its relative simplicity it can be applied to two and three-dimensional device structures. This especially becomes handy, when one has to account for complex geometrical device structures, material compositions, and doping profiles. However, due to the related computational high costs, three-dimensional simulations are only utilized in rare occasions, when the device structure can not be reduced to a set of simpler two-dimensional cuts.

Due to its closure relation $T_n = T_L$, the Drift–Diffusion Transport model neglects non-local effects and is, therefore, not able to accurately describe transport in short channel devices. This causes an accuracy decrease of the Drift–Diffusion Transport model for device feature lengths shorter than 100 nm [68], where one has to relax the restrictions of a constant carrier temperature in order to improve the description. Additionally, in the case of relevant temperature gradients the applied model has to cover heat flow and thermal diffusion as effects. In such situations, one has to add the energy flow to the Drift–Diffusion Transport model by taking the next higher moment equation into account.

5.4 *Energy Transport Model*

The Energy Transport model can be deduced by inserting the first three moments (q.v. (1.54) and (1.55)) into (1.48) and (1.49) [72]. In this way, an additional equation, the so-called energy balance equation, is gained. This extra equation incorporates the second even moment $\mathscr{E}$, while at the same time the energy flux abides as an unknown:

$$\partial_t \langle\!\langle \mathscr{E} \rangle\!\rangle + \nabla_{\mathbf{r}} \langle\!\langle \mathbf{v}\mathscr{E} \rangle\!\rangle - \langle\!\langle \mathbf{v}\nabla_{\mathbf{r}}\mathscr{E} \rangle\!\rangle + \langle\!\langle \nabla_{\mathbf{r}}\mathscr{E}\nabla_{\mathbf{p}}\mathscr{E} \rangle\!\rangle + s_\alpha q \langle\!\langle \nabla_{\mathbf{p}}\mathscr{E} \rangle\!\rangle \nabla_{\mathbf{p}}\tilde{\varphi} = -n \frac{\langle\!\langle \mathscr{E} \rangle\!\rangle - \langle\!\langle \mathscr{E}_0 \rangle\!\rangle}{\tau_1}. \tag{1.69}$$

After similar considerations as for the Drift–Diffusion Transport model, (1.69) can be simplified to:

$$\partial_t (n w_1) + \nabla_{\mathbf{r}} (n \mathbf{V}_1) + s_\alpha q n \mathbf{V}_0 \nabla_{\mathbf{r}} \tilde{\varphi} + n \frac{w_1 - w_{1,0}}{\tau_1} = 0. \tag{1.70}$$

$\mathbf{V}_1$ denotes the energy flux and w_{10} the equilibrium case of w_1. τ_1 represents the energy relaxation time. The Hydrodynamic Transport model is gained by incorporating the third moment $\mathbf{p}\mathscr{E}$ via (1.49) [72]. This yields an expression for the energy flux, which has been introduced in (1.70) and is up to now not defined in the Energy Transport model:

$$\underbrace{\nabla_{\mathbf{r}} \langle\!\langle \mathbf{v} \otimes \mathbf{p}\mathscr{E} \rangle\!\rangle}_{(1)} - \underbrace{\langle\!\langle \mathbf{v} \nabla_{\mathbf{r}} \otimes \mathbf{p}\mathscr{E} \rangle\!\rangle}_{(2)} + \underbrace{\langle\!\langle \nabla_{\mathbf{r}} \mathscr{E} \nabla_{\mathbf{p}} \otimes \mathbf{p}\mathscr{E} \rangle\!\rangle}_{(3)} + \underbrace{s_\alpha q \langle\!\langle \nabla_{\mathbf{p}} \otimes \mathbf{p}\mathscr{E} \rangle\!\rangle \nabla_{\mathbf{r}} \tilde{\varphi}}_{(4)} = -\frac{\langle\!\langle \mathbf{p}\mathscr{E} \rangle\!\rangle}{\tau_3}. \tag{1.71}$$

The first term on the left side of (1.71) can be approximated by:

$$\nabla_{\mathbf{r}} \langle\!\langle \mathbf{v} \otimes \mathbf{p}\mathscr{E} \rangle\!\rangle \approx \frac{1}{\mathrm{d}} \nabla_{\mathbf{r}} \langle\!\langle \mathrm{Tr}(\mathbf{v} \otimes \mathbf{v}\mathscr{E})\, \mathbb{1} \rangle\!\rangle = A\mathrm{H}_2 \nabla_{\mathbf{r}} (n w_2). \tag{1.72}$$

The second term of (1.71) can be reformulated, via the tensorial identity $\nabla_{\mathbf{x}} \otimes \mathbf{x} h(\mathbf{x}) = h(\mathbf{x}) \nabla_{\mathbf{x}} \otimes \mathbf{x} + \mathbf{x} \otimes \nabla_{\mathbf{x}} h(\mathbf{x})$, to:

$$\langle\!\langle \mathbf{v} \nabla_{\mathbf{r}} \otimes \mathbf{p}\mathscr{E} \rangle\!\rangle = \langle\!\langle \mathbf{v} (\mathscr{E} \nabla_{\mathbf{r}} \otimes \mathbf{p} + \mathbf{p} \otimes \nabla_{\mathbf{r}} \mathscr{E}) \rangle\!\rangle, \tag{1.73}$$

and the third term to:

$$\langle\!\langle \nabla_{\mathbf{r}} \mathscr{E} \nabla_{\mathbf{p}} \otimes \mathbf{p}\mathscr{E} \rangle\!\rangle = \langle\!\langle \nabla_{\mathbf{r}} \mathscr{E} (\mathscr{E} \nabla_{\mathbf{p}} \otimes \mathbf{p} + \mathbf{p} \otimes \nabla_{\mathbf{p}} \mathscr{E}) \rangle\!\rangle \approx \langle\!\langle \mathscr{E} \nabla_{\mathbf{r}} \mathscr{E} + \nabla_{\mathbf{r}} \mathscr{E} (\mathbf{p} \otimes \mathbf{v}) \rangle\!\rangle. \tag{1.74}$$

Taking a look at (1.73) and (1.74) reveals that they cancel each other. The fourth term on the left side of (1.71) is approximated with the same identity as in (1.72), which results in:

$$s_\alpha q \langle \nabla_{\mathbf{p}} \otimes \mathbf{p}\mathscr{E} \rangle \nabla_{\mathbf{r}} \tilde{\varphi} = s_\alpha q \langle\!\langle \mathscr{E} \nabla_{\mathbf{p}} \otimes \mathbf{p} + \mathbf{p} \otimes \nabla_{\mathbf{p}} \mathscr{E} \rangle\!\rangle \nabla_{\mathbf{r}} \tilde{\varphi} \tag{1.75}$$

$$= s_\alpha q n w_1 (1 + A\mathrm{H}_1) \nabla_{\mathbf{r}} \tilde{\varphi}.d \tag{1.76}$$

Merging all derived expressions gives the energy flux:

$$n\mathbf{V}_1 = -\frac{\mu_1}{q} \mathrm{H}_2 A \nabla_{\mathbf{r}} (n w_2) - s_\alpha n \mu_1 (1 + A\mathrm{H}_1) w_1 \nabla_{\mathbf{r}} \tilde{\varphi}. \tag{1.77}$$

The quantity μ_1 denotes the energy flux mobility defined as $\mu_1 = q\tau_3 / m^*_{n,p}$. Now the set of equations for the Hydrodynamic Transport models is complete and given by:

$$\partial_t (n w_0) + \nabla_{\mathbf{r}} (n \mathbf{V}_0) = -R, \tag{1.78}$$

$$n\mathbf{V}_0 = -\frac{\mu_0}{q}\mathsf{H}_1 A \nabla_{\mathbf{r}}(nw_1) - s_\alpha n\mu_0 w_0 \nabla_{\mathbf{r}}\tilde{\varphi}, \tag{1.79}$$

$$\partial_t(nw_1) + \nabla_{\mathbf{r}}(n\mathbf{V}_1) + s_\alpha qn\mathbf{V}_0\nabla_{\mathbf{r}}\tilde{\varphi} + n\frac{w_1 - w_{10}}{\tau_1} = 0, \tag{1.80}$$

$$n\mathbf{V}_1 = -\frac{\mu_1}{q}\mathsf{H}_2 A \nabla_{\mathbf{r}}(nw_2) - s_\alpha n\mu_1(1 + A\mathsf{H}_1)w_1\nabla_{\mathbf{r}}\tilde{\varphi}. \tag{1.81}$$

This set of equations is closed by assuming a heated Maxwellian distribution for the distribution function of the carriers. The concerning highest moment w_2 is then defined for the one-, two-, and three-dimensional electron gas by:

$$w_2^{1\mathrm{D}} = \frac{3}{4}(k_\mathrm{B}T_\mathrm{n})^2, \quad w_2^{2\mathrm{D}} = 2(k_\mathrm{B}T_\mathrm{n})^2, \quad \text{and} \quad w_2^{3\mathrm{D}} = \frac{15}{4}(k_\mathrm{B}T_\mathrm{n})^2. \tag{1.82}$$

Comparing (1.80) and (1.81) with Fourier's law (1.27) and the other additional thermal contributions (1.28)/(1.29) and (1.30) is not as straight forward as for the Drift–Diffusion Transport model, but can be carried out with some reasoning. Equation (1.80) is the so-called *energy flux conservation equation*. It contains a divergence term for the energy flux through the surface, which is analogous to the divergence term in Fourier's law. The time derivative of nw_1 is equivalent to the time derivative of the temperature T in Fourier's law and the remaining term with $n\mathbf{V}_0\nabla\tilde{\varphi}$ represents the source term H, which represents in the simplest case Joule heat.

One has to note, that during the derivation the diffusion approximation has been utilized and thus the so-called *convective* terms $\langle\mathbf{k}\rangle \otimes \langle\mathbf{k}\rangle$ and $\langle\mathbf{k}\rangle \cdot \langle\mathbf{k}\rangle$ were neglected against terms of the form $\langle\mathbf{k}\otimes\mathbf{k}\rangle$ and $\langle\mathbf{k}\cdot\mathbf{k}\rangle$. This causes the sole consideration of the thermal energy $k_\mathrm{B}T_\mathrm{n}$, ignoring the drift energy component of the carriers.

The limitation of this approach is that only the average energy is available to characterize the distribution function. This assumption is significantly violated in devices exhibiting lengths shorter than ≈50 nm [68].

Furthermore there is an arbitrary/synonymous use of the terms Hydrodynamic Transport model and Energy Transport model. The full transport model includes convective terms analog to the differential equations in fluid dynamics. These convective terms state a hyperbolic differential equation type which is hard to solve via numerical methods. Therefore, the diffusion approximation is introduced in order to get rid of these inconvenient terms. The resulting differential equations are of parabolic type and differ from the initial *hydrodynamic* problem. Hence all commonly employed four-moment models incorporating the diffusion approximation should be addressed as Energy Transport model.

A great variety of Hydrodynamic Transport and Energy Transport models have been developed [72]. They are deduced either by Bløtekjær's [28] or Stratton's [208] approach and yield with various assumptions a set of balance and flux equations.

These models are widely used in state of the art TCAD simulators. However, there are several crucial points due to the imposed assumptions during their derivation [72]:

- Band structure: Many employed models are based on the single effective parabolic band model. Due to its simplicity a closed-form solution for single effective parabolic band models exists, while even for the relatively simple non-parabolicity correction model by Kane [115] it is not possible to gain a closed-form solution.
- Non-homogeneous effects: The transport parameters (e.g. mobilities) are commonly gained by measurements or bulk simulations and described as function of the average carrier energy. This works fine for Bløtekjær's approach within, e.g., the channel region, where the absolute value of the electric field increases, while at the end of the channel, where the electric field decreases, these models can exhibit wrong results. A mixture of a cold and a hot-carrier population in this region leads to an inadequate description via the average carrier energy. This region is much smaller than the channel region for long-channel devices, and thus the incorporated error will be small. On the other hand, for devices with a channel length smaller $\approx$100 nm, the length of this region is in the order of the channel length, implying that the hot-carriers injected into the drain need a distance of about the channel length to relax. Therefore, the influence of this region is much more pronounced for future technologies.
- Closure: In order to obtain an amendable equation set, one has to transform the Boltzmann transport equation with the method of moments into an equivalent infinite set of equations and cut it at a certain moment. This set of equation has to be closed at its highest moment with a so-called *closure* relation, which is normally chosen by a heated Maxwellian distribution function. Concerning modern devices, this presents a rather crude approximation for the distribution function. One has to note, that this assumption leaves the lower order equations untouched, while the complete information of the higher order equations has to be bundled into the closure relation.
- Anisotropy: Equipartition of the energy is assumed for modeling of the temperature tensor. It has been demonstrated that such an approximation is invalid for $n^+ - n - n^+$ structures and MOS transistors. Due to the only indirect influence on the drain current in MOS transistors and a missing influence on the current in $n^+ - n - n^+$ structures, anisotropy has been addressed as an issue with negligible importance. However, the carriers penetrate much deeper into the bulk than predicted by Monte Carlo simulations, thus effecting the modeling of the energy-dependent parameters such as the mobility and impact ionization [76]. E.g., for partially depleted SOI transistors, this assumption does not hold and the Energy Transport models can not reproduce the transfer characteristics accurately. Despite the difficulty to treat the temperature tensor rigorously with additional equations for each temperature tensor component, empirical corrections offer promising results [76, 161].

- Drift energy: Due to the diffusion approximation, most Energy Transport models neglect the drift energy. Examination reveals that the drift energy can contribute up to 30% of the total energy inside the channel region [16, 204].
- Velocity overshoot: As a consequence of the afore mentioned approximations, like the truncation of the equations system, the applied closure relation, and the modeling of the transport parameters, the Energy Transport models tend to overestimate the velocity overshoot and expose a spurious velocity overshoot (SVO) at the end of the channel region of n^+-n-n^+ structures. Contrary, for MOS transistors the SVO coincides with the velocity overshoot at the end of the channel and is therefore not explicitly visible.
- Hot carrier effects: Since the Hydrodynamic Transport and Energy Transport models utilize only the first two moments of the energy distribution function it is hard to model hot-carrier effects. It can be demonstrated that the energy distribution function is not uniquely defined by the concentration and the average energy. Due to the dependence on the shape of the distribution function, hot-carrier effects like impact ionization, are destined to fail, if the employed model relies exclusively on the average energy. In such cases the extension of the Energy Transport model to a Six-Moments Transport model elevates the accuracy significantly.

5.5 *Six-Moments Transport Model*

In order to overcome the limitations of the Hydrodynamic Transport model, two further moments can be included in the equation system. The resulting model contains six moments and is therefore called Six-Moments Transport model. Substituting the fourth moment $\mathscr{E}^2$ into (1.48) delivers the second-order energy balance equation:

$$\partial_t \langle\!\langle \mathscr{E}^2 \rangle\!\rangle + \nabla_{\mathbf{r}} \langle\!\langle \mathbf{v}\mathscr{E}^2 \rangle\!\rangle - \langle\!\langle \mathbf{v}\nabla_{\mathbf{r}}\mathscr{E}^2 \rangle\!\rangle + \langle\!\langle \nabla_{\mathbf{r}}\mathscr{E}\nabla_{\mathbf{p}}\mathscr{E}^2 \rangle\!\rangle + \mathrm{s}_\alpha q \langle\!\langle \nabla_{\mathbf{p}}\mathscr{E}^2 \rangle\!\rangle \nabla_{\mathbf{p}}\tilde{\varphi}$$
$$= -n\frac{\langle\!\langle \mathscr{E}^2 \rangle\!\rangle - \langle\!\langle \mathscr{E}_0^2 \rangle\!\rangle}{\tau_2}. \tag{1.83}$$

Reexpressing $\nabla_{\mathbf{r}}\mathscr{E}^2 = 2\mathscr{E}\nabla_{\mathbf{r}}\mathscr{E}$, the second-order energy balance equation takes the following form:

$$\partial_t (n w_2) + \nabla_{\mathbf{r}} (n \mathbf{V}_2) + \mathrm{s}_\alpha q n \mathbf{V}_1 \nabla_{\mathbf{r}} \tilde{\varphi} + n \frac{w_2 - w_{2,0}}{\tau_2} = 0. \tag{1.84}$$

The second-energy flux equation can be deduced by inserting the fifth moment $\mathbf{p}\mathscr{E}^2$ into (1.49):

$$\underbrace{\nabla_{\mathbf{r}} \langle\!\langle \mathbf{v} \otimes \mathbf{p}\mathscr{E}^2 \rangle\!\rangle}_{(1)} - \underbrace{\langle\!\langle \mathbf{v}\nabla_{\mathbf{r}} \otimes \mathbf{p}\mathscr{E}^2 \rangle\!\rangle}_{(2)} + \underbrace{\langle\!\langle \nabla_{\mathbf{r}}\mathscr{E}^2\nabla_{\mathbf{p}} \otimes \mathbf{p}\mathscr{E}^3 \rangle\!\rangle}_{(3)} + \underbrace{\mathrm{s}_\alpha q \langle\!\langle \nabla_{\mathbf{p}} \otimes \mathbf{p}\mathscr{E}^2 \rangle\!\rangle \nabla_{\mathbf{r}}\tilde{\varphi}}_{(4)}$$
$$= -\frac{\langle\!\langle \mathbf{p}\mathscr{E}^2 \rangle\!\rangle}{\tau_4}, \tag{1.85}$$

Each term on the left hand side of (1.85) is gained by the same assumptions as for the energy flux equation. The first term can be approximated by:

$$\nabla_{\mathbf{r}} \langle\langle \mathbf{v} \otimes \mathbf{p}\mathscr{E}^2 \rangle\rangle \approx \frac{1}{d} \nabla_{\mathbf{r}} \langle\langle \mathrm{Tr}\left(\mathbf{v} \otimes \mathbf{p}\mathscr{E}^2\right) \mathbb{1} \rangle\rangle = A\mathrm{H}_3 \nabla_{\mathbf{r}} \left(n w_3\right), \tag{1.86}$$

while the second and third term can be neglected due to their mutual cancellation. The fourth term on the left hand side of (1.85) is substituted via the following expression:

$$s_\alpha q \langle\langle \nabla_{\mathbf{p}} \otimes \mathbf{p}\mathscr{E}^2 \rangle\rangle \approx \left(1 + 2A\mathrm{H}_2\right) n w_2 \nabla_{\mathbf{r}} \tilde{\varphi}. \tag{1.87}$$

Embracing now all contributions yields the second-order energy flux equation:

$$n\mathbf{V}_2 = -\frac{\mu_2}{q} \mathrm{H}_3 A \nabla_{\mathbf{r}} \left(n w_3\right) - s_\alpha n \mu_2 \left(1 + 2A\mathrm{H}_2\right) w_2 \nabla_{\mathbf{r}} \tilde{\varphi}. \tag{1.88}$$

μ_2 denotes the second-order flux mobility and is define via $q\tau_4/m^*_{n,p}$. The Six-Moments Transport model exhibits the following set of equations:

$$\partial_t \left(n w_0\right) + \nabla_{\mathbf{r}} \left(n\mathbf{V}_0\right) = -R, \tag{1.89}$$

$$n\mathbf{V}_0 = -\frac{\mu_0}{q} \mathrm{H}_1 A \nabla_{\mathbf{r}} \left(n w_1\right) - s_\alpha n \mu_0 w_0 \nabla_{\mathbf{r}} \tilde{\varphi}, \tag{1.90}$$

$$\partial_t \left(n w_1\right) + \nabla_{\mathbf{r}} \left(n\mathbf{V}_1\right) + s_\alpha q n \mathbf{V}_0 \nabla_{\mathbf{r}} \tilde{\varphi} + n \frac{w_1 - w_{10}}{\tau_1} = 0, \tag{1.91}$$

$$n\mathbf{V}_1 = -\frac{\mu_1}{q} \mathrm{H}_2 A \nabla_{\mathbf{r}} \left(n w_2\right) - s_\alpha n \mu_1 \left(1 + A\mathrm{H}_1\right) w_1 \nabla_{\mathbf{r}} \tilde{\varphi}, \tag{1.92}$$

$$\partial_t \left(n\mathbf{V}_2\right) + 2 s_\alpha q \mathbf{V}_1 \nabla_{\mathbf{r}} \tilde{\varphi} + n \frac{w_2 - w_{20}}{\tau_2} = 0, \tag{1.93}$$

$$n\mathbf{V}_2 = -\frac{\mu_2}{q} \mathrm{H}_3 A \nabla_{\mathbf{r}} \left(n w_3\right) - s_\alpha n \mu_2 \left(1 + 2A\mathrm{H}_2\right) w_2 \nabla_{\mathbf{r}} \tilde{\varphi}. \tag{1.94}$$

As before, one has to choose an extra closure relation to define the highest moment in the equation system. This is performed for the Six-Moments Transport model by the deviation of the carrier distribution function from a heated Maxwellian distribution, which is defined by the kurtosis β. The kurtosis β of a one-, two-, and three-dimensional electron gas is defined as:

$$\beta^{1\mathrm{D}} = \frac{1}{3} \frac{w_2}{w_1^2}, \quad \beta^{2\mathrm{D}} = \frac{1}{2} \frac{w_2}{w_1^2}, \quad \text{and} \quad \beta^{3\mathrm{D}} = \frac{3}{5} \frac{w_2}{w_1^2}. \tag{1.95}$$

The prefactors $1/3$, $1/2$, and $3/5$ serve as normalization factors, respectively. Assuming a heated Maxwellian distribution and parabolic bands the kurtosis is equal to unity. For realistic devices the kurtosis range is $[0.75, 3]$, portending strong deviations from a heated Maxwellian distribution. This leads to the following closure relation for the Six-Moments Transport model:

$$w_3^{1D} = \frac{15}{8}(k_B T_n)^3 \beta^c, \quad w_3^{2D} = 6\,(k_B T_n)^3 \beta^c, \quad \text{and} \quad w_3^{3D} = \frac{105}{8}(k_B T_n)^3 \beta^c. \tag{1.96}$$

c denotes a fit factor, where it has been found in [70, 125] that a value of $c = 2.7$ yields good results for w_3 in the source and channel regions.

6 The Analogy Between the Drift–Diffusion Transport Model and the Poisson–Nernst–Planck Model

The Poisson–Nernst–Planck [36, 41, 182] model describes the charge distribution and charge transport phenomena in electrolytes. It can be deduced by an averaging procedure from a Langevine model [182]. During the ensemble averaging process the many independent realizations of the stochastic system are bundled and lead to a continuous and steady state description of the system mathematically analog to the Drift–Diffusion Transport model. Instead of the electron and hole assisted charge transport in semiconductors, in electrolytes the ionic components are the charge carriers responsible for the transport and analogously to the Drift–Diffusion Transport model gradients in the electrostatic potential and the spatial concentration of the charged carriers raise forces, trying to extinguish the imbalance. Hence, similar physical conditions lead to a similar mathematical description of the system.

For a binary salt (e.g. *NaCl*) the Poisson–Nernst–Planck equation system may be written as [36]:

$$\mathbf{j}_\pm = -D_\pm \left(\nabla c_\pm + z_\pm c_\pm \nabla \varphi\right), \tag{1.97}$$

$$\nabla \cdot (\varepsilon \nabla \varphi) = \frac{F^2}{\varepsilon_0 RT}(c_- - c_+), \tag{1.98}$$

$$\nabla \cdot \mathbf{j}_\pm = 0. \tag{1.99}$$

$\mathbf{j}_\pm$ denotes the ionic flux, $D_\pm$ describe the diffusion coefficients, $c_\pm$ the ionic charge distributions, and $z_\pm$ the valency of the ion types, respectively. F stands for the Farrady constant, R describes the gas constant and T depicts the temperature of the liquid. In the equation system (1.97)–(1.99) has been assumed that the ionic components are fully dissolved and thus there is no generation-recombination term like in (1.15) and (1.16) and all transient effects are subsided.

Although, there is no doping profile like in common semiconductor devices, and, therefore, no C term in (1.97) like in (1.2), the boundaries of typical domains are charged, either due to differences in the work functions of the solute and the domain wall or through open binding sites at the surface of the boundaries (site-binding model by Yates [237]).

One of the application fields of the Poisson–Nernst–Plank model is the description of transport phenomena in natural and artificial nanopores/ion channels [36, 41, 182]. Biological ion channels constitute the key to understand and control the interaction between cells and their environment. They serve as gateways for various stimuli and the exchange of nutrition and secretion. Ion channels can be opened and closed to the flow of ions in a reliable and reversible manner by certain stimuli. In the open state many ion channels are restrictive to the conducted ion type: Some only conduct anions but not cations and vice versa, or are even more distinct and allow only one certain type to permeate. It was shown that artificial nanopores are feasible and exhibit similar behavior to biological ion channels. For instance, artificial ion channels are able to rectify electric current [13,194–197] or pump potassium ions against concentration gradients in response to a harmonically with time oscillating field [198]. Parameters like the amount of pores, their size and shape can be controlled within a few nanometers [198].

The analogy between the Drift–Diffusion Transport model and the Poisson–Nernst–Plank model stands out even more clearly by comparing the scaled equations of both descriptions [20, 184].

7 Modeling of Transport Parameters

The transport models presented in the previous sections exhibit various material parameters like mobilities. In order to obtain a sufficiently accurate and reliable device simulation one has to thoroughly describe these parameters. Most sufficiently accurate analytical models are derived from theoretical considerations and verified against data extracted from measurements.

7.1 Parameters for the Drift–Diffusion Transport Model

The carrier mobilities in semiconducting materials are determined by various physical mechanisms. The charge carriers experience scattering events by thermal lattice vibrations, ionized impurities, neutral impurities, vacancies, interstitials, dislocations, surfaces and with themselves. Furthermore mobility may depend on the driving electric field: There is a mobility reduction due to the saturation of the drift velocity of warm and hot carriers. Even though rigorous first principle models for the carrier mobilities are available, they are complicated and hard to implement and therefore often replaced by less demanding empirical expressions which are fitted to experimental data [185].

7.1.1 Carrier Mobilities

Due to the overwhelming complexity of rigorous models, we will also stick to the more appealing engineering approach handling the mobilities by fitted empirical models. Commonly it is assumed that the effective carrier mobility can be written as:

$$\mu_\nu^{\mathrm{LISF}} = \mu_\nu^{\mathrm{LISF}}\left(\mu_\nu^{\mathrm{LIS}}\left(\mu_\nu^{\mathrm{LI}}\left(\mu_\nu^{\mathrm{L}}\right)\right)\right). \tag{1.100}$$

ν denotes the charge carrier type (electrons or holes), and μ_ν^{LISF} depicts the effective mobility influenced by lattice scattering (L), ionized impurity scattering (I), surface roughness scattering (S) and carrier heating (F). This multi-level approach implies that the different scattering mechanisms can be separated and the effective mobilities can be obtained via consecutive sophistication of the model by including additional scattering mechanisms.

Lattice Scattering

Atoms in the semiconductor lattice vibrate around their equilibrium positions. Due to these oscillations, even in pure and perfectly ordered semiconductors, carriers are scattered by the vibrating lattice and the lattice mobility μ_ν^{L} depends on the lattice temperature. For simulation applications, an empirical power law is convenient [184]:

$$\mu_\nu^{\mathrm{L}} = \mu_\nu^0 \left(\frac{T}{300\,\mathrm{K}}\right)^{-\alpha_\nu}, \quad \nu = n, p. \tag{1.101}$$

The parameters μ_ν^0 and α_ν exhibit a certain spread of values [184]. For instance, the parameters for the electron mobility are frequently in the range $1{,}240\,\mathrm{cm}^2\,(\mathrm{Vs})^{-1} < \mu_n^0 < 1{,}600\,\mathrm{cm}^2\,(\mathrm{Vs})^{-1}$ and $2.2 < \alpha_n < 2.6$ for silicon. A possible explanation lies in the stochastic nature of the device fabrication process and the measurement itself. Corresponding parameters for *III–V* semiconductors can be found in [158].

Ionized Impurity Scattering

The mobility reduction in semiconductor devices due to scattering by charged impurities is a major effect. The influence of lattice and impurity scattering must be combined in an appropriate way in order to gain an effective mobility.

Caughey and Thomas introduced an empirical model which is able to fit the experimental data [34]. The exploited empirical expression is:

$$\mu_\nu^{\mathrm{LI}} = \mu_\nu^{\min} + \frac{\mu_\nu^{\mathrm{L}} - \mu_\nu^{\min}}{1 + \left(\frac{N_\mathrm{I}}{N_\nu^{\mathrm{ref}}}\right)^{\alpha_\nu}}, \tag{1.102}$$

where:

$$N_{\mathrm{I}} = \sum_i |Z_i| N_i \tag{1.103}$$

is the sum over all charged impurities and Z_i denotes the charge state of the impurity. For example, single ionized impurities (e.g. boron, phosphorus and aluminum in silicon) $|Z_i| = 1$. The Caughey and Thomas model requires three free parameters in order to fit the experimental data. Typical values for silicon at room temperature are $\mu_n^{\mathrm{min}} = 80\,\mathrm{cm}^2\,(\mathrm{Vs})^{-1}$, $N_n^{\mathrm{ref}} = 1.12\,10^{17}\,\mathrm{cm}^{-3}$ and $\alpha_n = 0.72$ for electrons and $\mu_p^{\mathrm{min}} = 45\,\mathrm{cm}^2\,(\mathrm{Vs})^{-1}$, $N_p^{\mathrm{ref}} = 2.23\,10^{17}\,\mathrm{cm}^{-3}$ and $\alpha_n = 0.72$ for holes.

Lombardi introduced an alternative mobility description for silicon in [136], based on the Matthiessen rule, and optimized for numerical simulations. The mobility model after Masetti [141] extends the description of Caughey and Thomas to high doping concentrations. For *III–V* semiconductors the required parameters can be found in the book by Palankovski and Quay [158].

Surface/Interface Scattering

The finite spatial dimensions of a semiconductor cause the perfect crystal periodicity to break at the crystal surfaces. The interfaces between different materials exhibit different lattice constants and thus lead to ineluctable imperfections. These imperfections have a huge impact, if the current is flowing primarily close to the interface, as commonly in modern MOSFETs. Usually, the mobility along a surface is significantly smaller than in the center of the crystal. The transition from the high mobility region in the bulk to the low mobility region at the surface is smooth.

An empirical model describing such a smooth transition depending on the depth has been proposed by [184]:

$$\mu_{\nu\mathrm{LIS}} = \frac{\mu_\nu^{\mathrm{ref}} + (\mu_\nu^{\mathrm{LI}} - \mu_\nu^{\mathrm{ref}})\,(1 - F(y))}{1 + F(y)\left(\frac{S_\nu}{S_\nu^{\mathrm{ref}}}\right)^{\gamma_\nu}}. \tag{1.104}$$

The depth dependence $F(y)$ is defined by:

$$F(y) = \frac{2\exp\left(-\frac{y^2}{y^{\mathrm{ref}2}}\right)}{1 + \exp\left(-2\frac{y^2}{y^{\mathrm{ref}2}}\right)}, \tag{1.105}$$

where the parameter y^{ref} is in the typical range from 2 to 10 nm. The pressing forces S_n and S_p are equal to the magnitude of the normal field strength at the interface, if the carriers are pulled by it otherwise they are zero. The parameters are fitted to experimental data.

Also the mobility model after Lombardi [136] can be employed due to the inclusion of surface acoustic phonon scattering and surface roughness scattering. An overview about the vast number utilized mobility models is documented in [100].

Field Dependent Mobility

The carrier energy can be split into two basic contributions, the thermal energy, which is related to the random thermal motion of the carriers, and the kinetic energy, describing the kinetic energy of the charge carriers $\frac{mv^2}{2}$. So the average energy per particle is given by:

$$w = \frac{3}{2}k_{\mathrm{B}}T_{\mathrm{n}} + \frac{1}{2}mv^2, \tag{1.106}$$

where T_n denotes the carrier temperature. Exerting the charged particles to an electric field, accelerates them and thus increases the kinetic energy, while scattering events convert kinetic energy to thermal energy and increase the carrier temperature. For weak electric fields the mobility is constant with respect to the field, and therefore the relation between the velocity and the electric field is linear.

Compared to the movement caused by the externally applied electric field the thermal velocity of electrons and holes is large and hence the carrier temperature is equal to the lattice temperature.

For large electric fields the relationship between the electric field and the carrier velocity begins to deviate from linear and saturates for very high fields. Within a simulation framework this effect is normally taken care of by a field dependent mobility.

Also here empirical mobility expressions are employed whose parameters are determined by fitting experimental data. A widely used expression was introduced by Caughey and Thomas [34]:

$$\mu_\nu^{\mathrm{LISF}}(E) = \frac{\mu_\nu^{\mathrm{LIS}}}{\left(1 + \left(\frac{\mu_\nu^{\mathrm{LIS}} E}{v_\nu^{\mathrm{sat}}}\right)^{\beta_\nu}\right)^{1/\beta_\nu}}, \tag{1.107}$$

or an alternative formulation by Jaggi [108, 109]:

$$\mu_\nu^{\mathrm{LISF}}(E) = \frac{2\mu_\nu^{\mathrm{LIS}}}{\left(1 + \left(\frac{2\mu_\nu^{\mathrm{LIS}} E}{v_\nu^{\mathrm{sat}}}\right)^{\beta_\nu}\right)^{1/\beta_\nu}}. \tag{1.108}$$

Both expressions contain the same parameters, the low-field mobility μ_ν^{LIS} and the saturation velocity v_ν^{sat}, respectively. These parameters pose the low-field and high-field limits of the carrier velocity as a function of the electric field:

$$\lim_{E\to 0} \mu_\nu^{\mathrm{LISF}}(E) = \mu_\nu^{\mathrm{LIS}}, \quad \lim_{E\to\infty} v_\nu(E) = \lim_{E\to\infty} \mu_\nu^{\mathrm{LISF}}(E) \times E = v_\nu^{\mathrm{sat}}. \tag{1.109}$$

Silicon at room temperature is characterized by the following parameters: $v_n^{\mathrm{sat}} = 10^7\,\mathrm{cm\,s^{-1}}$, $\beta_n = 2$, $v_p^{\mathrm{sat}} = 8 \times 10^6\,\mathrm{cm\,s^{-1}}$ and $\beta_p = 1$. For high electric fields both models (1.108) and (1.109) reach asymptotically $\mu_{\mathrm{LISF}}^{\nu} \sim 1/E$ as previously asserted.

For higher order transport models, the description of mobility becomes more complex due to the dependence on the carrier temperature [15, 16, 83, 84, 132, 213].

7.1.2 Carrier Generation and Recombination

Generation-recombination phenomena are involved in many fundamental effects like leakage current and device breakdown. In thermal equilibrium there is a dynamic balance between the generation and recombination of electron–hole pairs, which yields into an equilibrium concentration n_0 for electrons and p_0 for holes:

$$n_0 = N_\mathrm{c} \exp\left(\frac{\mathcal{E}_F - \mathcal{E}_C}{k_\mathrm{B} T_\mathrm{n}}\right) = n_i \exp\left(\frac{\mathcal{E}_F - \mathcal{E}_i}{k_\mathrm{B} T_\mathrm{n}}\right), \tag{1.110}$$

$$p_0 = N_\mathrm{v} \exp\left(\frac{\mathcal{E}_\mathrm{V} - \mathcal{E}_F}{k_\mathrm{B} T_\mathrm{n}}\right) = p_i \exp\left(\frac{\mathcal{E}_i - \mathcal{E}_F}{k_\mathrm{B} T_\mathrm{n}}\right). \tag{1.111}$$

$N_\mathrm{c}/N_\mathrm{v}$, and n_i/p_i denote the effective DOS for the conduction and valence band and the intrinsic concentrations for electrons and holes, respectively, while $\mathcal{E}_i$ describes the intrinsic energy. The product of the equilibrium concentrations for electrons and holes results in

$$n_0\, p_0 = N_\mathrm{c} N_\mathrm{v} \exp\left(\frac{\mathcal{E}_C - \mathcal{E}_\mathrm{V}}{k_\mathrm{B} T_\mathrm{n}}\right) = n_i^2, \tag{1.112}$$

with the introduction of the intrinsic concentration:

$$n_i = \sqrt{N_\mathrm{c} N_\mathrm{v}} \exp\left(-\frac{\mathcal{E}_g}{2 k_\mathrm{B} T_\mathrm{n}}\right). \tag{1.113}$$

Equations (1.110) and (1.111) are based on Boltzmann statistics and thus are only valid for non-degenerate semiconductors.

If the electron and hole concentrations differ from their equilibrium concentrations, the balance of generation and recombination rates is disturbed. Regions exhibiting excess carriers ($n\,p > n_i^2$) will experience mainly recombination while regions with a carrier deficiency ($n\,p < n_i^2$) will encounter a domination of the generation process.

Various physical mechanisms can cause the generation/recombination of an electron–hole pair. For instance, the absorption or emission of a photon, the absorption or emission of a phonon, three particle transitions, and transitions assisted by recombination centers. The impact of these mechanisms depends on the operation conditions and the properties of the employed materials.

The transition from the valence band to the conduction band requires energy. The needed amount of energy to lift an electron from the valence band to the conduction band or a hole from the conduction band to the valence band is at least the band gap energy $\mathscr{E}_g$. This energy can be gained by several means:

- Photons: Each photon carries an energy of $\hbar\omega$. If the energy of the photon is equal or greater than the band gap energy $\mathscr{E}_g$, an electron absorbing photon is able to raise into the conduction band.
- Phonons: Phonons represent the quantization of thermal lattice vibrations and are able to transfer energy to the charge carriers.
- Collisions: An electron in the conduction band with high energy is able to transfer enough energy to an electron in the valence band, so that it is elevated into the conduction band.

Trap Assisted Recombination and Generation

Silicon and Germanium are indirect semiconductors and it was experimentally found that these materials primarily generate and recombinate electron–hole pairs via trap centers. This so-called Shockley–Read–Hall generation-recombination mechanism is called after the authors who constituted the theory [81, 189]. The indirect generation-recombination process is a non-radiative process and can be separated into four independent processes (Fig. 1.2):

(a) Electron Capture: An electron jumps from the conduction band into an unoccupied trap state and fills it.
(b) Electron Emission: An electron occupying a trap site elevates into the conduction band and leaves the trap state empty.
(c) Hole Capture: An electron jumps from a trap site into an unoccupied valence band site, neutralizes a hole and leaves the trap site empty.
(d) Hole Emission: An electron from the valence band is lifted into the trap site occupies it and generates a hole in the valence band.

The reaction rates are given by:

$$\nu_a = k_a n N_t^0 \quad \text{(electron capture)}, \tag{1.114}$$

$$\nu_b = k_b N_t^- \quad \text{(electron emission)}, \tag{1.115}$$

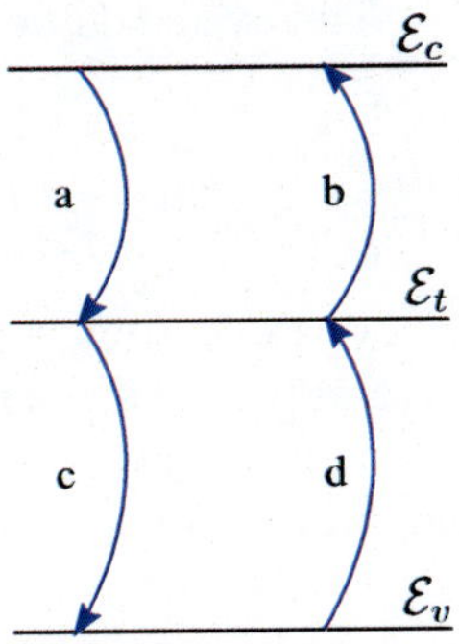

Fig. 1.2 The transition process can be split up into four partial processes

$$\nu_c = k_c\, p N_t^- \quad \text{(hole capture)}, \tag{1.116}$$

$$\nu_d = k_d N_t^0 \quad \text{(hole emission)}, \tag{1.117}$$

where Nt^0 denotes the concentration of neutral traps and N_t^- describes the concentration of occupied traps. The total trap concentration N_t is calculated by $N_t = N_t^0 + N_t^-$. The fraction of occupied traps is defined as $f_t = N_t^-/N_t$, $1 - f_t = N_t^0/N_t$. The rate equation ν_a describes the electron capture rate and assumes that the transmission rate is proportional to the number of carriers in the conduction band n and the number of neutral (free) traps N_t^0. The electron emission rate ν_b is expected to be proportional to the number of electrons N_t^- in the traps exclusively, due to the majority of empty states in the conduction band (i.e. the distribution function f is close to zero, hence $1 - f$ is close to 1). A consideration for holes is similar. The principle of detailed balance is valid for thermal equilibrium and allows the assumption of:

$$\nu_a^{eq} = \nu_b^{eq}, \quad \nu_c^{eq} = \nu_d^{eq}. \tag{1.118}$$

Thus, we obtain:

$$k_b = k_a \underbrace{n_0 \frac{1 - f_{t,0}}{f_{t,0}}}_{n_1}, \tag{1.119}$$

$$k_d = k_c \underbrace{p_0 \frac{f_{t,0}}{1 - f_{t,0}}}_{p_1}, \tag{1.120}$$

with the auxiliary concentrations n_1 and p_1. $f_{t,0}$ describes the fraction of occupied traps in thermal equilibrium. With the aid of the definitions (1.119) and (1.120) the net recombination rates can be expressed as:

$$R_n^{\mathrm{SRH}} = \nu_a - \nu_b = k_a N_t \left(n\,(1 - f_t) - n_1 f_t\right), \tag{1.121}$$

$$R_p^{\mathrm{SRH}} = \nu_c - \nu_d = k_c N_t \left(p\, f_t - p_1\,(1 - f_t)\right). \tag{1.122}$$

From a general viewpoint the recombination rates R_n^{SRH} and R_p^{SRH} are not automatically equal. This is taken into account by an additional conservation equation to the semiconductor equations:

$$\frac{\partial N_t^-}{\partial t} = R_n^{\mathrm{SRH}} - R_p^{\mathrm{SRH}}, \tag{1.123}$$

which has to be considered in the whole domain. Provided that the system is in steady state, the time derivative vanishes and the net recombination rate of electrons is equal to the net recombination rate for holes. Under these circumstances one can calculate the trap occupancy function explicitly:

$$f_t = \frac{k_a n + k_c p_1}{k_a (n + n_1) + k_c (p + p_1)}. \tag{1.124}$$

After introducing the carrier lifetimes $\tau_p^{-1} = k_a N_t$ and $\tau_n^{-1} = k_c N_t$ one is able to write down the recombination rate after Shockley and Read [189], and Hall [81]:

$$R^{\mathrm{SRH}} = \frac{np - n_i^2}{\tau_p (n + n_1) + \tau_n (p + p_1)}. \tag{1.125}$$

Traps are defined by defects with an energy level $\mathscr{E}_t$ and their concentration N_t. The interaction of carriers and trap centers is described by the capture cross section σ_n for electrons and σ_p for holes and linked to the rate constants and the carrier lifetimes by:

$$k_a = \sigma_n v_{th}^n, \;\; \tau_b^{-1} = \sigma_n v_{th}^n N_t, \;\; k_c = \sigma_p v_{th}^p, \;\; \tau_d^{-1} = \sigma_p v_{th}^p N_t, \tag{1.126}$$

with the thermal velocities v_{th}^n and v_{th}^p for electrons and holes, respectively.

Presuming that $np > n_i^2$ the recombination rate is larger than zero, thus recombination takes place until $np = n_i^2$. On the other hand if $np < n_i^2$ the generation is dominant which means that the carrier concentration increases until $np = n_i^2$ again. The carrier lifetimes τ_n and τ_p determine the transient response of the material in the non-equilibrium case. The smaller the carrier lifetime the bigger the recombination rate becomes and, hence, the faster the material gains equilibrium again.

Photon Transition

Direct generation/recombination can be associated with photon emission or absorption. Direct band to band transitions are only of importance for direct bandgap semiconductors like *GaAs*, due to the relatively small momentum of photons. In silicon and germanium the direct generation-recombination mechanism is therefore negligible. Starting with the electron–hole reaction:

$$e^- + h^+ \rightleftharpoons 0, \tag{1.127}$$

two distinct processes are available:

(a) Electron–Hole Recombination: An electron moves from the conduction band into the valence band and neutralizes a hole.
(b) Electron–Hole Generation: An electron from the valence band absorbs a photon which exhibits an energy larger than the bandgap energy and moves to the conduction band leaving a hole in the valence band.

The rate constants k_a^{opt} and k_b^{opt} allow a description of the rate equations for optical generation/recombination:

$$v_a = k_a^{\mathrm{opt}}(T)np, \tag{1.128}$$

$$v_b = k_b^{\mathrm{opt}}(T). \tag{1.129}$$

These two rates have to be equal in thermal equilibrium:

$$\nu_{a,0} = \nu_{b,0} \rightarrow k_a^{\text{opt}} n_i^2 = k_b^{\text{opt}} . \tag{1.130}$$

This way the net recombination rate results in:

$$R^{\text{opt}} = \nu_{a,0} - \nu_{b,0} = k_a^{\text{opt}} \left(n p - n_i^2\right) . \tag{1.131}$$

Here, once more the term $\left(n p - n_i^2\right)$ appears, which takes care of driving the system back into equilibrium.

Auger Generation-Recombination

The Auger generation-recombination is a three particle process, but only two move from one band to another. The third particle provides or receives the excess energy and moves to another energy level within the same band, where it releases its energy to thermal vibrations in the case of recombination. In the following we describe the direct band to band Auger process which is also known as phonon-assisted Auger process. This process is covered by four partial reactions:

$$2e^- + h^+ \rightleftharpoons e^- \tag{1.132}$$

$$e^- + 2h^+ \rightleftharpoons h^+ \tag{1.133}$$

(a) Electron Capture: An electron from the conduction band jumps into the valence band. The excess energy is transferred to another conduction band electron while the electron in the valence band neutralizes a hole.
(b) Electron Emission: A valence band electron gains energy from a high energetic conduction band electron and is lifted into the conduction band, leaving a hole behind.
(c) Hole Capture: An electron from the conduction band moves to the valence band. The excess energy is transferred to another hole. The new electron in the valence band neutralizes a hole.
(d) Hole Emission: A valence electron is lifted by a high energetic hole into the conduction band. A new hole remains in the valence band.

The reaction rates are written with the rate constants c_n, e_n, c_p and e_p as follows:

$$\nu_a = c_n n^2 p \text{ (electron capture),} \tag{1.134}$$

$$\nu_b = e_n n \text{ (electron emission),} \tag{1.135}$$

$$\nu_c = c_p p^2 n \text{ (hole capture),} \tag{1.136}$$

$$\nu_d = e_p p \text{ (hole emission).} \tag{1.137}$$

the n^2p term in (1.134) is caused by the need for two electrons from the conduction band and one hole from the valence band. On the other hand, although there are two electrons involved in the electron emission process in (1.135), only one electron from the conduction band participates.

Assuming thermal equilibrium, the principle of detailed balance demands:

$$\nu_{a,0} = \nu_{b,0} \rightarrow c_n n_i^2 = e_n, \tag{1.138}$$

$$\nu_{c,0} = \nu_{d,0} \rightarrow c_p n_i^2 = e_p. \tag{1.139}$$

The constants c_n and c_p denote the Auger coefficients and the net recombination rate for the Auger process is expressed as:

$$R^{\mathrm{Au}} = \nu_a - \nu_b + \nu_c - \nu_d = (c_n n + c_p p)\left(np - n_i^2\right). \tag{1.140}$$

Once more the $\left(np - n_i^2\right)$ term emerges and models the tendency of the system to reach an equilibrium. Commonly employed values for silicon at room temperature are $c_n = 2.9 \times 10^{-31}\,\mathrm{cm^6\,s^{-1}}$ and $c_p = 9.9 \times 10^{-32}\,\mathrm{cm^6\,s^{-1}}$.

Impact Ionization

Impact ionization is a process only generating electron–hole pairs via high energetic carriers. In the microscopic picture there is no difference between Auger generation and impact ionization. The difference is related to the energy sources. The Auger generation process was deduced with the aid of the principle of detailed balance, which is only valid in thermal equilibrium, while impact ionization is a typically non-equilibrium process requiring large fields.

For impact ionization two partial processes have to be taken into account:

$$e^- \rightharpoonup 2e^- + h^+, \tag{1.141}$$

$$h^+ \rightharpoonup 2h^+ + e^-. \tag{1.142}$$

(a) Electron Emission: A valence electron consumes energy from a high energetic electron in the conduction band and jumps into the conduction band, leaving a hole behind.

(b) Hole Emission: A valence electron moves to the conduction band by the energy from an high energetic hole in the valence band. A hole is generated in the valence band.

Even though these two partial processes are equivalent to the Auger processes (b) and (d), for modeling impact ionization, the reaction rates are differently expressed in the framework of the Drift–Diffusion Transport model:

$$\nu_a = \alpha_n \frac{\mathbf{j}_n}{q}, \tag{1.143}$$

$$\nu_b = \alpha_p \frac{\mathbf{j}_p}{q}. \tag{1.144}$$

α_n and α_p depict the ionization coefficients for electrons and holes. They are given by the reciprocal of the average distance carriers travel between successive ionization events. An electron generates on average one electron–hole pair, when it travels of $1/\alpha_n$. The total generation rate is determined by:

$$G^{\mathrm{II}} = v_a + v_b = \frac{\alpha_n}{q}\left|\mathbf{j}_n\right| + \frac{\alpha_p}{q}\left|\mathbf{j}_p\right|. \quad (1.145)$$

Thus the impact ionization rate is proportional to the current densities, while the Auger generation is proportional to the carrier concentrations (1.135) and (1.137). Therefore, Auger generation takes place in regions with high mobile carrier concentrations and not necessarily high current densities, while impact ionization requires a significant current flow. Theoretical and experimental surveys indicate an exponential dependence of the ionization coefficients on the electric field:

$$\alpha_n = A_n \exp\left(-(B_n/E)^{\beta_n}\right), \quad \alpha_p \exp\left(-(B_p/E)^{\beta_p}\right). \quad (1.146)$$

$E = \mathbf{E}\cdot\mathbf{j}/|\mathbf{j}|$ denotes the field component along the direction of the current flow. Chynoweth [40] found the exponents β_n and β_p to be unity on the basis of large experimental data sets. Shockley supports these findings by theoretical considerations [187], while Wolff predicts them to be two via a different approach [230].

Practically, β_n and β_p are adjusted between one and two in order to get a good matching to experimental data. Typical values for silicon at room temperature are $A_n = 7.03\times10^5\,\mathrm{cm}^{-1}$, $B_n = 1.231\times10^6\,\mathrm{V\,cm}^{-1}$, $\beta_n = 1$, $A_p = 6.71\times10^5\,\mathrm{cm}^{-1}$, $B_p = 1.693\times10^6\,\mathrm{V\,cm}^{-1}$ and $\beta_p = 1$.

In the case of *III–V* semiconductors the work from Palankovski and Quay [158] provides the necessary data. Due to the non-trivial dependence on several quantities, modeling of the parameters is much harder for higher order transport models [67, 69,72].

7.1.3 Modeling Biologically Sensitive Field-Effect Transistors

There are two common approaches for simulating biochemical systems. In the first, microscopic approach, every molecule is characterized by its electrostatic properties and free to move within the solute, trying to minimize the acting forces between them and, thus, minimizing the energy of the system. This is typically accomplished by a stochastic Monte Carlo process [134]. In the second macroscopic approach the system is characterized by a set of partial differential equations with well chosen boundary conditions. While the description of the system via its fundamental electrostatic interaction between single molecules is beneficial, the required amount of memory and the rather poor convergence rate in comparison to other methods ($\propto \frac{1}{\sqrt{N}}$, N is the sample size) pose computational problems. The vast amount of molecules/atoms in the solute is the reason for the high memory consumption. 1 ml of water contains about $\approx 3.35\,10^{22}$ of water molecules. Therefore, many simulations restrict to the molecules of interest and describe the surrounding water molecules by an average relative permittivity of $\sim$80, but the overall memory consumption is still quite high due to the fact that macromolecules regularly contain several thousand atoms. There are further ways to reduce the memory consumption, however, it remains an issue, so one is restricted to small volumes and/or short time scales ($\sim 10^{-15}$ s, [77, 241]).

On the contrary, the approach based on differential equations is less time and memory consuming, but neglects the quantized structure of the system and treats quantities as continuous. This complicates the description of the interaction between the molecules and also can lead to problems at low buffer concentrations [228]. We will follow the second approach below.

At first one has to identify the different parts of the simulation domain and classify them. There are: the zone where the macromolecules are contained, the region comprising the buffer, the dielectric and the semiconducting region (shown in Fig. 1.3). The devices are in the micrometer regime, even biologically sensitive field-effect transistors (BioFETs) utilizing nanowires commonly exhibit a length in the micrometer regime [56, 80, 203, 238] and therefore it is valid to model the semiconducting part via the Poisson equation, describing the charge distribution within the semiconductor, and the Drift–Diffusion Transport model taking care of the charge transport at least along the carrier transport direction [184, 214]. The dielectric is assumed to be a perfect isolator without charges modeled with the Laplace equation. The Stern layer[4] is covered by the Laplace equation and a relative permittivity of $\varepsilon_{\text{Ana}} \approx 80$ in order to guarantee a minimal distance of the charged zone holding the macromolecules to the oxide interface. Depending on the preparation of the device there can be charges at the oxide interface (q.v. site-binding model). Frequently the surface sites are passivated before the macromolecules are attached

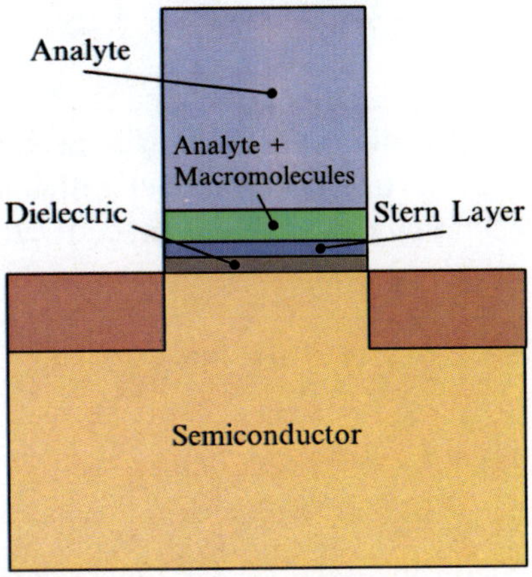

Fig. 1.3 Illustrating the different simulation zones

[4] Stern was the first to recognize, that the finite dimensions of dissolved ions cause a layer depleted from charges at interfaces (q.v. Sect. 8.2).

to avoid perturbing charge accumulation at the open oxide sites and to prepare the surface with certain functional groups allowing to attach (link) the macromolecules to the surface. The dimensions of the zone holding the charged macromolecules and their charge density can either be obtained by measurements or has to be estimated from the partial charge of a single macromolecule, derived from a protein data bank [167], and extrapolated. This zone and the rest of the electrolyte region are covered by several modeling approaches and will be explained in the sequel.

Poisson–Boltzmann Model

The Poisson–Boltzmann model is probably the most prominent one. For several mMol salt concentrations upwards it yields good results based on the assumption that the dissolved buffer ions are in thermodynamical equilibrium with their environment and only depend on the local potential. This premise allows to describe the buffer as sum over all ionic species weighted with Boltzmann type terms $e^{\frac{q\Psi}{k_B T}}$ and their valences:

$$\varepsilon_0 \nabla \cdot (\varepsilon_{\text{Ana}} \nabla \Psi(x,y)) = -\sum_{\xi \in S} \xi\, q\, c_\xi^\infty\, e^{-\xi \frac{q}{k_B T}(\Psi(x,y)-\Psi_\mu)} + \rho_{\text{Space}}(x,y). \tag{1.147}$$

ξ is the valence of the ions in the electrolyte, Ψ_μ is the chemical potential, c_σ^∞ is the ion concentration in equilibrium, while $\varepsilon_{Ana} \approx 80$ is the relative permittivity of water. ρ_{Space} represents the average space charge density in the simulation zone, where the charged macromolecules are contained.

Poisson–Boltzmann Model with Sheet Charge

If the charged macromolecules are directly linked to the surface and not dispensed in a gel, the zone height is typically in the deca- nanometer regime. Therefore, it will be extremely small compared to the rest of the device dimensions and it is justified to save mesh points by substituting this region by an equivalent sheet charge $\sigma_{\text{Sheet}}(x)$ at the surface y_0:

$$\varepsilon_0 \nabla \cdot (\varepsilon_{\text{Ana}} \nabla \Psi(x,y)) = -\sum_{\xi \in S} \xi\, q\, c_\xi^\infty\, e^{-\xi \frac{q}{k_B T}(\Psi(x,y)-\Psi_\mu)} + \sigma_{\text{Sheet}}(x)\, \delta(y-y_0). \tag{1.148}$$

Poisson–Boltzmann Model with Homogenized Interface Conditions

A similar but somewhat refined model is derived in [88, 89, 172]. The authors handled the multi-scale problem by exchanging the fast varying charge distribution at the surface (e.g. Proteins or DNA fragments scattered over the functionalized surface) by two interface conditions. These interface conditions describe the effects of the charge and the dipole moment of the biofunctionalized layer containing the charged macromolecules:

$$\varepsilon_{\mathrm{Oxid}}\, \partial_y \Psi(0-,x) - \varepsilon_{\mathrm{Ana}} \partial_y \Psi(0+,x) = -\frac{C(x)}{\varepsilon_0}, \tag{1.149}$$

$$\Psi(0-,x) - \Psi(0+,x) = -\frac{D_y(x)}{\varepsilon_{\mathrm{Ana}}\varepsilon_0}. \tag{1.150}$$

Here, $\Psi(0-)$ denotes the potential in the oxide, while $\Psi(0+)$ relates to the potential in the solute. The first equation describes the jump in the field, while the second introduces a dipole moment causing a shift of the potential (which can be accounted for by adjusting the potential in the analyte). $C(x)$ is the averaged (homogenized) charge density at the dielectric–electrolyte interface and can either be determined by experimental data or derived from first principle calculations via a data set from a protein data bank [167]. $D_y(x)$ expresses the averaged perpendicular dipole moment density and has to be gained from first principle calculations. For instance, the adaptive Poisson–Boltzmann Solver (APBS) [18,94,95] allows to assign partial charges to every atom for the desired macromolecule, and thus the calculation of the overall charge and in conjunction with the relative distances between the atoms also the dipole moment of the molecule. This charge and dipole moment can be extrapolated to the mean charge and mean dipole moment assuming an average distance between the macromolecules.

Extended Poisson–Boltzmann Model

The extended Poisson–Boltzmann model [228] is able to include the average closest possible approach of two ions in the liquid. This allows to include the Stern layer within this formulation without the need to add an ion free zone between the dielectric and the region where the Poisson–Boltzmann model is calculated. Furthermore, the minimal possible distance between two ions a is in this model a fit parameter and can therefore account for the varying screening behavior at different ionic concentrations:

$$\varepsilon_0 \nabla \cdot (\varepsilon_{\mathrm{ana}} \nabla \Psi) = 2qc_0^\infty \frac{\left(a-(a-1)\cosh\left(\frac{q\Psi}{2k_BT}\right)\right)\sinh\left(\frac{q\Psi}{2k_BT}\right)}{\left((1-a)+a\cosh\left(\frac{q\Psi}{2k_BT}\right)\right)^3}. \tag{1.151}$$

c_0^∞ denotes the bulk ion concentration for a $1:1$ salt, while a describes the closest possible approach between two ions. In the limit $\lim_{a\to 0}$ the Poisson–Boltzmann expression is recovered. One has to mention that this formulation is limited to $1:1$ electrolytes and therefore can not be applied to arbitrary buffers.

Debye–Hückel Model

The Poisson–Boltzmann equation represents a nonlinear differential equation for the electrostatic potential. Often there is a wish for a formulation which is numerically less demanding or offers quickly an analytical solution. This has been already achieved by Debye and Hückel [43] in 1923, deriving a linearized version of the Poisson–Boltzmann equation. Starting with the corresponding thermodynamical potential, they rigorously deduced the Poisson–Boltzmann model and their equation by Taylor expansion of the exponential terms, neglecting contributions higher than first order. This model is valid only for small potentials and relatively dilute electrolytes:

$$\varepsilon_0 \nabla \cdot (\varepsilon_{Ana} \nabla \Psi(x,y)) = \frac{q^2}{k_B T}(\Psi(x,y) - \Psi_\mu) \sum_{\xi \in S} \xi^2 c_\xi^\infty + \rho_{Space}(x,y). \tag{1.152}$$

From (1.152) two important properties can be gained. Firstly, the Debye length λ_D:

$$\lambda_\mathrm{D} = \sqrt{\frac{k_\mathrm{B} T \varepsilon_0 \varepsilon_\mathrm{Ana}}{q^2 \sum_{\xi \in S} \xi^2 c_\xi^\infty}} \quad \text{or in terms of ionic strength (see (1.155))} \tag{1.153}$$

$$= \sqrt{\frac{k_\mathrm{B} T \varepsilon_0 \varepsilon_\mathrm{Ana}}{2 q^2 I}}. \tag{1.154}$$

The Debye length λ_D states a characteristic length for the electrolytic system. It is the length at which the charge density and also the electric potential of an ion atmosphere reduces to $1/e$.

This approach offers the possibility to estimate the maximal distance of a charged macromolecule to the dielectric-electrolyte interface before its charge is entirely screened by counter ions, or in the case of very large macromolecules (e.g. DNA) to estimate the amount of charge coupled into the semiconductor. The Debye length λ_D influences the double layer thickness and increases the concentration of the counter ions[5] comprising the double layer.

The second parameter has already been introduced in (1.154) and describes the ionic strength of the electrolyte. The ionic strength of an electrolyte is defined as:

$$I(\mathbf{x}) = \frac{1}{2} \sum_{\xi \in S} \xi^2 c_\xi^\infty(\mathbf{x}). \tag{1.155}$$

The ionic strength describes the strength of a solution as a function of ion concentration and ion valence. It is one of the main characteristics of a solution

[5] A counter ion is the ion that accompanies an ionic species in order to gain charge neutrality. For instance, in sodium chloride, the sodium cation is the counter ion of the chlorine anion and vice versa.

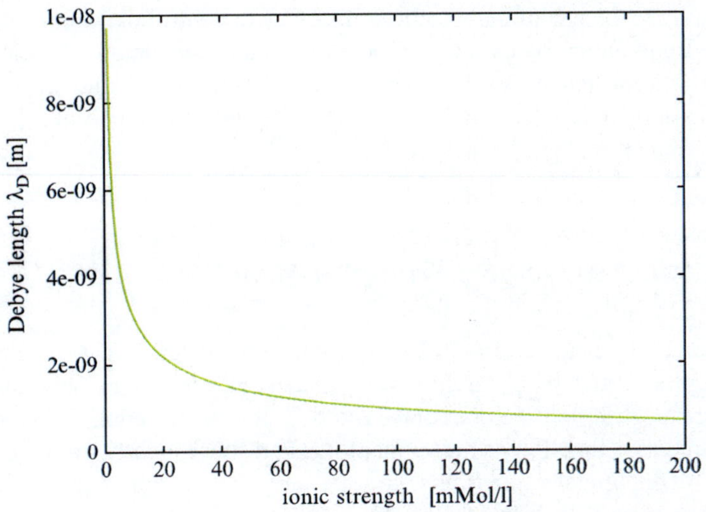

Fig. 1.4 Depicting the relation between the ion concentration for a 1 : 1 electrolytic solution and the Debye length λ_D. Increasing the salt concentration reduces the repulsion between complementary DNA strands and accelerates the hybridization events, but also decreases the Debye length λ_D and thus the device signal

containing dissolved ions and influences many important properties like the dissociation or solubility of different salts, and the double layer thickness (q.v. Sect. 8.2). The strong deviations from ideality which are typically experienced for ionic solutions described via the Debye–Hückel model are handled by the ionic strength. Furthermore, it is related to electrokinetic phenomena, electroacoustic phenomena in colloids and other heterogeneous systems and linked to the electric double layer. The Debye length λ_D is inversely proportional to the square root of the ionic strength (Fig. 1.4). Media with high ionic strength are employed to minimize the changes in the activity quotient of solutes during titration, which are more pronounced at lower concentrations. Natural waters such as seawater have a non-zero ionic strength due to the presence of dissolved salts, which significantly affects their properties.

Buffers and Ionic Strength

Commonly, an experiment is carried out in a so-called *buffer* solution. There are several reasons for this. Enzyme reactions are very sensitive to the local temperature, the local substrate concentration, and also to their *chemical* environment (e.g. pH). Here, the buffer fulfills the function of stabilizing the pH of the solution at a certain point and thus keeping the enzyme activity at its maximum. In the case of DNA hybridization, the ions in the buffer gather around the single DNA strands and screen partially the DNA charge. Therefore, the repulsion between two complementary negatively charged single DNA strands is reduced and they can approach each other close enough to enable the hybridization reaction.

In summary, the use of buffer solutes is a way to control the chemical properties of the environment in which the chemical reaction is conducted. Therefore, buffers are significant ingredients in the description of BioFETs and knowing the ion concentrations and the ionic strength for a buffer is of general importance [26].

7.2 *Parameters for the Energy Transport Model and the Six-Moment Transport Model*

A big advantage of the Drift–Diffusion Transport model is that, it only contains the carrier mobilities $\mu_{n,p}$. These parameter depend on various quantities such as the applied electric field, temperature, and doping concentration. The mobility can be measured as a function of these quantities and subsequently translated to fit parameters for analytical expressions. Unfortunately, this is not as easy for the higher order transport models. In the following we use μ_0 as abbreviation for $\mu_{n,p}$ and treat electrons and holes with structurally the same formulas. In the case of the Energy Transport model [132], two additional parameters are needed, the energy flux mobility μ_1 and the energy relaxation time τ_1. They can not be directly measured and therefore have either to be modeled [16, 83] or extracted from Monte Carlo simulations [111, 132, 213, 216]. The analytical models require parameters which are adjusted to fit the experimental data of a particular application. The problem is that there is no unique parameter set which fits all requirements.

For the sake of completeness we will start with the analytical description of the parameters. Due to the analog description of electrons and holes, we will restrict ourself to the modeling of electron parameters in the following.

Analytical Models for the Mobility

Two models are frequently used to describe the energy dependence of the mobility. There is the model after Baccarani [15, 16]:

$$\frac{\mu(T_n)}{\mu^{\mathrm{LIS}}} = \frac{T_L}{T_n} \tag{1.156}$$

and the model after Hänsch [83, 84]:

$$\frac{\mu(T_n)}{\mu^{\mathrm{LIS}}} = \left(1 - \frac{3}{2}\frac{\mu^{\mathrm{LIS}}}{\tau_1 v_s^2}\left(\frac{k_{\mathrm{B}} T_{\mathrm{L}}}{q} + \frac{2}{5}\frac{n\mathbf{s}}{\mathbf{j}}\right)\right)^{-1}. \tag{1.157}$$

Assuming homogeneous conditions the energy flux $\mathbf{s}$ is proportional to the particle current [72]:

$$\frac{\mathbf{s}}{\mathbf{j}} = -\frac{5 k_{\mathrm{B}} T_{\mathrm{n}}}{2q}. \tag{1.158}$$

Substituting (1.158) into (1.157) yields a simplified formulation:

$$\frac{\mu(T_n)}{\mu^{\mathrm{LIS}}} = \left(1 + \frac{3}{2}\frac{\mu^{\mathrm{LIS}} k_B}{q\tau_1 v_s^2}(T_n - T_L)\right)^{-1}. \tag{1.159}$$

As demonstrated in [130, 132], (1.158) reproduces the mobility quite reasonably in regions with increasing electric field, while for decreasing electric field, however, it is better to employ (1.157) cf. [132, 212].

Another approach to model the mobility has been introduced in [213] and is based on the separation of homogeneous and inhomogeneous parts of the mobility. It is proposed to describe the collision term C_p as:

$$n\mathsf{C}_p = \frac{\mathbf{j}}{\mu} = \frac{\mathbf{j}}{\mu^*} + \lambda_p n \nabla \cdot \hat{\mathsf{U}}, \tag{1.160}$$

with μ^* denoting the homogeneous mobility.

The ratio between the energy flux mobility μ_1 and the mobility μ_0 is usually expressed via constant values in the range $0.79 - 1.0$ps [132, 213]. Tang et al. [213] suggested to model the collision operator $\mathsf{C}_{p\mathscr{E}}$ as:

$$\mathsf{C}_{p\mathscr{E}} = -\frac{q\mathbf{s}}{\mu_1^*} + \lambda_{p\mathscr{E}} \nabla \cdot \hat{\mathsf{R}}, \tag{1.161}$$

which is analogous to (1.160). μ_1^* denotes the homogeneous energy flux mobility. The corresponding expressions for μ_1^* and $\lambda_{p\mathscr{E}}$ are given in [213].

Analytical Models for Relaxation Times

Commonly used values of the energy relaxation time τ_1 for silicon are in the range 0.3–0.4 ps, while in general values in the range 0.08–0.68 ps have been employed [98]. The Monte Carlo simulations demonstrate that the constant relaxation time assumption is quite reasonable [213]. However, there have been different energy expressions used. Baccarani et al. [15, 16] proposed for electrons:

$$\tau_1(T_n) = \frac{3}{2}\frac{k_B\mu_0}{qv_s^2}\frac{T_n T_L}{T_n + T_L} + \frac{m^*\mu_0}{2q}\frac{T_n}{T_L}. \tag{1.162}$$

Employing (1.156) and (1.161) together yields the correct homogeneous limit.

Hänsch's approach demands only an energy relaxation time τ_1 independent of the carrier temperature for (1.157) to reproduce the correct homogeneous limit. Defining:

$$\tau_{\mathscr{E}} = \frac{3 k_B \mu_0 T_L}{2qv_s^2} \tag{1.163}$$

and employing this to (1.157) and (1.159) results in a description equivalent to Baccarani's mobility model in the homogeneous case. A more detailed discussion about the inconsistencies arising, when combining an energy-dependent mobility and energy relaxation time models is found in [170]. On the basis of data from Fischetti [50], Agostinelli proposed a model for the energy relaxation time for silicon [4]:

$$\frac{\tau_1(W)}{1\,\mathrm{ps}} = \begin{cases} 0.172 + 2.656W - 3.448W^2, & \text{for } W \leq 0.4 \\ 0.68 & \text{, for } W > 0.4 \end{cases}, \tag{1.164}$$

with $W = w/(1\,\mathrm{eV})$. Another fit to newer data from Fischetti has been shown by Hasnat et al. [87] and is expressed via:

$$\frac{\tau_1(W)}{1\,\mathrm{ps}} = 0.27 + 0.62W - 0.63W^2 + 0.13W^3 + 0.01W^4, \tag{1.165}$$

exhibiting a maximum of approximately 0.42 ps. The effects of relaxation time and transport models on the performance of silicon bipolar transistors has been studied in [177] in more detail.

Parameter Extraction from Monte Carlo Simulations

For the Six-Moments Transport model the parameter set extends and includes μ_0, μ_1 and μ_2 for the mobilities, H_1, H_2 and H_3 as non-parabolicity factors in the flux equations (1.90), (1.92) and (1.94), and τ_1 and τ_2 relaxation times employed in the balance equations (1.89), (1.91) and (1.93). The parameters are difficult to model due to their dependence on the shape of the distribution function, on the band structure, and on hot carrier effects. Furthermore, the mobilities and relaxation times are scattering controlled. Simple empirical models are often non-satisfactory [71] and in particular hard to compare against Monte Carlo simulations, due to the non-matching results of the transport model with Monte Carlo data in the homogenous case and the questionable extension of these models into the inhomogeneous case.

In order to avoid these problems Grasser et al. [69] extracted all physical parameters as a function of the doping concentration and the average energy from homogenous Monte Carlo simulations. Due to the derivation of all model parameters from bulk Monte Carlo simulations, the resulting transport models are free of fit-parameters and yield a *no knobs to turn* description. Facing far too many parameters is an intrinsic property in many higher order transport models based on analytical models for the mobilities and relaxation times [72].

Figure 1.5 illustrates the non-parabolicity factor dependence on the energy at an electric field of $950\,\mathrm{kV\,cm^{-1}}$. As can be seen the non-parabolicity factors, gained from subband Monte Carlo simulations, head to unity for low energies and thus are consistent with the parabolic band case, where the non-parabolicity factors are equal to one [220].

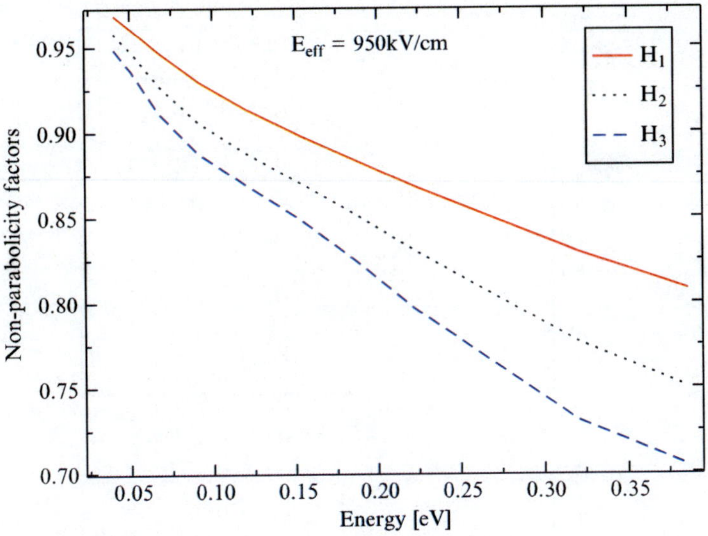

Fig. 1.5 H_1, H_2 and H_3 in relation to the energy at an electric field strength of $950\,\mathrm{kV\,cm^{-1}}$. At low energies, the non-parabolicity factors head to unity. The presented non-parabolicity factors have been extracted from subband Monte Carlo simulations [220]

Parameters for higher-order macroscopic transport models are displayed in Figs. 1.6 and 1.7. The carrier mobility μ_0 and the higher-order mobilities μ_1 and μ_2 are depicted as a function of the electric field $|\mathbf{E}|$ for different doping concentrations N_d (Fig. 1.6, [220]). For electric fields above $100\,\mathrm{kV\,cm^{-1}}$ the values of the mobilities exhibit no dependence on the doping concentration, while for low fields and low doping concentrations, the carrier mobility is very high compared to low fields and high doping concentrations. The energy flux mobility μ_1 and the second-order energy flux mobility μ_2 are smaller than the carrier mobility μ_0 at low doping concentrations and low fields, whereas at low fields and high doping concentrations the values of all mobilities are comparable.

Figure 1.7 depicts the relation between the relaxation times τ_1 and τ_2 for different doping concentrations and as a function of the kinetic carrier energy. Here, at high energies the relaxation times do not depend on the doping concentration and their decrease is caused by the increase of optical phonon scattering. At high doping concentrations N_d, the Monte Carlo simulations predict lower relaxation times in comparison to low N_d.

Avoiding fit-parameters is a crucial point for higher-order models, since their mutual influence is quite complex and the numerical stability of the whole transport model relies on an appropriate choice of these parameters. It has been demonstrated that the model based on the Monte Carlo data outperforms its counterparts based on analytical mobility models [71] substantially, in the quantitative agreement of the simulation results with Monte Carlo device simulations as well as in the numerical stability of the simulation.

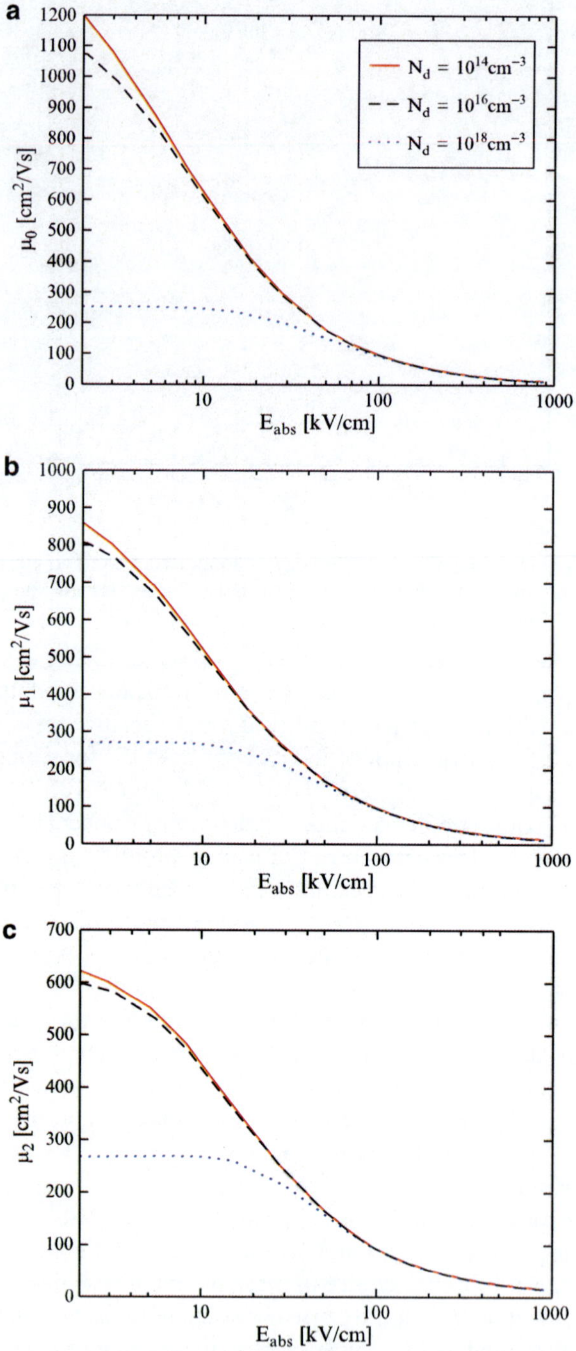

Fig. 1.6 Carrier mobility μ_0, energy flux mobility μ_1, and second-order energy flux mobility μ_2 as a function of driving field for different doping concentrations. While for low fields the values of the mobilities for the low doping case are high in comparison to the high doping case, for fields higher than $100\,\mathrm{kV\,cm^{-1}}$, the mobilities are independent of the doping concentration

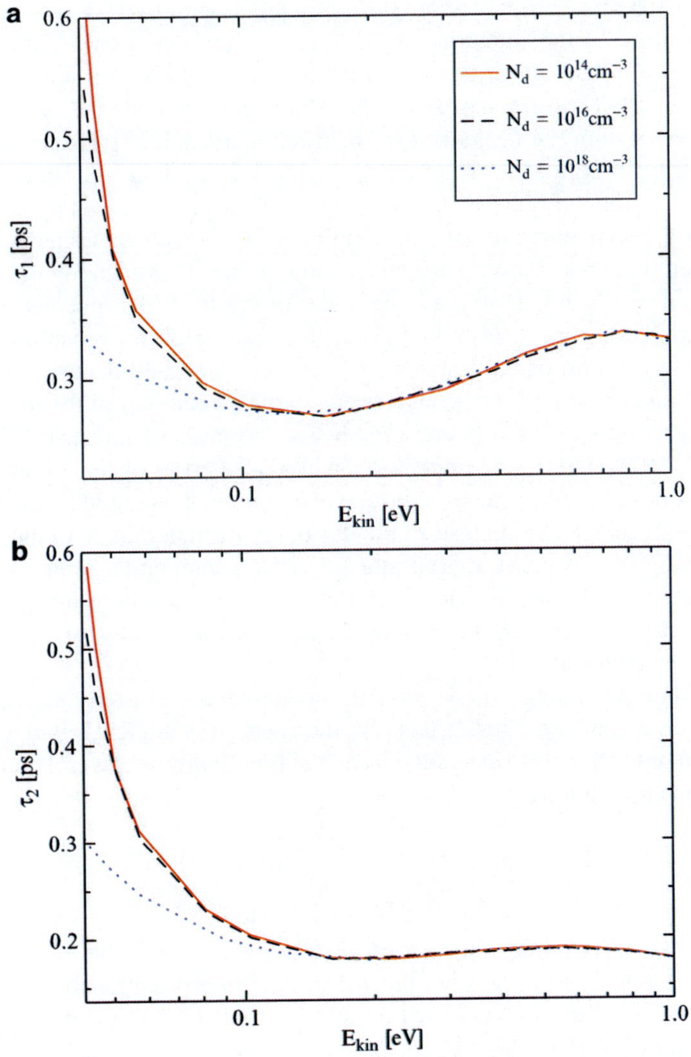

Fig. 1.7 Illustrating the energy-relaxation time τ_1 and the second-order energy relaxation time τ_2 as a function of the kinetic energy for different bulk dopings, extracted from bulk Monte Carlo simulations. At very high energies, the relaxation times decrease due to the increase in optical phonon scattering

7.2.1 Thermoelectric Phenomena

The advent of thermoelectric devices dates back to 1821, when Seebeck found the deviation of a compass needle due to two junctions of different metals at different temperatures [183]. This now called Seebeck effect was caused by the formation of a potential difference due to the temperature gradient. Thirteen years later Peltier discovered that an electrical current through a junction of two different metals alters the temperature at the junction [162]. Some years later, Lenz found that material combination and current direction determine uniquely, if a junction is cooled or heated [175]. Thomson explained the connection between the Seebeck and the Peltier effect within the framework of thermodynamics [217]. He was also able to predict a third thermoelectric effect, today known as Thomson effect. Altenkirch contributed significantly to the theory of thermoelectric materials by deducing that high quality thermoelectric materials exhibit high Seebeck coefficients and electrical conductivities but show low thermal conductivities [7,8]. Taking these attributes into account one is able to express the figure of merit for thermoelectric materials, which became an important part of the systematic search for novel thermoelectric materials. In the mid of the last century, Ioffe concentrated the research on semiconductor based thermoelectric devices due to the availability of the first artificially manufactured semiconductors and established the basis of modern thermoelectric theory [101,102]. Due to the improved material properties of semiconductors compared to metals, the efficiency of thermoelectric generators could be raised to about 5%. Intense research efforts lead to the discovery of materials with increased thermoelectric figures of merit appropriate for various temperature ranges. Today the basic structure of thermoelectric generators is a combination of n-type and p-type semiconductor rods, arranged thermally parallel and electrically serial, regardless of the employed materials.

In the sequel the three thermoelectric phenomena are briefly explained in the order of their discovery. These effects are the phenomenological foundations for the description of thermoelectric materials and the functioning of several thermoelectric devices and applications.

Seebeck Effect

The Seebeck effect relates the rise of an electrical voltage due to a temperature gradient. Seebeck not only gave the theoretical interpretation in his pioneering paper [183], but also an overview of several material combinations applicable in thermocouples (cf. Fig. 1.8).

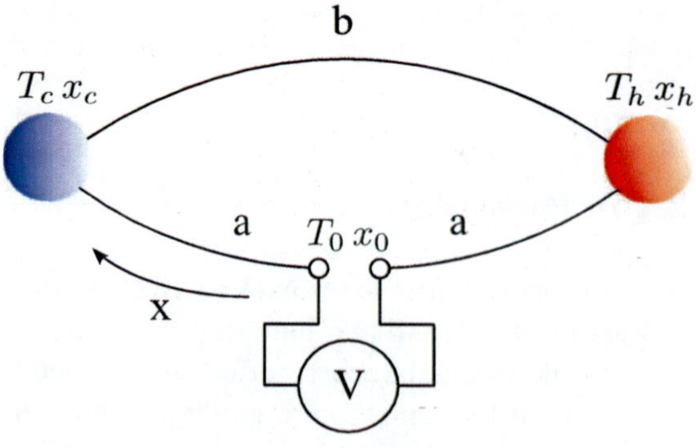

Fig. 1.8 Thermocouple scheme built with two metal rods

A thermocouple consists of two rods of different materials which are soldered together. The soldered points are held at the temperatures T_c and T_h, experiencing a temperature difference ΔT and thus exhibiting a temperature gradient along the rods. From the view point of a device, the given temperature step generates a certain voltage measured at the device contacts:

$$U_{\text{Seebeck}} \propto \Delta T. \tag{1.166}$$

On a microscopic level the Seebeck coefficient is defined via the limit at infinitesimal temperature differences:

$$\alpha(T) = \lim_{\Delta T \to 0} \frac{\Delta U}{\Delta T}. \tag{1.167}$$

The total voltage obtained on a rod is expressed by the path integral along the rod as:

$$U_{\text{Seebeck}} = \varphi_2 - \varphi_1 = \int_{x_1}^{x_2} \partial_x \varphi \, \mathrm{d}x = \int_{x_1}^{x_2} \alpha(T) \, \partial_x T \, \mathrm{d}x = \int_{T_1}^{T_2} \alpha(T) \, \mathrm{d}T. \tag{1.168}$$

In order to obtain the potential for the entire device one has to evaluate the path integral around both rods. Additionally to the contributions of the two rods, the contact potentials at the soldered points has to be added. However, due to the cancellation of the contact potentials the voltage is given by:

$$U_{\text{Seebeck}} = \int_{T_0}^{T_c} \alpha_a(T) \, \mathrm{d}T + \int_{T_c}^{T_h} \alpha_b(T) \, \mathrm{d}T + \int_{T_h}^{T_0} \alpha_a(T) \, \mathrm{d}T. \tag{1.169}$$

Averaging the temperature dependent Seebeck coefficient along the rods allows to express a combined coefficient for the material couple under given thermal conditions expressed as the difference of the single contribution of each rod:

$$U_{\text{Seebeck}} = (\bar{\alpha}_b - \bar{\alpha}_a) \int_{T_c}^{T_h} \mathrm{d}T = (\bar{\alpha}_b - \bar{\alpha}_a) \, \Delta T. \tag{1.170}$$

Two materials with opposite signs for their Seebeck coefficients must be chosen in order to maximize the gained output voltage. While most metals exhibit Seebeck coefficients in the range of $1-10\,\mu\mathrm{V\,K}^{-1}$, semiconductors offer values of mV and more. There are metals with positive and negative Seebeck coefficients. Depending on the intended application one has to choose an appropriate material combination. For instance, measurement applications do not necessarily need high total Seebeck coefficients, but a linear behavior in the desired temperature range is required. In semiconductors, the Seebeck coefficient is adjusted by appropriately varying the doping. While p-type materials posses positive Seebeck coefficients, n-type materials offer negative ones.

Peltier Effect

The Peltier effect states the phenomenological effect reverse to the Seebeck effect. Driving an electrical current through two connected rods generates a temperature difference between the two soldered points. Therefore, heat is absorbed at one end, while it is released at the other end. In conjunction, a heat flux through the rods is induced. The heat flux at the junctions can be interpreted as energy conservation within the junction and a change of total energy of the carriers when passing the junction.

The heat flux through the rods is governed by the charge current, and the Peltier coefficient and given by:

$$\mathbf{j}^{q}_{Peltier} = \pi_{ab}\mathbf{j}, \tag{1.171}$$

where π_{ab} denotes the Peltier coefficient of a junction, defined by the difference of the contributing materials $\pi_{ab} = \pi_a - \pi_b$. Thus, the direction of the heat flow at a junction is controlled by the choice of materials and the direction of the current flow. Furthermore, the Peltier coefficients are also temperature dependent like the Seebeck coefficients.

The Peltier coefficient and the Seebeck coefficient are not independent of each other. From phenomenological thermodynamics (first Kelvin relation [32]) as well as a systematic approach via the method of moments [224] the following relation is derived:

$$\pi_{ab} = \alpha_{ab} T. \tag{1.172}$$

Thomson Effect

Thomson (later Lord Kelvin) predicted and observed the third thermoelectric effect. For a homogeneous conductor exerted to a temperature gradient, the carriers crossing the temperature gradient will experience an energy gain or release depending on their relative direction to the temperature gradient. The energy change of the transiting carriers is absorbed or released as heat, respectively. The total Thomson heat absorbed or released along on rod is defined by:

$$\mathbf{j}^{q}_{\mathrm{Thomson}} = \int_{T_c}^{T_h} \chi(T)\,\mathbf{j}\,\mathrm{d}T, \tag{1.173}$$

where $\chi(T)$ denotes the temperature dependent Thomson coefficient. The Thomson and Seebeck coefficient are related by the second Kelvin law:

$$\chi = T\frac{\mathrm{d}\alpha}{\mathrm{d}T}. \tag{1.174}$$

Thermodynamic Relations

As already mentioned before the three thermoelectric coefficients are related to each other. In the following section, these relations will be discussed within the framework of fundamental thermodynamics [32, 166, 231].

Additionally to the three presented reversible phenomena, two irreversible processes occur in the structure. Firstly, each electrical current causes the dissipation of Joule heat, when passing a material with electrical resistance, and secondly heat is conducted in the device (Fourier's law).

In the following derivations the device illustrated in Fig. 1.8 is regarded as electrically short circuited. Therefore, no electric power is dissipated and no external voltage induced. Furthermore, the cold and the hot contact are connected to thermal reservoirs and energy losses by Joule heating are very small and negligible. The law of total energy conservation in the entire device including the reservoirs for a closed loop and all three thermoelectric effects is given by:

$$\underbrace{\mathbf{j}\alpha_{ab}\Delta T}_{\text{Seebeck}} = \underbrace{\mathbf{j}\pi_{ap}(T_h) - \mathbf{j}\pi_{ap}(T_c)}_{\text{Peltier}} + \underbrace{\mathbf{j}\left(\int_{T_c}^{T_h}\chi_b\,\mathrm{d}T - \int_{T_c}^{T_h}\chi_a\,\mathrm{d}T\right)}_{\text{Thomson}}. \tag{1.175}$$

The Seebeck effect generates the driving force for a current throughout the device. The current induces the Peltier and the Thomson effect. Substituting $T_h - T_c$ by ΔT and dividing (1.175) by $\mathbf{j}$ yields the following expression:

$$\alpha_{ab} = \frac{\pi_{ab}(T_c+\Delta T) - \pi_{ab}(T_c)}{\Delta T} + \frac{1}{\Delta T}\left(\int_{T_c}^{T_c+\Delta T}\chi_b\,\mathrm{d}T - \int_{T_c}^{T_c+\Delta T}\chi_a\,\mathrm{d}T\right). \tag{1.176}$$

Now, letting ΔT approach zero, the energy relation between the three effects is gained:

$$\alpha_{ab} = \frac{\mathrm{d}\pi_{ab}}{\mathrm{d}T} + \chi_b - \chi_a. \tag{1.177}$$

Neglecting irreversible processes allows to equate the net change of entropy of the entire structure including the reservoirs to zero. Hence, the contributions from all three effects anull:

$$\Delta S = -\mathbf{j}\frac{\pi_{ab}(T_c+\Delta T)}{T_c+\Delta T} + \mathbf{j}\frac{\pi_{ab}}{T_c} - \mathbf{j}\int_{T_c}^{T_h}\frac{\chi_b}{T}\,\mathrm{d}T + \mathbf{j}\int_{T_c}^{T_h}\frac{\chi_a}{T}\,\mathrm{d}T = 0. \tag{1.178}$$

Dividing (1.178) by $\mathbf{j}$ and extending the Peltier term by $\Delta t/\Delta T$ gives:

$$\left(-\frac{\pi_{ab}(T_c+\Delta T)}{T_c+\Delta T} + \frac{\pi_{ab}(T_c)}{T_c}\right)\frac{\Delta T}{\Delta T} = \int_{T_c}^{T_h}\frac{\chi_b}{T}\,\mathrm{d}T - \int_{T_c}^{T_h}\frac{\chi_a}{T}\,\mathrm{d}T. \tag{1.179}$$

In the limit of $\Delta T \rightarrow 0$, the relation between the Peltier and the Seebeck coefficient is gained:

$$-\frac{\mathrm{d}}{\mathrm{d}T}\left(\frac{\pi_{ab}}{T}\right) = \frac{\chi_b - \chi_a}{T}. \tag{1.180}$$

Expanding the derivative in (1.180) results in a more convenient form:

$$\frac{\pi_{ab}}{T} = \frac{\mathrm{d}\pi_{ab}}{\mathrm{d}T} + \chi_b - \chi_a. \tag{1.181}$$

Substituting the right hand side of (1.181) by (1.177) relates the Seebeck and the Peltier effect as observed by Thomson also known as the first Kelvin relation:

$$\frac{\pi_{ab}}{T} = \alpha_{ab}. \tag{1.182}$$

The second Kelvin relation, connecting the Seebeck and the Thomson coefficient, is derived by exchanging the Peltier term in (1.181) with (1.182):

$$T\frac{\mathrm{d}\alpha_{ab}}{\mathrm{d}T} = \chi_a - \chi_b. \tag{1.183}$$

The same results can be derived from Onsager's reciprocal relations [155].

7.2.2 Electrothermal Transport Model

It is important to correctly describe the energy relations in order to gain good results from thermoelectric device simulations. The contributions of the carrier subsystem and the lattice are combined to one heat-flux equation, incorporating rigorous treatment of the coupling mechanisms between the thermal and the electrical description.

Due to the rather low driving forces in electrothermal devices, it is safe to assume that the carrier gas is in local thermal equilibrium with the lattice, and the Electrothermal Transport model can be deduced from the Energy Transport model.

Besides the mandatory Poisson equation, the Electrothermal Transport model requires carrier balance equations and current equations for both carrier types. The energy relations are handled by the heat flow equation which can be deduced via systematic (method of moments) or phenomenological approaches (heuristic inclusion of heat transport).

In the following the Electrothermal Transport model is derived from the moment equations via the Bløtekjær approach (cf. Sect. 5.4). The according energy flux equation (1.81) expressed in terms of the particle flux is given in local thermal equilibrium by:

$$\mathbf{j}_{\nu,u} = \frac{5}{2}\frac{\mu_{\nu,1}}{\mu_{\nu,0}} k_{\mathrm{B}} T \mathbf{j}_\nu - \kappa_\nu \nabla_{\mathbf{r}} T, \tag{1.184}$$

with κ_ν denoting the thermal conductivity of the carrier subsystem in obedience to Wiedemann–Franz's laws:

$$\kappa_\nu = \frac{5}{2}\frac{k_\mathrm{B}^2}{q}\mu_{\nu,1}\nu T, \tag{1.185}$$

and ν as a placeholder for electrons n or holes p. Equation (1.184) shows the two distinct contributions to the energy flux, heat conduction, and the thermal energy of the moving carrier gas. For non-degenerate semiconductors, the thermal conductivities of the carrier subsystem can be neglected against the lattice contribution [142]. Substituting (1.184) into the energy balance equation (1.80) leads to:

$$\partial_t w + \frac{5}{2}\frac{\mu_{\nu,1}}{\mu_{\nu,0}}k_\mathrm{B}T\nabla\cdot\mathbf{j}_\nu + \frac{5}{2}\frac{\mu_{\nu,1}}{\mu_{\nu,0}}k_\mathrm{B}\mathbf{j}_\nu\cdot\nabla T - \nabla\cdot(\kappa_\nu\nabla T) + \mathrm{s}_\alpha q\mathbf{j}_\nu\cdot\nabla\tilde{\varphi} - G_\nu^{\mathscr{E}} = 0. \tag{1.186}$$

Here, $G_\nu^{\mathscr{E}}$ denotes the net generation rate. After a few rearrangements of (1.186), one is able to gain expressions for physical interpretation. In the first step the gradient of the electrochemical potential Φ_ν is substituted by the current relation [184, 224]:

$$\nabla\Phi_\nu = -\mathrm{s}_\alpha\frac{\mathbf{j}_\nu}{\mu_{\nu,0}\nu} - \mathrm{s}_\alpha\frac{k_\mathrm{B}}{q}\left(\frac{5}{2} - \ln\frac{\nu}{N_{c,\nu}}\right)\nabla T, \tag{1.187}$$

and the Seebeck coefficient is defined by:

$$\alpha_\nu = \mathrm{s}_\alpha\frac{k_\mathrm{B}}{q}\left(\frac{5}{2} - \ln\frac{\nu}{N_{c,\nu}}\right). \tag{1.188}$$

Rewriting (1.186) by insertion of (1.187) and (1.188) the following expression is obtained:

$$\begin{aligned}\partial_t\frac{3}{2}k_\mathrm{B}T \;-\; \nabla\cdot(\kappa_\nu\nabla T) + \mathrm{s}_\alpha q\frac{\mu_{\nu,1}}{\mu_{\nu,0}}\nabla\cdot\mathbf{j}_\nu\left(\alpha_\nu T + \Phi_\nu - \tilde{\varphi}\right) + \frac{\mu_{\nu,1}}{\mu_{\nu,0}}q\,\mathrm{s}_\alpha T\mathbf{j}_\nu\cdot\nabla\alpha_\nu \\ - \frac{\mu_{\nu,1}}{\mu_{\nu,0}}q\frac{|\mathbf{j}_\nu|^2}{\mu_{\nu,0}\nu} + \mathrm{s}_\alpha\left(1 - \frac{\mu_{\nu,1}}{\mu_{\nu,0}}\right)q\mathbf{j}_\nu\cdot\nabla\tilde{\varphi} - G_\nu^{\mathscr{E}} = 0.\end{aligned} \tag{1.189}$$

The energy balance equation (1.189) describes the electron and hole subsystem. The lattice contributes via an additional heat-flux term which represents the dominant contribution to heat conduction for most moderately doped semiconductors. This contribution is covered by Fourier's law with a corresponding lattice heat conductivity κ_L. Therefore, the energy balance equations for the three subsystems are given by:

$$\begin{aligned}\frac{3}{2}k_\mathrm{B}\partial_t T \;=\; \nabla\cdot(\kappa_n\nabla T) + \frac{\mu_{n,1}}{\mu_{n,0}}q\frac{|\mathbf{j}_n|^2}{\mu_{n,0}n} + \frac{\mu_{n,1}}{\mu_{n,0}}q\left(\alpha_n T + \Phi_n - \tilde{\varphi}\right)\nabla\cdot\mathbf{j}_n \\ + \frac{\mu_{n,1}}{\mu_{n,0}}qT\mathbf{j}_n\cdot\nabla\alpha_n + \left(1 - \frac{\mu_{n,1}}{\mu_{n,0}}\right)q\mathbf{j}_n\cdot\nabla\tilde{\varphi} + G_n^{\mathscr{E}},\end{aligned} \tag{1.190}$$

$$\frac{3}{2}k_{\mathrm{B}}\partial_t T = \nabla\cdot(\kappa_p \nabla T) + \frac{\mu_{p,1}}{\mu_{p,0}} q \frac{|\mathbf{j}_p|^2}{\mu_{p,0} p} - \frac{\mu_{p,1}}{\mu_{p,0}} q\,(\alpha_p T + \Phi_n - \tilde{\varphi})\,\nabla\cdot\mathbf{j}_p$$
$$- \frac{\mu_{p,1}}{\mu_{p,0}} q T\,\mathbf{j}_p\cdot\nabla\alpha_p - \left(1 - \frac{\mu_{p,1}}{\mu_{p,0}}\right) q\mathbf{j}_p\cdot\nabla\tilde{\varphi} + G_n^{\mathscr{E}}, \tag{1.191}$$

$$c_{\mathrm{L}}\partial_t T = \nabla\cdot(\kappa_{\mathrm{L}} \nabla T). \tag{1.192}$$

The cumulative heat-flow is defined by the sum of the contributions of all three subsystems. Specific heat as well as thermal conductivity are handled as parameters for the entire semiconductor. Thus, the heat-flow equation is given by:

$$c_{\mathrm{tot}}\partial_t T = \nabla\cdot(\kappa_{\mathrm{tot}} \nabla T) + H, \tag{1.193}$$

where the heat source term is expressed as:

$$H = \frac{\mu_{n,1}}{\mu_{n,0}} q \frac{|\mathbf{j}_n|^2}{\mu_{n,0} n} + \frac{\mu_{p,1}}{\mu_{p,0}} q \frac{|\mathbf{j}_p|^2}{\mu_{p,0} p} + \mathscr{E}_g R + \frac{\mu_{n,1}}{\mu_{n,0}} q\,(\alpha_n T + \Phi_n - \tilde{\varphi})\,\nabla\cdot\mathbf{j}_n$$
$$- \frac{\mu_{p,1}}{\mu_{p,0}} q\,(\alpha_p T + \Phi_p - \tilde{\varphi})\,\nabla\cdot\mathbf{j}_p + qT\left(\frac{\mu_{n,1}}{\mu_{n,0}}\mathbf{j}_n\cdot\nabla\alpha_n - \frac{\mu_{p,1}}{\mu_{p,0}}\mathbf{j}_p\cdot\nabla\alpha_p\right)$$
$$+ q\left(\left(1 - \frac{\mu_{n,1}}{\mu_{n,0}}\right)\mathbf{j}_n - \left(1 - \frac{\mu_{p,1}}{\mu_{p,0}}\right)\mathbf{j}_p\right)\nabla\tilde{\varphi}. \tag{1.194}$$

The divergence terms of the electron and hole currents can be substituted by the net recombination rate, for vanishing $\partial_t \nu$ terms in the carrier balance equation in stationary cases. The resulting source term is given by:

$$H = \frac{\mu_{n,1}}{\mu_{n,0}} q \frac{|\mathbf{j}_n|^2}{\mu_{n,0} n} + \frac{\mu_{p,1}}{\mu_{p,0}} q \frac{|\mathbf{j}_p|^2}{\mu_{p,0} p}$$
$$+ q\left(\frac{\mu_{n,1}}{\mu_{n,0}}(\alpha_n T + \Phi_n - \tilde{\varphi}) - \frac{\mu_{p,1}}{\mu_{p,0}}(\alpha_p T + \Phi_p - \tilde{\varphi}) - \mathscr{E}_g\right) G$$
$$+ qT\left(\frac{\mu_{n,1}}{\mu_{n,0}}\mathbf{j}_n\cdot\nabla\alpha_n - \frac{\mu_{p,1}}{\mu_{p,0}}\mathbf{j}_p\cdot\nabla\alpha_p\right) + q\left(\left(1 - \frac{\mu_{n,1}}{\mu_{n,0}}\right)\mathbf{j}_n - \left(1 - \frac{\mu_{p,1}}{\mu_{p,0}}\right)\mathbf{j}_p\right)\nabla\tilde{\varphi}. \tag{1.195}$$

A not fully justifiable but frequently used assumption is to set the mobility ratios to unity for electrons and holes [72]. The heat source term simplifies then to:

$$H = q\frac{|\mathbf{j}_n|^2}{\mu_n n} + q\frac{|\mathbf{j}_p|^2}{\mu_p p} + q\left(T\,(\alpha_n - \alpha_p) + \Phi_n - \Phi_p - \mathscr{E}_g\right) G + qT\,(\mathbf{j}_n\cdot\nabla\alpha_n - \mathbf{j}_p\cdot\nabla\alpha_p). \tag{1.196}$$

Equation (1.196) contains a contribution from Joule heat losses due to the current flow through the structure, heat transferred to the lattice by the carrier recombination, and Thomson heat.

7.2.3 Seebeck Coefficient

While up to now the Seebeck coefficient has been treated on a phenomenological basis, its inclusion in the semiconductor current equations will be studied in the sequel.

For a non-zero temperature gradient between the two ends of a homogeneous and solid material, a thermoelectric voltage can be measured. The Seebeck coefficient is defined by the ratio of the resulting voltage and the temperature difference. The temperature gradient times the Seebeck coefficient is equal to the negative gradient of the electrochemical potential:

$$-\nabla \Phi_\nu = \alpha_\nu \nabla T. \tag{1.197}$$

This equation is only valid for zero current at open circuit conditions. The current equation (1.187) deduced via the Bløtekjær approach is:

$$\mathbf{j}_\nu = -s_\alpha \mu_{\nu,0} \nu \nabla \Phi_\nu - \mu_{\nu,0} \nu \frac{k_\mathrm{B}}{q} \left(\frac{5}{2} - \ln \frac{\nu}{N_{c,\nu}} \right) \nabla T = 0. \tag{1.198}$$

Since an assumption that the carrier gas is in local equilibrium with the lattice was used, the carrier temperature is equal to the lattice temperature and can be expressed by a single temperature in the current relations:

$$T_\nu = T_\mathrm{L} = T. \tag{1.199}$$

The Seebeck coefficient in (1.197) is identified as:

$$\alpha_\nu = s_\alpha \frac{k_\mathrm{B}}{q} \left(\frac{5}{2} - \ln \frac{\nu}{N_{c,\nu}} \right). \tag{1.200}$$

The resulting current relations for electrons and holes are expressed by:

$$\begin{aligned} \mathbf{j}_n &= \mu_{n,0} n \nabla \Phi_n - \mu_{n,0} n \frac{k_\mathrm{B}}{q} \left(\frac{5}{2} - \ln \frac{n}{N_c} \right) \nabla T \\ &= \mu_{n,0} n \left(\nabla \Phi_n + \alpha_n \nabla T \right), \end{aligned} \tag{1.201}$$

$$\begin{aligned} \mathbf{j}_p &= -\mu_{p,0} p \nabla \Phi_p - \mu_{p,0} p \frac{k_\mathrm{B}}{q} \left(\frac{5}{2} - \ln \frac{p}{N_v} \right) \nabla T \\ &= -\mu_{p,0} p \left(\nabla \Phi_p + \alpha_p \nabla T \right), \end{aligned} \tag{1.202}$$

and the corresponding Seebeck coefficients are defined as:

$$\alpha_n = -\frac{k_\mathrm{B}}{q}\left(\frac{5}{2} - \ln\frac{n}{N_c}\right), \tag{1.203}$$

$$\alpha_p = \frac{k_\mathrm{B}}{q}\left(\frac{5}{2} - \ln\frac{p}{N_v}\right). \tag{1.204}$$

The opposing signs of the Seebeck coefficients in (1.203) and (1.204) are the reason for the basic thermoelectric device behavior, exhibiting two legs with p and n doping, respectively. These devices are commonly built electrically in serial but thermally in parallel, thus yielding a constructive interference of the contributions from both legs (see Fig. 1.20).

Several physical mechanisms causing an additional driving force for carriers by a temperature gradient are incorporated in the Seebeck coefficient model. In the following the expressions (1.203) and (1.204) are reformulated to depend on energy levels in the semiconductor. The carrier concentrations are expressed assuming Boltzmann statistics:

$$n = N_c \exp\left(\frac{\mathcal{E}_f - \mathcal{E}_c}{k_\mathrm{B}T}\right), \tag{1.205}$$

$$p = N_v \exp\left(\frac{\mathcal{E}_v - \mathcal{E}_f}{k_\mathrm{B}T}\right). \tag{1.206}$$

and substituted into (1.203) and (1.204):

$$\alpha_n = -\frac{k_\mathrm{B}}{q}\left(\frac{5}{2} - \frac{\mathcal{E}_f - \mathcal{E}_c}{k_\mathrm{B}T}\right), \tag{1.207}$$

$$\alpha_p = \frac{k_\mathrm{B}}{q}\left(\frac{5}{2} - \frac{\mathcal{E}_v - \mathcal{E}_f}{k_\mathrm{B}T}\right). \tag{1.208}$$

The temperature dependence of the Fermi level itself raises a gradient along the thermoelectric device and thus a carriers driving force. Additionally the positions of the band edges are temperature dependent and they therefore contribute to the driving force in the semiconductor. The previously assumed Boltzmann statistics is only valid for low doping concentrations, while at high doping concentrations the Fermi–Dirac statistics has to be taken into account. The phonon system, which acts as scattering centers for the carriers, has been assumed in local thermal equilibrium. This is not valid in electrothermal devices, due to the strong temperature gradients and the phonon movement through the structure. Caused by the phonons transiting from the hot side to the cold side of the device, the carriers gain additional momentum, which is also known as *phonon-drag* effect [57, 58, 144, 226] and can be modeled by adding an extra driving force for the carriers in the expression for the Seebeck coefficients [90–92]. A theoretical approach incorporating the phonon-drag effect has been presented in [233]. For silicon the phonon-drag effect is significant in the temperature range from 10 to 500 K [92].

The correction terms ζ_n and ζ_p are introduced into (1.209) and (1.210) in order to account for the deviation from Boltzmann statistics in the degenerate case and the phonon-drag effect:

$$\alpha_n = -\frac{k_B}{q}\left(\frac{5}{2} - \ln\frac{n}{N_c} + \zeta_n\right), \tag{1.209}$$

$$\alpha_p = \frac{k_B}{q}\left(\frac{5}{2} - \ln\frac{p}{N_v} + \zeta_p\right). \tag{1.210}$$

7.3 *Comparing the Six-Moments Transport Model with Spherical Harmonics Expansion*

Spherical Harmonics Expansion (SHE) is a numerical method for the solution of the Boltzmann Transport equation. By expanding the distribution function $f(\mathbf{r},\mathbf{k},t)$ in the **k**-space into spherical harmonics functions $Y_{lm}(\theta,\Phi)$ one can obtain an approximate solution of the Boltzmann Transport equation [212]. The SHE procedure is able to reproduce the results calculated by Monte Carlo methods quite well while at the same time exhibiting less computational costs. Spherical harmonics functions are defined via [2, 129]:

$$Y_{lm}(\theta,\Phi) = \sqrt{\frac{2l+1}{4\pi}\frac{(l-m)!}{(l+m)!}}\,P_l^m(\cos(\theta))\,e^{\mathrm{i}m\Phi}, \tag{1.211}$$

with $P_l^m(\cos(\theta))$ commonly known as Legendre polynomials and the indices l and m defined in the ranges $l \in [0,\infty)$ and $m \in [-l,l]$, respectively. The spherical harmonics functions are orthogonal [113]:

$$\int_0^{\pi}\int_0^{2\pi} \mathrm{d}\Omega\, Y_{lm}(\theta,\Phi)\,Y^*_{l'm'}(\theta,\Phi) = \delta_{ll'}\,\delta_{mm'}. \tag{1.212}$$

The asterisk in the term $Y^*_{l'm'}$ in (1.212) represents the complex conjugate of $Y_{l'm'}$, while $\mathrm{d}\Omega$ is defined by $\mathrm{d}\Omega = \sin(\theta)\,\mathrm{d}\theta\,\mathrm{d}\Phi$. In order to give an impression of the structure of theses functions Y_{00}, Y_{10}, Y_{11} and Y_{20} are given below:

$$Y_{00} = \sqrt{\frac{1}{4\pi}}, \qquad Y_{10} = -\sqrt{\tfrac{3}{8\pi}}\sin(\theta)e^{\mathrm{i}m\Phi}, \qquad Y_{11} = \sqrt{\frac{3}{4\pi}}\cos(\theta)$$

$$\text{and} \quad Y_{20} = \sqrt{\tfrac{5}{16\pi}}\left(3\cos^2(\theta) - 1\right). \tag{1.213}$$

Under the prerequisite of rotational symmetry along the Φ direction, the spherical harmonics functions reduce to the associated Legendre polynomials.

The distribution function can be expanded into spherical harmonics functions:

$$f(r,k) = \sum_{l=0}^{\infty} \sum_{m=-l}^{l} f_{lm}(r,k) Y_{lm}(\theta, \Phi). \tag{1.214}$$

The coefficients $f_{lm}(r,k)$ are defined as:

$$f_{lm}(r,k) = \int_0^{\pi} \int_0^{2\pi} \mathrm{d}\Omega f(\theta, \Phi) Y^*_{l'm'}(\theta, \Phi). \tag{1.215}$$

Thus, the fluxes in the three-dimensional case (q.v. (1.55)) are given by:

$$n\mathbf{V}_i = \sum_{l=0}^{\infty} \sum_{m=-l}^{l} \int_0^{\pi} \int_0^{2\pi} \mathrm{d}\Omega \mathbf{v} \mathscr{E}^i f_{lm}(\theta, \Phi) Y_{lm}(\theta, \Phi). \tag{1.216}$$

In the next step the SHE method is applied to the stationary Boltzmann transport equation. In order to simplify the following derivation, we will restrict to the transport direction of the carriers and assume it to be along the z-axis in conjunction with parabolic bands. Thus, (1.214) reduces to:

$$f(z,k) = \sum_{l=0}^{N} f_l(z,k) P_l(\cos(\theta)), \tag{1.217}$$

where θ denotes the direction of the electric field and $P_l(\cos(\theta))$ describe the Legendre polynomials. Before substituting the distribution function via spherical harmonics functions, the **k**-space is transformed into the $\mathscr{E}$-space, offering advantages such as an isotropic distribution function on energy surfaces in equilibrium [113]. By expanding the Boltzmann transport equation via (1.217), one yields the SHE [169,221]. The two lowest order expansions are defined as:

$$l = 0 \longrightarrow \quad \partial_z f_1 - qE\left(\partial_{\mathscr{E}} f_1 + \Gamma_B f_1\right) = \frac{1}{v}\left(\partial_t f_0\right)_{\mathrm{coll}}, \tag{1.218}$$

$$l = 1 \longrightarrow \quad \partial_z f_0 - 2\partial_z f_2 - qE\left(\partial_{\mathscr{E}} f_0 + 2\partial_{\mathscr{E}} f_2 + 3\Gamma_B f_2\right) = \frac{1}{v}\left(\partial_t f_1\right)_{\mathrm{coll}}. \tag{1.219}$$

In the limit $N \rightarrow \infty$ the resulting description is an exact solution of the Boltzmann transport equation. Vasicek [220] demonstrated for a velocity profile of an n^+nn^+ structure, that already for considering the first nine Legendre polynomials there is a good agreement between the SHE results and data from Monte Carlo simulations.

In the next step, the relation between the SHE and the macroscopic transport models is derived (e.g. Drift–Diffusion Transport model). Presuming a homogeneous, stationary system under an externally applied electric field **E**, parabolic bands, and the validity of the diffusion approximation, leads to a description of the Boltzmann transport equation with a macroscopic relaxation time approximation as follows:

$$-q\mathbf{E}\nabla_{\mathbf{p}}f = -\frac{f-f_0}{\tau_0}. \tag{1.220}$$

The distribution function f can be split into a symmetric part f_s and an anti-symmetric part f_a, which incorporates the non-equilibrium conditions. The diffusion assumption [67] states that:

$$f_s \gg f_a, \tag{1.221}$$

which demands that the system is not far from equilibrium and hence (1.220) can be exploited to deduce an expression for the anti-symmetric part f_a [138]:

$$f_a = q\tau_0\mathbf{E}\nabla_{\mathbf{p}}f_0 = \frac{q\tau_0 f_0}{k_B T_L}\mathbf{E}\,\mathbf{v} = \frac{q\tau_0\hbar}{k_B T_L m^*} f_0\,|\mathbf{E}|\,|\mathbf{k}|\,P_1\left(\cos(\theta)\right). \tag{1.222}$$

By inserting (1.222) into (1.216) one obtains the drift term of the Drift–Diffusion Transport model. Hence, just considering the first Legendre polynomial of the SHE for low-fields yields the same results as the Drift–Diffusion Transport model.

Therefore, the SHE is an appealing alternative to the solution of the Boltzmann transport equation via the Monte Carlo method and can be used as a reference solution for the previously derived three-dimensional higher-order macroscopic transport models.

As an example, Figs. 1.9 and 1.10 compare the Drift–Diffusion Transport model, the Energy Transport model and the Six-Moments Transport model for different channel lengths against results obtained by SHE as a reference and the corresponding relative error of the applied transport model [220]. As expected at a channel length of 1,000 nm all transport models yield the same results in conjunction with a small relative error of ~1% (Fig. 1.10). For a channel length of 250 nm the errors of the Energy Transport model and the Six-Moments Transport model stay within reasonable 2.5%, while the Drift–Diffusion Transport model starts with ever increasing errors and reaches a relative error of ~16% at a channel length of 100 nm. Below 250 nm the error of the Energy Transport model continuously increases, while the Six-Moments Transport model stays close to the SHE reference.

Simulating short channel devices with the Drift–Diffusion Transport model gives only poor results, as expected. However, for devices exhibiting a channel length of 1 μm, the Drift–Diffusion Transport model, the Energy Transport model, the Six-Moments Transport model, and the SHE model yield the same current value with a relative accuracy of 1%. The reason for the failure of the Drift–Diffusion Transport model lies in its closure assumption. By setting the charge carriers' temperature equal to the lattice temperature, the corresponding distribution function constantly underestimates the amount of available charge carriers and thus yields too small currents. On the other hand, the Energy Transport model assumes a heated Maxwellian distribution function, which is not valid in devices with channel lengths below ~250 nm.

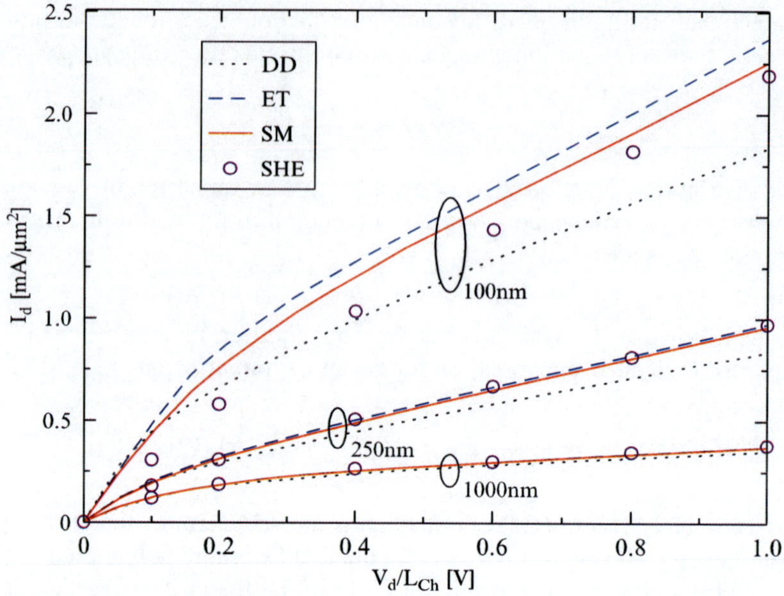

Fig. 1.9 Output currents for different $n^+ - n - n^+$ structures calculated with Drift–Diffusion Transport model, the Energy Transport model, and the Six-Moments Transport model. The SHE results are employed as a reference. For 1,000 nm, all models predict the same current, while the Drift–Diffusion Transport model underestimates the current for a channel length of 100 nm

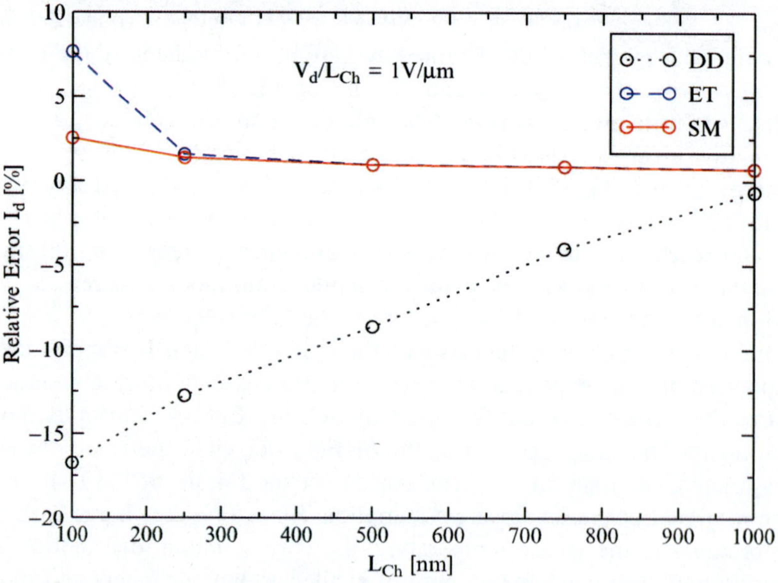

Fig. 1.10 Illustrating the relative error of the current calculated with the Drift–Diffusion Transport model, Energy Transport model, and the Six-Moments Transport model as a function of the channel length at a voltage of 1 V. While the Energy Transport model and the Six-Moments Transport model is below 7.5%, the Drift–Diffusion Transport model heads to 16% at a channel length of 100 nm

8 Applications

Despite the nowadays readily available nanometer technology (the semiconductor industry is entering the 22 nm node [103]), there is still plenty of room for the application of classical transport models. In the subsequent section three examples of up to date devices covered by classical device simulation will be given. These examples treat solar cells, BioFETs, and thermovoltaic elements.

8.1 Solar Cells

The French physicist Becquerel was the first to recognize the photovoltaic effect in 1839, but it took until 1883 to build the first solar cell, which was realized by semiconducting selenium coated with an extremely thin layer of gold to create the junction. This device possessed a poor conversion efficiency of about 1%. After Stoletow [205–207], who built the first solar cell based on the outer photoelectric effect, and Einstein explaining the photoelectric effect in 1905, Ohl patented the modern junction semiconductor solar cell in 1946 [154].

The present research in the field of photovoltaics can be divided into three main topics: reducing the cost of state of the art solar cells and/or increase their efficiency, so that they can compete with other energy sources; developing new solar cells with new technologies and new solar cell architectural designs; and advancing materials serving as light absorbers and charge carriers.

8.1.1 Working Principle of Solar Cells

Solar cells are similar to photo diodes [63] (Fig. 1.11). The distinction between both applications is that solar cells are designed to convert photons into electric power and not just detect photons like the photo diode. Therefore, in order to increase the amount of photons penetrating the solar cell and generating electron–hole pairs, the area accessible to light must be as large as possible and the coupling of the photons in the cell efficient. Under optimum conditions a photon is able to enter the solar cell and to generate an electron–hole pair. The electron and the hole start to diffuse and reach the space charge region of the pn-diode. There, the electron is pushed into the n-doped region and the hole moves to the p-doped zone and later on into the p+-doped zone, due to the built-in electric field created by the space charges. Furthermore photons generate electron–hole pairs and the electrons accumulate at the front, while the holes aggregate at the back contacts, thus generating an electric potential difference.

Consequently, different demands arise in order to optimize the conversion efficiency of solar cell. One way to enhance the yield from the generated electron–hole pairs is to design the depletion zone as large as possible and keep at the same time the average diffusion length of electrons and holes at a high level. Thus the electrons

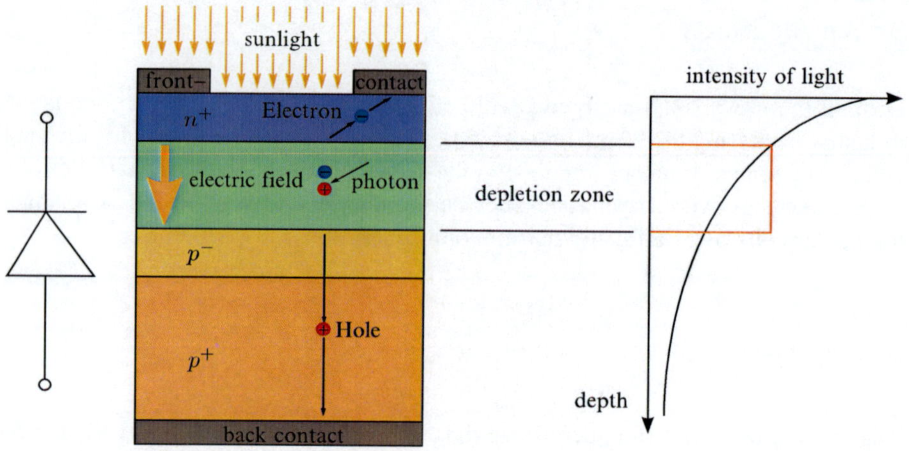

Fig. 1.11 Illustration of the working principle of a solar cell

and holes, which do not originate in the depletion region, increase their chance to reach the junction region and become swept to the corresponding side of the pn-diode. Furthermore, the number of photons exponentially decreases with increasing depth and thus it is beneficial to situate the depletion layer close to the surface. This is commonly achieved by a thin highly n-doped layer at the surface and a thick low p-doped substrate. If electrons are generated in the p-doped layer they are able to get to the back contact and recombine there. This effect is suppressed by an additional highly doped p-layer (also called p+-layer) at the back contact. The p+-layer induces a so-called *back surface field* between the p- and the p+-region and reflects electrons on their way to the back contact giving the electrons a second chance to reach the depletion layer. On the other hand also holes arriving at the surface of the front side are able to recombine with surface trap states, reducing the number of holes on the back side of the cell. This is the reason for an extra antireflection coating which decreases the reflectivity of the surface and also saturates surface traps, but may require an extra passivation step [1]. Another way of coupling light more efficiently into the solar cell is to texture the surface [63,73]. The front contacts represent a trade off between minimizing the energy losses in the fingers and busbars of the contact and maximizing the accessible area for the incoming photons. Commonly, this is realized by two thick busbars connected to many thin fingers perpendicular (see Fig. 1.12).

Despite many efforts to introduce alternative solar cell designs, the results can not compete with established solar cells either in efficiency or over all costs [78]. For instance, nanostructures exhibit lower thermal conductivities than their bulk materials, due to increased acoustic phonon scattering, which causes issues related to heat removal and reliability [118,135,192] (unlike in thermovoltaic applications where it is actually beneficial. cf. Sect. 8.3). Therefore, wafer based silicon (e.g. single crystal, polycrystalline and multicrystalline) solar cells and thin film solar cells manufactured with amorphous silicon, *CdTe*, *CuInGaSe*$_2$, and *III–V* semiconductors rule the photovoltaic manufacturing [193]. Shockley and Queisser estimated the thermodynamic limit of maximum efficiency for a pn-junction silicon solar cell at 300 K

Fig. 1.12 Example for a monocrystalline solar cell exhibiting the commonly employed contact pattern

and *AM* 1.5 illumination around 30% [188]. Photon energies higher than the energy band gap are converted into heat. Heat loss is the major effect for efficiency degradation in silicon solar cells. Therefore, major efforts are being put on exploiting hot carriers created by absorbing photons with an energy higher than the energy band gap and generating higher output currents or voltages, and introducing energy states within the band gap to trap carriers originating from photons smaller than the band gap energy [78].

The class of high-efficiency solar cells is characterized by the ability to generate more electric power per incident solar power unit. The industry is interested in the most cost efficient technologies in the sense of cost per generated Watt. The two major solutions to reduce costs of photovoltaic electricity are enhancement of cell efficiency and reduction of the costs per unit area. Thus, it is highly desirable that the efficiency of the solar cell is increased and the total cost per kilowatt-hour is reduced at the same time.

8.1.2 Multiple Junction Solar Cells

Multiple junction photovoltaic cells consist of many layers of epitaxially deposited films. The band gap of each layer is adjusted by a different alloy composition of *III–V* semiconductors, enabling every layer to absorb a specific band of the solar spectrum. The optimization of the respective band gaps of the various junctions is aggravated by the constraint of matching lattice constants for all layers. Beginning with the highest band gap material on top, all layers are optically in series. The first junction receives all of the incoming spectrum and photons with energies higher than the first band gap are absorbed in the first layer. Photons with energies below the first band gap travel to the next layer and are subsequently absorbed.

Currently available commercial cells are electrically connected in series. Due to the series connection, the generated current through each junction must be equal. Therefore, current match for each junction is an important design criterion for these devices.

For multiple junction solar cells the highest reported efficiencies are claimed to be 42.8, 41.1, and 40.8% from the university of Delaware [96], the Fraunhofer institute for solar energy systems [52], and the US national renewable energy research facility (NREL) [152], respectively.

Various multiple junction solar cells and their properties have been studied by simulations [6, 114, 218, 222].

8.1.3 Thin-Film Solar Cells

There are several thin-film technologies currently in development. The goal is to reduce the amount of light absorbing material required for producing a solar cell. This decreases the processing cost compared to using bulk materials, but at the same time also reduces the energy conversion efficiency to about 7–10% [75]. However, many multi-layer thin-film cells exhibit efficiencies above those fabricated on bulk silicon wafers. Their advantage, in addition to bulk silicon, lies in lower costs, flexibility, lighter weights, and ease of integration.

An efficiency of 19.9% for solar cells based on copper indium gallium selenide thin films (CIGS) was achieved by the NREL [153]. The CIGS films were grown by physical vapor deposition via a three-stage co-evaporation process. During this process indium, gallium and selenium are evaporated and afterwards copper and Se co-evaporated followed by In, Ga and Se evaporation at the end.

Thin film solar products have about 14% marketshare, while the other 86% are held by crystalline silicon [107]. The biggest amount of commercially produced thin-film solar cells is based on $CdTe$ with a typical efficiency of 11%.

Pieters et al. introduced a new version of their simulation tool, suitable for thin-film solar cells [163], Song et al. numerically studied CIGS tandem solar cells [201], Malm et al. studied CIGS thin-film solar cells with the aid of the finite element method [139], and Iwata et al. studied the influence of solar cell thickness and surface roughness on the conversion efficiency [104].

8.1.4 Crystalline Silicon

Crystalline silicon is currently the material of choice for solar cells, also known as *solar grade silicon*. Bulk silicon can be further distinguished into multiple categories according to crystallinity and crystal size in the resulting ingot, ribbon, or wafer. Monocrystalline silicon is frequently produced by the Czochralski process which tends to be expensive. Poly- or multicrystalline silicon, which is cheaper in production, but also exhibits smaller efficiency, and ribbon silicon [120], which is a subtype of multicrystalline silicon formed by drawing flat thin films from molten silicon, which further reduces production costs and silicon waste, but at the same time causes efficiencies smaller than polysilicon.

Monocrystalline silicon cells yield the highest efficiencies in silicon. The highest commercially available efficiency (22%) is manufactured by SunPower by utilizing expensive, high quality silicon wafers. An efficiency of 25% has been reported on monocrystalline silicon under laboratory conditions [219]. Crystalline silicon devices achieve an energy payback period of 1–6 years depending on cell type and environmental conditions [54, 168], and they are heading to the theoretical limiting efficiency of 30% [74, 188].

Klusak et al. [123] presented a modeling and optimization study of industrial n-type high-efficiency back-contact back-junction silicon solar cells. Ghargi et al. [59] numerically studied spherical silicon solar cells as a cost effective alternative to planar silicon solar cells. Campa et al. [33] found an increase up to 45% in current density for an optimized structure with periodic sinusoidal textured interfaces in comparison to that of the cell with flat interfaces. Tasaki et al. [215] investigated the influence of interface states on high performance amorphous silicon solar cells.

From the simulation point of view, feasible photovoltaic devices posses areas from several cm^2 to several hundred cm^2 and thicknesses around $\sim 100-400\,\mu m$ for standard cells and several μm for thin film cells. Therefore, the simulation of these devices is well covered by the Drift–Diffusion Transport model and, if necessary (e.g. inhomogeneous illumination of the surface and other effects causing inhomogeneous heating of the cell) by the Energy Transport model.

8.2 Biologically Sensitive Field-Effect Transistors

Today's technology for detecting pathogens, antigen-antibody complexes, and tumor markers is a timeconsuming, complex and expensive task [164, 186]. For instance, a typically workflow to detect, e.g., a certain deoxyribonucleic acid (DNA) molecule involves several process steps. First, one has to increase the number of DNA samples either by polymerase chain reaction (PCR) or reverse transcription (RT), followed by a subsequent process step which marks the DNA with a so-called label. The label enables the sensing of the DNA via radiation or light. This sample is afterwards applied to a microarray. Which consists of an array of spots. Each of the spots is prepared differently and thus offers the detection of a defined molecule. When the reaction took place, the array is read by a costly microarray reader.

At this point the introduction of a field-effect transistor which replaces the optical sensing mechanism by an electrical signal detection, offers several advantages. First, there is no need for the expensive readout device anymore. Furthermore, utilization of field-effect transistors allows the integration of analyzing and amplifying circuits on the same chip and, hence, enables a further cost reduction due to cheaper equipment. The outstanding development of semiconductor process technology allows mass production for this kind of devices in combination with a reduction of the price per piece.

Various reaction pairs are accessible and extensively studied for detecting DNA [53, 56, 80], cancer markers [238], proteins like biotin-streptavidin [42, 79, 99, 203], albumin [160], and transferrin [62], representing only a small sample to illustrate the diversity of the investigated device types and materials. It also shows, that there are many high-potential researches on BioFETs, which is still in its initial phase.

The concept of a BioFET is extremely versatile. Nearly all molecules posses a charge, when they are dissolved in a solute. This charge can be exploited by binding it on the surface of a BioFET, thus enabling its detection.

Among the possibility of exploiting BioFETs for DNA sequencing in a lab, these devices could enable a family doctor to screen for diseases on his own and decide faster and with less effort, which treatment is best for a patient, supported by the integrated analyzing and amplifying circuits in conjunction with the shrinking device size and easier handling. Furthermore, the amount of multi-resistance germs could be reduced by the consideration of a patients genetic profile, and a choice at deciding for a certain treatment or medication could be optimized, yielding the best possible results.

The integration of a BioFET into a chip environment is a manageable task. Either by putting a microfluidic channel above the functionalized gate of the BioFET or by isolating the surrounding areas by a thick oxide or polymere, the chip can be transformed into a mini-laboratory also known as lab on chip. This kind of systems improve the control over environmental parameters like local pH or detects the amount of a special protein, and facilitates local measurements (e.g. how a cell reacts to certain stimuli), thus offering a complete lab on chip. However, even though great advances have been made, far more research is needed to overcome the many obstacles.

The Working Principle of a BioFET

A BioFET consists of several parts: a reference electrode (analog to the gate contact for a MOSFET), the analyte, a biofunctionalized surface layer, a dielectric layer, and a semiconductor transducer (as depicted in Fig. 1.13). The semiconductor transducer is implemented as a conventional field-effect transistor. The dielectric is commonly an oxide (e.g. SiO_2) and serves two purposes: the first is to electrically isolate the channel of the field-effect transistor from the liquid and the second is to couple the surface layer charge to the channel electrostatically. The biofunctionalized surface layer exhibits immobilized biomolecule receptors able to bind a certain molecule. The sample molecules are dissolved in a solution which is also known as analyte. The sensitivity of the device is adjusted by the reference electrode (the optimum sensitivity lies around moderate inversion [44]).

If the desired molecules bind to the receptors, the surface charge density changes. This modifies the potential in the semiconductor and thus the conductivity in the channel of the field-effect transducer. The scale of the chemical reaction between sample and receptor molecules lies in the Angstrom regime, while the BioFET size is in the micrometer regime. Hence, it is significant to employ a proper multiscale mathematical description of the solution/semiconductor interface.

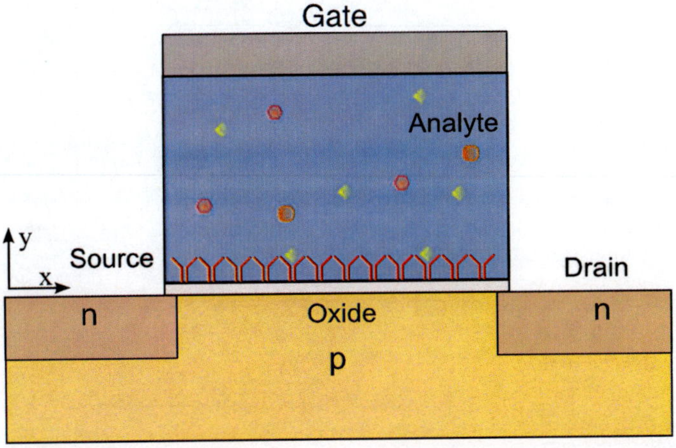

Fig. 1.13 When a charged sample molecule reaches a matching receptor at the biofunctionalized surface, it binds to it. This event alters the surface potential and also the potential distribution in the semiconductor, which results in a resistance change of the field-effect transistor's channel

Modeling Electrolytic Interfaces

Chemical and biological experiments are commonly carried out in ionic solutions [227]. The non-vanishing dipole moment of polar solution molecules (e.g. water) enables the reduction of the electric field between bound ions and, hence, to break up initially strong ionic bonds leading to more chemically active reactants. $NaCl$ and KCl are common salts for buffer solutions and exhibit valences for their anions and cations equal in absolute value (so-called $1:1$ solution/salt). Without external forces the charges are homogeneously distributed across the electrolyte and each ion is surrounded by an aggregate of water molecules. The water shell around an ion influences the relative dielectric constant around the ion and reduces the effect of electric fields stemming from other ions. Therefore, ions can freely move in the solvent and facilitate the conduction of an electrical current.

Insulator Surface Charge (Double Layer)

Charges gather at the electrode and in the surface area of the electrolyte, either due to an externally applied field or a difference of the chemical potentials between the electrode and the electrolyte. In the electrode (assuming a metal) the majority of charge carriers resides on the electrodes surface due to their mutual repulsion. In the electrolyte, the dissolved ions of opposing charge will be attracted by the electrodes surface charge. Unlike the electrons in the electrode, the ions exhibit a water shell and therefore a larger radius. Thus, a single layer of ions is not able to compensate the surface charge of the electrode and a diffusive layer of ions at the

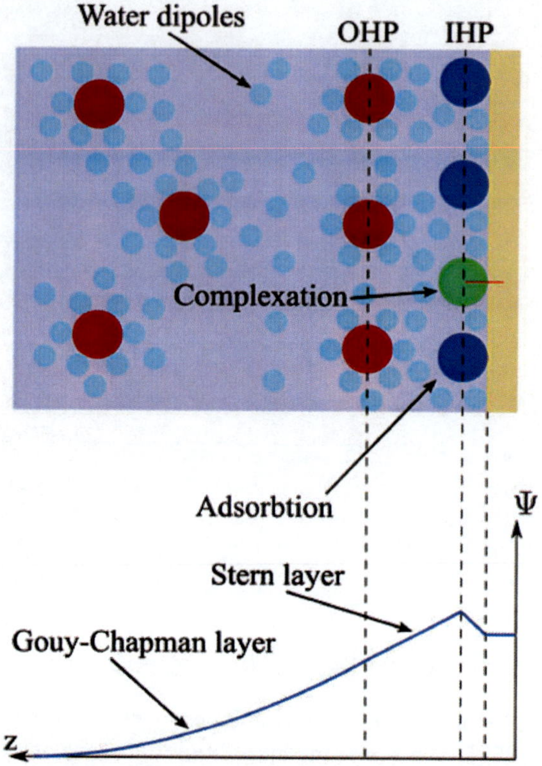

Fig. 1.14 The different contributions to the potential profile are: the inner Helmholtz plane (IHP) due to (non-) specific adsorption (caused by partial release of the solvation shell and therefore closer distance to the interface, *blue circles*), surface complexation, caused by the high affinity of attracting counter ions (*green circle*), the outer Helmholtz plane (OHP), and end of the Stern layer (zone without counter ions possessing their whole water shell, *red circles* with small *blue circles*) and the Gouy–Chapman layer [186]

electrodes surface will rise. In this *double layer* a potential drop will occur which has to obey the Poisson equation. Combining the Poisson equation and assuming thermal equilibrium of the ions with their environment the Poisson–Boltzmann equation is deduced. The diffusive layer is also known as Gouy–Chapman layer or electric double layer (depicted in Fig. 1.14).

Stern Modification

The potential distribution in the vicinity of the electrode–electrolyte interface is calculated by the Poisson–Boltzmann equation. Nevertheless, experimental data exhibit deviations from the predicted values for the double layer charge and capacitance [45]. It was shown that the Gouy–Chapman model overestimates the interface charge and the capacitance for high-concentration electrolytes. Stern was the first to recognize that the ions in the electrolyte posses a certain dimension and can not approach the electrode surface closer than their ionic radius. This led to the introduction of the outer Helmholtz plane (OHP) [61], taking care of the closest possible distance. The water molecules aggregated around the ion are included in this distance. The release of the aqueous shell of the ion would require an extensive amount of energy and thus a zone close to the electrode surface exists, which is depleted of ionic charges giving rise to an additional contribution to the total capacitance. This Stern capacitance has a typical value of $\sim 20\,\mu F\,cm^{-2}$. However, there are further contributions to the potential profile. Usually these effects are small and therefore negligible. Some of the effects are summarized in Fig. 1.14:

- Specific adsorption of ions on the surface: If ions (partially) release their solvation shell, they are able to move closer to the interface than the OHP. The new emerging radius of closest approach is called inner Helmholtz plane (IHP). The resulting model treating IHP and OHP, is called Gouy–Chapman–Stern–Graham model [186].
- Non-specific adsorption: Instead of releasing water molecules from the solvation shell, they are adsorbed on the surface due to distant coulombic attraction.
- Polarization of solvent: Commonly, the effects caused by the electric field weakening through the dipole moment of water is covered via adjustment of the relative permittivity. In the case of bulk applications this works well, while in the neighborhood of a surface many water molecules are not able to polarize with the electric field and the effective relative permittivity will not be the same as for the bulk.
- Surface complexation: Several charged surfaces posses an increased attraction and support the formation of complex compounds at the surface, influencing the potential in their vicinity.

Insulator Surface Charge: Site-Binding Model

The Gouy–Chapmen–Stern model expresses the main contributions to the electric double layer. It relates the accumulated charge at the surface of the electrochemical interface to the applied potential. Up to now, only electrostatic interactions were considered and chemical reactions at the interface were neglected. However, there are potentially significant chemical reactions at the interface leading to a net charge aggregation at the insulators interface [237]. The site-binding model allows to include chemical reactions at the insulator's interface into the simulation. Chemical reactions, unlike electrostatic forces which interact over long ranges, are restricted to molecular distances. This encourages the assumption for the site-binding model, that chemical reactions are limited to the region between the surface and the OHP. Ionic species assigned to the dissolved salt exhibit a water shell and thus are restricted to stay outside the OHP. Hence, no ions can contribute to the chemical reactions at the insulator interface (neglecting the possibility of specific adsorption of salt ions). On the other hand, the much smaller hydrogen ions are not blocked by the OHP due to their smaller ionic radius and can approach the interface close enough to allow chemical reactions.

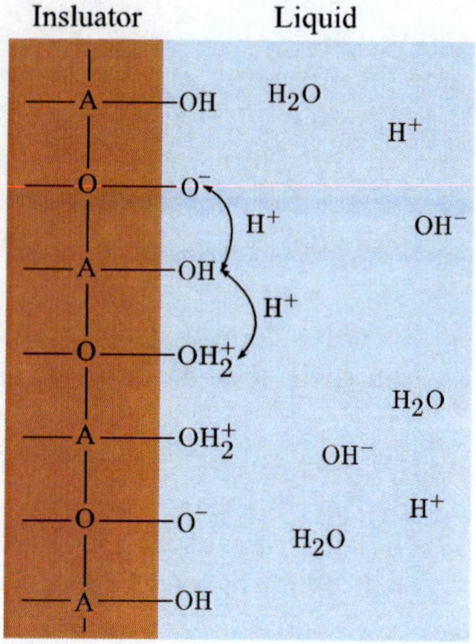

Fig. 1.15 Due to the lack of further bonding partners at the insulator surface there are open binding sites left. These binding sites may be either positively/negatively charged or neutral, depending on the properties of the liquid covering the surface. The surface charge density is related to the surface potential Ψ_0, material properties, and the local hydrogen concentration $[\mathrm{H}]_b^+$

Table 1.1 Parameters needed for the site-binding model commonly ($pK_i = -\log_{10}(K_i)$ analog to the definition of $pH = -\log_{10}([\mathrm{H}^+])$)

Oxide	pK_a	pK_b	N_s (cm^{-2})	Reference
SiO_2	−2	6	5×10^{14}	[29]
Si_3N_4	−8.1	6.2	5×10^{14}	[85]
Al_2O_3	6	10	8×10^{14}	[29]
Ta_2O_5	2	4	1×10^{15}	[30]
Gold surface	4.5	4.5	1×10^{18}	[60]

Figure 1.15 illustrates the open bonds of an insulator surface. Without unspecific adsorption, the only ions capable of binding to these sites are the hydrogen and hydroxyl ions [29, 128, 237]. The relation between the surface charge density, local potential and hydrogen concentration is given by [237]:

$$\sigma_{\mathrm{Ox}} = qN_s \frac{\frac{[\mathrm{H}^+]_b}{K_a} e^{-\frac{q\Psi_0}{qk_\mathrm{B}T}} - \frac{K_b}{[\mathrm{H}^+]_b} e^{\frac{q\Psi_0}{qk_\mathrm{B}T}}}{1 + \frac{[\mathrm{H}^+]_b}{K_a} e^{-\frac{q\Psi_0}{qk_\mathrm{B}T}} + \frac{K_b}{[\mathrm{H}^+]_b} e^{\frac{q\Psi_0}{qk_\mathrm{B}T}}}. \tag{1.223}$$

As an example some parameter sets for common materials are given in Table 1.1. The maximum amount of surface charge is directly proportional to the number of surface sites per unit area N_s, while the steepness and width of the two appearing steps is related to the difference between the reaction rates pK_a and pK_b. This additional effect influences the charge distribution in the double layer and in the semiconductor. Adding the site-binding model to the system of equations, the description of the ion-sensitive field-effect transistor (ISFET) is able to cover chemical reactions at the insulators surface. The charge aggregation at the oxide surface raises problems for the design of biosensors, while at the same time, it can be exploited to build highly efficient pH sensors [234].

The hydrogen concentration is properly taken into account at the oxide interface but outside of the OHP only the salt ions concentration is included and the hydrogen concentration is ignored. This contradiction can be resolved as follows: At the oxide surface the hydrogen concentration strongly influences the equilibrium constants, while outside the OHP, the hydrogen diffusive layer is much smaller than the ion diffusive layer. For instance, in a dilute solution containing 1 mM of $NaCl$, the salt is fully dissolved into 1 mM Na^+ and 1 mM Cl^-. Assuming at the same time a pH of 7, the hydrogen concentration in the electrolyte will be about 100 nM. This states a concentration discrepancy of four orders of magnitude between the hydrogen and the sodium concentration. Thus, the hydrogen diffusive layer will only have negligible influence on the potential in the Gouy–Chapman layer in comparison to the site-binding region of the electrolyte.

BioFET Examples

There are two principle operation modes for detecting molecules: the first one exploits the change in surface charge density due to the pH sensitivity caused by the open binding sites at the oxide surface (also known as ISFET [23–25]). The second one utilizes the field-effect [165] discussed in this section. For the field-effect the intrinsic charge of the desired molecule is sensed directly without the intermediate step of generating H^+ or OH^- molecules. Due to the specific binding of the macromolecules via the *lock-key* principle, information about the macromolecules structure is contained.

At first an example for the modeling of a DNA-FET will be given. Several models have been investigated in order to find the best suited for a suspend gate FET (SGFET) at low salt concentrations [228]. The second example studies the detection of a streptavidin-biotin reaction depending on the molecules orientation and the employed dielectrics [229].

The ability of ISFET structures to detect the charge in deoxyribonucleic acid (DNA) can be utilized to build biosensors capable of detecting specific DNA sequences [128, 181, 227]. This application offers huge opportunities for many areas like food and environmental monitoring, development of patient specific drugs, and gene expression experiments. Hence, the simulation of so-called DNA-FETs is currently a topic of great interest. DNA and proteins are commonly considered as the main active components in all living organisms [210]. The DNA stores all the genetic information via molecular sequences within its polymere structure. Watson and Crick were the first to find that DNA consists of a double helix structure and each helix is build from a repeating structure, containing a sugar polymer, a nitrogen base, and a phosphate ion. The nitrogen base can be distinguished between four select bases. Namely, adenine (A), thymin (T), cytosine (C), and guanine (G). The repeating structure, also known as DNA strand, often consists of several millions of these base pairs and the specific order of bases in the DNA strand encodes the specific genetic information concerning an organism. A unique genetic sequence can be generated from particular subsequences of an organisms DNA, allowing a genetic finger print [210, 227]. The two helical strands are bound by weak hydrogen bonds. The thermodynamically favorable and therefore stable bonds are found between adenine and thymin, and between cytosine and guanine. Only helical DNA strands with complementary bases are thermodynamically stable and form stable complexes. The process of double helix formation is called hybridization.

In [228], the experimental data of a SGFET have been studied via three different modeling approaches. The SGFET design is the same as for a standard MOSFET with one exception: it exhibits an elevated gate with an empty space under it. The bare gate-oxide layer is biofunctionalized with single stranded DNA and able to hybridize with a complementary strand. The intrinsic charge of DNA stems from its phosphate groups, with minus one elementary charge per group. The phosphate groups are fundamental building blocks of the DNAs nucleotides. Every base contained in the DNA is charged by minus one elementary charge. Therefore DNA possesses a high intrinsic charge, and big shifts in the transfer characteristics of BioFETs are induced. Thus, label-free, time-resolved, and in-situ detection of DNA is possible.

Harnois et al. [86] prepared a SGFET with 60 oligo-deoxynucleotides (ODN), also known as single stranded DNA, which were embedded onto a glutaraldehyd coated nitride layer. Test runs proved the specificity of the device. Their experimental data demonstrates two interesting properties. One is the fairly high threshold voltage shift of $\sim$800 mV and the other is that the probe transfer curve lies centered between the target and the reference curve. Typical threshold voltage shifts are in a range from several mV to $\sim$100 mV [165] and depend on the applied buffer concentration. Furthermore, the data display a big shift between the reference curve and the probe/target curve ($\sim$100 mV), but a much smaller shift between the probe and the target curve ($\sim$10–20 mV) [186].

Three models were employed, trying to reproduce the device behavior: the Poisson–Boltzmann model in combination with a space charge equivalent to the charged DNA (60 base pairs probe and 120 base pairs target), the Poisson–Boltzmann model with a sheet charge describing the DNAs, and the Debye–Hückel model with a corresponding space charge. Figure 1.16a–c show the transfer characteristics for the unprepared SGFET (reference), the prepared but unbound (probe), and after the DNA has bound to functionalized surface (target), respectively. The experimental data are displayed in discrete grey tones to enable better comparison to the simulated curves. Even at a very low salt concentration of 0.6 mV, Fig. 1.16b, c exhibit a bigger shift between the reference curve and the probe/target curve than between probe and target curve. This behavior complies with observations in [186] and is caused by the nonlinear screening of the Poisson–Boltzmann model. Figure 1.17b, c show that doubling the charge does not lead to twice the curve shift. Even though there is a bigger shift for the sheet charge model due to the less screening in comparison to the model with the space charge, the overall trend is a bigger shift between the reference curve and the probe/target curve and a much smaller shift between the probe and the target curve.

On the other hand the employed Debye–Hückel model offers good agreement for the same parameter set as for the Poisson–Boltzmann models given in Fig. 1.16a. Figure 1.17a illustrates that for the Debye–Hückel model, doubling the charge is connected to twice the potential shift, due to the linear screening characteristics of the model.

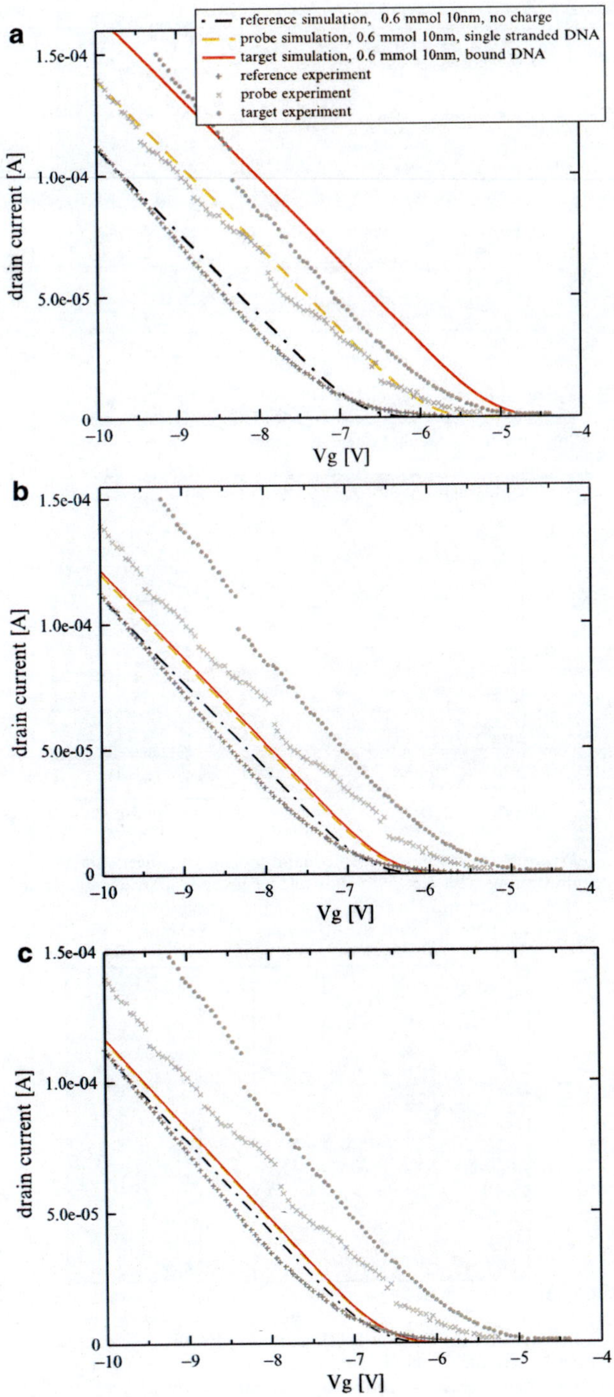

Fig. 1.16 The transfer characteristics for the Debye–Hückel model and DNA charge modeled via space charge density is given by (**a**), (**b**) shows the transfer characteristics for the same SGFET for the Poisson–Boltzmann model and DNA charge described via sheet charge density, and (**c**) illustrates the transfer characteristics of a SGFET for the Poisson–Boltzmann model and DNA charge modeled via space charge density

a

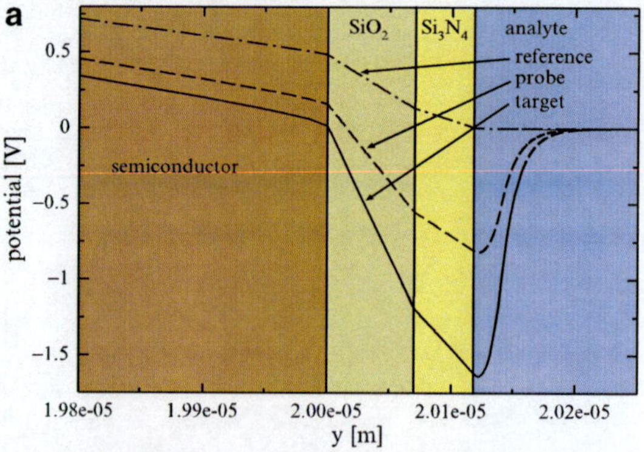

Potential for the Debye-Hückel model with space charge, showing that doubling the charge leads to twice the potential shift due to the weaker linear screening.

b

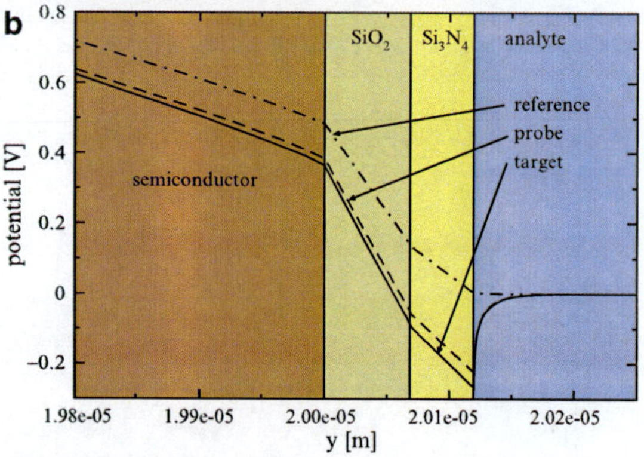

Potential for the Poisson-Boltzmann model with sheet charge, illustrating a bit increased shift but far away from the values from the measurement. However, doubling the charge does not lead to twice the potential shift due to nonlinear screening.

c

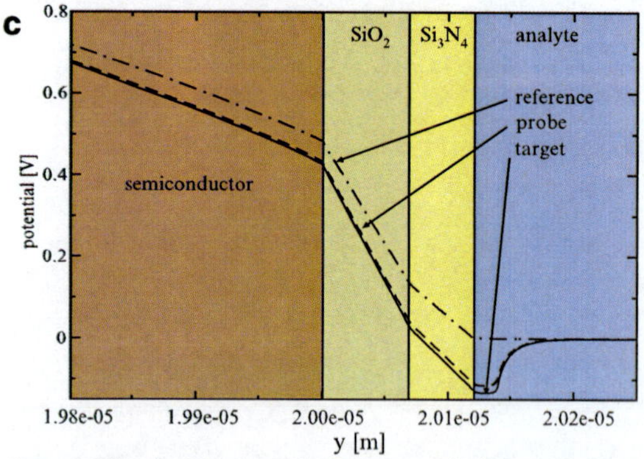

Potential for the Poisson-Boltzmann model with space charge, demonstrating doubling the charge also here does not lead to twice the potential shift due to nonlinear screening.

Fig. 1.17 The potential profile for the different modeling approaches, starting from the semiconductor (*left*) and ending in the analyte (*right*)

Assuming a single 60 bases DNA strand, it will occupy a volume of about $V_0 = 10 \times 10 \times 20\,\mathrm{nm}^3$. Taking this volume and multiplying it with 1 mM sodium-chloride bulk concentration results into approximately one sodium/chloride ion on average per V_0. Therefore, strong nonlinear screening at low salt concentrations is extremely unlikely. Furthermore, the Poisson–Boltzmann model is a continuum model and thus describes the salt concentration as a continuous quantity. This causes the Poisson–Boltzmann model to overestimate the screening and, therefore, to fail at small salt concentrations. The Debye–Hückel model is derived from the Poisson–Boltzmann model by expanding the exponential terms into a Taylor series and neglecting all terms higher than second-order [43]. Due to the laws of series expansion $q\Psi/k_B T \ll 1$ the potential has to be much smaller than the thermal energy. However, even though this constraint is not fulfilled, the Debye–Hückel model is able to reproduce the data. One reasonable explanation might be that in this case the extended Poisson–Boltzmann model and the Debye–Hückel model coincide as shown in [228] and thus the screening depends on the average closest possible distance between the ions.

The second example studies a BioFET equipped with a biotin-streptavidin reaction for different dielectric materials and different molecule orientations at the surface.

Streptavidin is a tetrameric protein purified from the bacterium streptomyces avidini. Each subunit is able to bind biotin with equal affinity (Fig. 1.18). It is frequently used in molecular biology due to its very strong affinity for biotin which represents one of the strongest non-covalent interactions known in nature. It is commonly used for purification or detection of various biomolecules. With the aid of the strong streptavidin-biotin bond various biomolecules can be attached to one another or onto a solid support.

Here, the biotin-streptavidin reaction pair is modeled with the Poisson–Boltzmann model with homogenized interface conditions (1.149) and (1.150). The charge and dipole moment for a single molecule (biotin/streptavidin) is ob-

Fig. 1.18 The tetrameric protein streptavidin (*black*) and its four binding sites for biotin (*white*)

tained from a protein data bank [167] and extrapolated to the mean charge density and the mean dipole moment density of the boundary layer.

Three different oxide types were used as dielectric. SiO_2 served as a reference, Al_2O_3, and Ta_2O_5 were studied as possible high-k materials, with relative permittivities of 3.9, 10, and 25, respectively. The solute was sodium chloride at $pH = 7$. Several simulation runs were performed for each dielectric, such as the unprepared state (only water and salt), the prepared but unbound state (water, salt and biotin) and the bound state, when the chemical reaction took place (water, salt, and biotinstreptavidin), for two different molecule orientations (0°...perpendicular to the surface and 90°...parallel to the surface). The output characteristics were gained for several parameter combinations, under the prerequisite of 100% binding efficiency. The reference electrode was set to 0.4 V in order to shift the FET into moderate inversion as proposed in [44].

Figure 1.19 compares the output characteristics for the different employed dielectrics and molecule orientations. The following trends are recognizable: the lowest output characteristics are found for 0°, followed by 90°, and finally without dipole moment, for each group. This is caused by the inhomogeneously charged biomolecules and the related dipole moment entering the boundary conditions (1.150), hence, resulting in different output characteristics of the BioFET for different orientation angles in relation to the surface. Furthermore, higher ε_r leads to higher output currents, due to the better coupling of the surface charge to the channel.

There are several conclusions which can be deduced by considering the molecules electrostatic properties. The signal-to-noise ratio can be improved by exploiting a only minimally charged or better a neutral linker. Therefore, in the

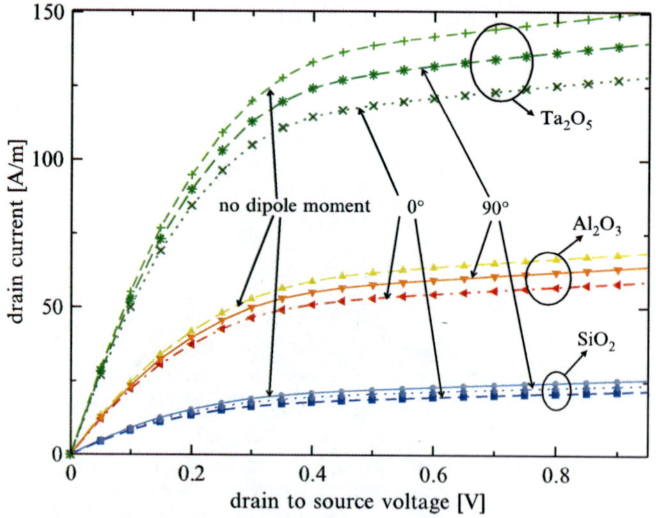

Fig. 1.19 Output characteristics for SiO_2, Al_2O_3, and Ta_2O_5 for calculation without dipole moment, 0° (perpendicular to surface), and 90° (parallel to surface)

case of detecting strepatividin via biotin, biotin should be attached to the surface by a neutral linker. Streptavidin is negatively charged with minus four elementary charges and biotin with minus one elementary charge. By attaching streptavidin after biotin to the surface the relative change in surface charge will be quite big, even when partial screening of the intrinsic charges is taken into account. Due to the tetrameric nature of streptavidin (four binding sites for biotin, Fig. 1.18), the linker utilized should be short enough to prevent binding several biotin molecules to a single streptavidin protein. Furthermore, if there is freedom of choice in deciding, wether biotin or streptavidin is initially attached to the surface, biotin is a better choice. In this case the relative change in charge will be bigger (from minus one elementary charge to minus five elementary charges) yielding a more pronounced change in the output signal.

Summarizing the results gained by the Poisson–Boltzmann model with homogenized interface conditions show a strong dependence on surface charges and indicate a detectable shift in the threshold voltage depending on the molecule orientation relative to the surface.

8.3 Thermovoltaic Elements

In the last decades enormous efforts in engineering and science were taken to increase the fuel efficiency, but unfortunately it has not been possible to keep up with the economical growth. Thermoelectric energy conversion is one among several technologies with the potential to break through in future energy technology. The underlying physical effect has been well known for about 200 years and is based on the direct energy conversion from temperature gradients into electrical energy. Despite extensive research efforts, the usage of the thermoelectric energy is still restricted to few highly specialized fields, due to a low conversion efficiency. The enormous efforts on material research over the last years introduced novel materials for thermoelectric devices as well as a better understanding of the prerequisites for higher conversion efficiencies.

8.3.1 Materials for Thermoelectric Devices

The goal of research and engineering of materials for thermoelectric devices is to maximize their efficiency. A potentially promising thermoelectric material is characterized by a high Seebeck coefficient, good electric transport properties, and an impeded thermal transport [150]. Commonly, these material properties are accompanied with a pronounced temperature dependence yielding ideal thermal operation conditions for a certain material.

In this section, thermoelectric materials with technological importance and their operational range are introduced. Silicon-germanium alloys are appealing due to their low thermal conductivity compared to pure materials. Furthermore, *SiGe* is attractive due to its importance in mainstream electronics and the availability of an elaborate physical description.

Lead telluride ($PbTe$) is used in the intermediate temperature range with a maximum operation temperature of approximately 900 K. Apart from its application in thermoelectric devices, lead telluride is employed for optical devices in the infrared wavelength regime. Additionally to doping, the material properties can be altered by deviating its stoichiometric composition. Ternary alloys are also part of ongoing research efforts. In principle, lead telluride is avaliable as p-type as well as n-type material but, contrary to the n-type material, the p-type material suffers from degradation of stability at high temperatures and devices are difficult to bond and exhibit poorer mechanical properties [199]. Therefore, the p-doped leg is frequently exchanged by alloys containing silver antimony and germanium telluride, also known as *TAGS*.

Bismuth telluride stands at the lower end of the temperature scale. Because of its good thermoelectric properties at room temperature, it is frequently utilized for cooling applications. By analogy to lead telluride, the material type and number of excess carriers can be controlled by adjusting the stoichiometry of the alloy.

Apart from these classical thermoelectric materials frequently employed in generation and cooling applications, there are ongoing research efforts on novel materials [21, 22, 149] and specially designed nanostructures for thermoelectric applications [37, 38, 119, 121, 145, 232, 235, 236, 240].

Material Characterization

The performance of thermoelectric generators is judged by their characteristic numbers as efficiency, total power output, and power density. The efficiency is limited by several parameters. Apart from the geometrical construction, several material properties as the Seebeck coefficient, the thermal conductivity, and the electrical conductivity, influence the transport of charge carriers and phonons and thus the overall device characteristics. Here the figure of merit for thermoelectric materials comes into play. It embraces the material parameters influencing the device behavior as well as the device efficiency.

Ioffe [101] showed that the maximum conversion efficiency $\eta_{\max}$ of a thermoelectric generator at matched load conditions is obtained via the product of the ideal reversible thermodynamic process efficiency and a factor describing the energy losses in the device due to Joule heating and non-ideal thermal conductivity [231]

$$\eta_{\max} = \frac{T_h - T_c}{T_h} \frac{M-1}{M + \frac{T_c}{T_h}}, \tag{1.224}$$

where T_h and T_c describe the temperature of the heated and the cooled side of the device, respectively and M is given by:

$$M = \sqrt{1 + \frac{1}{2} Z \left(T_c + T_h \right)}. \tag{1.225}$$

The averaged thermoelectric figure of merit for each leg (different subscripts for each leg) and matched geometry is defined by:

$$Z = \frac{(\alpha_1 - \alpha_2)^2}{\left(\sqrt{\frac{\kappa_1}{\sigma_1}} + \sqrt{\frac{\kappa_2}{\sigma_2}}\right)^2}. \tag{1.226}$$

Equation (1.226) incorporates all significant material parameters like the Seebeck coefficient α, thermal conductivity κ, and the electric conductivity σ. Since the figure of merit exhibits a strong dependence on temperature as well as on the concentration of free carriers, inherited from its input quantities, every material possesses an optimum range of operation. However, common devices have legs with similar material properties and the so-called bulk figure of merit for a given material can be used:

$$Z = \frac{\alpha^2 \sigma}{\kappa}. \tag{1.227}$$

From a microscopic viewpoint, the figure of merit is affected by charge and heat transport as well as their interaction in the semiconductor. Thus, the figure of merit depends on band structure, lattice dynamics, and charge carriers scattering mechanisms.

While higher doping levels commonly have an adverse effect on the Seebeck coefficient, the electric conductivity increases due to the increased number of carriers. The electric part of the thermal conductivity κ_ν becomes significant at high carrier concentrations and even the dominant thermal conductivity mechanism on the transition to metals. Insulators and metals exhibit superior conditions for single parameters, but at the same time poor conditions for others. For instance, metals are known for their low Seebeck coefficients and relatively high thermal conductivities, which can not be counterbalanced by their high electric conductivities. Semiconductors lie approximately in the middle of the competing parameter ranges and, thus, are able to yield a maximal thermoelectric figure of merit. This maximum is reinforced by still moderate Seebeck coefficients and low electrical resistance but restricted electron thermal conductivity in the regions of high carrier concentrations. Furthermore the optimum carrier concentration can be adjusted by an appropriate doping.

Silicon-Germanium

SiGe thermogenerators have been effectively employed in several applications. Among these, the utilization as thermogenerators powered by radioisotopes (RTG), as reliable power source on space missions and in remote weather stations is probably the most impressive one [166]. Due to its high reliability and high operating temperatures, *SiGe* is also a good candidate to meet the conditions in nuclear reactors.

SiGe also constitutes an important material system due to it's use in mainstream microelectronics. Initiated by the introduction of strain techniques to commercially avaliable devices, the research efforts on the properties and processability of *Si/SiGe* were intensified [10, 39, 176]. A detailed analysis and material characterization of *SiGe* alloys with emphasis on physical modeling for device simulation has been presented in [157], and a review on the mobility modeling at high temperatures can be found in [179]. *SiGe* alloys are attractive due to their differing influence on thermal conductivity and mobility for varying composition, in comparison to their pure constitutes. Vining [223] and Slack et al. [200] studied the theoretical maximal figure of merit. While in [223] a two-band model has been employed, in [200] the second conduction band has been considered, yielding a broader temperature range of the validity of the model.

With increasing germanium content (up to 50%) the lattice thermal conductivity of *SiGe* reduces significantly. For increasing germanium content this trend first halts, than begins to reverse, and finally reaches the value of pure germanium content. The reason for this characteristic is alloy disorder scattering of phonons, created by the random distribution of silicon and germanium in the alloy [17]. A further reduction in thermal conductivity has been reported for sintered samples, due to extra phonon scattering at the grain boundaries [143].

Sintered composites exhibit a low sensitivity of the thermal conductivity on the material composition over a wide range of germanium content and, therefore, posses a good figure of merit. This proves beneficial for inhomogeneous samples, where clustering causes relatively large localized deviations in the material parameters. Compared to the thermal conductivity, the mobility decreases more slowly with rising germanium content yielding a range with favorable figures of merit. The Seebeck coefficient for pure silicon, germanium, and several of their alloys have been determined and documented in the literature [9, 55, 57, 58, 174].

Despite the excellent reliability performance of *SiGe* alloys, there are degradation effects for *SiGe* thermoelectric generators, reducing the figure of merit over the device lifetime [223]. High temperature conditions may cause sublimation and result in thermal and/or electrical shortcuts. Furthermore, erosion can appear under extreme conditions and induce device failure by open circuits or mechanical damage. Additionally, high doping concentrations, intentionally introduced to raise the figure of merit, are prone to build up local accumulations. These accumulations reduce the free carrier concentration, decrease the electric conductivity, and degrade the figure of merit. Due to the lower diffusion rate of boron doped p-type samples compared to n-type samples doped with phosphorus, the p-samples are less sensitive to this phenomenon.

Lead Telluride and Its Alloys

Lead telluride ($PbTe$) and lead tin telluride ($Pb_{1-x}Sn_xTe$) devices operate with regard to temperature between those of bismuth telluride and silicon-germanium. Despite the slightly smaller maximum figure of merit compared to bismuth telluride, lead telluride exhibits an equivalently good efficiency over a large temperature window. By changing the stoichiometry of the material composition the electrical properties of the alloy can be adjusted. Utilizing an excess of tellurium results in a p-type semiconductor, while a raised plumbum content has the adverse effect and gives a n-type semiconductor. However, this processing path restricts a maximum carrier concentration to about $10^{18}\,\mathrm{cm}^{-3}$, which is lower than the ideal doping for thermoelectric applications [166]. Therefore, higher carrier concentrations are realized by doping with PbI_2, $PbBr_2$, or Ge_2Te_3 for an increased donor concentration and Na_2Te or K_2Te for elevated acceptor concentrations.

Lead telluride and lead tin telluride are available as sintered materials as well as single crystals. Sintered samples exhibit a lower thermal and electrical conductivity compared to single crystals due to the additional scattering at the grain boundaries [49].

Bismuth Telluride and Its Alloys

Bismuth telluride (Bi_2Te_3) and some related alloys are frequently used for cooling applications in commercial Peltier elements due to a good thermoelectric figure of merit at room temperature. Common ternary alloys are bismuth telluride either with bismuth selenide (Bi_2Se_3) or antimony telluride (Sb_2Te_3) [46]. Their crystal structure is characterized as hexagonal [126], but also has been described as rhombohedral [35]. The temperature range for thermoelectric applications of Bi_2Te_3 is limited by its melting point at 858 K [48].

By analogy to lead telluride the free carrier concentration can be controlled, either by changing the material composition or by extra dopants. Contrary to lead telluride, stoichiometric bismuth telluride is of p-type with a free carrier concentration of about $10^{19}\,\mathrm{cm}^{-3}$. Raising the tellurium concentration converts the material to n-type.

Bismuth telluride belongs to the group of narrow gap semiconductors and possesses an indirect band gap of 160 mV at 300 K. The population of higher energy levels is relatively high due to a low DOS. Therefore, the large non-parabolicity of the bandstructure is of significance [27]. Founding on the theoretical work of [131,178] experimental work has been accomplished, serving as a basis for future performance optimizations, e.g. introduction of low-dimensional structures [191].

Reducing the thermal conductivity is also a possible solution in order to increase the figure of merit. Ternary alloys show a dependence of the lattice thermal conductivity on the additional phonon scattering by alloy disordering. Bismuth antimony telluride, in the form of $(Bi_{0.5}Sb_{0.5})_2\,Te_3$, features the highest lattice disorder and therefore the lowest thermal conductivity [180]. However, related to the adverse effect on the evolution of the electrical conductivity and the carrier contribution to the total thermal conductivity, the maximum figure of merit is gained at higher antimony content [64, 65]. In analogy to the previous materials, sintered samples show a reduced lattice thermal conductivity due to extra grain boundary scattering [211]. The influence of dopants on the thermal conductivity has been studied in [239].

The overall device performance of a thermoelectric generator is lower than the theoretical maximum, due to the narrow temperature range of about 50 K for the maximum figure of merit. A possible solution to circumvent this limitation is to incorporate graded or segmented materials along the temperature gradient in order to meet the optimum material properties [127].

Optical, transport, and several mechanical parameters exhibit a strong anisotropy. Despite the relatively isotropic Seebeck coefficient with deviations of about 10% between the opposing extrema, the electrical resistivity and the thermal conductivity show anisotropy ratios of 4–6 and 2–2.5, respectively [66, 93, 180]. p-type as well as n-type samples show Seebeck coefficients between 100 and 250 $\mu V K^{-1}$ depending on the material composition [5,51]. The maximum observed figure of merit is in direction parallel to the cleavage plain and is superior to the normal direction by a factor of 2.

8.3.2 Examples

The first introduced thermoelectric generator type is commonly used in commercial energy conversion applications and is based on the classical design of thermocouples and temperature sensors. It is built from two semiconducting legs, one consisting of n-type semiconductor and the other of p-type semiconductor (Fig. 1.20). The p-type leg features a larger cross section than the n-type leg in order to compensate the lower hole mobility. The two legs are arranged thermally in parallel and electrically in series, exhibiting an electrical contact at the heated side of the device. Since the signs of the Seebeck coefficients of the two legs are opposing, their voltage contributions add up to the total voltage of the device.

The second example represents an alternative thermoelectric device and is built from a large area pn-junction [202]. The principle of its design is illustrated in Fig. 1.21. The contacts are situated at the cooled end of the device structure

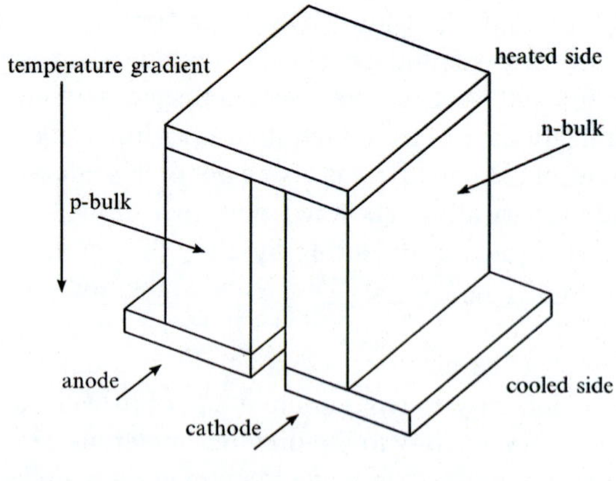

Fig. 1.20 Scheme of a thermoelectric device

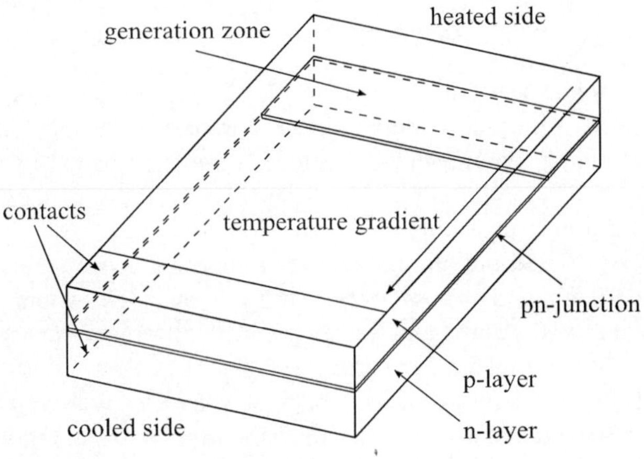

Fig. 1.21 Scheme of a large pn-junction thermoelectric generator

and the temperature gradient is brought into play along the pn-junction. Contrary to conventional thermoelectric devices, the thermal electron–hole pair generation is exploited in a large area pn-junction. Applying a temperature gradient within the structure, induces the generation of an electric current which is related to the temperature effect on the electrostatic potential of the pn-junction. Higher temperatures lead to a smaller energy step from the potential of the n-layer to the p-layer compared to the step at lower temperatures. Due to the temperature gradient in the large area pn-junction both conditions exist in the same device and thus carriers at the higher potential experience a driving force into the colder region. Both carrier types move into the same direction (ambipolar drift and diffusion). They leave the pn-junction at the high temperature, which therefore becomes depleted and induces a disturbance in the local thermal equilibrium. This shifts the local generation-recombination balance to a raised generation of carriers in order to compensate the off-drifting carriers, while at the end of the device with the lower temperature the opposite effect takes place. Therefore, a circular current is driven from the hot region with increased generation to the cold region with enhanced recombination. Using selective contacts for the n- and the p-layer, the circular current can be exploited for an external load, and a power source in the form of a thermoelectric element is established.

Depending on the environmental conditions, the devices can either be connected in series for a higher output voltage or in parallel for higher output currents. Similar reasoning is valid for the thermal circuit. Multiple single elements on the same temperature level increase the heat flux through the entire module in order to exhaust relatively strong temperature reservoirs at temperature differences suitable for a single stage. For an ambience with a higher temperature difference it is beneficial to apply modules with multiple stages, where each stage is optimized for a certain temperature.

Reduced Thermal Conductivity by Alloys

The enhanced phonon scattering rates for *SiGe* alloys lead to a strongly pronounced reduction in phonon thermal conductivity in comparison to pure silicon, but also influence several other parameters. Therefore, a trade-off has to be found between the beneficial lower thermal conductivity and the decrease of carrier mobility to achieve the optimum improvement of conversion efficiency.

Wagner [224] carried out simulations for a thermoelectric generator exhibiting a leg length of 20 mm and a cross section of $5 \times 1\,\text{mm}^2$. The dopings for both legs were set constant to $10^{19}\,\text{cm}^{-3}$ and the temperature difference was considered to be situated 600 K above room temperature.

With increasing germanium content, the Seebeck voltages as well as the mobility decrease over a wide range and lead to a drop in output current. Figure 1.22 demonstrates that the highest electric output is obtained in pure silicon. Low mobility in *SiGe* causes a reduction in the absolute maximum of the power output and shifts it to higher resistances.

However, the influence of the material composition on the thermal conductivity and thus the heat flux through the device outweighs the negative impact on the electrical properties [224]. The heat flux decreases to a minimum at approximately 50% germanium concentration which is an order of magnitude lower than for pure silicon. The resulting maximum of conversion efficiency is expected at about 30% germanium content, where the optimum between thermal and electrical properties

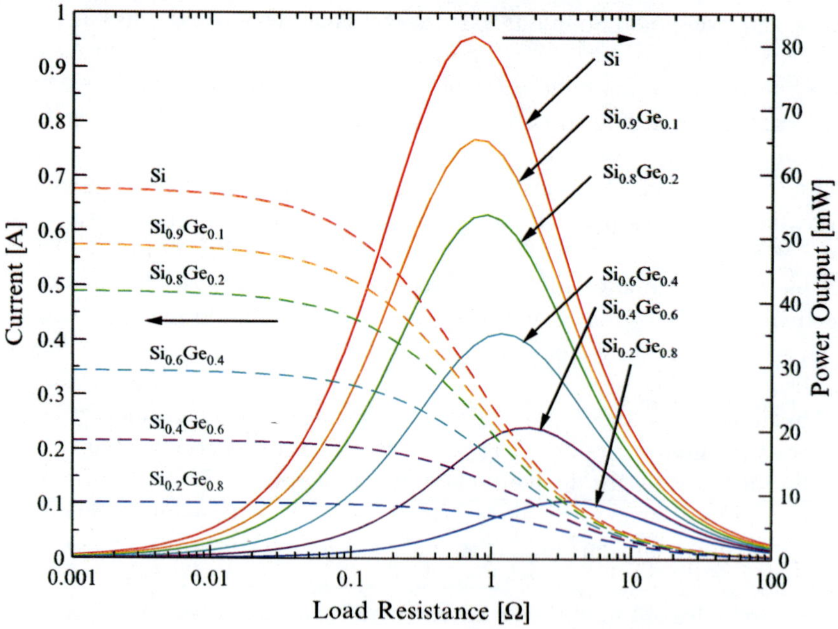

Fig. 1.22 Electric current and power output as a function of load resistance for several *SiGe* alloys

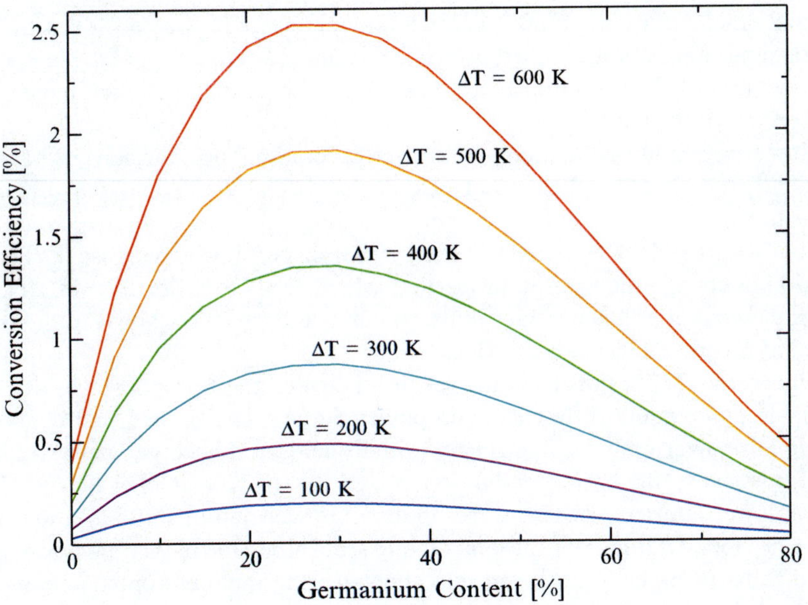

Fig. 1.23 Illustrating the conversion efficiency in relation to the material composition and temperatures

is situated. At higher germanium content, the thermal conductivity still decreases, but can not compensate the degrading mobility and Seebeck voltages any longer. Figure 1.23 depicts the conversion efficiency in relation to the material composition at match load conditions. While the power output continuingly decreases with increasing germanium content, the conversion efficiency reaches its maximum at approximately 30% germanium.

pn-Junctions as Thermoelectric Devices

The device characteristics of a pn-junction thermoelectric generator are mainly controlled by carrier generation. Due to the strong influence of the lattice temperature on the carrier generation, the zone of high generation rates is restricted to the hottest parts of the structure.

Therefore it is advantageous to keep large parts of the device at high temperatures. In the case of a pure material, the temperature distribution along the pn-junction is concave because of the decreasing thermal conductivity with increasing temperature. This yields a steep temperature gradient at the heated end and a relatively short zone at high temperature, restricting the carrier generation.

By engineering the spatial distribution of the thermal conductivity it is possible to increase the dimension of the zone at high temperatures. The implementation of graded material alloys is a pathway to achieve this. *SiGe* alloys decrease their thermal conductivity up to 50% germanium content. A device profile with higher germanium content at the cooled side of the device leads to a shift of the temperature drop to the cooled side of the device, in analogy to a potential divider in the electric counterpart of the model.

However, in addition to high total carrier generation rates, the carriers have to be efficiently transferred to the contacts. Doping as well as the geometrical dimensions of the transport layers have to be chosen accordingly in order to keep recombination as small as possible. Geometrically oversized transport layers increase the heat flux but do not change the electric properties, which leads to a decrease in efficiency. Thus, the goal of efficient device optimization is the careful analysis of the interrelation of several effects in the device.

Figure 1.24 [224] illustrates the relation between the transport layer thickness, available temperature difference, and power output. The dashed line denotes the maximum power output curve for an initially chosen device geometry, while the solid line shows the corresponding optimized device with thicker layers. Caused by the lower internal resistance, the optimum power output shifts to a lower load resistance too. Additionally, the temperature scale along the maximum power output curve shifts to higher values and thus the same thermal environment results in a significantly enhanced power output.

Due to the dependence of the device on the carrier generation rate, one can further improve the device performance by introducing trap states in the forbidden energy gap. In accordance with the Shockley–Read–Hall formalism [189] the carrier generation rate is controlled by the local temperature, the amount of traps, and the energy levels of the traps. Trap energy levels situated in the middle of the band

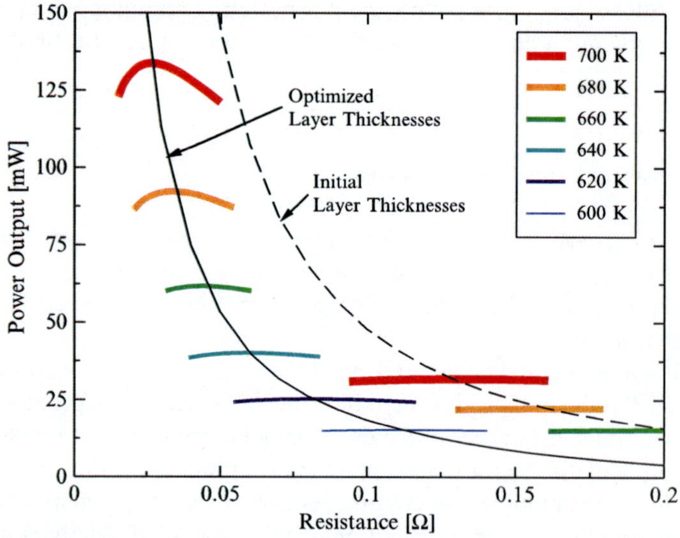

Fig. 1.24 Power output for a pn-junction thermoelectric generator in relation to load resistance, for several temperatures and two layer thicknesses

gap yield the highest thermal generation rates. For instance, for silicon, gold can be exploited as an additional dopant in the generation region of the device to add deep levels close to the mid band gap [171]. Due to the ability of the impurity state to absorb differences in momentum between the carriers, this carrier generation process is dominant in silicon and other indirect semiconductors. Hence, in a certain regime, the device performance of a pn-junction thermoelectric generator at a specific temperature can be shifted to lower temperatures by adjusting the extra trap concentration and distribution.

Acknowledgements Special thanks go to Prof. Tibor Grasser, Prof. Hans Kosina, and Neophytos Neophytou for their support in questions related to higher order transport models and modeling transport in thermovoltaic elements. Also the various discussions about higher order transport models and nice pictures regarding higher order transport models and SHE from Martin Vasicek, and the examples related to thermovoltaic elements from Martin Wagner are highly appreciated. This work was partly funded by the Austrian Science Fund project P18316-N13 and partly by the "Klima- und Energiefonds" Austria, project No. 825467.

References

1. Aberle, A.: Crystalline Silicon Solar Cells: Advanced Surface Passivation and Analysis. Centre for Photovoltaic Engineering, University of New South Wales, Sydney (1999)
2. Abramowitz, M., Stegun, I.: Handbook of Mathematical Functions. Dover Publications Inc. (1965)
3. Adler, M.: Accurate Calculations of the Forward Drop and Power Dissipation in Thyristors. IEEE Trans.Electron Devices **25**(1), 16–22 (1978)
4. Agostinelli, V.J., Bordelon, T., Wang, X., Yeap, C., Maziar, C., Tasch, A.: An Energy-Dependent Two-Dimensional Substrate Current Model for the Simulation of Submicrometer MOSFET's. IEEE Electron Device Lett. **13**(11), 554–556 (1992). DOI 10.1109/55.192837
5. Ainsworth, L.: Single Crystal Bismuth Telluride. Proc.Phys.Soc.B **69**(6), 606–612 (1956)
6. Allen, C., Jeon, J.H., Woodall, J.: Simulation Assisted Design of a Gallium Phosphide n-p Photovoltaic Junction. Solar Energy Materials and Solar Cells **94**(5), 865–868 (2010). DOI 10.1016/j.solmat.2010.01.009
7. Altenkirch, E.: Über den Nutzeffekt der Thermosäule. Physikalische Zeitschrift **10**, 560–580 (1909)
8. Altenkirch, E.: Elektrothermische Kälteerzeugung und reversible elektrische Heizung. Physikalische Zeitschrift **12**, 920–924 (1911)
9. Amith, A.: Seebeck Coefficient in n-Type Germanium-Silicon Alloys: "Competition" Region. Physical Review **139**(5A), A1624–A1627 (1965). DOI 10.1103/PhysRev.139.A1624
10. Andrieu, F., Ernst, T., Faynot, O., Rozeau, O., Bogumilowicz, Y., Hartmann, J., Brévard, L., Toffoli, A., Lafond, D., Ghyselen, B., Fournel, F., Ghibaudo, G., Deleonibus, S.: Performance and Physics of sub-50 nm Strained Si on $Si_{1-x}Ge_x$-On-Insulator (SGOI) nMOSFETs. Solid-State Electron. **50**(4), 566–572 (2006). DOI 10.1016/j.sse.2006.03.029
11. Anile, A., Pennisi, S.: Extended Thermodynamics of the Blotekjaer Hydrodynamical Model for Semiconductors. Continuum Mechanics and Thermodynamics **4**(3), 187–197 (1992). DOI 10.1007/BF01130290
12. Anile, A.M., Muscato, O.: Improved Hydrodynamical Model for Carrier Transport in Semiconductors. Physical Review B **51**(23), 16,728–16,740 (1995). DOI 10.1103/PhysRevB.51.16728

13. Apel, P., Korchev, Y., Siwy, Z., Spohr, R., Yoshida, M.: Diode-Like Single-Ion Track Membrane Prepared by Electro-Stopping. Nucl. Instrum. and Meth. in Phys. Res. Sect. B: Beam Interactions with Materials and Atoms **184**(3), 337–346 (2001). DOI 10.1016/S0168-583X(01)00722-4
14. Ashcroft, N., Mermin, N.: Solid State Physics. Cengage Learning Services (1976)
15. Baccarani, G., Rudan, M., Guerrieri, R., Ciampolini, P.: Physical Models for Numerical Device Simulation, vol. 1. North-Holland Publishing Co., Amsterdam, The Netherlands (1986)
16. Baccarani, G., Wordeman, M.: An Investigation of Steady-State Velocity Overshoot in Silicon. Solid-State Electron. **28**(4), 407–416 (1985). DOI 10.1016/0038-1101(85)90100-5
17. Bahandari, C.: CRC Handbook of Thermoelectrics, Chapter: Minimizing the Thermal Conductivity. CRC Press LLC (1994)
18. Baker, N.A., Sept, D., J., S., Holst, M.J., McCammon, J.A.: Electrostatics of Nanosystems: Application to Microtubules and the Ribosome. Proc. of the National Academy of Sciences of the United States of America **98**(18), 10,037–10,041 (2001). DOI 10.1073/pnas.181342398
19. Bandyopadhyay, S., Klausmeier-Brown, M., Maziar, C., Datta, S., Lundstrom, M.: A Rigorous Technique to Couple Monte Carlo and Drift-Diffusion Models for Computationally Efficient Device Simulation. IEEE Trans.Electron Devices **34**(2), 392–399 (1987)
20. Barcilon, V., Chen, D.P., Eisenberg, R.S., Jerome, J.W.: Qualitative Properties of Steady-State Poisson-Nernst-Planck Systems: Perturbation and Simulation Study. SIAM J. Appl. Math. **57**(3), 631–648 (1997). DOI 10.1137/S0036139995312149
21. Bentien, A., Christensen, M., Bryan, J., Sanchez, A., Paschen, S., Steglich, F., Stucky, G., Iversen, B.: Thermal Conductivity of Thermoelectric Clathrates. Physical Review B **69**(4) (2004). DOI 10.1103/PhysRevB.69.045107
22. Bentien, A., Johnsen, S., Madsen, G., Iversen, B., Steglich, F.: Colossal Seebeck Coefficient in Strongly Correlated Semiconductor $FeSb_2$. EPL (Europhysics Letters) **80**(1), 17,008 (2007). stacks.iop.org/0295-5075/80/17008
23. Bergveld, P.: Development of an Ion-Sensitive Solid-State Device for Neurophysiological Measurements. IEEE Trans. Biomed. Eng. **17**(1), 70–71 (1970). DOI 10.1109/TBME.1970.4502688
24. Bergveld, P.: The Development and Application of FET-Based Biosensors. Biosensors **2**(1), 15–34 (1986)
25. Bergveld, P.: Thirty Years of ISFETOLOGY: What Happened in the Past 30 Years and What May Happen in the Next 30 Years. Sensors and Actuators B: Chemical **88**(1), 1–20 (2003). DOI 10.1016/S0925-4005(02)00301-5
26. Beynon, R., Easterby, J.: Buffer Solutions the Basiscs. Oxford University Press, Oxford New York Tokyo (1996)
27. Bhandari, C., Agrawal, V.: Thermal and Electrical Transport in Bismuth Telluride. Indian Journal of Pure and Applied Physics **28**, 448–451 (1990)
28. Bløtekjær, K.: Transport Equations for Electrons in Two-Valley Semiconductors. IEEE Intl. Electron Devices Meeting **17**, 38–47 (1969)
29. Bousse, L.: Single Electrode Potentials Related to Flat-Band Voltage Measurements on EOS and MOS Structures. J.Chem.Phys. **76**(10), 5128–5133 (1982). DOI 10.1063/1.442812
30. Bousse, L., Mostarshed, S., Van Der Shoot, B., De Rooij, N.F., Gimmel, P., Gopel, W.: Zeta Potential Measurements of Ta_2O_5 and SiO_2 Thin Films. Journal of Colloid and Interface Science **147**(1), 22–32 (1991)
31. Brugger, S., Schenk, A., Fichtner, W.: Moments of the Inverse Scattering Operator of the Boltzmann Equation: Theory and Applications. SIAM Journal on Applied Mathematics **66**(4), 1209–1226 (2006). DOI 10.1.1.72.3270
32. Callen, H.: Thermodynamics. John Wiley & Sons Inc. (1966)
33. Campa, A., Krc, J., Topic, M.: Analysis and Optimisation of Microcrystalline Silicon Solar Cells with Periodic Sinusoidal Textured Interfaces by Two-Dimensional Optical Simulations. J.Appl.Phys. **105**(8), 083,107–083,107–5 (2009). DOI 10.1063/1.3115408
34. Caughey, D., Thomas, R.: Carrier Mobilities in Silicon Empirically Related to Doping and Field. Proc.IEEE **55**(12), 2192–2193 (1967)

35. Caywood, L.P., Miller, G.R.: Anistropy of the Constant-Energy Surfaces in n-type Bi_2Te_3 and Bi_2Se_3 from Galvanomagnetic Coefficients. Physical Review B **2**(8), 3209–3220 (1970). DOI 10.1103/PhysRevB.2.3209
36. Cervera, J., Schiedt, B., Ramírez, P.: A Poisson/Nernst-Planck Model for Ionic Transport Through Synthetic Conical Nanopores. Europhys. Lett. **71**(1), 35–41 (2005). DOI 10.1209/epl/i2005-10054-x
37. Chen, G., Narayanaswamy, A., Dames, C.: Engineering Nanoscale Phonon and Photon Transport for Direct Energy Conversion. Superlattices and Microstructures **35**(3-6), 161–172 (2004). DOI 10.1016/j.spmi.2003.08.001. Eurotherm 75 'Microscale Heat Transfer 2'
38. Chen, G., Zeng, T., Borca-Tasciuc, T., Song, D.: Phonon Engineering in Nanostructures for Solid-State Energy Conversion. Mat.Sci.Eng.A **292**(2), 155–161 (2000). DOI 10.1016/S0921-5093(00)00999-0
39. Cheng, Z., , Currie, M., Leitz, C., Taraschi, G., Fitzgerald, E., Hoyt, J., Antoniadas, D.: Electron Mobility Enhancement in Strained-Si n-MOSFETs Fabricated on $SiGe$-On-Insulator (SGOI) Substrates. IEEE Electron Device Lett. **22**, 321–323 (2001). DOI 10.1109/55.930678
40. Chynoweth, A.: Ionization Rates for Electrons and Holes in Silicon. Physical Review **109**(5), 1537–1540 (1958). DOI 10.1103/PhysRev.109.1537
41. Coalson, R., Kurnikova, M.: Poisson-Nernst-Planck Theory Approach to the Calculation of Current Through Biological Ion Channels. IEEE Trans. NanoBioscience **4**(1), 81–93 (2005). DOI 10.1109/TNB.2004.842495
42. Cui, Y., Wei, Q., Park, H., Lieber, C.M.: Nanowire Nanosensors for Highly Sensitive and Selective Detection of Biological and Chemical Species. Science **293**(5533), 1289–1292 (2001)
43. Debye, P., Hückel, E.: Zur Theorie der Elektrolyte: I. Gefrierpunktserniedrigung und verwandte Erscheinungen. Physikalische Zeitschrift **24**(9), 185–206 (1923)
44. Deen, M.J., Shinwari, M.W., Ranuárez, J.C., Landheer, D.: Noise Considerations in Field-Effect Biosensors. J.Appl.Phys. **100**(7), 074,703–1 –074,703–8 (2006)
45. Delahay, P.: Double Layer and Electrode Kinetics. New York: Interscience Publishers (1965)
46. Ding, Z., Huang, S., Marcus, D., Kaner, R.: Modification of Bismuth Telluride for Improving Thermoelectric Properties. In: Intl. Conf. on Thermoelectrics, pp. 721–724 (1999). DOI 10.1109/ICT.1999.843487
47. van Dort, M., Woerlee, P., Walker, A.: A Simple Model for Quantisation Effects in Heavily-Doped Silicon MOSFETs at Inversion Conditions. Solid-State Electron. **37**(3), 411–414 (1994). DOI 10.1016/0038-1101(94)90005-1
48. Drabble, J.: Progress in Semiconductors, vol. 7. John Wiley & Sons Inc., New York (1963)
49. Fano, V.: CRC Handbook of Thermoelectrics, Chapter: Lead Telluride and Its Alloys. CRC Press LLC (1994)
50. Fischetti, M.: Monte Carlo Simulation of Transport in Technologically Significant Semiconductors of the Diamond and Zinc-Blende Structures. I. Homogeneous Transport. IEEE Trans.Electron Devices **38**(3), 634–649 (1991). DOI 10.1109/16.75176
51. Fleurial, J., Gailliard, L., Triboulet, R., Scherrer, H., Scherrer, S.: Thermal Properties of High Quality Single Crystals of Bismuth Telluride Part I: Experimental Characterization. J.Phys.Chem.Solids **49**(10), 1237–1247 (1988). DOI 10.1016/0022-3697(88)90182-5
52. World Record: 41.1% Efficiency Reached for Multi-Junction Solar Cells at Fraunhofer ISE. Press Release (2009). www.ise.fraunhofer.de
53. Fritz, J., Cooper, E., Gaudet, S., Soger, P., Manalis, S.: Electronic Detection of DNA by its Intrinsic Molecular Charge. PNAS **99**(22), 1412–1416 (2002). DOI 10.1073/pnas.232276699
54. Fthenakis, V., Kim, H., Alsema, E.: Emissions from Photovoltaic Life Cycles. Environmental Science & Technology **42**(6), 2168–2174 (2008). DOI 10.1021/es071763q
55. Fulkerson, W., Moore, J.P., Williams, R.K., Graves, R.S., McElroy, D.L.: Thermal Conductivity, Electrical Resistivity, and Seebeck Coefficient of Silicon from 100 to 1300° K. Physical Review **167**(3), 765–782 (1968). DOI 10.1103/PhysRev.167.765
56. Gao, Z., Agarwal, A., Trigg, A., Singh, N., Fang, C., Tung, C.H., Fan, Y., Buddharaju K.D., and Kong, J.: Silicon Nanowire Arrays for Label-Free Detection of DNA. Analytical Chemistry **79**(9), 3291–3297 (2007). DOI 10.1021/ac061808q

57. Geballe, T.H., Hull, G.W.: Seebeck Effect in Germanium. Physical Review **94**(5), 1134–1140 (1954). DOI 10.1103/PhysRev.94.1134
58. Geballe, T.H., Hull, G.W.: Seebeck Effect in Silicon. Physical Review **98**(4), 940–947 (1955). DOI 10.1103/PhysRev.98.940
59. Gharghi, M., Bai, H., Stevens, G., Sivoththaman, S.: Three-Dimensional Modeling and Simulation of $p-n$ Junction Spherical Silicon Solar Cells. IEEE Trans.Electron Devices **53**(6), 1355–1363 (2006). DOI 10.1109/TED.2006.873843
60. Giesbers, M., Kleijn, J., Stuart, M.: The Electrical Double Layer on Gold Probed by Electrokinetic andSurfaceForce Measurements. Journal of Colloid and Interface Science **248**(1), 88–95 (2002). DOI 10.1006/jcis.2001.8144
61. Gileadi, E., Kirowa Eisner, E., Penciner, J.: Interfacial Electrochemistry: An Experimental Approach. Addison-Wesley Publishing Company (1975)
62. Girard, A., Bendria, F., Sagazan, O.D., Harnois, M., Bihan, F.L., Salaün, A., Mohammed-Brahim, T., Brissot, P., Loréal, O.: Transferrin Electronic Detector for Iron Disease Diagnostics. In: IEEE Conf. on Sensors, pp. 474–477 (2006). DOI 10.1109/ICSENS.2007.355509
63. Goetzberger, A., Voß, B., Knobloch, J.: Sonnenenergie: Photovoltaik, vol. 2. Teubner Studienbücher (1997)
64. Goldsmid, H.: The Thermal Conductivity of Bismuth Telluride. Proc.Phys.Soc.B **69**(2), 203–209 (1956). stacks.iop.org/0370-1301/69/203
65. Goldsmid, H.: Heat Conduction in Bismuth Telluride. Proc.Phys.Soc. **72**(1), 17–26 (1958). stacks.iop.org/0370-1328/72/17
66. Goldsmid, H.: Recent Studies of Bismuth Telluride and Its Alloys. J.Appl.Phys. **32**(10), 2198–2202 (1961). DOI 10.1063/1.1777042
67. Grasser, T.: Non-Parabolic Macroscopic Transport Models for Semiconductor Device Simulation. Physica A: Statistical Mechanics and its Applications **349**(1-2), 221–258 (2005). DOI 10.1016/j.physa.2004.10.035
68. Grasser, T., Jungemann, C., Kosina, H., Meinerzhagen, B., Selberherr, S.: Advanced Transport Models for Sub-Micrometer Devices. In: Intl. Conf. on Simulation of Semiconductor Processes and Devices, pp. 1–8. Springer-Verlag (2004)
69. Grasser, T., Kosik, R., Jungemann, C., Kosina, H., Selberherr, S.: Nonparabolic Macroscopic Transport Models for Device Simulation Based on Bulk Monte Carlo Data. J.Appl.Phys. **97**(9), 093710 (2005). DOI 10.1063/1.1883311
70. Grasser, T., Kosik, R., Jungemann, C., Meinerzhagen, B., Kosina, H., Selberherr, S.: A Non-Parabolic Six Moments Model for the Simulation of Sub-100 nm Semiconductor Devices. J.Comp.Electronics **3**(3), 183–187 (2004). DOI 10.1007/s10825-004-7041-1
71. Grasser, T., Kosina, H., Gritsch, M., Selberherr, S.: Using Six Moments of Boltzmann's Transport Equation for Device Simulation. J.Appl.Phys. **90**(5), 2389–2396 (2001). DOI 10.1063/1.1389757
72. Grasser, T., Tang, T.W., Kosina, H., Selberherr, S.: A Review of Hydrodynamic and Energy-Transport Models for Semiconductor Device Simulation. Proc.IEEE **91**(2), 251–274 (2003). DOI 10.1109/JPROC.2002.808150
73. Green, M.: Solar Cells : Operating Principles, Technology, and System Applications. Prentice-Hall, Englewood Cliffs, NJ (1982)
74. Green, M.: Third Generation Photovoltaics: Solar Cells for 2020 and Beyond. Physica E: Low-Dimensional Systems and Nanostructures **14**(1-2), 65–70 (2002). DOI 10.1016/S1386-9477(02)00361-2
75. Green, M.: Consolidation of Thin-Film Photovoltaic Technology: The Coming Decade of Opportunity. Progress in Photovoltaics: Research and Applications **14**(5), 383–392 (2006). DOI 10.1002/pip.702
76. Gritsch, M., Kosina, H., Grasser, T., Selberherr, S.: Revision of the Standard Hydrodynamic Transport Model for SOI Simulation. IEEE Trans.Electron Devices **49**(10), 1814–1820 (2002). DOI 10.1109/TED.2002.803645
77. Grubmüller, H.: Force Probe Molecular Dynamics Simulations. Springer Protocols **305** (2005). DOI 10.1007/978-1-59259-912-7_23

78. Gupta, N., Alapatt, G., Podila, R., Singh, R., Poole, K.: Prospects of Nanostructure-Based Solar Cells for Manufacturing Future Generations of Photovoltaic Modules. Intl. Journal of Photoenergy p. 13 (2009). DOI 10.1155/2009/154059
79. Gupta, S., Elias, M., Wen, X., Shapiro, J., L.Brillson: Detection of Clinical Relevant Levels of Protein Analyte UnderPhysiologicBuffer Using Planar Field Effect Transistors. Biosensors and Bioelectronics **24**, 505–511 (2008). DOI 10.1016/j.bios.2008.05.011
80. Hahm, J., Lieber, C.: Direct Ultrasensitive Electrical Detection of DNA and DNA Sequence Variations Using Nanowire Nanosensors. Nano Letters **4**(1), 51–54 (2004)
81. Hall, R.: Electron-Hole Recombination in Germanium. Physical Review **87**(2), 387 (1952). DOI 10.1103/PhysRev.87.387
82. Hänsch, W.: Carrier Transport Near the Si/SiO_2 Interface of a MOSFET. Solid-State Electron. **32**, 839–849 (1989). DOI 10.1016/0038-1101(89)90060-9
83. Hänsch, W.: The Drift Diffusion Equation and its Application in MOSFET Modeling. XII. Springer-Verlag (1991)
84. Hänsch, W., Miura-Mattausch, M.: The Hot-Electron Problem in Small Semiconductor Devices. J.Appl.Phys. **60**(2), 650–656 (1986). DOI 10.1063/1.337408
85. Harame, D., Bousse, L., Shott, J., Meindl, J.: Ion-Sensing Devices with Silicon Nitride and Borosilicate Glass Insulators. IEEE Trans.Electron Devices **34**(8), 1700–1707 (1987)
86. Harnois, M., Sagazan, O., Girard, A., Salaün, A.C., Mohammed-Brahim, T.: Low Concentrated DNA Detection by SGFET. In: Transducers & Eurosensors, pp. 1983–1986. Lyon, France (2007)
87. Hasnat, K., Yeap, C.F., Jallepalli, S., Hareland, S., Shih, W.K., Agostinelli, V., Tasch, A., Maziar, C.: Thermionic Emission Model of Electron Gate Current in Submicron NMOSFETs. IEEE Trans.Electron Devices **44**(1), 129–138 (1997). DOI 10.1109/16.554802
88. Heitzinger, C., Kennell, R., Klimeck, G., Mauser, N., McLennan, M., Ringhofer, C.: Modeling and Simulation of Field-Effect Biosensors (BioFETs) and Their Deployment on the nanoHUB. J. Phys.: Conf. Ser. **107**, 012,004/1–12 (2008). DOI 10.1088/1742-6596/107/1/012004
89. Heitzinger, C., Klimeck, G.: Computational Aspects of the Three-Dimensional Feature-Scale Simulation of Silicon-Nanowire Field-Effect Sensors for DNA Detection. J.Comp.Electronics **6**(1-3), 387–390 (2007). DOI 10.1007/s10825-006-0139-x
90. van Herwaarden, A.: The Seebeck Effect in Silicon ICs. Sensors and Actuators **6**(4), 245–254 (1984). DOI 10.1016/0250-6874(84)85020-9
91. van Herwaarden, A., van Duyn, D., van Oudheusden, B., Sarro, P.: Integrated Thermopile Sensors. Sensors and Actuators A: Physical **22**(1-3), 621–630 (1989). DOI 10.1016/0924-4247(89)80046-9
92. van Herwaarden, A., Sarro, P.: Thermal Sensors Based on the Seebeck Effect. Sensors and Actuators **10**(3-4), 321–346 (1986). DOI 10.1016/0250-6874(86)80053-1
93. Hodgson, D.: Anisotropy of Thermoelectric Power in Bismuth Telluride. Tech. Rep. 377, Massachusetts Institute of Technology, Research Laboratory of Electronics (1961). hdl.handle.net/1721.1/4445
94. Holst, M., Saied, F.: Multigrid Solution of the Poisson-Boltzmann Equation. J. Comput. Chem. **14**, 105–113 (1993)
95. Holst, M., Saied, F.: Numerical Solution of the Nonlinear Poisson-Boltzmann Equation: Developing More Robust and Efficient Methods. J. Comput. Chem. **16**, 337–364 (1995)
96. Honsberg, C., Barnett, A.: UD-Led Team Sets Solar Cell Record, Joins DuPont on $100 Million Project. Press Release (2007). www.udel.edu/PR/UDaily/2008/jul/solar072307.html
97. Huang, C., Wang, T., Chen, C., Chang, M., Fu, J.: Modeling Hot-Electron Gate Current in Si MOSFET's Using a Coupled Drift-Diffusion and Monte Carlo Method. IEEE Trans.Electron Devices **39**(11), 2562–2568 (1992). DOI 10.1109/16.163464
98. Ieong, M.K.: A Multii-Valley Hydrodynamic Transport Model for *GaAs* Extracted from Self-Consistent Monte Carlo Data. M.S. Thesis, University of Massachusetts (1993)
99. Im, H., Huang, X.., Gu, B., Choi, Y..: A Dielectric-Modulated Field-Effect Transistor for Biosensing. Nature Nanotechnology **2**(7), 430–434 (2007)

100. Institute for Microelectronics, TU Wien, Gußhausstraße 2729, 1040 Wien, Austria/Europe: Minimos-NT Device and Circuit Simulator Release 2.1. www.iue.tuwien.ac.at/software/minimosnt
101. Ioffe, A.: Semiconductor Thermoelements and Thermoelectric Cooling. Infosearch London (1957). Originally published-U.S.S.R. Academy of Sciences, 1956.
102. Ioffe, A.: The Revival of Thermoelectricity. Scientific American (1958)
103. International Technology Roadmap for Semiconductors: 2009 Edition (2009). www.itrs.net/Links/2009ITRS/Home2009.htm
104. Iwata, H., Ohzone, T.: Numerical Solar Cell Simulation Including Multiple Diffused Reflection at the Rear Surface. Solar Energy Materials and Solar Cells **61**(4), 353 – 363 (2000). DOI 10.1016/S0927-0248(99)00119-1
105. Jacoboni, C., Lugli, P.: The Monte Carlo Method for Semiconductor Device Simulation. Springer-Verlag (1989)
106. Jacoboni, C., Reggiani, L.: The Monte Carlo Method for the Solution of Charge Transport in Semiconductors with Applications to Covalent Materials. Rev. Mod. Phys. **55**(3), 645–705 (1983). DOI 10.1103/RevModPhys.55.645
107. Jäger-Waldau, A.: Research, Solar cell Production and Market Implementation of Photovoltaics. Tech. rep., European Commission, DG Joint Research Centre, Institute for Energy, Renewable Energy Unit (2009). www.jrc.ec.europa.eu
108. Jaggi, R.: High-Field Drift Velocities in Silicon and Germanium. Helvetia Physica Acta **42**, 941–943 (1969)
109. Jaggi, R., Weibel, H.: High-Field Electron Drift Velocities and Current Densities in Silicon. Helvetia Physica Acta **42**, 631–632 (1969)
110. Jüngel, A.: Transport Equations for Semiconductors XVII, vol. 773. Springer-Verlag (2009)
111. Jungemann, C., Meinzerhagen, B.: Hierachical Device Simulation: The Monte Carlo Perspective. Springer-Verlag (2003)
112. Jungemann, C., Nguyen, C., Neinhüs, B., Decker, S., Meinerzhagen, B.: Improved Modified Local Density Approximation for Modeling of Size Quantization in NMOSFETs. In: Intl. Conf. on Modeling and Simulation of Microsystems, vol. 1, pp. 458–461 (2001)
113. Jungemann, C., Pham, A., Meinerzhagen, B., Ringhofer, C., Bollhöfer, M.: Stable Discretization of the Boltzmann Equation Based on Spherical Harmonics, Box Integration, and a Maximum Entropy Dissipation Principle. J.Appl.Phys. **100**(2), 024502 (2006). DOI 10.1063/1.2212207
114. Kabir, M., Ibrahim, Z., Sopian, K., Amin, N.: Effect of Structural Variations in Amorphous Silicon Based Single and Multi-Junction Solar Cells from Numerical Analysis. Solar Energy Materials and Solar Cells (2010). DOI 10.1016/j.solmat.2009.12.031
115. Kane, E.: Band Structure of Indium Antimonide. J.Phys.Chem.Solids **1**(4), 249–261 (1957). DOI 10.1016/0022-3697(57)90013-6
116. Karner, M., Gehring, A., Holzer, S., Pourfath, M., Wagner, M., Goes, W., Vasicek, M., Baumgartner, O., Kernstock, C., Schnass, K., Zeiler, G., Grasser, T., Kosina, H., Selberherr, S.: A Multi-Purpose Schrödinger-Poisson Solver for TCAD Applications. J.Comp.Electronics **6**(1), 179–182 (2007). DOI 10.1007/s10825-006-0077-7
117. Karner, M., Wagner, M., Grasser, T., Kosina, H.: A Physically Based Quantum Correction Model for DG MOSFETs. In: Materials Research Society Spring Meeting (MRS), pp. 104–105 (2006)
118. Khitun, A., Balandin, A., Liu, J., Wang, K.: In-Plane Lattice Thermal Conductivity of a Quantum-Dot Superlattice. J.Appl.Phys. **88**(2), 696–699 (2000). DOI 10.1063/1.373723
119. Khitun, A., Liu, J., Wang, K.: Thermal Conductivity of Si/Ge Quantum Dot Superlattices. In: Conf. on Nanotechnology, pp. 20–22 (2004). DOI 10.1109/NANO.2004.1392236
120. Kim, D., Gabor, A., Yelundur, V., Upadhyaya, A., Meemongkolkiat, V., Rohatgi, A.: String Ribbon Silicon Solar Cells with 17.8% Efficiency. In: 3rd World Conf. on Photovoltaic Energy Conversion. Georgia Institute of Technology (2003). hdl.handle.net/1853/26154
121. Kim, W., Singer, S., Majumdar, A., Zide, J., Gossard, A., Shakouri, A.: Role of Nanostructures in Reducing Thermal Conductivity Below Alloy Limit in Crystalline Solids. In: Intl. Conf. on Thermoelectrics, pp. 9–12 (2005). DOI 10.1109/ICT.2005.1519874

122. Kittel, C.: Einführung in die Festkörperphysik, vol. 14. Oldenburg Wissenschaftsverlag (2006)
123. Kluska, S., Granek, F., Rüdiger, M., Hermle, M., Glunz, S.: Modeling and Optimization Study of Industrial n-Type High-Efficiency Back-Contact Back-Junction Silicon Solar Cells. Solar Energy Materials and Solar Cells **94**(3), 568–577 (2010). DOI 10.1016/j.solmat.2009.11.025
124. Knaipp, M.: Modellierung von Temperatureinflüssen in Halbleiterbauelementen. Dissertation, Technische Universität Wien (1998). www.iue.tuwien.ac.at/phd/knaipp
125. Kosik, R., Grasser, T., Entner, R., Dragosits, K.: On The Highest Order Moment Closure Problem [Semiconductor Device Modelling Applications]. In: Electronics Technology: Meeting the Challenges of Electronics Technology Progress, 2004, vol. 1, pp. 118–121 (2004)
126. Kullmann, W., Geurts, J., Richter, W., Lehner, N., Rauh, H., Steigenberger, U., Eichhorn, G., Geick, R.: Effect of Hydrostatic and Uniaxial Pressure on Structural Properties and Raman Active Lattice Vibrations in Bi_2Te_3. Phys.Stat.Sol.(b) **125**(1), 131–138 (1984). DOI 10.1002/pssb.2221250114
127. Kuznetsov, V.L., Kuznetsova, L.A., Kaliazin, A.E., Rowe, D.M.: High Performance Functionally Graded and Segmented Bi_2Te_3-Based Materials for Thermoelectric Power Generation. Journal of Material Science **37**(14), 2893–2897 (2002). DOI 10.1023/A:1016092224833
128. Landheer, D., Aers, G., McKinnon, W., Deen, M., Ranuárez, J.: Model for the Field Effect from Layers of Biological Macromolecules on the Gates of Metal-Oxide-Semiconductor Transistors. J.Appl.Phys. **98**(4), 044,701–1 – 044,701–15 (2005)
129. Lang, C., Pucker, N.: Mathematische Methoden in der Physik, zweite erweiterte Auflage. Spektrum Akademischer Verlag (2005)
130. Lee, C.S., Tang, T.W., Navon, T.: Transport Models for MBTE. In: J. Miller (ed.) Numerical Analysis of Semiconductor Devices and Integrated Circuits, pp. 261–265. Dublin Ireland: Boole (1989)
131. Lee, S., von Allmen, P.: Tight-Binding Modeling of Thermoelectric Properties of Bismuth Telluride. Appl.Phys.Lett. **88**(2), 022107 (2006). DOI 10.1063/1.2162863
132. Lee, S.C., Tang, T.W.: Transport Coefficients for a Silicon Hydrodynamic Model Extracted from Inhomogeneous Monte-Carlo Calculations. Solid-State Electron. **35**(4), 561–569 (1992). DOI 10.1016/0038-1101(92)90121-R
133. Levermore, C.: Moment Closure Hierarchies for Kinetic Theories. Journal of Statistical Physics **83**(5), 1021–1065 (1996). DOI 10.1007/BF02179552
134. Liu, J.S.: Monte Carlo Strategies in Scientific Computing. Springer-Verlag, New York Berlin Heidelberg (2001)
135. Liu, W., nd G. Chen J.L. Liu, T.B.T., Wang, K.: Anisotropic Thermal Conductivity of Ge Quantum-Dot and Symmetrically Strained Si/Ge Superlattices. Journal of Nanoscience and Nanotechnology **1**, 39–42(4) (2001). DOI 10.1166/jnn.2001.013
136. Lombardi, C., Manzini, S., Saporito, A., Vanzi, M.: A Physically Based Mobility Model for Numerical Simulation of Nonplanar Devices. IEEE Trans.Computer-Aided Design of Integrated Circuits and Systems **7**(11), 1164–1171 (1988). DOI 10.1109/43.9186
137. Ludwig, W.: Festkörperphysik. Akademische Verlagsgesellschaft Wiesbaden (1978)
138. Lundstrom, M.: Fundamentals of Carrier Transport. Cambridge University Press (2000)
139. Malm, U., Edoff, M.: 2D Device Modelling and Finite Element Simulations for Thin-Film Solar Cells. Solar Energy Materials and Solar Cells **93**(6-7), 1066–1069 (2009). DOI 10.1016/j.solmat.2008.11.058
140. Markowich, P., Ringhofer, C., Schmeiser, C.: Semiconductor Equations. Springer-Verlag (1990)
141. Masetti, G., Severi, M., Solmi, S.: Modeling of Carrier Mobility Against Carrier Concentration in Arsenic-, Phosphorus-, and Boron-Doped Silicon. IEEE Trans.Electron Devices **30**, 764–769 (1983)
142. Maycock, P.: Thermal Conductivity of Silicon, Germanium, III-V Compounds and III-V Alloys. Solid-State Electron. **10**(3), 161–168 (1967). DOI 10.1016/0038-1101(67)90069-X
143. Meddins, H., Parrott, J.: The Thermal and Thermoelectric Properties of Sintered Germanium-Silicon Alloys. J.Phys.C:Solid State Phys. **9**(7), 1263–1276 (1976). stacks.iop.org/0022-3719/9/1263

144. Miele, A., Fletcher, R., Zaremba, E., Feng, Y., Foxon, C.T., Harris, J.J.: Phonon-Drag Thermopower and Weak Localization. Physical Review B **58**(19), 13,181–13,190 (1998). DOI 10.1103/PhysRevB.58.13181
145. Mingo, N.: Thermoelectric Figure of Merit and Maximum Power Factor in III–V Semiconductor Nanowires. Appl.Phys.Lett. **84**(14), 2652–2654 (2004). DOI 10.1063/1.1695629
146. Muscato, O., Romano, V.: Simulation of Submicron Silicon Diodes with a Non-Parabolic Hydrodynamical Model Based on the Maximum Entropy Principle. In: Intl. Workshop on Computational Electronics, pp. 94–95 (2000). DOI 10.1109/IWCE.2000.869941
147. Neinhüs, B.: Hierarchische Bauelementsimulationen von Si/SiGe Hochfrequenztransistoren. Dissertation, Universität Bremen (2002)
148. Nguyen, C., Jungemann, C., Meinerzhagen, B.: Modeling of Size Quantization in Strained Si-nMOSFETs with the Improved Modified Local Density Approximation. In: NSTI Nanotechnology Conference and Trade Show, vol. 3, pp. 33–36 (2005)
149. Nolas, G., Cohn, J., Slack, G., Schujman, S.: Semiconducting Ge Clathrates: Promising Candidates for Thermoelectric Applications. Appl.Phys.Lett. **73**(2), 178–180 (1998). DOI 10.1063/1.121747
150. Nolas, G., Sharp, J., Goldsmid, H.: Thermoelectrics: Basic Principles and New Materials Developments. Springer-Verlag (2001)
151. Nolting, W.: Grundkurs Theoretische Physik 6: Statistische Physik, vol. XII. Springer-Verlag (2007)
152. NREL Solar Cell Sets World Efficiency Record at 40.8 Percent. Press Release (2008). www.nrel.gov/news/press/2008/625.html
153. Golden, CO, USA: NREL Sets New CIGS Thin Film Efficiency Record. Press Release (2008). www.solarbuzz.com/news/NewsNATE50.htm
154. Ohl, R.: Light-Sensitive Electric Device (1946). www.freepatentsonline.com/2402662.html
155. Onsager, L.: Reciprocal Relations in Irreversible Processes. I. Physical Review **37**(4), 405–426 (1931). DOI 10.1103/PhysRev.37.405
156. Paasch, G., Übensee, H.: A Modified Local Density Approximation. Electron Density in Inversion Layers. Phys.Stat.Sol.(b) **113**(1), 165–178 (1982). DOI 10.1002/pssb.2221130116
157. Palankovski, V.: Simulation of Heterojunction Bipolar Transistors. Dissertation, Technische Universität Wien (2000). www.iue.tuwien.ac.at/phd/palankovski
158. Palankovski, V., Quay, R.: Analysis and Simulation of Heterostructure Devices. Springer-Verlag (2004)
159. Palestri, P., Mastrapasqua, M., Pacelli, A., King, C.: A Drift-Diffusion/Monte Carlo Simulation Methodology for $Si_{1-x}\,Ge_x$ HBT Design. IEEE Trans.Electron Devices **49**(7), 1242–1249 (2002). DOI 10.1109/TED.2002.1013282
160. Park, K., Lee, S., Sohn, Y., Choi, S.: BioFET Sensor for Detection of Albumin in Urine. Electronic Letters **44**(3) (2008)
161. Pejčinović, B., Tang, H., Egley, J., Logan, L., Srinivasan, G.: Two-Dimensional Tensor Temperature Extension of the Hydrodynamic Model and its Applications. IEEE Trans.Electron Devices **42**(12), 2147–2155 (1995). DOI 10.1109/16.477773
162. Peltier, J.: Nouvelles Experiences sur la Caloriecete des Courans Electrique. Ann. Chim. **LVI**, 371–387 (1834)
163. Pieters, B., Krc, J., Zeman, M.: Advanced Numerical Simulation Tool for Solar Cells - ASA5. In: Photovoltaic Energy Conversion, vol. 2, pp. 1513–1516 (2006). DOI 10.1109/WCPEC.2006.279758
164. Pirrung, M.: How to Make a DNA Chip. Angew. Chem. Intl. Ed. **41**, 1276–1289 (2002)
165. Poghossian, A., Cherstvy, A., Ingebrandt, S., Offenhäusser, A., Schöning, M.J.: Possibilities and Limitations of Label-Free Detection of DNA Hybridization with Field-Effect-Based Devices. Sensors and Actuators, B: Chemical **111-112**, 470–480 (2005)
166. Pollock, D.: CRC Handbook of Thermoelectrics, Chapter: Thermoelectric Phenomena. CRC Press LLC (1994)
167. protein data bank: www.pdb.org
168. PV FAQ's: What is the Energy Payback for PV? Press Release (2004). www.nrel.gov/docs/fy05osti/37322.pdf

169. Rahmat, K.: Simulation of Hot Carriers in Semiconductor Devices. Technical Report 591, Research Laboratory of Electronics, Massachusetts Institute of Technology (1995)
170. Rahmat, K., White, J., Antoniadis, D.: Computation of Drain and Substrate Currents in Ultra-Short-Channel nMOSFET's Using the Hydrodynamic Model. IEEE Trans.Computer-Aided Design of Integrated Circuits and Systems **12**(6), 817–824 (1993). DOI 10.1109/43.229756
171. Richou, F., Pelous, G., Lecrosnier, D.: Thermal Generation of Carriers in Gold-Doped Silicon. J.Appl.Phys. **51**(12), 6252–6257 (1980). DOI 10.1063/1.327611
172. Ringhofer, C., Heitzinger, C.: Multi-Scale Modeling and Simulation of Field-Effect Biosensors. ECS Transactions **14**(1), 11–19 (2008). DOI 10.1149/1.2956012
173. Roosbroeck, W.V.: Theory of Flow of Electrons and Holes in Germanium and Other Semiconductors. Bell Syst. Techn. J. **29**, 560–607 (1950)
174. Rowe, D.: Electrical Properties of Hot-Pressed Germanium-Silicon-Boron Alloys. J.Phys.D:Appl. Phys. **8**(9), 1092–1103 (1975). stacks.iop.org/0022-3727/8/1092
175. Rowe, D.: CRC Handbook of Thermoelectrics. CRC Press LLC (1994)
176. Sadaka, M., Thean, A., Barr, A., Tekleab, D., Kalpat, S., White, T., Nguyen, T., Mora, R., Beckage, P., Jawarani, D., Zollner, S., Kottke, M., Liu, R., Canonico, M., Xie, Q., Wang, X., Parsons, S., Eades, D., Zavala, M., Nguyen, B., Mazure, C., Mogab, J.: Fabrication and Operation of sub-50 nm Strained-Si on $Si_{1-x}Ge_x$ Insulator (SGOI) CMOSFETs. In: IEEE Intl. SOI Conf., pp. 209–211 (2004)
177. Sadovnikov, A., Roulston, D.: A Study of the Influence of Hydrodynamic Model Effects on Characteristics of Silicon Bipolar Transistors. COMPEL: The Intl. Journal for Computation and Mathematics in Electrical and Electronic Engineering **12**(4), 245–262 (1993). DOI 10.1108/eb051803
178. Scheidemantel, T.J., Ambrosch-Draxl, C., Thonhauser, T., Badding, J.V., Sofo, J.O.: Transport Coefficients from First-Principles Calculations. Physical Review B **68**(12), 125,210 (2003). DOI 10.1103/PhysRevB.68.125210
179. Schenk, A.: Re-Examination of Physical Models in the Temperature Range 300 K – 700 K. Tech. Rep. 99, German Bundesministerium für Bildung und Forschung (2001)
180. Scherrer, H., Scherrer, S.: CRC Handbook of Thermoelectrics, Chapter: Bismulth Telluride, Antimony Telluride, and Their Solid Solutions. CRC Press LLC (1994)
181. Schöning, M.J.: "Playing Around" with Field-Effect Sensors on the Basis of EIS Structures, LAPS and ISFETs. Sensors **5**(3), 126–138 (2005). DOI 10.3390/s5030126
182. Schuss, Z., Nadler, B., Eisenberg, R.: Derivation of Poisson and Nernst-Planck Equations in a Bath and Channel from a Molecular Model. Physical Review E **64**(3), 036,116 (2001). DOI 10.1103/PhysRevE.64.036116
183. Seebeck, T.J.: Über die magnetische Polarisation der Metalle und Erze durch Temperatur-Differenz. Annalen der Physik **82**(2), 133–160 (1826). DOI 10.1002/andp.18260820202
184. Selberherr, S.: Analysis and Simulation of Semiconductor Devices. Springer-Verlag (1984)
185. Selberherr, S., Hänsch, W., Seavey, M., Slotboom, J.: The Evolution of the MINIMOS Mobility Model. Solid-State Electron. **33**(11), 1425–1436 (1990). DOI 10.1016/0038-1101(90)90117-W
186. Shinwari, M., Deen, M., Landheer, D.: Study of the Electrolyte-Insulator-Semiconductor Field-Effect Transistor (EISFET) with Applications in Biosensor Design. Microelectronics Reliability **47**(12), 2025–2057 (2007). DOI 10.1016/j.microel.2006.10.003
187. Shockley, W.: Problems Related to p-n Junctions in Silicon. Solid-State Electron. **2**(1), 35–67 (1961). DOI 10.1016/0038-1101(61)90054-5
188. Shockley, W., Queisser, H.: Detailed Balance Limit of Efficiency of p-n Junction Solar Cells. J.Appl.Phys. **32**(3), 510–519 (1961). DOI 10.1063/1.1736034
189. Shockley, W., Read, W.T.: Statistics of the Recombinations of Holes and Electrons. Physical Review **87**(5), 835–842 (1952). DOI 10.1103/PhysRev.87.835
190. Singh, J.: Physics of Semiconductors and their Heterostructures. McGraw-Hill, Inc. (1993)
191. Singh, M., Bhandari, C.: Thermoelectric Properties of Bismuth Telluride Quantum Wires. Solid State Communications **127**(9-10), 649–654 (2003). DOI 10.1016/S0038-1098(03)00520-9

192. Singh, R., Chandran, P., Grujicic, M., Poole, K., Vingnani, U., Ganapathi, S., Swaminathan, A., Jagannathan, P., Iyer, H.: Dominance of Silicon CMOS Based Semiconductor Manufacturing Beyond International Technology Roadmap. Semiconductor Fabtech **30**, 104–113 (2006)
193. Singh, R., Gupta, N., Poole, K.: Global Green Energy Conversion Revolution in 21st Century Through Solid State Devices. In: Intl. Conf. on Microelectronics, pp. 45–54 (2008). DOI 10.1109/ICMEL.2008.4559221
194. Siwy, Z., Apel, P., Baur, D., Dobrev, D., Korchev, Y., Neumann, R., Spohr, R., Trautmann, C., Voss, K.O.: Preparation of Synthetic Nanopores with Transport Properties Analogous to Biological Channels. Surface Science **532-535**, 1061–1066 (2003). DOI 10.1016/S0039-6028(03)00448-5
195. Siwy, Z., Apel, P., Dobrev, D., Neumann, R., Spohr, R., Trautmann, C., Voss, K.: Ion Transport Through Asymmetric Nanopores Prepared by Ion Track Etching. Nucl. Instrum. and Meth. in Phys. Res. Sect. B: Beam Interactions with Materials and Atoms **208**, 143–148 (2003). DOI 10.1016/S0168-583X(03)00884-X. Ionizing Radiation and Polymers
196. Siwy, Z., Dobrev, D., Neumann, R., Trautmann, C., Voss, K.: Electro-Responsive Asymmetric Nanopores in Polyimide with Stable Ion-Current Signal. Appl.Phys.A **76**(5), 781–785 (2003). DOI 10.1007/s00339-002-1982-7
197. Siwy, Z., Gu, Y., Spohr, H.A., Baur, D., Wolf-Reber, A., Spohr, R., Apel, P., Korchev, Y.E.: Rectification and Voltage Gating of Ion Currents in a Nanofabricated Pore. EPL (Europhysics Letters) **60**(3), 349–355 (2002)
198. Siwy, Z., Fuliński, A.: Fabrication of a Synthetic Nanopore Ion Pump. Physical Review Letters **89**(19), 198,103 (2002). DOI 10.1103/PhysRevLett.89.198103
199. Skrabek, E., D.Trimmer: CRC Handbook of Thermoelectrics, Chapter: Properties of the General TAGS System. CRC Press LLC (1994)
200. Slack, G., Hussain, M.: The Maximum Possible Conversion Efficiency of Silicon-Germanium Thermoelectric Generators. J.Appl.Phys. **70**(5), 2694–2718 (1991). DOI 10.1063/1.349385
201. Song, J., Li, S., Huang, C., Anderson, T., Crisalle, O.: Modeling and Simulation of a $CuGaSe_2/Cu(In_{1-x},Ga_x)Se_2$ Tandem Solar Cell. In: Photovoltaic Energy Conversion, vol. 1, pp. 555–558 (2003). DOI 10.1109/WCPEC.2003.1305344
202. Span, G.: Thermoelectric Element (2008). Austrian Patent AT 410 492 B. Intl. Patent Application PCT/AT01/00123, Granted in USA, Russia, Europe
203. Stern, E., Klemic, J., Routenberg, D., Wyrembak, P., Turner Evans, D., Hamilton, A., LaVan, D., Fahmy, T., Reed, M.: Lable-free Immunodetection with CMOS-compatible Semiconducting Nanowires. Nature Letters **445**(1), 519–522 (2007). DOI 10.1038/nature05498
204. Stettler, M., Alam, M., Lundstrom, M.: A Critical Examination of the Assumptions Underlying Macroscopic Transport Equations for Silicon Devices. IEEE Trans.Electron Devices **40**(4), 733–740 (1993). DOI 10.1109/16.202785
205. Stoletow, A.: Suite des recherches actino-electriques. Comptes Rendus **CVII**, 91 (1888)
206. Stoletow, A.: Sur une sorte de courants electriques provoques par les rayons ultraviolets. Comptes Rendus **CVI** (1888)
207. Stoletow, A.: Sur les courants actino-électriques dans l'air raréfié. Journal de Physique **9**, 468 (1890)
208. Stratton, R.: Diffusion of Hot and Cold Electrons in Semiconductor Barriers. Physical Review **126**(6), 2002–2014 (1962). DOI 10.1103/PhysRev.126.2002
209. Stratton, R.: Semiconductor Current-Flow Equations (Diffusion and Degeneracy). IEEE Trans.Electron Devices **19**(12), 1288–1292 (1972)
210. Stryer, L.: Biochemistry 4th Edition. New York: W.H. Freeman and Company (1995)
211. Sugihara, S., Tomita, S., Asakawa, K., Suda, H.: High Performance Properties of Sintered Bi_2Te_3-Based Thermoelectric Material. In: Intl. Conf. on Thermoelectrics, pp. 46–51 (1996). DOI 10.1109/ICT.1996.553254
212. Tang, T., Gan, H.: Two Formulations of Semiconductor Transport Equations Based on Spherical Harmonic Expansion of the Boltzmann Transport Equation. IEEE Trans.Electron Devices **47**(9), 1726–1732 (2000). DOI 10.1109/16.861583

213. Tang, T., Ramaswamy, S., Nam, J.: An Improved Hydrodynamic Transport Model for Silicon. IEEE Trans.Electron Devices **40**(8), 1469–1477 (1993). DOI 10.1109/16.223707
214. Tang, T.W., Ieong, M.K.: Discretization of Flux Densities in Device Simulations Using Optimum Artificial Diffusivity. IEEE Trans.Computer-Aided Design of Integrated Circuits and Systems **14**(11), 1309–1315 (1995)
215. Tasaki, H., Kim, W.Y., Hallerdt, M., Konagai, M., Takahashi, K.: Computer Simulation Model of the Effects of Interface States on High-Performance Amorphous Silicon Solar Cells. J.Appl.Phys. **63**(2), 550–560 (1988). DOI 10.1063/1.340085
216. Thoma, R., Emunds, A., Meinerzhagen, B., Peifer, H.J., Engl, W.: Hydrodynamic Equations for Semiconductors with Nonparabolic Band Structure. IEEE Trans.Electron Devices **38**(6), 1343–1353 (1991). DOI 10.1109/16.81625
217. Thomson, W.: On a Mechanical Theory of Thermoelectric Currents. Proc. of the Royal Society of Edinburgh pp. 91–98 (1851)
218. Tsutagawa, M., Michael, S.: Triple Junction $InGaP/GaAS/Ge$ Solar Cell Optimization: The Design Parameters for a 36.2% Efficient Space Cell Using Silvaco ATLAS Modeling & Simulation. In: Photovoltaic Specialists Conf., pp. 001,954–001,957 (2009). DOI 10.1109/PVSC.2009.5411544
219. University of New South Wales: Highest Silicon Solar Cell Efficiency Ever Reached. Press Release (2008). www.sciencedaily.com/releases/2008/10/081023100536.htm
220. Vasicek, M.: Advanced Macroscopic Transport Models. Dissertation, Technische Universität Wien (2009). www.iue.tuwien.ac.at/phd/vasicek
221. Ventura, D., Gnudi, A., Baccarani, G.: A Deterministic Approach to the Solution of the BTE in Semiconductors. La Rivista del Nuovo Cimento (1978-1999) **18**(6), 1–33 (1995). DOI 10.1007/BF02743029
222. Villanueva, J., Diaz, V., Bolivar, S., Tejada, T., Rodriguez, E.: A Multijunction Solar Cell Simulation Program for the Development of Concentration Systems. IEEE Trans.Electron Devices pp. 262–265 (2007). DOI 10.1109/SCED.2007.384042
223. Vining, C.: A Model for the High-Temperature Transport Properties of Heavily Doped n-type Silicon-Germanium Alloys. J.Appl.Phys. **69**(1), 331–341 (1991). DOI 10.1063/1.347717
224. Wagner, M.: Simulation of Thermoelectric Devices. Dissertation, Technische Universität Wien (2007). www.iue.tuwien.ac.at/phd/wagner
225. Wagner, M., Karner, M., Cervenka, J., Vasicek, M., Kosina, H., Holzer, S., Grasser, T.: Quantum Correction for DG MOSFETs. J.Comp.Electronics **5**(4), 397–400 (2006). DOI 10.1007/s10825-006-0032-7
226. Weinberg, I.: Phonon-Drag Thermopower in Cu-Al and Cu-Si Alloys. Physical Review **139**(3A), A838–A843 (1965). DOI 10.1103/PhysRev.139.A838
227. Willner, I., Katz, E.: Bioelectronics: From Theory to Applications, 1 ed. Wiley-VCH (2005). www.worldcat.org/isbn/3527306900
228. Windbacher, T., Sverdlov, V., Selberherr, S.: Modeling of Low Concentrated Buffer DNA Detection with Suspend Gate Field-Effect Transistors (SGFET). In: Intl. Workshop on Computational Electronics, pp. 169–172 (2009)
229. Windbacher, T., Sverdlov, V., Selberherr, S., Heitzinger, C., Mausern, N., Ringhofer, C.: Study of the Properties of Biotin-Streptavidin Sensitive BioFETs. In: Intl. Conf. on Biomedical Electronics and Devices, pp. 24–30 (2009)
230. Wolff, P.: Theory of Electron Multiplication in Silicon and Germanium. Physical Review **95**(6), 1415–1420 (1954). DOI 10.1103/PhysRev.95.1415
231. Wood, C.: Materials for Thermoelectric Energy Conversion. Reports on Progress in Physics **51**(4), 459–539 (1988). stacks.iop.org/0034-4885/51/459
232. Wu, D.L., Fan, R., Yang, P., Majumdar, A.: Thermal Conductivity of $Si/SiGe$ Superlattice Nanowires. Appl.Phys.Lett. **83**(15), 3186–3188 (2003). DOI 10.1063/1.1619221
233. Wu, M.W., Horing, N.J.M., Cui, H.L.: Phonon-Drag Effects on Thermoelectric Power. Physical Review B **54**(8), 5438–5443 (1996). DOI 10.1103/PhysRevB.54.5438
234. Xu, J., Luo, X., Chen, H.: Analytical Aspects of FET-Based Biosensors. Frontiers in Bioscience **10**, 420–430 (2005)

235. Yang, R., Chen, G.: Thermal Conductivity Modeling of Periodic Two-Dimensional Nanocomposites. Physical Review B **69**(19), 195,316 (2004). DOI 10.1103/PhysRevB.69.195316
236. Yang, R., Chen, G.: Nanostructured Thermoelectric Materials: From Superlattices to Nanocomposites. Materials Integration **18**, 1–12 (2006)
237. Yates, D., Levine, S., Healy, T.: Site-Binding Model of the Electrical Double Layer at the Oxide/Water Interface. Journal of the Chemical Society **70**, 1807–1818 (1974). DOI 10.1039/F19747001807
238. Zheng, G., Patolsky, F., Cui, Y., Wang, W.U., Lieber, C.M.: Multiplexed Electrical Detection of Cancer Markers with Nanowire Sensor Arrays. Nature Biotechnology **23**(10), 1294–1301 (2005)
239. Zhitinskaya, M., Nemov, S., Svechnikova, T., Luk'yanova, L., Konstantinov, P., Kutasov, V.: Thermal Conductivity of Bi_2Te_3: Sn and the Effect of Codoping by Pb and I Atoms. Physics of the Solid State **45**(7), 1251–1253 (2003). DOI 10.1134/1.1594237
240. Zhou, J.: Thermal and Thermoelectric Transport Measurements of One-Dimensional Nanostructures. Ph.D. thesis, University of Texas at Austin (2005)
241. Zhou, J., Zhang, L., Leng, Y., Tsao, H.K., Sheng, Y.J., Jiang, S.: Unbinding of the Streptavidin-Biotin Complex by Atomic Force Microscopy: A Hybrid Simulation Study. J.Chem.Phys. **125**(10), 104905 (2006). DOI 10.1063/1.2337629

Chapter 2
Quantum and Coulomb Effects in Nano Devices

Dragica Vasileska, Hasanur Rahman Khan, Shaikh Shahid Ahmed, Gokula Kannan, and Christian Ringhofer

Abstract In state of the art devices, it is well known that quantum and Coulomb effects play significant role on the device operation. In this book chapter we demonstrate that a novel effective potential approach in conjunction with a Monte Carlo device simulation scheme can accurately capture the quantum-mechanical size quantization effects. Inclusion of tunneling within semi-classical simulation schemes is discussed in details. We also demonstrate, via proper treatment of the short-range Coulomb interactions, that there will be significant variation in device design parameters for devices fabricated on the same chip due to the presence of unintentional dopant atoms at random locations within the channel of alternative technology devices.

Keywords Nanoscale devices · Quantum confinement · SCHRED · Random dopants

1 Introduction

As semiconductor devices are being scaled into nanometer dimensions (Fig. 2.1), significant number of effects start to become important and they can be classified into quantum and classical reliability effects. In general, there are three manifestations of quantum effects in nanodevices: (1) quantum-mechanical size quantization, (2) tunneling and (3) quantum interference. Quantum-mechanical size quantization effects and gate leakage can be easily incorporated into classical simulators, but quantum interference effects require fully quantum-mechanical treatment. In this book chapter we focus on the inclusion of quantum-mechanical size quantization and tunneling effects into particle-based device simulators. Several separate book chapters in this book are devoted to quantum transport. In addition

D. Vasileska (✉)
School of Electrical, Computer and Energy Engineering, Arizona State University, Tempe, AZ, USA
e-mail: vasileska@asu.edu

D. Vasileska and S.M. Goodnick (eds.), *Nano-Electronic Devices: Semiclassical and Quantum Transport Modeling*, DOI 10.1007/978-1-4419-8840-9_2,

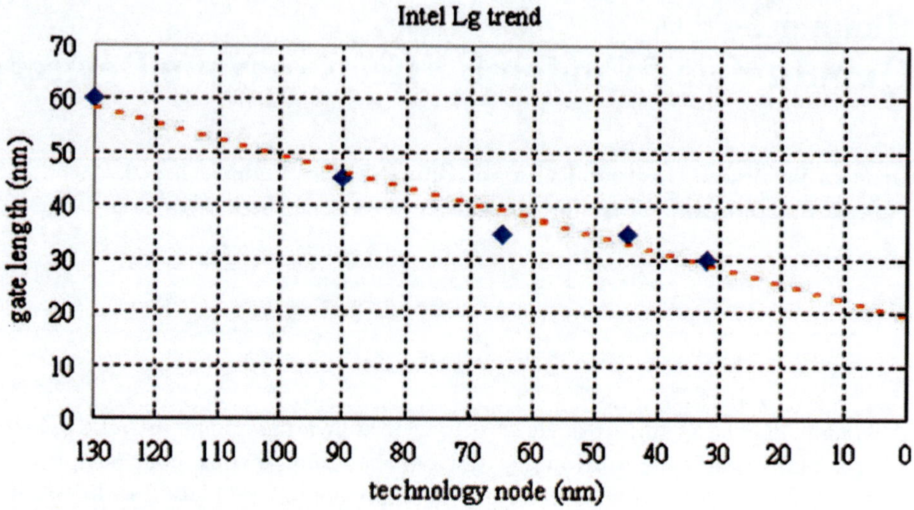

Fig. 2.1 Intel trend in transistor channel length scaling

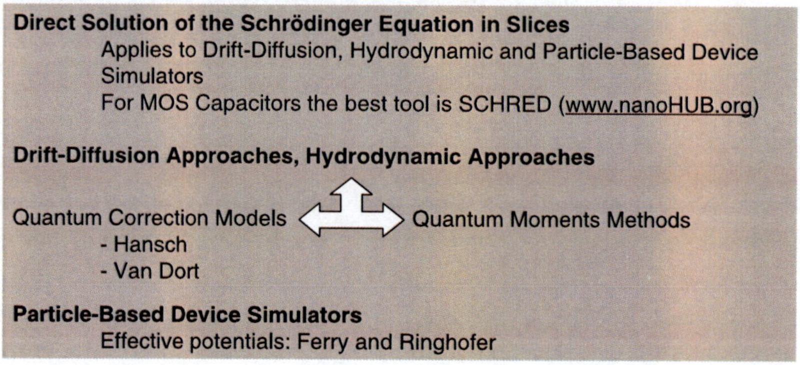

Fig. 2.2 Inclusion of Quantum Mechanical Space/Size Quantization effects in classical device simulators

to this, in this book chapter we also address in detail the issue of transistor reliability due to random dopant effects or due to unintentional dopants in alternative technology devices.

The inclusion of quantum-mechanical size quantization effects in drift-diffusion, hydrodynamic and particle-based device simulators is schematically illustrated in Fig. 2.2 and explained in more detail later in the text.

Quantum correction models try to incorporate quantum-mechanical description of carrier behavior via modification of certain device parameters within the standard drift-diffusion or hydrodynamic model. For example, the Hansch model [1] modifies the effective density of states function using,

$$N_C^* = N_C[1 - \exp(-z/\text{LAMBDA})]^2 \tag{2.1}$$

where LAMBDA is a parameter.

On the other hand, the very popular Van Dort model [2] modifies the intrinsic carrier concentration by taking into account the effective band-gap increase due to quantum-mechanical size quantization effects. Namely, the surface potential is modified according to:

$$\psi_s^{QM} = \psi_s^{CONV} + \Delta\varepsilon/q + E_n\Delta z,\ \Delta z = \langle z^{QM}\rangle - \langle z^{CONV}\rangle \tag{2.2}$$

The second term on the RHS of the above expression accounts for the band-gap widening effect because of the upward shift of the lowest allowed state. The third term accounts for the larger displacement of the carriers from the interface and the extra band-bending needed for given population that is expressed with

$$qE_n\Delta z \approx \frac{4}{9}\Delta\varepsilon \tag{2.3}$$

The energy shift that appears in the above equation is calculated using the variational approach of Fang and Howard [3]. With these modifications, one arrives at the following expression for the effective band-gap

$$E_g^{QM} = E_g^{CONV} + \frac{13}{9}\Delta\varepsilon,\ \Delta\varepsilon \approx \beta\left(\frac{\varepsilon_{Si}}{4qk_BT}\right)^{1/3} E_\perp^{2/3} \tag{2.4}$$

where β is a parameter. The modification in the effective bandgap leads to modification of the intrinsic carrier concentration

$$\begin{aligned} n_i^{QM} &= n_i^{CONV} \exp\left[(E_g^{QM} - E_g^{CONV})/2k_BT\right] \\ n_i &= n_i^{CONV}[1 - F(y)] + F(y)n_i^{QM} \end{aligned} \tag{2.5}$$

where the function $F(y)$ defined with

$$F(y) = 2\exp(-a^2)/\left[1+\exp(-2a^2)\right],\quad a = y/y_{ref} \tag{2.6}$$

enables a smooth transition between the intrinsic carrier density in the quantum region (towards the semiconductor-oxide interface) and the semiclassical region (towards the bulk portion of the device). The meaning of the various parameters that appear in the expressions of the Van Dort model is graphically represented in Fig. 2.3 below.

The quantum moment methods for inclusion of size quantization effects into drift-diffusion and hydrodynamic simulators are discussed in Sect. 2.1 below. SCHRED First and Second Generation are discussed in Sect. 2.2. SCHRED First Generation (or SCHRED V1.0) is a tool developed by Prof. Vasileska from Arizona State University back in 1992 and it was further developed in 1998 and installed on PUNCH (in fact, SCHRED was the first tool installed on Purdue University Network Computational Hub). When the Network for Computational Nanotechnology (NCN) was formed, SCHRED V1.0 was immediately transferred on the

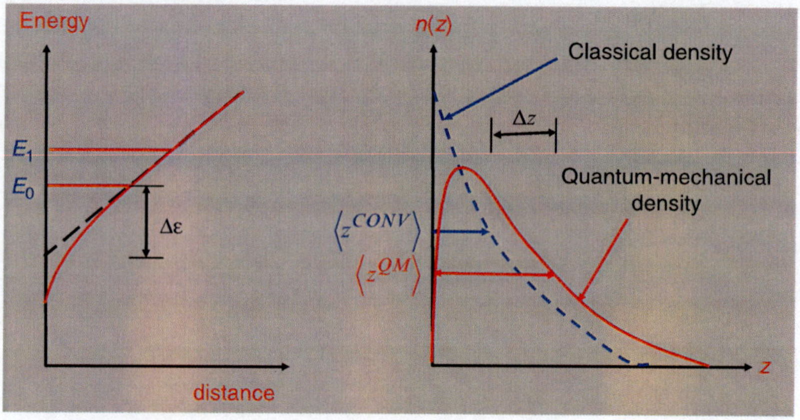

Fig. 2.3 Graphical description of the idea of the Van Dort model

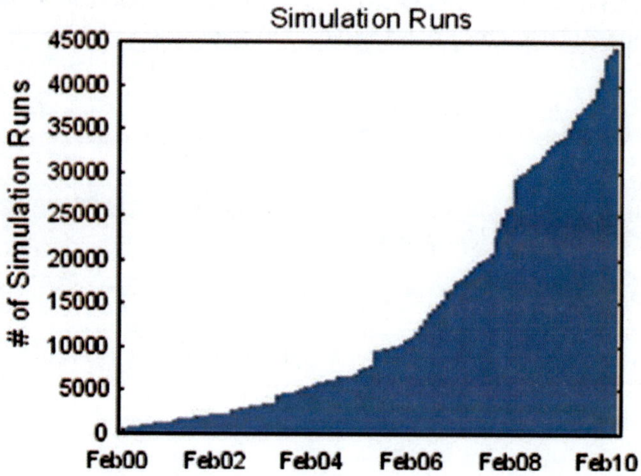

Fig. 2.4 SCHRED usage statistics

nanoHUB portal. In the meantime SCHRED V1.0 went through several revisions made by Prof. Vasileska and Dr. Zhibin Ren (Currently at IBM T. J. Watson), the most important being the introduction of quantization of holes using a heavy-hole and a light-hole band model and calculation of the tunneling current through the gate oxide. After being installed on PUNCH, and more so after its installment on the nanoHUB, SCHRED V1.0 gained enormous popularity. In fact, it was not only popular for educators to help teach students principles of operation of MOS capacitors, it was also heavily used in research work all around the world and is at the moment cited in 108 research papers (www.nanoHUB.org). The usage statistic of SCHRED v1.0 is depicted in Fig. 2.4 and its world-wide usage is illustrated in Fig. 2.5.

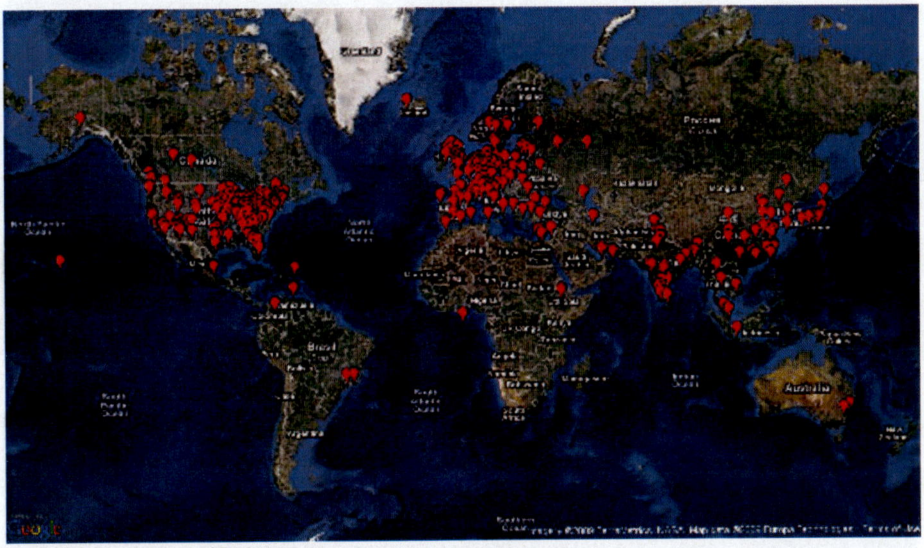

Fig. 2.5 SCHRED worldwide usage

The trend in transistors channel length scaling shown in Fig. 2.1 also requires oxide thickness reduction to improve the device transconductance and achieve better control of the charge in the channel with the gate. Since 1 nm oxide have shown to be very leaky, Intel in its 45 nm technology node already introduced high-k dielectrics, thus eliminating the gate leakage problem.

However, the gate leakage is still a big issue in Schottky transisors like MESFETs and HEMTs. The calculation of the gate leakage current in these structures can be accomplished by the use of either the WKB approximation or the transfer matrix approach. With regard to the injection between the Schottky gate and the device channel, it is best handled by using transmission probabilities, which are obtained as solutions of the Schrödinger equation along paths perpendicular to the semiconductor/metal interface. The potential along these paths is taken from the solution of the Poisson equation at each self-consistent step of the Monte Carlo procedure. The transmission probability is calculated using standard Airy function approach based on the 1 D Schrödinger equation on the propagating path. A transfer matrix approach is then applied, where the potential is interpolated linearly between the grid points on which the Poisson equation is solved in the Monte Carlo region. The unique solution is calculated with the application of the boundary conditions for the continuity of the wavefunction and its derivative at each grid point. The use of the Airy functions approach is better than the simple WKB approximation, because WKB model neglects quantum-mechanical reflections for the thermionic emission and is typically inaccurate for tunneling near the top of the potential barrier. Direct solution of the Schrödinger equation, as implemented via the Airy function formalism, also has the advantage of treating on an equal footing both thermionic emission and field-emission tunneling.

To compute the current injected by the metal contact, we calculate transmission coefficient as a ratio of the transmitted and incident probability current densities. At each iteration step, a table of transmission probabilities is generated for each mesh location along the contact interface. Then, the injected current density is obtained by integrating the product between carrier distribution and transmission probability. In its actual implementation within the Monte Carlo scheme, the transmission probability is evaluated separately for each particle and a random number technique is used to decide whether the particle is absorbed or not. Note that a similar version of the above-described approach has been successfully applied in simulations of Schottky barrier MOSFETs, as described in more detail in [4]. The WKB approximation and the transfer matrix approach that employs Airy function solutions for piecewise linear potential barrier are explained in Sect. 3 of this book chapter.

Yet another issue that we discuss in this book chapter in great details is transistor mismatch due to random number and random position of the impurity atoms in the active region of the device. These statistical fluctuations of the channel dopant number were predicted by Keyes [5] as a fundamental physical limitation of MOSFET down-scaling. Entering into the nanometer regime results in a decreasing number of channel impurities whose random distribution leads to significant fluctuations of the threshold voltage and off-state leakage current. These effects are likely to induce serious problems on the operation and performances of logical and analog circuits. It has been experimentally verified by Mizuno and co-workers [6] that threshold voltage fluctuations are mainly caused by random fluctuations of the number of dopant atoms and that other contributions such as fluctuations of the oxide thickness are comparably very small. It follows from these remarks that impurities cannot be considered anymore using the continuum doping model in advanced semiconductor device modeling but the precise location of each individual impurity within a full Coulomb interaction picture must be taken into account.

In the past, the effect of discrete dopant random distribution in MOSFET channel has been assessed by analytical or drift-diffusion (DD) approaches. The first DD study consisted in using a stochastically fluctuating dopant distribution obeying Poisson statistics [7]. 3D *atomistic* simulators have also been developed for studying threshold voltage fluctuations [8,9]. Even though the DD/HD methods are very useful because of their simplicity and fast computing times, it is not at all clear whether such macroscopic simulation schemes can be exploited into the atomistic regime. In fact, it is not at all clear how such discrete electrons and impurities are modeled in macroscopic device simulations due to the long-range nature of the Coulomb potential.

Three-dimensional (3D) Monte Carlo (MC) simulations should provide a more realistic transport description in ultra-short MOSFETs. The MC procedure gives an exact solution of the Boltzmann transport equation. Thus it correctly describes the non-stationary transport conditions. Even where the microscopic simulations such as the MC method are considered, the treatment of the electrons and impurities is not straightforward which is again due to the long-range nature of the Coulomb potential. The incorporation of the long-range Coulomb potential in the MC method has been a long-standing issue [10,11]. This problem is, in general, avoided by assuming that the electrons and the impurities are always screened by the other carriers so that the long-range part of the Coulomb interaction is effectively suppressed. The complexity of the MC simulation increases as one takes into account more complicated screening processes by using the dynamical and wave-vector dependent dielectric function obtained from, for example, the random phase approximation. However, the screening is a very complicated many-body matter [12].

This situation is also complicated in the MC *device simulations* in which the BTE is self-consistently coupled with the Poisson equation [13]. The Coulomb potential due to electrons and impurities is then separated into the long-range and the short-range parts. The long-range part is taken into account by the solution of the Poisson equation, whereas the short-range part is usually included in the BTE through the scattering kernel. In other words, the Coulomb potential is separated into the long-range and short-range parts by the size of the mesh employed in the Poisson equation. However, the choice of the mesh size is not trivial. For example, the mesh cannot be arbitrarily small as the Coulomb potential would then be double-counted by the Poisson equation and the BTE. Since the long-range part of the Coulomb potential is responsible for the many-body effects, the mesh size has to be determined consistently with, say, the renormalized electron (kinetic) energy calculated from the many-body theory [14]. This is of course not an easy task, especially for the case of small device structures. On the other hand, since the size of localized electrons in the MC device simulations is roughly given by the size of the mesh, this is not consistent with the concept of the electron wave packet. The BTE (or equivalently, the microscopic simulation) assumes that the electrons are localized and described by the wave packet whose size is comparable to the de Broglie wavelength. However, the size of the active device region is now comparable with the size of the wave packet in nanoscale MOSFETs and so it is not clear how the localized electrons in the channel should be interpreted in such microscopic simulations.

2 Inclusion of Quantum-Mechanical Size Quantization and Tunneling Effects in Particle-Based Device Simulators

2.1 Quantum-Mechanical Size Quantization Effects in Conjunction with Device Simulators

Successful scaling of MOSFETs towards shorter channel lengths requires thinner gate oxides and higher doping levels to achieve high drive currents and minimized short-channel effects [15, 16]. For these nanometer devices it was demonstrated a long time ago that, as the oxide thickness is scaled to 10 nm and below, the total gate capacitance is smaller than the oxide capacitance due to the comparable values of the oxide and the inversion layer capacitances. As a consequence, the device transconductance is degraded relative to the expectations of the scaling theory [17]. The inversion layer capacitance was also identified as being the main cause

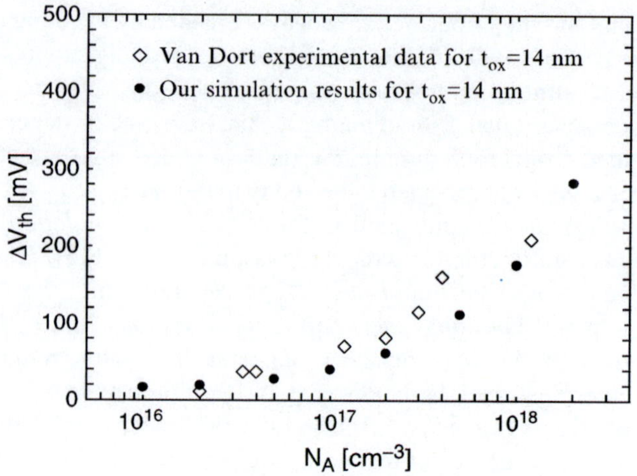

Fig. 2.6 SCHRED simulation data for the shift in the threshold voltage compared to the experimental values provided by van Dort and co-workers [20, 21]

of the second-order thickness dependence of MOSFET's *IV*-characteristics [18]. The finite inversion layer thickness was estimated experimentally by Hartstein and Albert [19]. The high levels of substrate doping, needed in nano-devices to prevent the punch-through effect has lead to quasi-two-dimensional (Q2D) nature of the carrier transport which is found responsible for the increased threshold voltage and decreased channel mobility, and a simple analytical model that accounts for this effect was proposed by van Dort and co-workers [20, 21]. Later on, Vasileska and Ferry [22] confirmed these findings by investigating the doping dependence of the threshold voltage in MOS capacitors. The experimental data for the doping dependence of the threshold voltage shift and our simulation results from [22] are shown in Fig. 2.6.

These results clearly demonstrate the influence of quantum-effects on the operation of nano-scale MOSFETs in both the off- and the on-state. The two physical origins of the inversion layer capacitance due to the finite density of states and due to the finite inversion layer thickness were demonstrated experimentally by Takagi and Toriumi [23]. A computationally efficient three-subband model that predicts both the quantum-mechanical effects in the electron inversion layer and the electron distribution within the inversion layer was proposed and implemented into the PICSEC simulator [24]. The influence of the image and many-body exchange-correlation effects on the inversion layer and the total gate capacitance was studied by Vasileska et al. [25]. It was also pointed out that the depletion of the poly-silicon gates considerably affects the magnitude of the total gate capacitance [26].

The above examples outline the advances during the two decades of research on the influence of quantum-effects on the operation on nano-devices. The conclusion is that any state-of-the-art device simulator must take into consideration the quantum-mechanical nature of the carrier transport and the poly-depletion effects to correctly predict the device off- and on-state behavior. As noted by many of these authors, to account for the quantum-mechanical effects, one in principle has to solve the 2D/3D Schrödinger-Poisson problem in conjunction with an appropriate transport kernel. (For devices in which velocity overshoot is strongly pronounced, minimum that one can do is to solve the Boltzmann transport equation using the Ensemble Monte Carlo (EMC) technique.) Since the exact solution of the 2D/3D Schrödinger–Poisson problem is time-consuming even with present state-of-the-art computers, alternative paths have been sought for device simulators that utilize quantum potentials.

The idea of quantum potentials originates from the hydrodynamic formulation of quantum mechanics, first introduced by de Broglie and Madelung [27–29], and later developed by Bohm [30, 31]. In this picture, the wave function is written in complex form in terms of its amplitude $R(\mathbf{r},t)$ and phase $\psi(\mathbf{r},t) = R(\mathbf{r},t)\exp[iS(\mathbf{r},t)/\hbar]$. These are then substituted back into the Schrödinger equation to obtain the following coupled equations of motion for the density and phase

$$\frac{\partial \rho(\mathbf{r},t)}{\partial t} + \nabla \cdot \left(\rho(\mathbf{r},t) \frac{1}{m} \nabla S(\mathbf{r},t) \right) = 0, \tag{2.7}$$

$$-\frac{\partial S(\mathbf{r},t)}{\partial t} = \tfrac{1}{2m} [\nabla S(\mathbf{r},t)]^2 + V(\mathbf{r},t) + Q(\rho,\mathbf{r},t), \tag{2.8}$$

where $\rho(\mathbf{r},t) = R^2(\mathbf{r},t)$ is the probability density. By identifying the velocity as $\frac{1}{m}\nabla S$, and the flux as $\mathbf{j} = \rho \mathbf{v}$, (2.7) becomes the continuity equation. Hence, (2.7) and (2.8) arising from this so-called *Madelung transformation* to the Schrödinger equation have the form of classical hydrodynamic equations with the addition of an extra potential, often referred to as the *quantum* or *Bohm potential*, written as

$$V_Q = -\frac{\hbar^2}{2mR} \nabla^2 R \rightarrow -\frac{\hbar^2}{2m\sqrt{n}} \nabla^2 \sqrt{n} \tag{2.9}$$

where the density n is related to the probability density as $n(\mathbf{r},t) = N\rho(\mathbf{r},t) = NR^2(\mathbf{r},t)$, where N is the total number of particles. The Bohm potential essentially represents a field through which the particle interacts with itself. It has been used, for example, in the study of wave packet tunneling through barriers [32], where the effect of the quantum potential is shown to lower or smoothen barriers, and hence *allow* for the particles to leak through.

An alternate form of the quantum potential was proposed by Iafrate, Grubin and Ferry [33], who derived a form of the quantum potential based on moments of the *Wigner–Boltzmann equation*, the kinetic equation describing the time evolution of the Wigner distribution function [34]. Their form is based on moments of the Wigner function in the pure state, and involve an expansion of order $O(\hbar^2)$, which is given by

$$V_Q = -\frac{\hbar^2}{8m} \nabla^2 (\ln n), \tag{2.10}$$

this is sometimes referred to as the Wigner potential, or as the density gradient correction. Such quantum potentials have been extensively used in *density-gradient* and *quantum-hydrodynamic* methods. Their use in particle-based simulation schemes becomes questionable due to the presence of statistical noise in the representation of the electron density and the considerable difficulty to calculate the second derivative of the density on a completely unstructured mesh given by the particle discretization.

To avoid this problem, Ferry and Zhou derived a form for a smooth quantum potential [35], based on the effective classical partition function of Feynman and Kleinert [36]. More recently, Gardner and Ringhofer [37] derived a smooth quantum potential for hydrodynamic modeling, valid to all orders of $\hbar^2$, which involves a smoothing integration of the classical potential over space and temperature. There, it was shown that close to the equilibrium regime, the influence of the potential on the ensemble can be replaced by the classical influence of a smoothed non-local barrier potential. While this effective potential depends non-locally on the density, it does not directly depend on its derivatives. Through this effective quantum potential, the influence of the barriers on an electron is felt at quite some distance from the barrier. The smoothed effective quantum potential has been used successfully in quantum-hydrodynamic simulations of resonant tunneling effects in one-dimensional double-barrier structures [38].

In analogy to the smoothed potential representations discussed above for the quantum hydrodynamic models, it is desirable to define a smooth quantum potential for use in quantum particle-based simulations. Ferry [40] has suggested an *effective potential scheme* that emerges from a wave packet description of the particle motion, where the extent of the wave packet spread is obtained from the range of wavevectors in the thermal distribution function (characterized by an electron temperature). The effective potential, V_{eff}, is related to the self-consistent Hartree potential V, obtained from the Poisson equation, through an integral smoothing relation

$$V_{\text{eff}}(\mathbf{x}) = \int V(\mathbf{x}+\mathbf{y})\, G(\mathbf{y},a_0)\, d\mathbf{y} \tag{2.11}$$

where G is a Gaussian with standard deviation a_0. The effective potential V_{eff} is then used to calculate the electric field that accelerates the carriers in the transport kernel of the Monte Carlo particle-based device simulator discussed in [39]. The calculation of V_{eff} has a fairly low computational cost, but the requirement that the electric field is updated every 0.01 fs to get physically accurate particle trajectories and to eliminate the artificial heating of the carriers in the vicinity of the Si/SiO_2 interface (where the fields are the strongest), adds to the computational cost. Note also that within this approach the parameter a_0 has to be adjusted in the initial stages of the simulation via comparisons of the sheet/line density of the Q2D/Q1D structure being investigated using the effective potential approach and the 1D/2D Schrödinger–Poisson simulations.

In this book chapter, in addition to the effective potential approach due to Ferry [40], we present a new form of the effective quantum potential for use in Monte Carlo device simulators. The proposed approach is based on perturbation theory around thermodynamic equilibrium and leads to an effective potential which depends on the energy and wavevector of each individual electron, thus effectively lowering step-function barriers for high-energy carriers [41]. The quantum potential is derived from the idea that the Wigner and the Boltzmann equation with the quantum corrected potential should possess the same steady state. The resultant quantum potential is in general two-degrees smoother than the original Coulomb and barrier potentials, i.e. possesses two more classical derivatives which essentially eliminate the problem of statistical noise. The computation of the quantum potential involves only the evaluation of pseudo-differential operators and can therefore, be effectively facilitated using Fast Fourier Transform (FFT) algorithms. The approach is quite general and can easily be modified to modeling of, for example, triangular quantum wells. The above-described approach has been used in simulation of 25 nm MOSFET device with oxide thickness of 1.2 nm.

2.1.1 Thermodynamic Effective Potential

The basic idea of the thermodynamic approach to effective quantum potentials is that the resulting semiclassical transport picture should yield the correct thermalized equilibrium quantum state. Using quantum potentials, one generally replaces the quantum Liouville equation

$$\partial_t \rho + \frac{i}{\hbar}[H,\rho] = 0 \tag{2.12}$$

for the density matrix $\rho(x,y)$ by the classical Liouville equation

$$\partial_t f + \frac{\hbar}{2m^*} k \cdot \nabla_x f - \frac{1}{\hbar} \nabla_x V \cdot \nabla_k f = 0, \tag{2.13}$$

for the classical density function $f(x,k)$. Here, the relation between the density matrix and the density function f is given by the Weyl quantization,

$$f(x,k) = W[\rho] = \int \rho(x+y/2, x-y/2) \exp(ik \cdot y) dy. \tag{2.14}$$

The thermal equilibrium density matrix in the quantum mechanical setting is given by $\rho^{eq} = \mathrm{e}^{-\beta H}$, where $\beta = 1/k_B T$ is the inverse energy and the exponential is understood as a matrix exponential, i.e. $\rho^{eq}(x,y) = \sum_\lambda \psi_\lambda(x) \exp(-\beta\lambda) \psi_\lambda(y)^*$ holds with $\{\psi_\lambda\}$ the orthonormal eigensystem of the Hamiltonian H. On the other hand, in the semiclassical transport picture, the thermodynamic equilibrium density function f_{eq} is given by the Maxwellian $f_{eq}(x,k) = \exp\left(-\frac{\beta\hbar^2|k|^2}{2m^*} - \beta V\right)$. Consequently, to obtain the quantum mechanically correct equilibrium states in the semiclassical Liouville equation with the effective quantum potential V^Q, we set

$$\begin{aligned} f_{eq}(x,k) = \exp\left(-\frac{\beta\hbar^2|k|^2}{2m^*} - \beta V^Q\right) &= W[\rho^{eq}] \\ &= \int e^{-\beta H} \rho(x+y/2, x-y/2) \exp(ik \cdot y) dy. \end{aligned} \tag{2.15}$$

This basic concept was originally introduced by Feynman and Kleinert [36]. Different forms of the effective quantum potential arise from different approaches to approximate the matrix exponential $e^{-\beta H}$.

In the approach presented in this paper, we represent $e^{\beta H}$ as the Green's function of the semigroup generated by the exponential. Introducing an artificial dimensionless parameter γ and defining $\rho(x,y,\gamma) = \sum_\lambda \psi_\lambda(x)\exp(-\gamma\beta\lambda)\psi_\lambda(y)^*$, we obtain a heat equation for ρ by differentiating ρ w.r.t. γ and using the eigenfunction property of the wave functions ψ_λ. This heat equation is referred to as the Bloch equation

$$\partial_\gamma \rho = -\frac{\beta}{2}(H\cdot\rho + \rho\cdot H),\ \rho(x,y,\gamma=0) = \delta(x-y), \tag{2.16}$$

and $\rho^{eq}(x,y)$ is given by $\rho(x,y,\gamma=1)$. Under the Weyl quantization this becomes with the usual Hamiltonian $H = -\frac{\hbar^2}{2m^*}\Delta_x + V$ and defining the effective energy E by $f = W[\rho] = e^{-\beta E}$,

$$\begin{aligned}\partial_\gamma E = & \frac{\beta\hbar^2}{8m^*}(\Delta_x E - \beta|\nabla_x E|^2) + \frac{\hbar^2|k|^2}{2m^*} \\ & + \frac{1}{2(2\pi)^3}\sum_{\nu=\pm1}\int V(x+\nu y/2)\exp[\beta E(x,k,\gamma) - \beta E(x,q,\gamma) \\ & + iy(k-q)]dqdy, E(x,k,\gamma=0) = 0.\end{aligned} \tag{2.17}$$

The effective quantum potential in this formulation is given by $E(x,k,\gamma=1) = V^Q + \frac{\hbar^2|k|^2}{2m^*}$. The logarithmic Bloch equation is now solved 'asymptotically' using the *Born approximation*, i.e. by iteratively inverting the highest order differential operator (the Laplacian). This involves successive solution of a heat equation for which the Green's function is well known, giving (see [42] for the details),

$$V^Q(x,k) = \frac{1}{(2\pi)^3}\int \frac{2m^*}{\beta\hbar^2 k\cdot\xi}\sinh\left(\frac{\beta\hbar^2 k\cdot\xi}{2m^*}\right)\exp\left(-\frac{\beta\hbar^2}{8m^*}|\xi|^2\right)V(y)e^{i\xi\cdot(x-y)}dyd\xi. \tag{2.18}$$

Note that the effective quantum potential V^Q now depends on the wave vector k. For electrons at rest, i.e. for $k=0$, the effective potential V^Q reduces to the Gaussian smoothing given in (2.11) and [40]. Also note that there are no fitting parameters in this approach, i.e. the size of the wavepacket is determined by the particle's energy.

The potential $V(y)$ that appears in the integral of (2.18) can be represented as a sum of two potentials: the barrier potential $V_B(x)$, which takes into account the discontinuity at the Si/SiO_2 interface due to the difference in the semiconductor and the oxide affinities and the Hartree potential $V_H(x)$ that results from the solution of the Poisson equation. Note that the barrier potential is 1D and independent of time and needs to be computed only once in the initialization stage of the code. On the other hand, the Hartree potential is 2D and time-dependent it describes the evolution of charge from quasi-equilibrium to a non-equilibrium state. Since the evaluation of the effective Hartree potential as given by (2.18), is very time consuming and CPU intensive, approximate solution methods have been pursued to resolve this term within a certain level of error tolerance.

We recall from the above discussion that the barrier potential is just a step-function. Under these circumstances $e\nabla_x V_B(x) = B(1,0,0)^T\delta(x_1)$, where B is the barrier height (in the order of 3.2 eV) and x_1 is a vector perpendicular to the interface. We actually need only the gradient of the potential so that using the pseudo-differential operators, we compute

$$\nabla_x V_B^Q(x,p) = \exp\left[\frac{\beta\hbar^2|\nabla_x|^2}{8m^*}\right]\frac{2m^*\sin\left(\frac{\beta\hbar p\cdot\nabla_x}{2m^*}\right)}{\beta\hbar p\cdot\nabla_x}\nabla_x V_B(x). \tag{2.19}$$

This gives

$$e\nabla_x V_B^Q(x,p) = \frac{B}{2\pi}(1,0,0)^T\int\exp\left[-\beta\frac{\hbar^2|\xi_1|^2}{8m^*}\right]\frac{2m^*\sinh\left(\frac{\beta\hbar p_1\cdot\xi_1}{2m^*}\right)}{\beta\hbar p_1\cdot\xi_1}e^{i\xi_1\cdot x_1}d\xi_1 \tag{2.20}$$

Note that V_B^Q is only a function of (x_1,p_1), i.e. it remains to be strictly one-dimensional, where x_1 and p_1 are the position and the momentum vector perpendicular to the interface. This when combined with the fact that we have to calculate this integral only once is a reason why we have decided to tabulate the result given by (2.20) on a mesh.

The Hartree potential, as computed by solving the d-dimensional Poisson equation depends in general upon d particle coordinates. For example, on a rectangular mesh the 2D Hartree potential is given by $V_H(x_1,x_2,t)$, and one has to evaluate $V_H^Q(x_1,x_2,p_1,p_2,t)$ using (2.18) N times each time step for all particles position and momenta: $x^n,p^n,n=1,\ldots,N$ (where N is the number of electrons, which is large). Of course, this is an impossible task to be accomplished in finite time on present state-of-the-art computers. We, therefore, suggest the following scheme. According to (2.18), we evaluate the quantum potential by multiplying the Hartree potential by a function of $\hbar\nabla_x$, or by multiplying the Fourier transform of the Hartree potential by a function of $\hbar\xi$. We factor the expression in (2.18) into

$$\begin{aligned}V_H^Q(x,k) &= \frac{2im^*}{\beta\hbar^2 k\cdot\nabla_x}\sinh\left(\frac{\beta\hbar^2 k\cdot\nabla_x}{2im^*}\right)\exp\left(\frac{\beta\hbar^2}{8m^*}|\nabla_x|^2\right)V_H(x)\\ &= \frac{2im^*}{\beta\hbar^2 k\cdot\nabla_x}\sinh\left(\frac{\beta\hbar^2 k\cdot\nabla_x}{2im^*}\right)V_H^0(x),\end{aligned} \tag{2.21}$$

with

$$V_H^0(x) = \exp\left(\frac{\beta\hbar^2}{8m^*}|\nabla_x|^2\right)V_H(x). \tag{2.22}$$

The evaluation of the potential $V_H^0(x)$, which is a version of the Gaussian smoothed potential due to Ferry [40]. This is computationally inexpensive since it does not depend on the wavevector k. On the other hand because of the Gaussian smoothing, $V_H^0(x)$ will be a smooth function of position, even if the Hartree potential $V_H(x)$ is computed via the Poisson equation where the electron density is given by a particle discretization. Therefore, the Fourier transform of the potential $V_H^0(x)$ will decay rapidly as a function of ξ, and it is admissible to use a Taylor expansion for small values of $\hbar\xi$ in the rest of the operator. This gives

$$\frac{2im^*}{\beta\hbar^2 k\cdot\nabla_x}\sinh\left(\frac{\beta\hbar^2 k\cdot\nabla_x}{2im^*}\right)\approx 1-\frac{\beta^2\hbar^4(k\cdot\nabla_x)^2}{24(m^*)^2}, \tag{2.23}$$

or

$$\partial_{x_r}V_H^Q(x^n,p^n)=\partial_{x_r}V_H^0(x^n)-\frac{\beta^2\hbar^2}{24m^{*2}}\sum_{j,k=1}^{2}p_j^n p_k^n\partial x_j\partial x_k\partial x_r V_H^0(x^n),\quad n=1,\ldots,N \tag{2.24}$$

for all particles. This is done simply by numerical differentiation of the sufficiently smooth grid function V_H^0 and interpolation. The evaluation of (2.24) is the price we have to pay when we compare the computational cost of this approach as opposed to the Ferry approach [40] which uses simple forward, backward or centered difference scheme for the calculation of the electric field. However, with this novel effective potential approach we avoid the use of adjustable parameters.

Example: Quantum Effects in a Conventional 25 nm MOSFET

As a first example to which we apply the Ringhofer's effective potential approach we take conventional MOSFET device with 25 nm channel length. The parameters of the device structure being simulated are as follows: the average channel/substrate doping is $10^{19}\,\mathrm{cm}^{-3}$, the doping of the source and drain regions is $10^{19}\,\mathrm{cm}^{-3}$, the junction depth is 30 nm, the oxide thickness is 1.2 nm and the gates are assumed to be metal gates with work-function equal to the semiconductor affinity. The gate/channel length is 25 nm. First in Fig. 2.7, the carrier confinement within the triangular potential well with and without the inclusion of the quantum-mechanical size-quantization effects is shown for the bias conditions $V_\mathrm{G}=V_\mathrm{D}=1\mathrm{V}$. From the results shown in this figure, it is evident that the low-energy electrons are displaced little more than the high-energy electrons; the reason being the fact that the high-energy electrons tend to behave as classical particles and hence are displaced relatively less. Also note that there is practically no carrier heating for the case when the effective potential is used in calculating the driving electric field. The carrier displacement from the interface proper is also seen from the results presented in Fig. 2.8. Notice that there is approximately 2 nm average shift of the electron density distribution near the source end of the channel when quantization effects are included in the model.

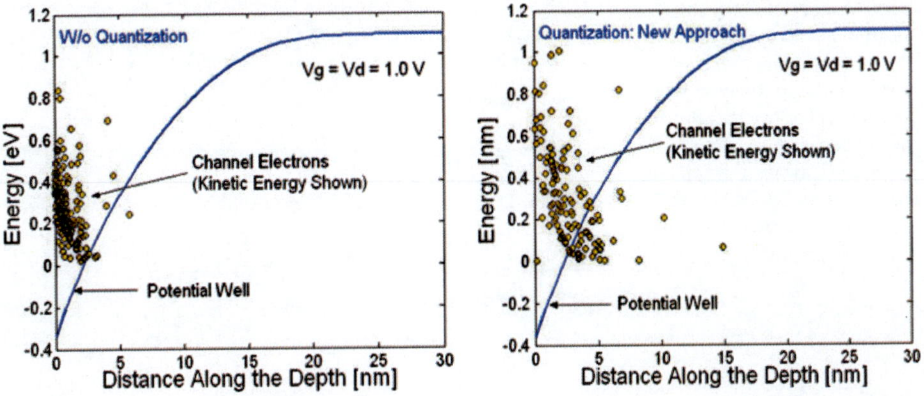

Fig. 2.7 Electron localization within the triangular potential barrier for the case when quantization effects are not included in the model *(left panel)* and for the case when we include quantum-mechanical space-quantization effects by using the effective potential approach presented in this paper *(right panel)*. The potential profile is taken in the middle portion of the channel, not at the drain end, and because of that some electrons seem to be in regions where they should not, but that is just an artifact of presenting the results. The triangular potential at the drain end of the channel is much wider

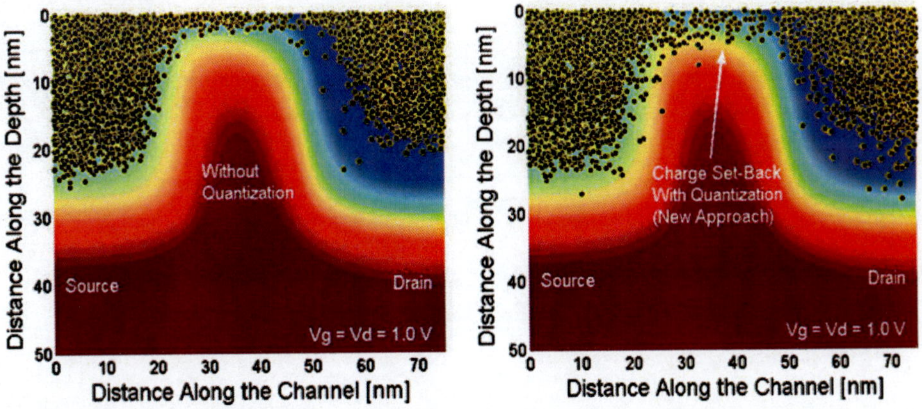

Fig. 2.8 Electron distribution in the device without *(left panel)* and with *(right panel)* the incorporation of quantum-mechanical size-quantization effects

Also note that carriers behave more like bulk carriers at the drain end of the channel and are displaced in the same manner when using both the classical and the quantum-mechanical model.

The channel length variation of the sheet electron density is shown in Fig. 2.9 for classical, fully-quantum ($V_H^Q + V_B^Q$) and quantum-barrier field (V_B^Q) models [43]. Also compared are the simulation results for the sheet electron density from the new method with those utilizing the approach due to Ferry [44]. There are several noteworthy features to be observed in this figure. First, the pinch-off of the sheet

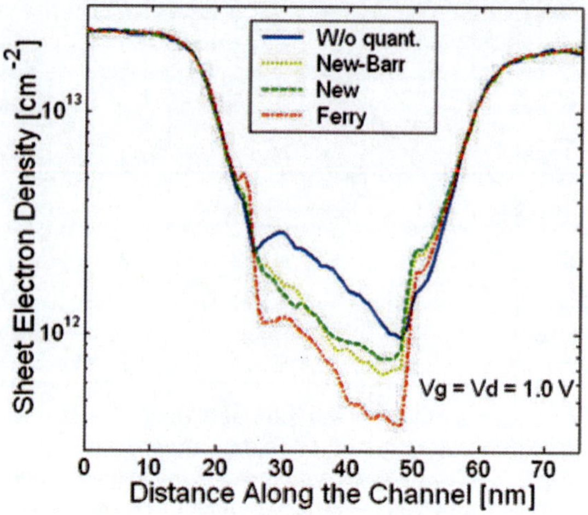

Fig. 2.9 Variation of the sheet electron density along the channel. *New-barr* corresponds to the case when we only include the influence of the barrier field. *New* represents the case when we include both the barrier and the Hartree contributions to the total electric field

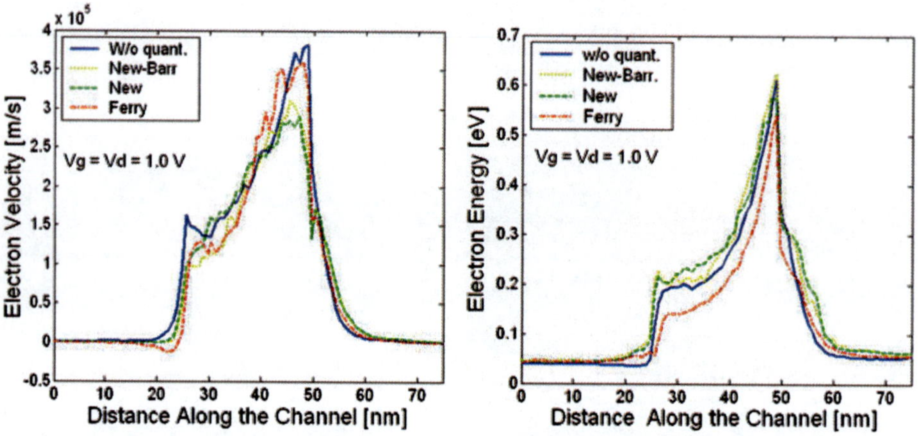

Fig. 2.10 Average electron velocity *(left panel)* and average electron energy *(right panel)* variation along the channel

electron density near the drain end of the channel is evident in all models used. Second, the barrier and the full-effective potential scheme give almost the same value for the sheet electron density, which suggests that the repulsive barrier field dominates over the attractive field due to the Hartree potential. Third, the method due to Ferry leads to significantly lower value for the sheet electron density which can be improved by choosing lower values of the Gaussian smoothing parameter.

The average electron velocity and the average electron energy are shown in the left and the right panels of Fig. 2.10, respectively. Comparing the results for the average carrier energy on the right panel, one can see that the data for the case when one has not included the effective potential and the case when one has used the new model for the effective potential agree very well with each other. The slight increase in the carrier energy in the channel region (which is non-physical) when one uses the new effective potential approach is because of the very high value of the quantum field being present in the vicinity of the Si/SiO_2 interface proper. The situation can be improved by using a sufficiently small time-step (for example 0.01 fs) during Monte Carlo simulation. The approach due to Ferry gives significantly lower value for the carrier energy near the source end of the channel which has been explained to be due to the bandgap widening effect. Also, here we do not observe the non-physical carrier heating because of the fact that Ferry's effective potential is calculated from the mesh potential which depends on both the meshing and the Gaussian parameter used in the model. The quantum field is calculated from direct differentiation of the effective mesh-potential and has every possibility of being underestimated due to the finite size of the meshing used in simulations. It also is independent on carrier energy (according to the current implementation of the model). When one confronts these data with the results for the average electron velocity, its east to say that in the low-energy region near the source end of the channel the velocity is almost the same for all cases considered. At the drain end, one finds degradation of the velocity due to the smearing introduced by the quantum potential. Again, the inclusion of the barrier field and of the quantum-corrected Hartree term give similar values, which suggests that for the device being considered in this study only the barrier field has significant impact [45].

The device transfer characteristics are shown in the left panel of Fig. 2.11. Again, it becomes clear that the proposed full quantum potential and the barrier potential give similar values for the current. Looking more in detail the device transfer characteristics one finds that the quantization effects lead to threshold voltage increase of about 220 mV. When properly adjusted for the oxide thickness difference, this result is consistent with previously published data [20]. Evidently, as deduced from the output characteristics shown in the right panel of Fig. 2.11, the shift in the threshold voltage leads to a decrease in the on-state current by 30%. The later observation confirms earlier findings that one must include quantum effects into the theoretical model to be able to properly predict the device threshold voltage and its on-state current.

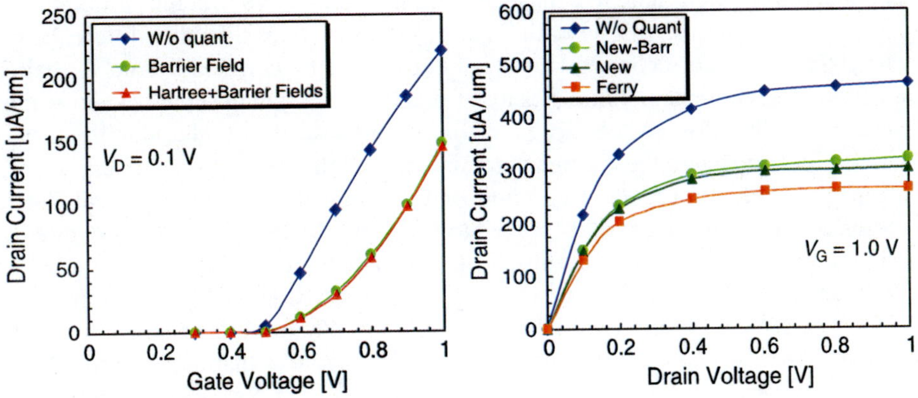

Fig. 2.11 Device transfer characteristic for $V_D = 0.1\,V$ *(left panel)*. Device output characteristics for $V_G = 1.0\,V$ *(right panel)*

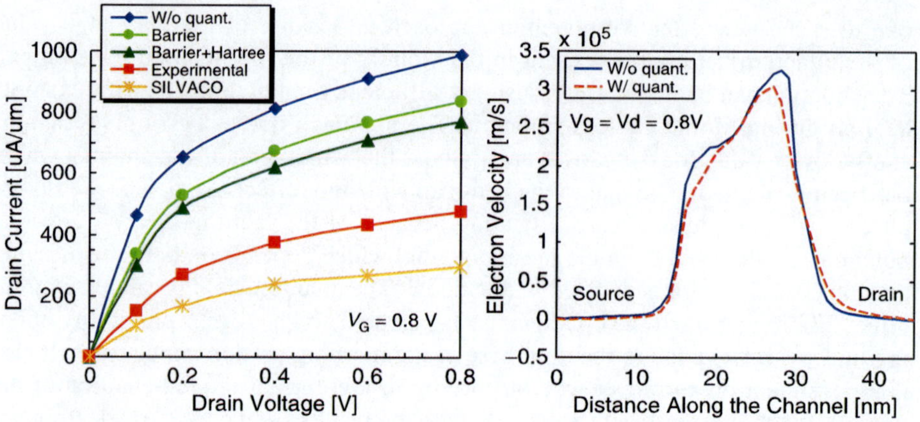

Fig. 2.12 *Left panel*: Conventional 15 nm MOSFET device output characteristics. *Right panel*: Average electron velocity along the channel

Next, the simulation results of a 15 nm conventional *n*-channel MOSFET device are discussed. Similar devices have been fabricated by Intel Corporation [46]. The physical gate length of the device used is 15 nm. The source/drain length equals 15 nm and the junction depth is also 15 nm. The bulk substrate thickness used for simulations is 45 nm. The height of the fabricated polysilicon gate electrode for this device is 25 nm. The gate oxide used was SiO_2 with physical thickness of only 0.8 nm. The source/drain doping density is $2 \times 10^{19}\,\text{cm}^{-3}$ and the channel doping is $1.5 \times 10^{19}\,\text{cm}^{-3}$. The substrate doping used is $1 \times 10^{18}\,\text{cm}^{-3}$. The simulated device output characteristics are shown in Fig. 2.12.

There are again several noteworthy features in these results: (1) Quantum-mechanical size quantization increases the threshold voltage as observed from the decrease in the slope in the linear region and hence degrades the device transconductance. (2) Drain current degradation due to the quantum effects is not uniform rather decreases with the increase in drain bias. The reason may be attributed again to the fact that the electrons tend to behave as classical particles as average carrier energy increases with the increase in drain bias, (3) there is a considerable difference between the barrier-correction and the barrier-Hartree (full) correction which is mainly due to the use of higher doping density ($1.5 \times 10^{19}\,\text{cm}^{-3}$) in the channel region than was used in the 25 nm MOSFET ($1 \times 10^{19}\,\text{cm}^{-3}$) case. The higher doping density has a direct impact on the Hartree potential making the triangular channel potential steeper and hence introducing a pronounced quantum effects. But the overall degradation of the drain current as compared to the 25 nm MOSFET device structure has reduced in the 15 nm device because of the ballistic nature of the carrier motion in the latter case. This fact becomes clear if one observes the velocity profile of the device as depicted in the right panel of Fig. 2.12. What is important in this figure is that the carriers attain a velocity which is comparable to that in the 25 nm device structure even with a lesser biases applied i.e. $V_G = V_D = 0.8\,\text{V}$. Also, the gate oxide thickness is lesser in the 10 nm device which means that the gate oxide capacitance constitutes the major portion of the total effective gate capacitance thereby reducing the impact of the quantum capacitance. (4) The discrepancy between the experimental and the simulated results is attributed mainly to two reasons: (a) the series resistance coming from the finite width of the actual device structure and the contact resistances, and (b) the gate polysilicon depletion effects which as previously mentioned, can introduce further degradation of the drain current on the order of 10–30% depending on the doping density and the height of the polysilicon gate used. The limited data as supplied by the Intel Corporation shows that the polysilicon gate is of 25 nm height which can indeed contribute to a significant degradation of the drain current. (5) The use of a commercial simulator like the drift-diffusion based SILVACO Atlas fails considerably to predict the device behavior mainly because of the ballistic and quantized nature of the carriers in these nanoscale device structures.

Example: Size-Quantization in Nanoscale SOI Devices

Because of using lightly/nearly undoped channel region, size-quantization effects in nanoscale fully-depleted SOI devices find a major source in the very physical nature of the confined region which remains sandwiched between the two oxide layers. In order to verify the applicability of the quantum potential approach developed in this work, a single gated SOI device structure will be studied first. Simulations will be carried out to calculate the threshold voltage as a function of the silicon film thickness and the results will be compared to other available methods. The SOI device used here has the following specifications: gate length is 40 nm, the source/drain length is 50 nm each, the gate oxide thickness is 7 nm with a 2 nm source/drain overlap, the box oxide thickness is 200 nm, the channel doping is uniform at $1 \times 10^{17}\,\mathrm{cm}^{-3}$, the doping of the source/drain regions equals $2 \times 10^{19}\,\mathrm{cm}^{-3}$, and the gate is assumed to be a metal gate with workfunction equal to the semiconductor affinity. There is a 10 nm spacer region between the gate and the source/drain contacts. The silicon (SOI) film thickness is varied over a range of 1–10 nm for the different simulations that were performed to capture the trend in the variations of the device threshold voltage. Similar experiments were performed in [47, 48] using the Schrödinger–Poisson solver and Ferry's effective potential approaches, respectively. For comparison purposes, threshold voltage is extracted from the channel inversion density vs. gate bias profile and extrapolating the linear region of the characteristics to a zero value. This method also corresponds well to the linear extrapolation technique using the drain current-gate voltage characteristics.

The results showing the trend in the threshold voltage variation with respect to the SOI film thickness are depicted in Fig. 2.13. One can see that Ferry's effective potential approach overestimates the threshold voltage for a SOI thickness of 3 nm due to the use of a rather approximate value for the standard deviation of the Gaussian wave packet which results in a reduced sheet electron density. As the silicon film thickness decreases, the resulting confining potential becomes more like rectangular from a combined effects of both the inversion layer quantization and the SOI film (physical) quantization, which also emphasizes the need for using a more

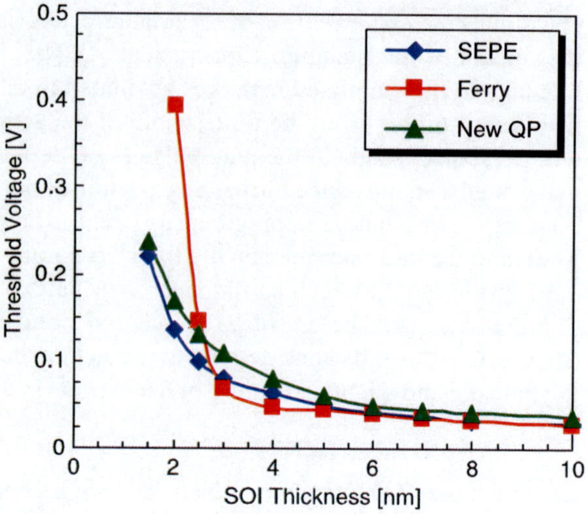

Fig. 2.13 Threshold voltage variation with SOI film thickness. SEPE stands for Schrödinger-Poisson, Ferry stands for Ferry's effective potential approach and New QP stands for new quantum potential

realistic quantum-mechanical wavepacket description for the confined electrons. Of most importance in this figure is the very fact that the new quantum potential approach is free from this large discrepancy and can capture the trend in the threshold voltage as it is obtained from the more accurate 2D Schrödinger-3D Poisson solver. These results indicate that the new quantum potential method can be applied to the simulations of SOI devices with a greater accuracy and predictive capability as it will be seen from the results presented in the next section.

Example: Size-Quantization in Nanoscale DG SOI Devices

Figure 2.14 shows the simulated DG SOI device structure used in this work, which is similar to the devices reported in [49]. For quantum simulation purposes only the dotted portion of the device, termed as the *intrinsic* device is taken into considerations. The device was originally designed in order to achieve the ITRS performance specifications for the year 2016.

The effective intrinsic device consists of two gate stacks (gate contact and SiO_2 gate dielectric) above and below a thin silicon film. For the intrinsic device, the thickness of the silicon film is 3 nm. Use of a thicker body reduces the series resistance and the effect of process variation but it also degrades the short channel effects (SCE). From the SCE point of view, a thinner body is preferable but it is harder to fabricate very thin films of uniform thickness, and the same amount of process variation ($\pm 10\%$) may give intolerable fluctuations in the device characteristics. A thickness of 3 nm seems to be a reasonable compromise, but other body thicknesses are also examined. The top and bottom gate insulator thickness is 1 nm, which is expected to be near the scaling limit for SiO_2. As for the gate contact, a metal gate with tunable workfunction, Φ_G, is assumed, where Φ_G is adjusted to

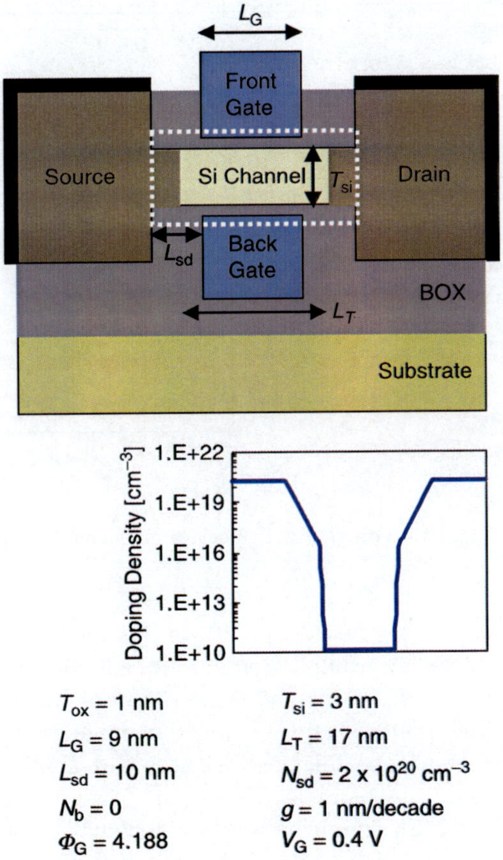

Fig. 2.14 DG device structure being simulated

4.188 eV to provide a specified off-current value of $4\,\mu A/\mu m$. The background doping of the silicon film is taken to be intrinsic, however due to diffusion of the dopant ions, the doping profile from the heavily doped S/D extensions to the intrinsic channel is graded with a coefficient of g which equals to 1 nm/dec. For convenience, the doping scheme is also shown in Fig. 2.14. According to the roadmap, the high performance (HP) device should have a gate length of $L_G = 9$ nm at the year 2016. At this scale, two-dimensional (2D) electrostatics and quantum mechanical effects both play an important role and traditional device simulators may not provide reliable projections. The length L_T, is an important design parameter in determining the on-current, while gate metal workfunction Φ_G, directly controls the off-current. The doping gradient g, affects both on-current and off-current. Values of all the structural parameters of the device are shown in Fig. 2.14 as well.

The intrinsic device is simulated using the new quantum potential approach in order to gauge the impact of size-quantization effects on the DG SOI performance. The results are then compared to that from a full quantum approach based on the non-equilibrium Green's function (NEGF) formalism (NanoMOS–2.5) developed

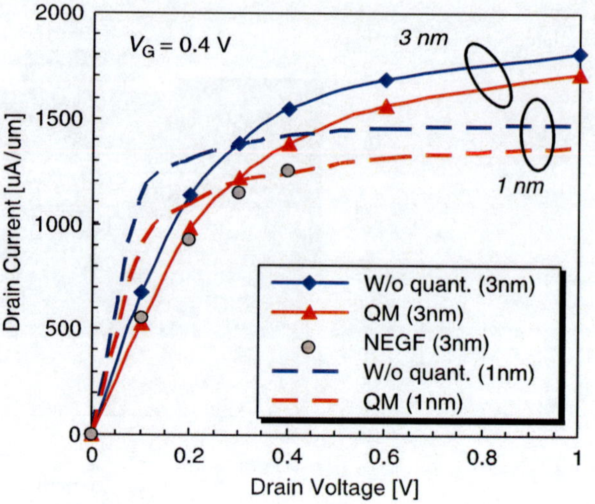

Fig. 2.15 Generic DG SOI device output characteristics

at Purdue University [50]. In this method, scattering inside the intrinsic device is treated by a simple Büttiker probe model, which gives a phenomenological description of scattering and is easy to implement under the Greens' function formalism. The simulated output characteristics are shown in Fig. 2.15. Devices with both 3 and 1 nm channel thickness are used with applied gate bias of 0.4 V. The salient features of this figure are as follows: (1) Even with an undoped channel region, the devices achieve a significant improvement with respect to the SCEs as depicted in flatness of the saturation region. This is due to the use of the two gate electrodes and an ultrathin SOI film which makes the gates gain more control on the channel charge. (2) Reducing the channel SOI film thickness to 1 nm further reduces the SCEs and improves the device performance. However, the reduction in the drive current at higher drain biases is due to series resistance effect pronounced naturally when the drain current increases. (3) Regarding the quantum effects, one can see that quantum-mechanical size quantization does not play a very dominant role in degrading the device drive current mainly because of use of an undoped channel region. Also, looking at the 3 nm (or 1 nm) case alone one can see that the impact of quantization effects reduces as the drain voltage increases because of the growing bulk nature of the channel electrons. (4) Percentage reduction in the drain current is more pronounced in 1 nm case throughout the range of applied drain bias because of the stronger physical confinement arising from the two SiO_2 layers sandwiching the silicon film. (5) Finally, the comparison between the quantum potential formalism and the NEGF approach for the device with 3 nm SOI film thickness shows reasonable agreement which further establishes the applicability of this method in the simulations of different technologically viable nanoscale classical and non-classical MOSFET device structures.

2.2 SCHRED First and Second Generation

Proper inclusion of the quantum-mechanical size quantization effects in device simulators is achieved by solving the Schrödinger–Poisson–Boltzmann problem. This approach was discussed in details in [51]. Here we only focus on solving the 1D Schrödinger–Poisson problem for proper description of charge quantization in MOS capacitors. This can be achieved with SCHRED First Generation tool that is installed on the Network for Computational Nanotechnology (www.nanoHUB. org). However, in the past 2–3 years many users of the existing SCHRED expressed wishes for increasing the present capabilities of SCHRED tool in terms of making it capable to study MOS capacitors made of silicon or strained silicon with arbitrary crystallographic transport directions and to be able to simulate MOS capacitors fabricated of other materials. To satisfy user needs, an effort was undertaken at ASU and SCHRED Second Generation was developed that has all the required features that were on the wish list of SCHRED First Generation. The tool was developed by a M.S. student of Prof. Vasileska at Arizona State University Gokula Kannan. In what follows, we will first explain the capabilities of the SCHRED First Generation Tool and then we will describe SCHRED Second Generation Tool in details.

2.2.1 SCHRED First Generation Capabilities

The periodic crystal potential in the bulk of semiconducting materials is such that, for a given energy in the conduction band, the allowed electron wavevectors trace out a surface in **k**-space. In the effective-mass approximation for silicon, these constant energy surfaces can be visualized as six equivalent ellipsoids of revolution (Fig. 2.16), whose major and minor axes are inversely proportional to the effective masses. A collection of such ellipsoids for different energies is referred to as a valley.

In this framework, the bulk Hamiltonian for an electron, residing in one of these valleys is of the form

$$H_o(\mathrm{R}) = -\left(\frac{\hbar^2}{2m_x^*}\frac{\partial^2}{\partial x^2} + \frac{\hbar^2}{2m_y^*}\frac{\partial^2}{\partial y^2} + \frac{\hbar^2}{2m_z^*}\frac{\partial^2}{\partial z^2}\right) + V_{eff}(z) = H_{o||}(\mathrm{r}) + H_{o\perp}(z), \tag{2.25}$$

where $\mathbf{R} = (\mathbf{r}, z)$, $V_{eff}(z) = V_H(z) + V_{exc}(z)$ is the effective potential energy profile of the confining potential, $V_H(z)$ is the Hartree potential which is nothing more but a solution of the 1D Poisson equation introduced later in the text, $V_{exc}(z)$ is the exchange-correlation potential also discussed later in the text, $H_{o||}$ is the parallel part of H_o, and the transverse part is defined as

$$H_{o\perp}(z) = -\frac{\hbar^2}{2m_z^*}\frac{\partial^2}{\partial z^2} + V_{eff}(z). \tag{2.26}$$

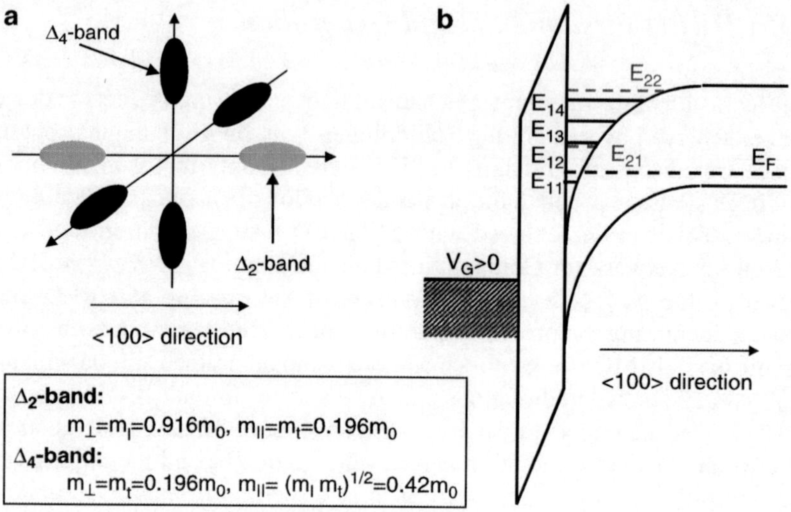

Fig. 2.16 *Right panel* – Potential diagram for inversion of *p*-type semiconductor. In this first notation E_{ij} refers to the jth subband from either the Δ_2-band (i = 1) or Δ_4-band (i = 2). *Left panel* – Constant-energy surfaces for the conduction-band of silicon showing six conduction-band valleys in the <100> direction of momentum space. The band minima, corresponding to the centers of the ellipsoids, are 85% of the way to the Brillouin-zone boundaries. The long axis of an ellipsoid corresponds to the longitudinal effective mass of the electrons in silicon, $m_l = 0.916m_o$, while the short axes correspond to the transverse effective mass, $m_t = 0.190m_o$. For <100> orientation of the surface, the Δ_2-band has the longitudinal mass (m_l) perpendicular to the semiconductor interface and the Δ_4-band has the transverse mass (mt) perpendicular to the interface. Since larger mass leads to smaller kinetic term in the Schrödinger equation, the unprimed ladder of subbands (as is usually called), corresponding to the Δ_2-band, has the lowest ground state energy. The degeneracy of the unprimed ladder of subbands for <100> orientation of the surface is 2. For the same reason, the ground state of the primed ladder of subbands corresponding to the Δ_4-band is higher than the lowest subband of the unprimed ladder of subbands, The degeneracy of the primed ladder of subbands for (100) orientation of the interface is 4

The basis-states of the unperturbed Hamiltonian are assumed to be of the form

$$\Psi_n(\mathrm{R}) = \frac{1}{\sqrt{A}} e^{i\mathbf{k}\cdot\mathbf{r}} \psi_n(z), \tag{2.27}$$

where **k** is a wavevector in the *xy*-plane and *A* is the area of the sample interface. The subband wavefunctions satisfy the one-dimensional Schrödinger equation,

$$H_{o\perp}(z)\psi_n(z) = \varepsilon_n \psi_n(z) \tag{2.28}$$

subject to the boundary conditions that $\psi_n(z)$ are zero for $z = 0$ and approach zero as $z \to \infty$. In (2.28), ε_n is the subband energy and $\psi_n(z)$ is the corresponding wavefunction. In the parabolic band approximation, the total energy of the electrons is given by

$$E_n(\mathrm{k}) = \frac{\hbar^2 \mathrm{k}^2}{2m_{xy}^*} + \varepsilon_n = \varepsilon_\mathrm{k} + \varepsilon_n, \tag{2.29}$$

where $\varepsilon_{\mathbf{k}}$ is the kinetic energy and m_{xy}^* is the density of states mass along the xy-plane. An accurate description of the charge in the inversion layer of deep-submicrometer devices and, therefore, the magnitude of the total gate capacitance C_{tot} requires a self-consistent solution of the 1D Poisson

$$\frac{\partial}{\partial z}\left[\varepsilon(z)\frac{\partial\phi}{\partial z}\right] = -e[N_D^+(z) - N_A^-(z) + p(z) - n(z)], \tag{2.30}$$

and the 1D Schrödinger equation

$$\left[-\frac{\hbar^2}{2m_i^\perp}\frac{\partial^2}{\partial z^2} + V_{eff}(z)\right]\psi_{ij}(z) = E_{ij}\psi_{ij}(z). \tag{2.31}$$

In (2.30) and (2.31), $\varphi(z)$ is the electrostatic potential [the Hartree potential $V_H(z) = -e\varphi(z)$], $\varepsilon(z)$ is the spatially dependent dielectric constant, $N_D^+(z)$ and $N_A^-(z)$ are the ionized donor and acceptor concentrations, $n(z)$ and $p(z)$ are the electron and hole densities, $V_{eff}(z)$ is the effective potential energy term that equals the sum of the Hartree and exchange-correlation corrections to the ground state energy of the system, $m_i^\perp$ is the effective mass normal to the semiconductor-oxide interface of the ith valley, and E_{ij} and $\psi_{ij}(z)$ are the energy level and the corresponding wavefunction of the electrons residing in the jth subband from the ith valley. The electron-density is calculated using

$$n(z) = \sum_{i,j} N_{ij}\psi_{ij}^2(z) \tag{2.32}$$

where N_{ij} is the sheet electron concentration in the ith subband from the jth valley is given by

$$N_{ij} = g_i\frac{m_{xy}^*}{\pi\hbar^2}k_BT\ln\left\{1 + \exp[(E_F - E_{ij})/k_BT]\right\} \tag{2.33}$$

where g_i is the valley degeneracy factor and E_F is the Fermi energy. When evaluating the exchange-correlation corrections to the chemical potential, we have relied on the validity of the density functional theory (DFT) of Hohenberg and Kohn [52], and Kohn and Sham [53]. According to DFT, the effects of exchange and correlation can be included through a one-particle exchange-correlation term $V_{exc}[n(z)]$, defined as a functional derivative of the exchange-correlation part of the ground-state energy of the system with respect to the electron density $n(z)$. In the local density approximation (LDA), one replaces the functional $V_{exc}[n(z)]$ with a function $V_{exc}[n(z)] = \mu_{exc}[n_0 = n(z)]$, where μ_{exc} is the exchange-correlation contribution to the chemical potential of a homogeneous electron gas of density n_0, which is taken to be equal to the local electron density $n(z)$ of the inhomogeneous system. In our model, we use the LDA and approximate the exchange-correlation potential energy term $V_{exc}(z)$ by an interpolation formula developed by Hedin and Lundqvist [54]

$$V_{exc}(z) = -\frac{e^2}{8\pi\varepsilon_{sc}b}\left[1 + 0.7734x\ln\left(1 + \frac{1}{x}\right)\right]\left(\frac{2}{\pi\alpha r_s}\right), \tag{2.34}$$

which is accurate over a large density range. In (2.34), $\alpha = (4/9\pi)^{1/3}$, $x = x(z) = r_s/21$, $r_s = r_s(z) = [4\pi b^3 n(z)/3]^{-1/3}$, and $b = 4\pi\varepsilon_{sc}\hbar^2/m^* e^2$. Exchange and correlation effects tend to lower the total energy of the system and lead to non-uniform shift of the energy levels and repopulation of the various subbands. The enhancement of the exchange-correlation contribution to the energy predominantly affects the ground subband of the occupied valley; the unoccupied subbands of the same valley are essentially unaffected. As a result, noticeable increase in the energy of the inter-subband transitions can be observed at high electron densities.

Similarly, the valence band is represented by the heavy hole band and light hole band, the spit-off band is ignored because the spit-off energy is large enough to exclude any hole staying there. In treating holes quantum mechanically, the same effective mass based Schrodinger equation is solved with the masses quoted from references [55, 56]. Due to their different perpendicular masses, the heavy holes form the first set of energy levels which are relatively low, and the light holes form the second set with higher confined energies. SCHRED V1.0 also has the capability of treating the electron/hole density in the inversion layer classically by using either Maxwell–Boltzmann or Fermi–Dirac statistics.

In doing bulk structure quantum mode simulation, SCHRED V1.0 can not only solve the effective mass based Schrödinger equation for inversion layer carriers, but also can solve the equation for accumulation layer carriers, for example, if the bulk is *p*-type silicon, in the inversion range, electrons are treated quantum mechanically, whereas in the accumulation range, holes are treated quantum mechanically. This is a feature that many other simulators do not offer.

In doing SOI quantum mode simulation, both electrons and holes are treated quantum mechanically at the same time. This is because in most cases, the SOI bodies are undoped or lightly doped, and the two dielectric gates confine the carriers in both inversion and accumulation regimes, therefore, the quantum effects can be equally important for both electrons and holes at low biases.

For both simulation modes, (classical or quantum mechanical) if the gate contacts are polysilicon, the charge density on the gates will always be computed classically. The gate dielectric constant can be specified different from SiO_2. The latest version also allows different dielectrics for the top and bottom gates in a SOI structure. This eases the simulations of effects of exotic insulator materials on device performance. Typical outputs of the solver are the spatial variations of the conduction-band edge and 3D charge density in the body; 2D surface charge density, average distance of the carriers from the interface; inversion layer capacitance C_{inv}, depletion layer capacitance C_{depl}, total gate capacitance C_{tot} and in the case of capacitors with poly-silicon gates, it also calculates the poly-gate capacitance C_{poly}. When choosing quantum-mechanical description of the electron density in the channel, it also provides the subband energies, the subband population, and the wavefunction variations in the body.

Schred is written in Fortran 77. The program is more efficient compared to other 1D Schrödinger–Poisson self-consistent simulators. A simplified flow-chart of the SCHRED V1.0 code is given in Fig. 2.17.

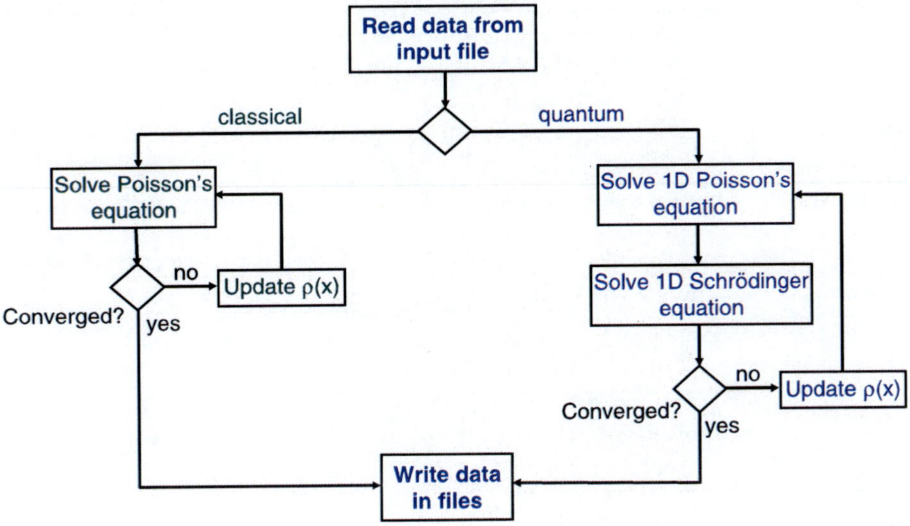

Fig. 2.17 Flow-chart of Schred V1.0

Examples of the application of SCHRED V1.0 can be found in [57–59] and in the sub-sections below.

Representative Simulation Results Obtained with SCHRED V1.0

Example 1: Semiclassical Versus Quantum Behavior

A first set of important simulation results that can be obtained with SCHRED V1.0 is the comparison between the semi-classical and quantum-mechanical models and how that affects the shape of the electron density and the magnitude of the sheet charge density. For that purpose we simulate an MOS capacitor with oxide thickness $t_{ox} = 1$ nm, substrate doping $N_A = 10^{18}\,cm^{-3}$ and applied gate bias of 1 V. The metal workfunction is assumed to be equal to the semiconductor affinity.

The simulation results for the sheet electron density obtained with SCHRED V1.0 are: $N_s(\text{semi} - \text{classical}) = 1.43 \times 10^{13}\,cm^{-2}$ and $N_s(\text{quantum}) = 1.08 \times 10^{13}\,cm^{-2}$. These results indicate that the semiclassically calculated sheet electron density is about 30% higher than the quantum-mechanically calculated sheet electron density. There are two reasons for this: (1) the bandgap widening effect in the case of the quantum-mechanical model due to the shift of the first allowed state in the conduction band by 200.47 meV, and (2) the charge set-back from the interface because the wavefunction vanishes right at the interface, which leads to effective oxide thickness larger than the physical oxide thickness, thus leading to transconductance degradation. The charge set-back is clearly seen from the results shown in Fig. 2.18 where we plot the semi-classically calculated total electron density and the quantum-mechanically calculated total electron density. We see that

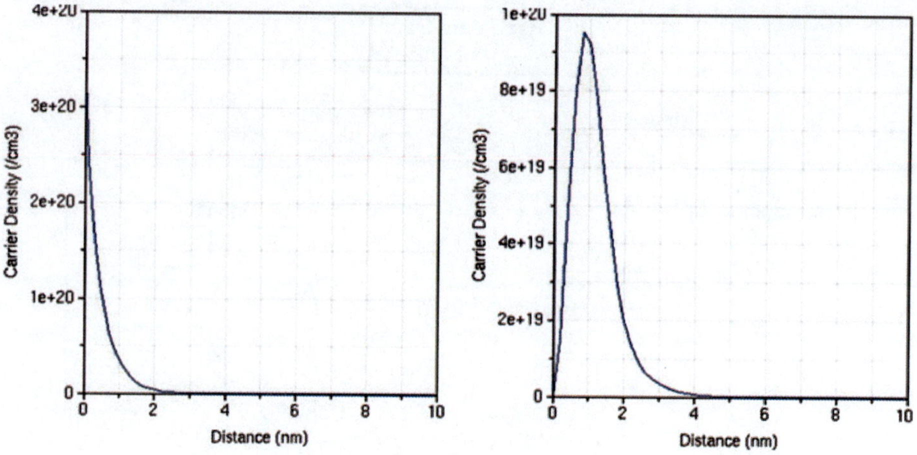

Fig. 2.18 Semiclassical *(left panel)* and quantum-mechanical *(right panel)* electron density

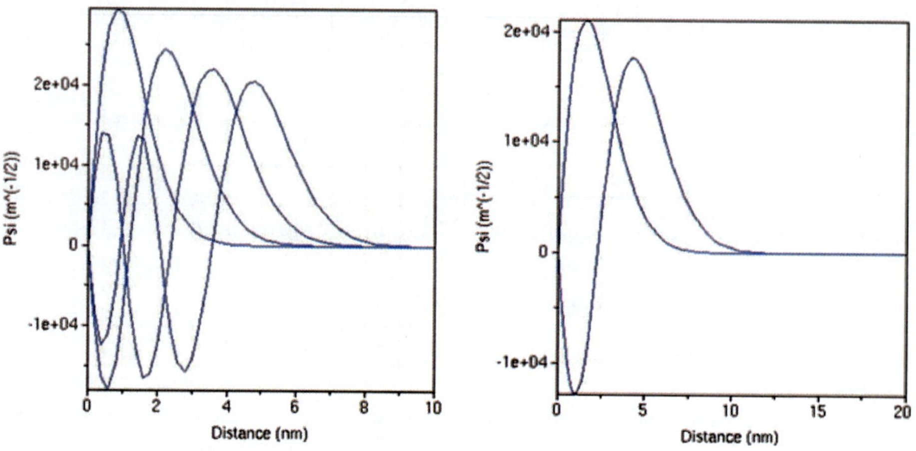

Fig. 2.19 Wavefunctions of the unprimed *(left)* and primed *(right)* ladder of subbands

the semiclassical charge density peaks at the interface as it is exponentially dependent of the negative of the potential, whereas the quantum-mechanically calculated electron density is zero at the interface and peaks at few angstroms away from the interface.

For the case of the quantum-mechanical model we have taken 4 subbands from the unprimed ladder of subbands and 2 subbands from the primed ladder of subbands. The spatial variation of the corresponding wavefunctions is shown in Fig. 2.19. There are several important things that can be observed from the results shown in Fig. 2.19. First, the shape of the wavefunctions resembles Airy functions that are solution to the 1D Schrödinger equation with linear potential energy term. Second, if we compare the first two wavefunctions from both the unprimed and

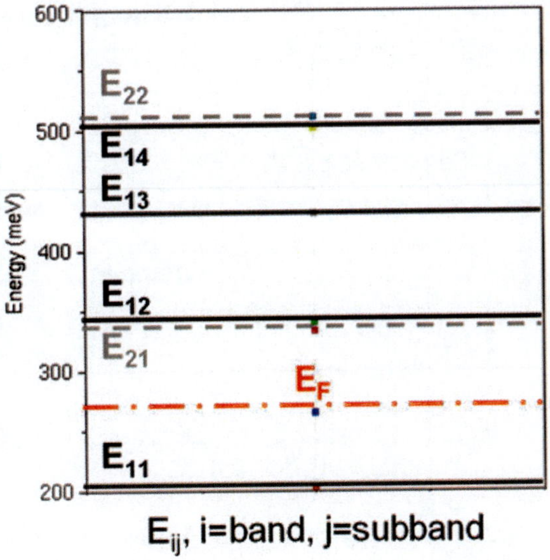

Fig. 2.20 Energy levels values from the unprimed and primed ladder of subbands

primed latter of subbands, then we see that the unprimed wavefunctions are more squeezed as the energies are lower and for those energies (see Fig. 2.20) the well is squeezed, therefore there exists larger localization of the carriers. Third, the first wavefunction has zero intersections with the x-axis, the second one has one, the third one has two, etc.

The corresponding energy levels of the unprimed and primed ladder of subbands are shown in Fig. 2.20. We see that the Fermi-level is above the first subband, therefore the semiconductor is degenerate. More importantly, we see that as we go higher in energy, the well widens and the energy level separation becomes smaller and smaller.

Example 2: Total Capacitance Degradation for Old and New Technology Nodes

In this second example we examine degradation of the total gate capacitance as a function of technology node. We consider what we call state of the art device technology, which is essentially the MOS capacitor discussed in Sect. 2.2.1. Regarding the older device technology MOS capacitor, its parameters are as follows: $N_A = 10^{16}\,\mathrm{cm}^{-3}$ and $t_{ox} = 40\,\mathrm{nm}$. The results of the simulations are presented in Figs. 2.21 and 2.22. There are several noteworthy features that can be deduced from the results shown.

For the case of state-of-the-art MOS capacitors, looking at the capacitances obtained for the case when the electron density is treated classically and quantum-mechanically, we observe two very important things: (1) there is a threshold voltage shift due to the quantum-mechanical size-quantization effect, and (2) there is a significant degradation of the total gate capacitance when using the quantum charge model that effectively degrades the device transconductance. The total capacitance degradation can be explained by examining the results for the average distance of the

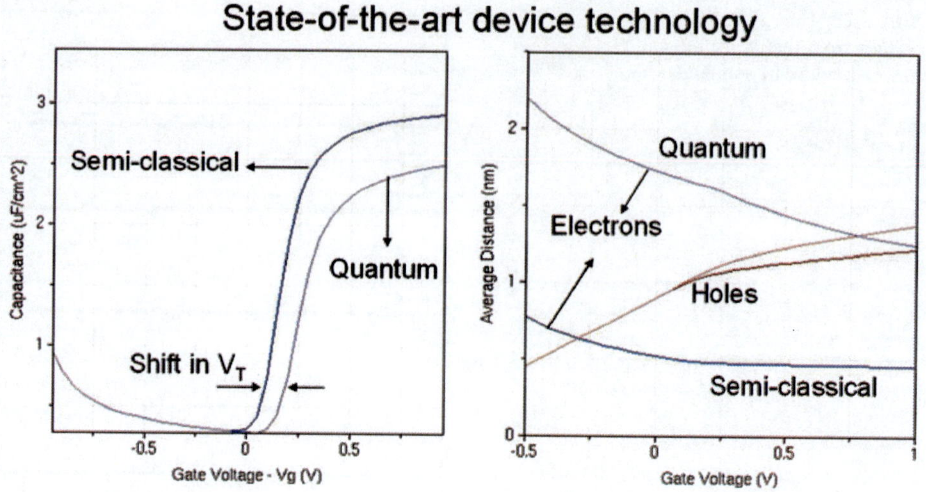

Fig. 2.21 *Left panel* – Total gate capacitance vs. gate voltage for state of the art device technology. *Right panel* – Average distance of the carriers from the interface

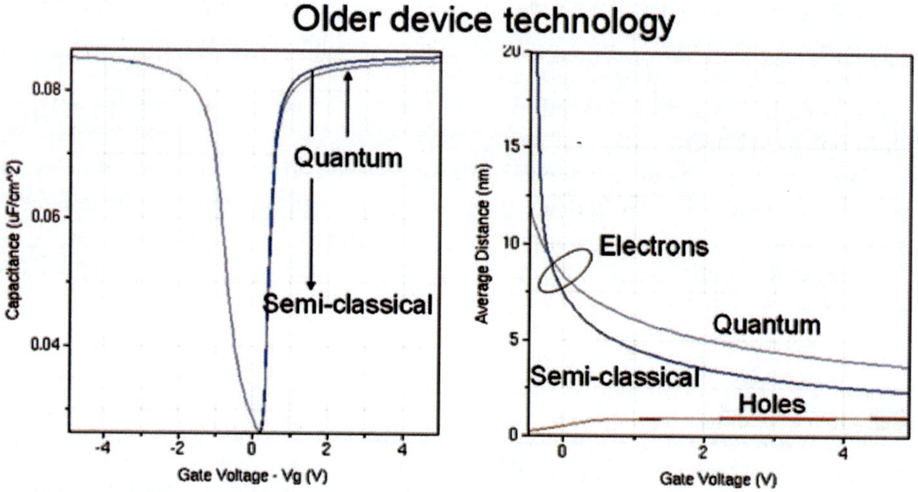

Fig. 2.22 *Left panel* – Total gate capacitance vs. gate voltage for older device technology. *Right panel* – Average distance of the carriers from the interface

electrons from the interface (Fig. 2.21 – Right panel). We see that classically carriers are about three times closer to the semiconductor/oxide interface when compared to the quantum case. The average distance in a way is a measure of the effective oxide thickness and quantum charge model leads to larger effective oxide thickness; therefore smaller transconductance.

For the case of older technology devices, looking at the results for the total gate capacitance shown in the left panel of Fig. 2.22, we might safely say that quantum effects are not important as the total capacitance degradation is negligible. This can be attributed to the lower energy levels due to the wider well because of two orders of magnitude lower doping. As the well is wider, the average distance of the electrons from the interface is larger but that does not lead to transconductance degradation because the oxide thickness is 40 nm (40 times larger than in state-of-the-art devices).

From these two examples we might conclude that when modeling novel technology devices, quantum effects must be accounted for to properly determine the threshold voltage and total gate capacitance.

Example 3: Single Versus Dual Gate Capacitors

One of the primary reasons for device degradation at shorter channel lengths in FD SOI devices is the encroachment of drain electric field in the channel region. As shown in Fig. 2.23, the gate electrode shields the channel region from those lines at the top of the device, but electric field lines penetrate the device laterally and from underneath, through the buried oxide and the silicon wafer substrate causing the undesirable DIBL for the charge carriers.

To prevent the encroachment of electric field lines from the drain on the channel region, special gate structures can be used as shown in Fig. 2.24. Such "multiple-gate" devices include double-gate transistors, triple-gate devices such

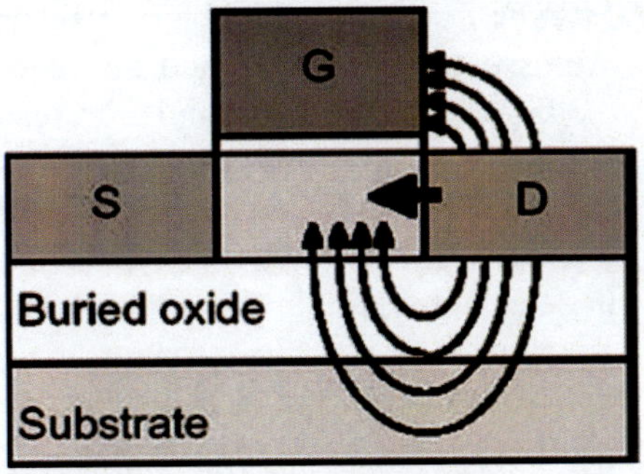

Fig. 2.23 Electric field lines from the drain

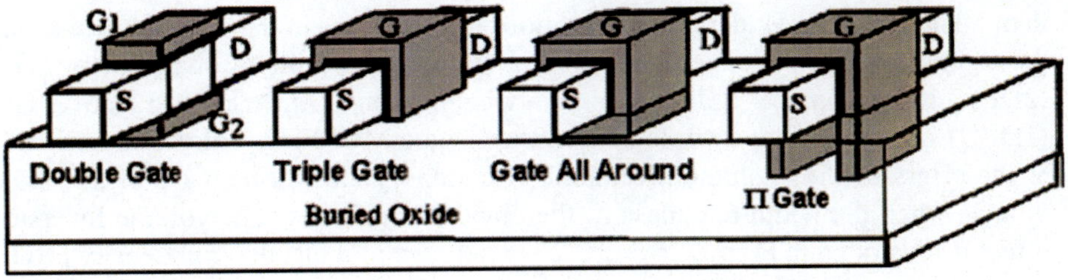

Fig. 2.24 Double-gate, triple-gate, gate all around (GAA), and Π-gate SOI MOSFETs

as the quantum wire [60], the FinFET [61] and Π-channel SOI MOSFET [62], and quadruple-gate devices such as the gate-all-around device [63], the DELTA transistor [64], and vertical pillar MOSFETs [65].

The double-gate device structure allows for termination of the drain electric field at the gates and leads to a more scalable FET. The double-gate concept was first reported in 1984 [66] and has been fabricated by several groups since then. The salient features of the DG FET (Fig. 2.24) are: (1) control of short-channel effects by device geometry, as compared to bulk FET, where the short-channel effects are controlled by doping (channel doping and/or halo doping); and (2) a thin silicon channel leading to tight coupling of the gate potential with the channel potential. These features provide potential DG FET advantages that include: (1) reduced 2D short-channel effects leading to a shorter allowable channel length compared to bulk FET; (2) a sharper subthreshold slope (60 mV/dec compared to 80 mV/dec for bulk FET) which allows for a larger gate overdrive for the same power supply and the same off-current; and (3) better carrier transport as the channel doping is reduced (in principle, the channel can be undoped). Reduction of channel doping also relieves a key scaling limitation due to the drain-to-body band-to-band tunneling leakage current. A further potential advantage is more current drive (or gate capacitance) per device area; however, this density improvement depends critically on the specific fabrication methods employed and is not intrinsic to the device structure. The most common mode of operation of the DG FET is to switch the two gates simultaneously.

In this exercise, we compare the performance of single-gate vs. double-gate MOSFET device structure by considering the double-gate option in SCHRED V1.0. We assume metal gates and the second gate is set to $V_{G2} = 1\,V$, and we sweep the first gate V_{G1}. The simulation results of the sheet electron density in the channel for single-gate and double-gate MOS capacitor are shown in Fig. 2.25. We use $t_{ox} = 1\,nm$ and $N_A = 10^{18}\,cm^{-3}$. For the double-gate MOS capacitor the body thickness is 10 nm. Evidently, we have almost twice the number of electrons in the channel region in the double-gate structure when compared to the single-gate structure.

Example 4: Dual Gate Capacitors – Volume Inversion

The thickness and/or width of multi-gate FETs are reaching values that are less than 10 nanometers. Under these conditions the electrons in the channel (if we take the example of an n-channel device) form either a two-Dimensional Electron Gas (2DEG) if we consider a double-gate device or a one-Dimensional Electron Gas (1DEG) if we consider a triple or quadruple-gate MOSFET. This confinement is at the origin of the "volume inversion" effect and yields an increase of threshold voltage when the width/thickness of the devices is reduced. The volume inversion effect is illustrated in Figs. 2.26 and 2.27, where we plot the electron density profile vs. gate voltage and the sheet electron density vs. body thickness, respectively.

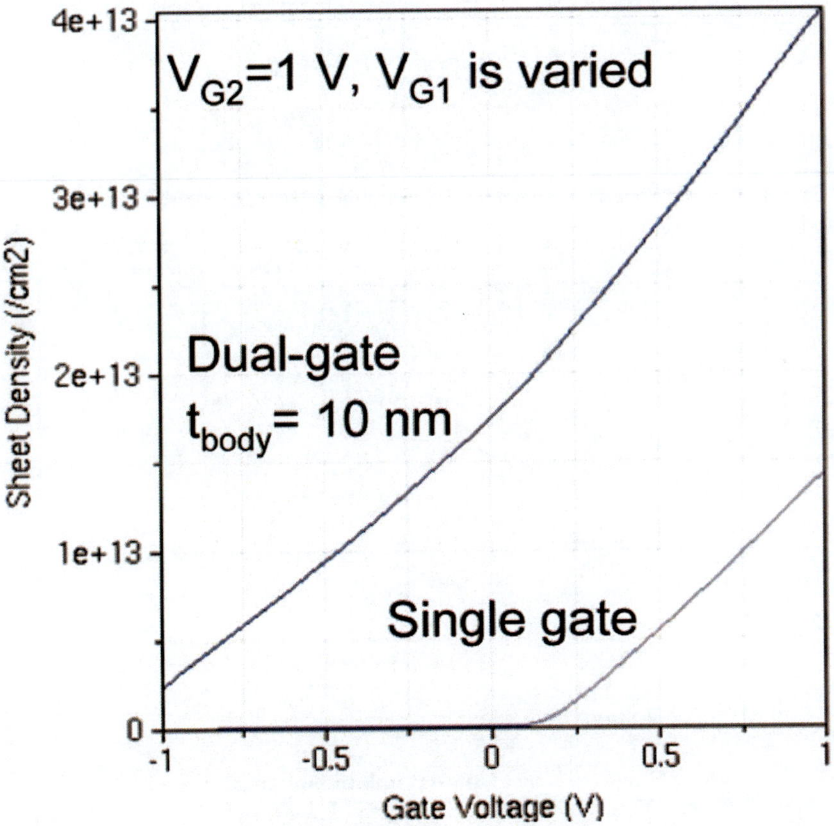

Fig. 2.25 Sheet electron density in a single-gate and double-gate structure as a function of the front gate voltage

2.2.2 SCHRED Second Generation Capabilities

Theoretical Model and Implementation Details

The theoretical model implemented is as follows. First user chooses one of the material systems described below. Then user specifies how many conduction bands are going to be taken into consideration. Then, for each specified conduction band (or pair of bands in the case of Si or strained-Si) the user specifies the effective masses. For the case of materials different than Si, the masses are taken to be isotropic. In the case of Si or strained-Si material system, the mass is assumed to be anisotropic, therefore crystallographic directions become important. Following the nomenclature of Rahman and co-workers [67], the user specifies the device, the crystal and the transport direction based on which one calculates the width, the confinement and the transport mass for each of the three pairs of ellipsoids of revolution

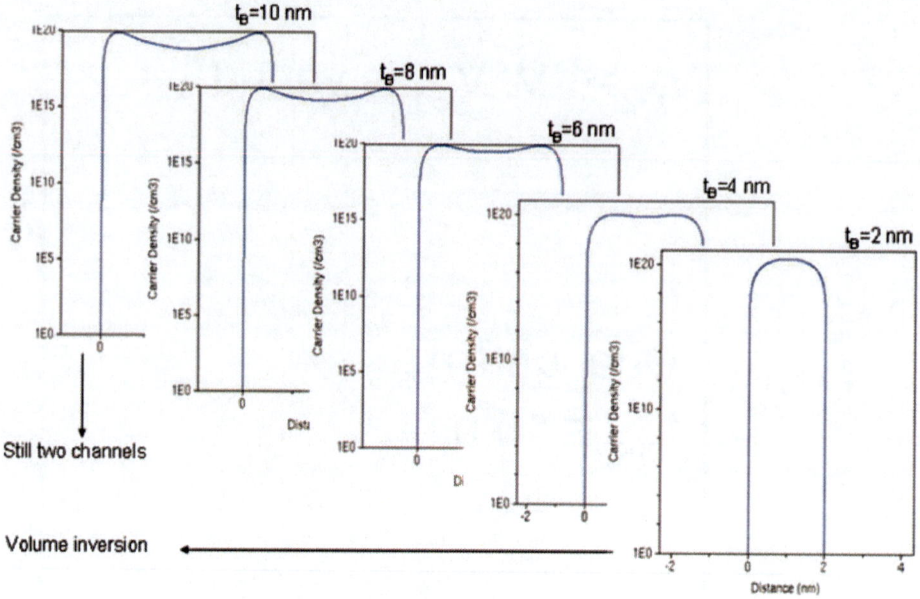

Fig. 2.26 Electron density profile for $V_{G1} = V_{G2} = 1V$

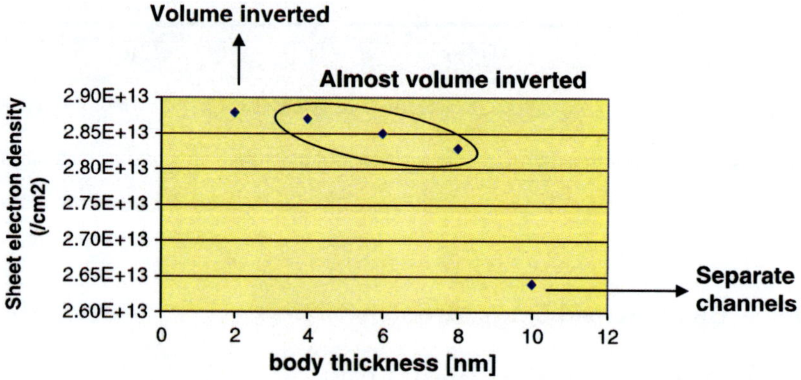

Fig. 2.27 Sheet electron density vs. silicon body thickness in the dual-gate structure

for the conduction band. Thus for a general conduction band ellipsoid (assuming 3 valleys) in the ellipse coordinate system (ECS),

$$E = \frac{\hbar^2 k_{||}^2}{2m_1} + \frac{\hbar^2 k_{\perp 1}^2}{2m_2} + \frac{\hbar^2 k_{\perp 2}^2}{2m_3} \tag{2.35}$$

For a given crystal coordinate system (CCS) and the ellipsoidal effective masses, we can write rotation matrix R_{E-C} for transforming components of an arbitrary vector in CCS to its components in the ellipse co-ordinate system (ECS). Similarly we can write a rotation matrix R_{C-D} for transforming wave vector in the device co-ordinate system (DCS) to CCS. Thus we can write the inverse effective mass in the DCS as [68],

$$(M_D^{-1}) = R_{E\leftarrow D}^T (M_E^{-1}) R_{E\leftarrow D} \tag{2.36}$$

where

$$R_{E\leftarrow D} = R_{E\leftarrow C} R_{C\leftarrow D}, \tag{2.37}$$

and M_E^{-1} is a 3×3 diagonal matrix with m_l^{-1}, m_t^{-1}, m_t^{-1} along the diagonal. As a result, we can effectively model different orientations of Si or strained Si based on this approach for the effective mass calculation.

The valley offset in the conduction band in strained Si can be modeled using our three valley conduction band model. The various different effective masses for these three valleys can also be taken into consideration while solving the coupled system of Schrödinger–Poisson equations. The change in effective masses in the valence band of strained Si can also be included for the simulation.

As shown in Fig. 2.28, any material that can be expressed using a three valley conduction band system can be modeled by using our three valley conduction band model. This would enable us to model even those materials that are being researched at present. We can thus include in our simulation the different effective masses for the various conduction band and valence bands.

Because in some regimes of operation of the MOS capacitor there is no quantum-mechanical confinement and charge has to be treated classically, the effective density of states of the conduction band is calculated. Note that in SCHRED Second Generation holes at the moment are treated classically. In near future k.p method

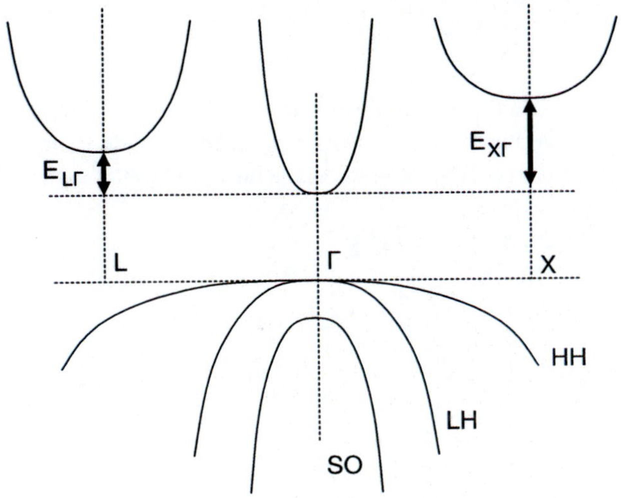

Fig. 2.28 General 3 valley conduction band model of a material

will be implemented to properly account for the warped valence bands and how they change under the influence of strain. User can choose whether to use semi-classical or quantum-mechanical charge description for the electrons. For the case of classical charge description the user has the option of Maxwell–Boltzmann and Fermi–Dirac statistics. The gate electrode can be treated as either a metal with user-defined workfunction or polysilicon. For simulations at low temperatures the users can also include partial ionization of the impurity atoms.

For the case of semiclassical charge description of the electrons and holes, only the linearized Poisson equation is solved using the LU decomposition method. When the electrons are treated quantum mechanically then a self-consistent solution of the 1D Poisson and the 1D Schrödinger equation is obtained. Note that the 1D Schrödinger equation is solved separately for each conduction band valley/valley pair. It is important to note that when finite difference approximation is applied to the 1D Schrodinger equation, a tri-diagonal non-symmetric coefficient matrix is obtained. Since the EISPACK routines that solve the eigenvalue problem are designed for symmetric coefficient tridiagonal matrices, a symmetrization procedure is necessary. This is achieved in the following manner. The discretized 1D Schrodinger equation is given by,

$$\sum_{j=1}^{n} A_{ij}\psi_j = \lambda\psi_i \tag{2.38}$$

where

$$A_{ij} = \begin{cases} -\dfrac{\hbar^2}{m^* x_i (x_i + x_{i-1})} & j = i+1 \\ \dfrac{\hbar^2}{m^* x_i (x_i + x_{i-1})} + \dfrac{\hbar^2}{m^* x_{i-1} (x_i + x_{i-1})} + V_i & j = i \\ -\dfrac{\hbar^2}{m^* x_{i-1} (x_i + x_{i-1})} & j = i-1 \\ 0 & otherwise \end{cases}$$

Thus, with the finite difference discretization of the 1D Schrödinger equation on a non-uniform mesh one arrives at a tridiagonal matrix that is not symmetric. The symmetrization of the coefficient matrix is achieved with the matrix transformation technique detailed below [69].

Let $x_i + x_{i-1}$ be L_i^2. Then, we have

$$A_{ij} = \begin{cases} -\dfrac{\hbar^2}{m^* x_i}\dfrac{1}{L_i^2} & j = i+1 \\ \left(\dfrac{\hbar^2}{m^* x_i} + \dfrac{\hbar^2}{m^* x_{i-1}}\right)\dfrac{1}{L_i^2} + V_i & j = i \\ -\dfrac{\hbar^2}{m^* x_{i-1}}\dfrac{1}{L_i^2} & j = i-1 \\ 0 & otherwise \end{cases}$$

Let $B_{ij} = L_i^2 A_{ij}$ or in matrix notation, $B = MA$, where M is the diagonal matrix with elements L_i^2, and B is tridiagonal and symmetric matrix. Thus the eigenvalue matrix (2.39) becomes,

$$B\psi = MA\psi = \lambda M\psi \tag{2.39}$$

The matrix M can be written as: $M = LL$, where L is a diagonal matrix with elements L_i. One can show that

$$L^{-1}BL^{-1}L\psi = L^{-1}LLA\psi = \lambda L^{-1}LL\psi, \tag{2.40}$$

or

$$H\varphi = \lambda\varphi, \tag{2.41}$$

where

$$H = L^{-1}BL{-}1, \tag{2.42}$$

and

$$\psi = L^{-1}\varphi. \tag{2.43}$$

Thus we can now solve using the symmetric matrix H, obtain the value of the φ matrix and from that obtain the value of ψ matrix – the eigenvectors.

Simulation Results

This section is divided into three parts. The first Sub-Section details the results from SCHRED Second Generation for the Silicon case. The following Sub-Section explains the results of SCHRED Second Generation in comparison with experimental results for a multi-valley semiconductor such as GaAs. The last Sub-Section compares experimental results of Strained Silicon for <100> transport orientation with the results of SCHRED Second Generation.

Example 1: Simulations of Regular Silicon for Specific Crystallographic Orientations

As shown in Table 2.1, the following orientations (wafer/transport/width directions) are simulated using SCHRED Second Generation.

We simulate MOS Capacitor with the following parameters: metal gate, substrate doping concentration of $10^{17}\,\text{cm}^{-3}$, and oxide thickness of 4 nm. Two subbands are assumed for each of the three pairs of valleys. The resultant plots are then discussed. The effective masses for the different conduction band valley pairs are shown in Table 2.2 [67]. The mass m_z refers to the confinement effective mass and the mass m_{xy} refers to the product of the transport and width direction masses. This product contributes to the 2D density of states (DOS) mass.

Table 2.1 Different crystallographic orientations of silicon

(Wafer)/[Transport]/[Width]
(001) /[100]/[010]
(111)/[$\bar{2}$11]/[0$\bar{1}$1]
(110)/[001]/[00]

Table 2.2 Transport, width and confinement effective masses

Confinement direction	Transport, width and confinement effective mass	Valley 1	Valley 2	Valley 3
(001)	m_z	0.19	0.19	0.98
(110)	m_z	0.3189	0.3189	0.19
(111)	m_z	0.2598	0.2598	0.2598
(001)	m_{xy}	1.17	1.17	0.0361
(110)	m_{xy}	0.2223	0.2223	0.3724
(111)	m_{xy}	0.13604	0.13572	0.13572

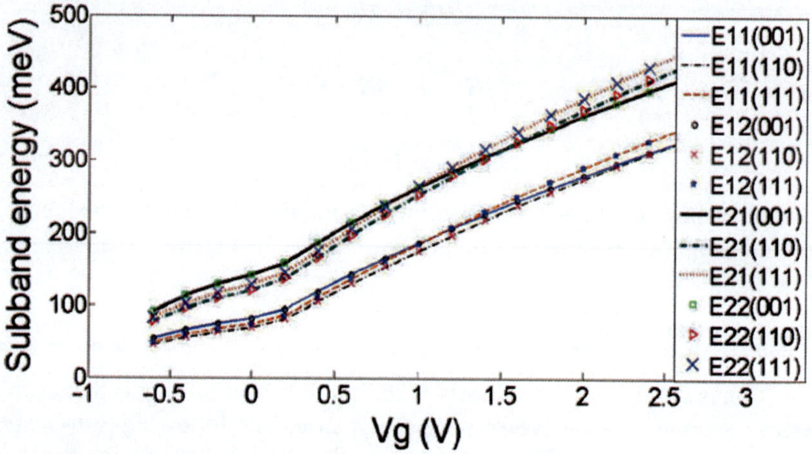

Fig. 2.29 Subband energy vs. applied voltage for valleys 1 and 2 (for various subband energy Eij, where i – denotes the subband, j – denotes the valley)

From the result shown in Fig. 2.29, it is evident that conduction band valley pair 1 has the lowest confinement mass for (001) confinement direction (see Table 2.2) and highest for (110) direction. Thus, the subband energies are lowest for the (110) direction and highest for the (001) direction. (The kinetic energy term in the Schrödinger equation will be the highest for the lowest mass, hence higher total subband energy). The valley pair 2 subband energies follow the same variation as the valley pair 1 subbands as they have the same set of masses in given directions and hence are equivalent to valley pair 1. The lower subband energies of valley pair 3 (unprimed set of subbands) as shown in Fig. 2.30, and are lower due to their higher confinement mass m_z (Table 2.2). As we increase the applied voltage, the potential well deepens, and the subband energies increase.

As shown in Fig. 2.31, the 2D sheet charge density is highest for the (001) orientation due to its lowest subband energy values. Thus we have lower sheet charge densities for the case of (110) which has higher subband energy than (001). In Fig. 2.32, the capacitance variation is presented for the three crystallographic directions. There is slight degradation for the total gate capacitance for orientations different than [001]. The most prominent result is shown in Fig. 2.33 where we

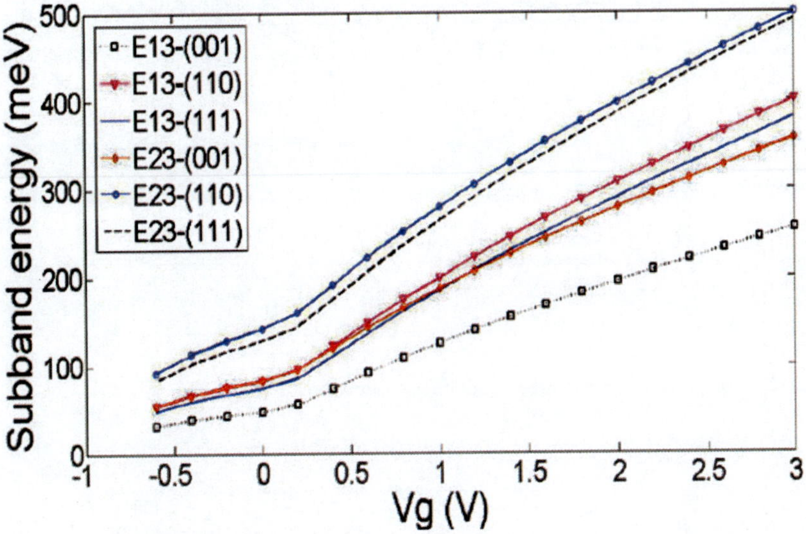

Fig. 2.30 Subband energy vs. applied voltage for valley3 (for various subband energy Eij, where i – denotes the subband, j – denotes the valley)

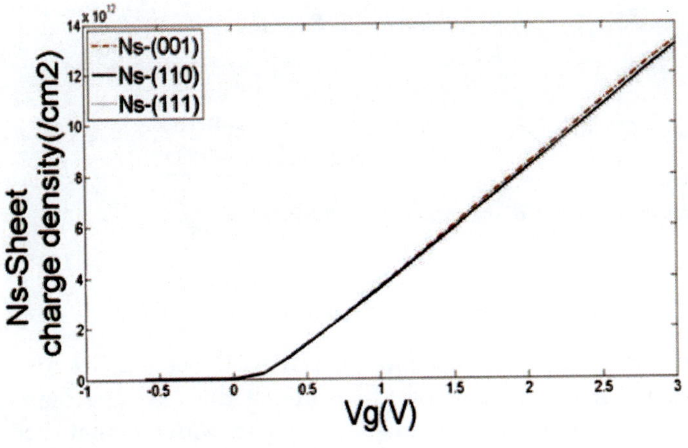

Fig. 2.31 Sheet charge density (Ns) vs. voltage

plot the average distance of the carriers from the interface as a function of the gate bias. We see that for [001] orientation we have the smallest average distance which means that in these devices interface roughness will play much higher role when compared to the other two crystallographic directions. This can significantly affect the on-current of the device fabricated in this material system.

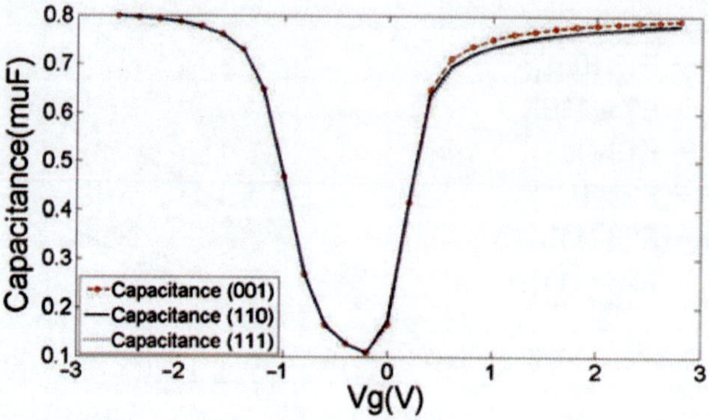

Fig. 2.32 Capacitance for the three confinement directions

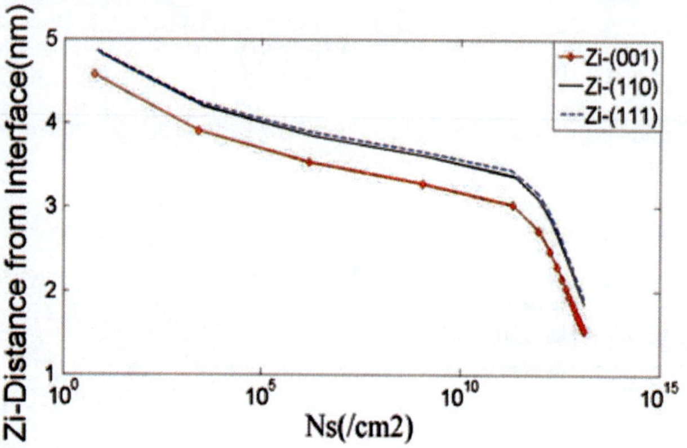

Fig. 2.33 Average distance of the carriers from the interface

Example 2: Gallium Arsenide MOS Capacitors

In order to verify the actual capability of SCHRED Second Generation in solving for multi-valley semiconductors, we had simulated MOS capacitors for a specific case of GaAs and compared our simulation results with the published data [70]. A substrate doping concentration of $10^{18}\,\mathrm{cm}^{-3}$ is used together with an oxide thickness of $t_{ox} = 16\,\mathrm{nm}$. The simulation runs have been performed for voltages in the range (−4 to 4 V). We use three conduction band valleys (gamma, X and L valleys). We use two subbands for each of these valleys. The offsets between the valleys are included in the simulation.

From the results shown in Fig. 2.34 it can be seen that our results match much closer to the experimentally determined capacitance than the simulation results of [70]. The capacitance values match in the inversion and accumulation regions. We also observe that our results indicate a higher value of accumulation capacitance because we have not included hole confinement in the negative bias region of the simulation.

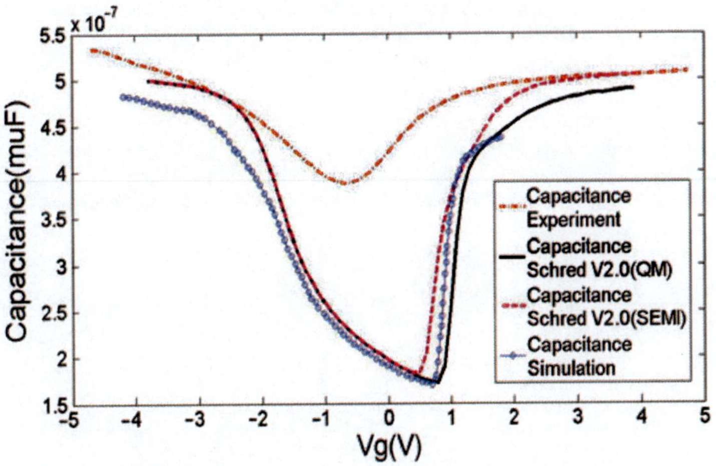

Fig. 2.34 GaAs capacitance for quantum mechanical (QM) and semi-classical case with experimental and simulation results from [70]

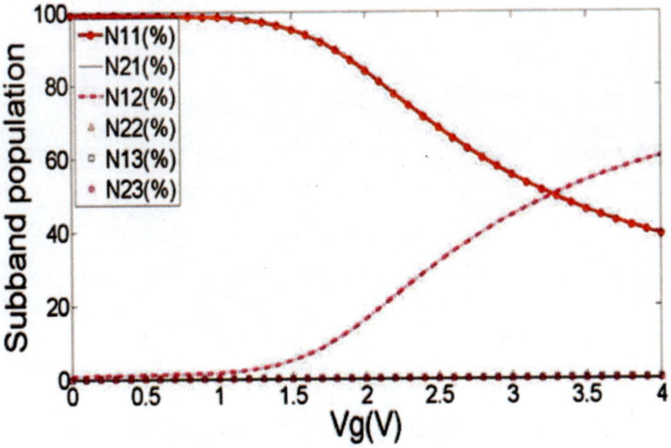

Fig. 2.35 Subband population

From the results presented in Figs. 2.35 and 2.36, we can clearly see that the subband population shifts from valley 1(gamma) to valley 2(L valley) as the gate voltage increases. More carriers are being excited to higher valleys, namely the L valley, as the applied voltage increases, thus increasing their population density. From the results presented in Fig. 2.36, it is also observed that only the lower subbands contribute to the majority of the population in a given valley, whereas the higher subbands are relatively unoccupied. This can be explained with the plot of the energy levels variation shown in Fig. 2.37.

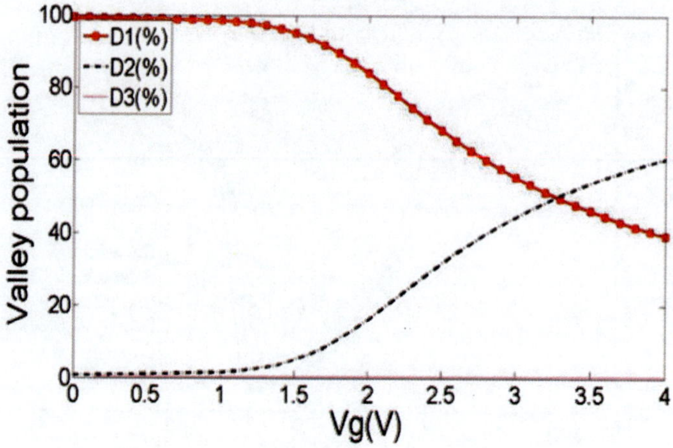

Fig. 2.36 Valley population

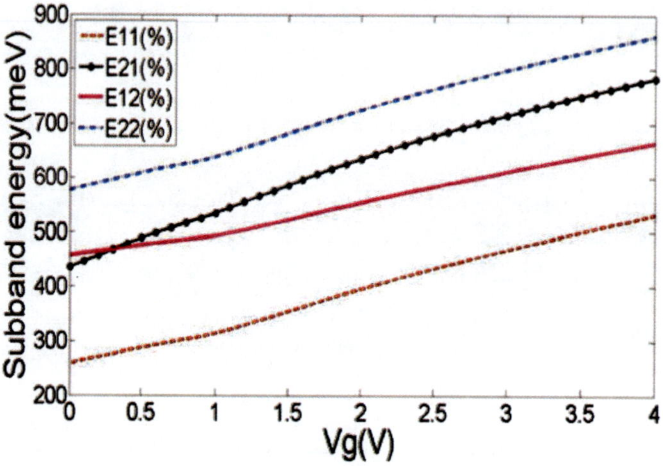

Fig. 2.37 Lowest two subband energies variation of the gamma and L valleys

Example 3: Strained Silicon

In the case of strained Si, strain on the Si material forces the valence bands degenerate levels to split; the heavy hole band crosses the light hole band and also the equi-energy Δ valleys are split into Δ_4 and Δ_2 conduction bands. This leads to change in the effective masses of the heavy hole and light hole valence bands (Figs. 2.38 and 2.39) and a change in the bandgap of the material.

Here we simulate to match experimental results of tensile strained Si (Silicon on silicon germanium). The experiment uses a polysilicon gate on a bi-axial strained Si layer on $Si_{0.8}Ge_{0.2}$. The experimental values are: polysilicon gate with doping concentration of $10^{20}\,cm^{-3}$ oxide thickness $t_{ox} = 1.33\,nm$, temperature $T = 300\,K$, substrate doping $N_A = 9 \times 10^{19}\,cm^{-3}$.

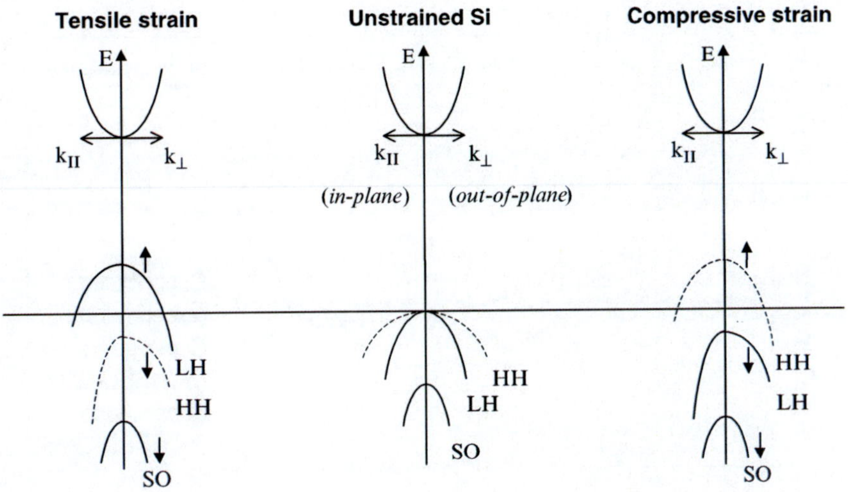

Fig. 2.38 Schematic band representation in strained layers under tensile and compressive strain, along with the unstrained case as a reference

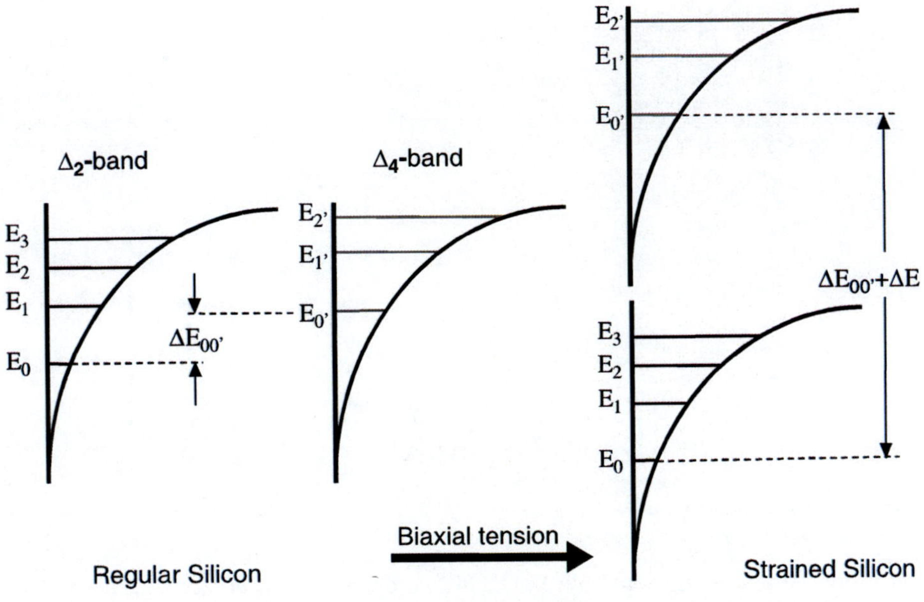

Fig. 2.39 Subband structure in the inversion layer of regular and surface-channel strained-Si layer

Our results in Fig. 2.40 closely match with the experimental results of [71]. The quantum capacitance matches with the experimental values in the inversion region, but differs in the accumulation and the depletion region due to the omission of the hole confinement in this work.

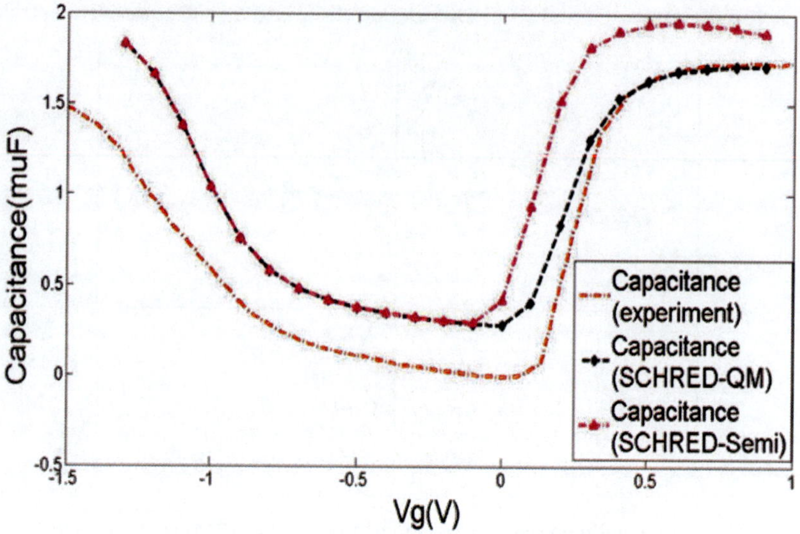

Fig. 2.40 Bi-axial strained on silicon (100) capacitance, experimental results from [71]

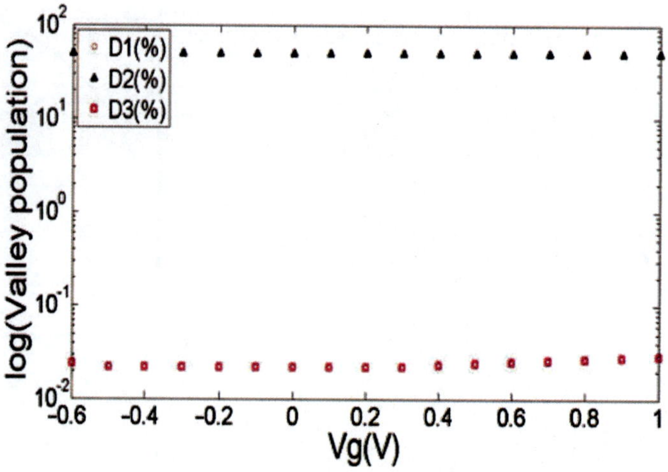

Fig. 2.41 Valley population

From the results presented in Fig. 2.41, we observe that, contrasting to the case of normal Si the population now shifts to the Δ_2 band(D2 valleys) from the Δ_4 band (D1 and D2 valleys) due to the application of the bi-axial strain (see Fig. 2.41), which makes the Δ_2 band to have a lower energy than the Δ_4 band.

Conclusions

This part of the research work presented in this book chapter has successfully created a nano-device simulator that can model MOS/SOS capacitors with the inclusion of quantum effects, poly gate depletion, uniform/non-uniform doping, and user defined number of valleys, partial/complete ionization of carriers and several other features.

The simulator is built with a fast direct LU-decomposition Poisson solver that is coupled with the Schrödinger equation. The Schrödinger equation is solved in the bulk region using three point finite difference scheme, which results in a non-symmetric matrix (due to the non-uniform mesh used). This matrix is then transformed to a symmetric matrix using a matrix transformation technique. This transformed symmetric matrix is used to solve for eigenvalues and wavefunctions using the EISPACK routine.

2.3 Inclusion of Tunneling in Particle-Based Device Simulators

Tunneling is an important phenomenon in the operation of some devices in both the positive and the negative sense. For example, the negative differential characteristics in an Esaki diode (heavily doped p + /n+ junction – see Fig. 2.42) or in resonant tunneling diode are due to tunneling/resonant tunneling in these structures respectively. The peak to valley current is an important indicator on the quality of the device and larger the ratio, better is the device usability in oscillators.

Also, tunneling into the floating gate is necessary for the operation of EEPROM memories. Tunneling is the basic principle on which the operation of scanning tunneling microscopes is based, which revolutionized the understanding of surfaces and surface reconstructions in different semiconductor materials.

There are also instances in which tunneling is an undesired phenomenon, such as gate leakage in FET devices (see Fig. 2.43) or transistors with Schottky gate. In the case of FET devices, if the carriers tunnel through the tip of the barrier, then we call this tunneling process as Fowler–Nordheim tunneling. In small structures with thin oxides, carriers tunnel through the whole thickness of the oxide and in that case we have direct tunneling process.

The WKB (Wentzel, Kramers, Brillouin) approximation is a quasi-classical method for solving the one-dimensional (and effectively one-dimensional, such as radial) time-independent Schrödinger equation. The nontrivial step in the method is the connection formulas, that problem was first solved by Lord Rayleigh [72] and as Jeffries notes [73] "it has been rediscovered by several later writers" presumably referring to Wentzel, Kramers and Brillouin (WKB). A more accurate method for the calculation of the transmission coefficient in 1D tunneling structures is the transfer matrix approach which sometimes suffers from numerical overflow problems. To avoid these issues, a variant of this approach, the so-called scattering matrix approach is typically used. For 2D and 3D problems, the Usuki method [74] is the method of choice alongside with the Green's function approaches [75]. In what follows here, we first describe the WKB approximation on the example of tunneling through a triangular barrier, and then we discuss the transfer matrix approach on the example of a piecewise linear approximation of the potential barrier and its application in calculation of tunneling current in SOI Schottky MESFET.

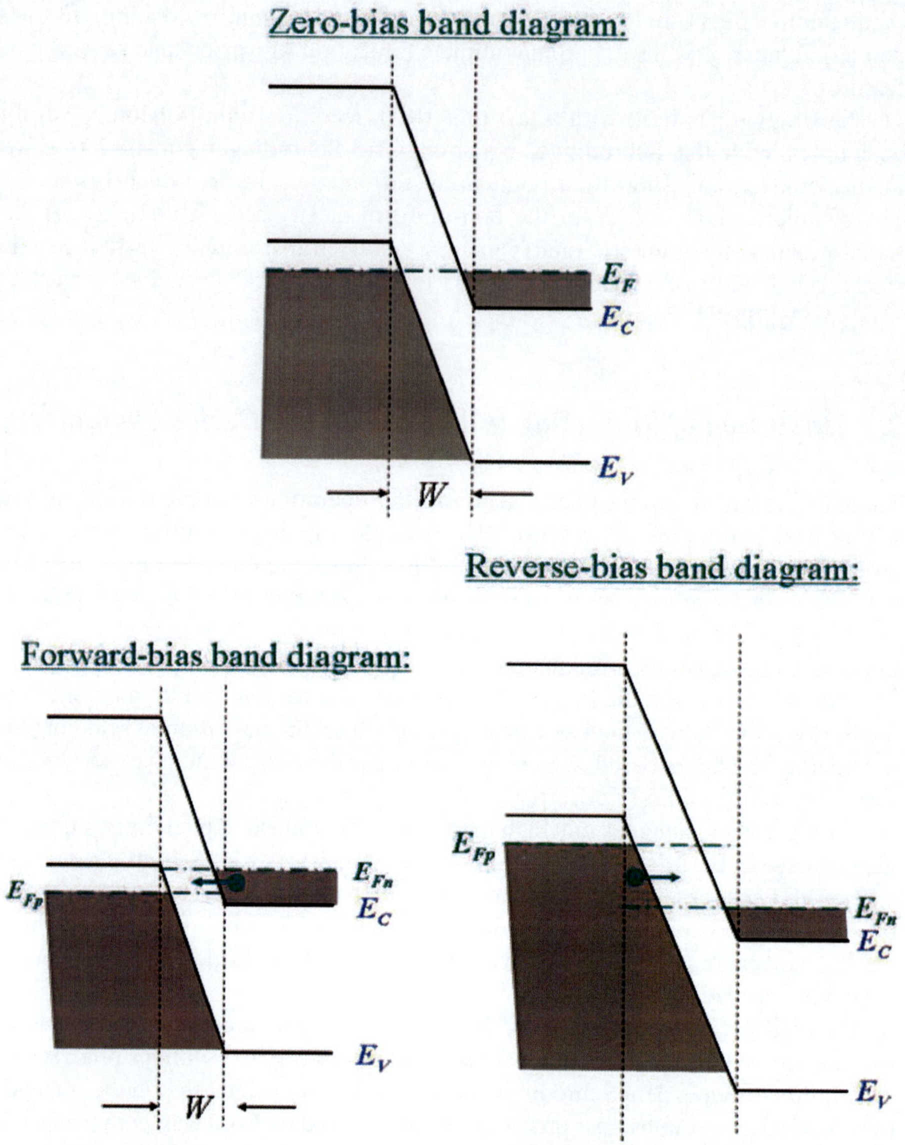

Fig. 2.42 Forward and reverse tunneling in heavily-doped PN (Esaki diodes). *Top panel* – Equilibrium band diagram, *bottom left panel* – forward bias conditions and *bottom right panel* – reverse bias conditions

2.3.1 WKB Approximation Used in Tunneling Coefficient Calculation

Consider a particle of mass m^* and energy $E > 0$ moving through some *slowly varying* potential $V(x)$. The particle's wave-function satisfies

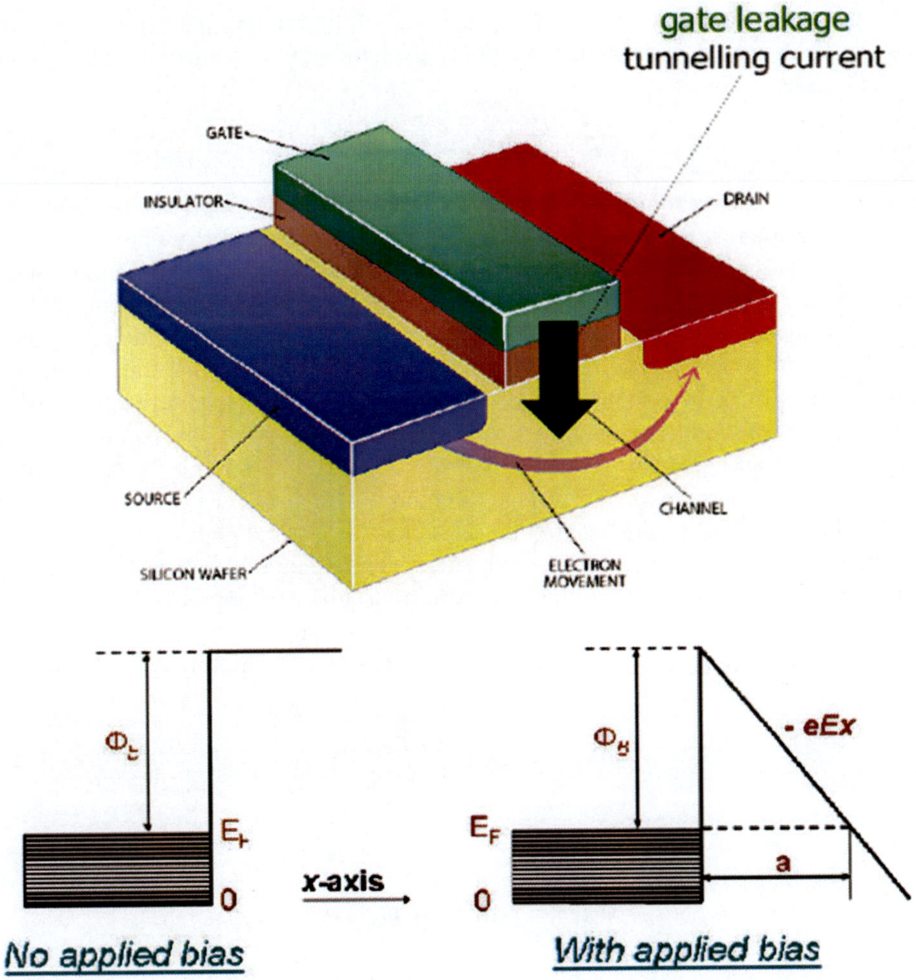

Fig. 2.43 Tunneling (gate leakage) limiting device miniaturization and leading to the introduction of gate stacks with high-k dielectrics (*top panel*). *Bottom panel* – Schematics of a tunnel barrier and the concept of Fowler–Nordheim tunneling

$$\frac{d^2\psi}{dx^2} = -k^2(x)\psi(x) \tag{2.44}$$

where

$$k^2(x) = \frac{2m^*[E - V(x)]}{\hbar^2} \tag{2.45}$$

Let us try a solution to (2.44) of the form

$$\psi(x) = \psi_0 \exp\left[\int_0^x ik(x')dx'\right] \tag{2.46}$$

where ψ_0 is a complex constant. Note that this solution represents a particle moving to the right with the continuously varying wavenumber $k(x)$. Substituting (2.46) into (2.44) gives

$$\frac{d^2\psi}{dx}^2 = ik'(x)\psi(x) - k^2(x)\psi(x) \tag{2.47}$$

where $k' = dk/dx$. From (2.44–2.47) it follows that (2.46) is a solution to (2.44) provided that the first term on its right-hand side is negligible compared to the second. This yields the validity criterion $|k'| << k^2$. In other words, the variation length-scale of $k(x)$ (which is approximately the same as the variation length-scale of $V(x)$) must be *much greater* than the particle's de Broglie wave-length (which is of order k^{-1}). Let us suppose that this is the case. Incidentally, the approximation involved in dropping the first term on the right-hand side of (2.47) is generally known as the *WKB approximation*. Similarly, (2.46) is termed a WKB solution. According to the WKB solution (2.46), the probability density remains constant: *i.e.*, $|\psi(x)|^2 = |\psi_0|^2$ as long as the particle moves through a region in which $E > V(\mathrm{x})$ and k(x) is consequently real (i.e., an allowed region according to classical physics).

Suppose, however, that the particle encounters a potential barrier (i.e., a region from which the particle is excluded according to classical physics). By definition, $E < V(x)$ inside such a barrier, and $k(x)$ is consequently imaginary. Let the barrier extend from $x = x_1$ to x_2, where $0 < x_1 < x_2$. The WKB solution inside the barrier is written

$$\psi(x) = \psi_1 \exp\left[-\int_{x_1}^{x} |k(x')|\, dx'\right] \tag{2.48}$$

where

$$\psi_1(x) = \psi_0 \exp\left[\int_{0}^{x_1} ik(x')dx'\right]. \tag{2.49}$$

Here, we have neglected the unphysical exponentially growing solution. According to the WKB solution, the probability density *decays exponentially* inside the barrier: i.e.,

$$|\psi(x)|^2 = |\psi_1|^2 \exp\left[-2\int_{x_1}^{x} |k(x')|\, dx'\right], \tag{2.50}$$

where $|\psi_1|^2$ is the probability density at the left-hand side of the barrier (i.e., $x = x_1$). It follows that the probability density at the right-hand side of the barrier (i.e., $x = x_2$) is

$$|\psi_2|^2 = |\psi_1|^2 \exp\left[-2\int_{x_1}^{x_2} |k(x')|\, dx'\right]. \tag{2.51}$$

Note that $|\psi_2|^2 < |\psi_1|^2$. Of course, in the region to the right of the barrier (i.e., $x > x_2$), the probability density takes the constant value $|\psi_2|^2$. We can interpret the ratio of the probability densities to the right and to the left of the potential barrier as the probability $|T|^2$, that a particle incident from the left will tunnel through the barrier and emerge on the other side: i.e.,

$$T = \frac{|\psi_2|^2}{|\psi_1|^2} = \exp\left[-2\int_{x_1}^{x_2} |k(x')|dx'\right] \tag{2.52}$$

It is easily demonstrated that the probability of a particle incident from the right tunneling through the barrier is the same.

Note that the criterion for the validity of the WKB approximation implies that the above transmission probability is *very small*. Hence, the WKB approximation only applies to situations in which there is very little chance of a particle tunneling through the potential barrier in question. Unfortunately, the validity criterion breaks down completely at the edges of the barrier (i.e., at $x = x_1$ and x_2), since $k(x) = 0$ at these points. However, it can be demonstrated that the contribution of those regions, around $x = x_1$ and x_2, in which the WKB approximation breaks down to the integral in (2.52) is fairly negligible. Hence, the above expression for the tunneling probability is a reasonable approximation provided that the incident particle's de Broglie wave-length is much smaller than the spatial extent of the potential barrier.

Let us now apply the result given in (2.52) to the triangular barrier shown in Fig. 2.44. Upon the calculation of the integral in the exponent given by (2.52), one gets the transmission coefficient as,

$$T = \exp\left(-\frac{\pi m^{*1/2} E_G^{3/2}}{2\sqrt{2}\hbar eE}\right)\exp\left(-\frac{2E_z}{\bar{E}}\right), \tag{2.53}$$

where

$$\bar{E} = \frac{4\sqrt{2}\hbar eE}{3\pi\sqrt{m^* E_G}}, \tag{2.54}$$

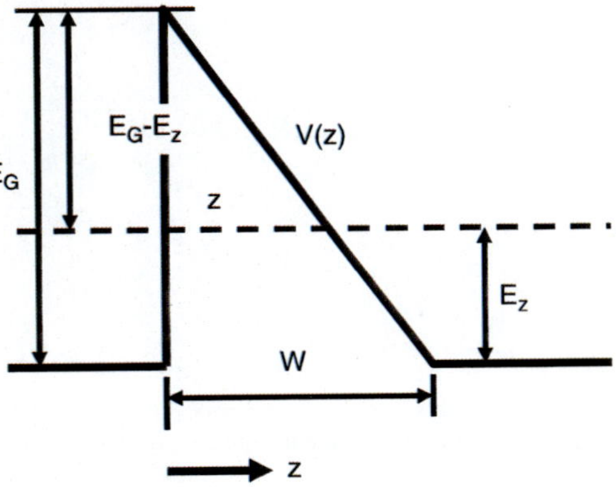

Fig. 2.44 Triangular potential barrier encountered by the electrons in an Esaki diode from Fig. 2.42 under forward and reverse bias conditions

and E is the electric field along the transport direction. The result given in (2.53) is then substituted in the Tsu–Esaki Formula for the current to get:

$$J_t = \frac{e^3 m^{*1/2} \xi V_a}{4\sqrt{2}\pi^2 \hbar^2 E_g^{1/2}} \exp\left(\frac{-4\sqrt{2m^*} E_G^{3/2}}{3e\hbar\xi} \right). \tag{2.55}$$

2.3.2 Transfer Matrix Approach for Piece-Wise Linear Approximation of the Potential Barrier

We next discuss the methodology for the calculation of the transmission probability and apply the technique for the calculation of the transmission coefficient through an arbitrary varying potential barrier. The exact method [76] that we use is based on the analytical solution of the Schrödinger equation across a linearly varying potential. In this case, the solution can be expressed as linear combination of Airy functions. Proper boundary conditions are imposed at the interface between adjacent linear intervals of the potential using a transfer matrix [77] procedure. The method for the calculation of the transmission coefficient is outlined below.

Let us consider a piecewise linear potential function such that the potential energy profile varies linearly in the region (a_{i-1}, a_i) (Fig. 2.45).

$$V(x) = V(a_{i-1}) + \frac{x - a_{i-1}}{a_i - a_{i-1}}[V(a_i) - V(a_{i-1})] = V_{i-1} + \frac{V_i - V_{i-1}}{a_i - a_{i-1}}(x - a_{i-1}) \tag{2.56}$$

The electric field profile is given by,

$$F_i = -\left.\frac{d\phi}{dx}\right|_i = \frac{1}{e}\left.\frac{dV}{dx}\right|_i = -\frac{V_i - V_{i-1}}{a_i - a_{i-1}}, \tag{2.57}$$

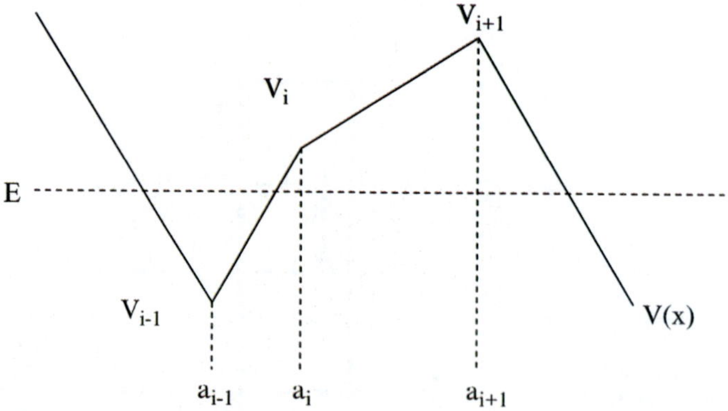

Fig. 2.45 Piecewise linear potential barrier

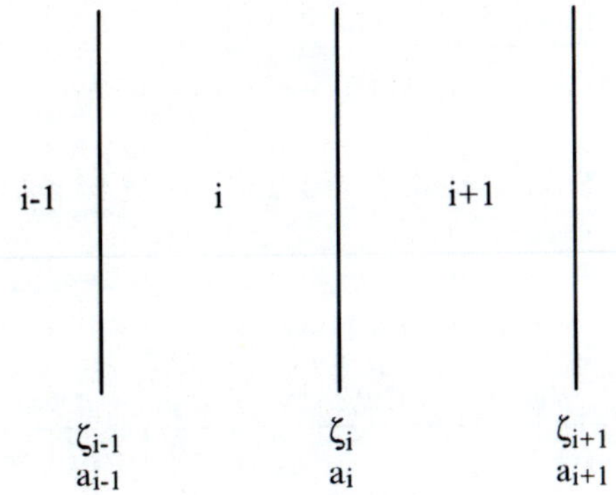

Fig. 2.46 Slicing of the region and corresponding variables in the slices

where V_i is in eV. Therefore,

$$V(x) = V_{i-1} + F_i(x - a_{i-1}) \tag{2.58}$$

Substituting back into the time-independent Schrödinger Wave Equation (TISE) gives (Fig. 2.46),

$$\begin{aligned} &-\frac{\hbar^2}{2m}\frac{d^2\Psi}{dx^2} + V(x)\psi = E\psi, \\ &\Rightarrow -\frac{\hbar^2}{2m}\frac{d^2\Psi}{dx^2} + [V_{i-1} + F_i(x - a_i)]\psi = E\psi, \\ &\Rightarrow -\frac{\hbar^2}{2m}\frac{d^2\Psi}{dx^2} + F_i x\psi = (E + F_i a_i - V_{i-1})\psi, \\ &\Rightarrow -\frac{\hbar^2}{2m}\frac{d^2\Psi}{dx^2} + F_i x\psi = \varepsilon'\psi. \end{aligned} \tag{2.59}$$

We now define a dimensionless variable ξ such that

$$\xi = \left(\frac{2mF_i}{\hbar^2}\right)^{1/3} x - \frac{2m\varepsilon'}{\hbar^2}\left(\frac{\hbar^2}{2mqF_i}\right)^{2/3}. \tag{2.60}$$

Substituting (2.60) into (2.59) leads to

$$\frac{d^2\psi}{d\xi^2} - \xi\psi(\xi) = 0, \tag{2.61}$$

where $\varepsilon' = E + qF_i a_i - V_{i-1}$. The solutions of the reduced equation are the Airy functions and the modified Airy functions. Thus,

$$\psi_i = C_i^{(1)} A_i(\xi) + C_i^{(2)} B_i(\xi), \tag{2.62a}$$

and

$$\psi_{i+1}(\xi) = C_{i+1}^{(1)} A_i(\xi) + C_{i+1}^{(2)} B_i(\xi) \tag{2.62b}$$

From the continuity and the smoothness conditions for the wave function at $x = a_i$ we get

$$\psi_i(\xi_i) = \psi_{i+1}(\xi_i), \tag{2.63a}$$

$$\left.\frac{d\psi_i}{dx}\right|_{a_i} = \left.\frac{d\psi_{i+1}}{dx}\right|_{a_i} \Rightarrow \frac{d\psi_i}{dx} = \left.\frac{d\psi_i}{d\xi}\right|_{\xi_i} \frac{d\xi}{dx} = r_i \frac{d\psi_i}{d\xi}, \quad \left.\frac{d\psi_{i+1}}{dx}\right|_{a_i} = r_{i+1} \left.\frac{d\psi_{i+1}}{dx}\right|_{\xi_i} \tag{2.63b}$$

Therefore,

$$C_i^{(1)} A_i(\xi_i) + C_i^{(2)} B_i(\xi_i) = C_{i+1}^{(1)} A_i(\xi_i) + C_{i+1}^{(2)} B_i(\xi_i), \tag{2.64a}$$

$$r_i C_i^{(1)} A_i'(\xi_i) + r_i C_i^{(2)} B_i'(\xi_i) = r_{i+1} C_{i+1}^{(1)} A_i'(\xi_i) + r_{i+1} C_{i+1}^{(2)} B_i'(\xi_i). \tag{2.64b}$$

Rearranging (2.64a) and (2.64b) and writing them in a matrix form gives,

$$\begin{bmatrix} A_i(\xi_i) & B_i(\xi_i) \\ r_i A_i'(\xi_i) & r_i B_i'(\xi_i) \end{bmatrix} \begin{bmatrix} C_i^{(1)} \\ C_i^{(2)} \end{bmatrix} = \begin{bmatrix} A_i(\xi_i) & B_i(\xi_i) \\ r_{i+1} A_i'(\xi_i) & r_{i+1} B_i'(\xi_i) \end{bmatrix} \begin{bmatrix} C_{i+1}^{(1)} \\ C_{i+1}^{(2)} \end{bmatrix}$$

$$\Rightarrow \begin{bmatrix} C_i^{(1)} \\ C_i^{(2)} \end{bmatrix} = M^{-1} \begin{bmatrix} A_i(\xi_i) & B_i(\xi_i) \\ r_{i+1} A_i'(\xi_i) & r_{i+1} B_i'(\xi_i) \end{bmatrix} \begin{bmatrix} C_{i+1}^{(1)} \\ C_{i+1}^{(2)} \end{bmatrix}$$

where

$$M^{-1} = \frac{1}{\det M} \begin{bmatrix} r_i B_i'(\xi_i) & -r_i A_i'(\xi_i) \\ -B_i(\xi_i) & A_i(\xi_i) \end{bmatrix}^T, \tag{2.65}$$

and $\det(M) = r_i [A_i(\xi_i) B_i'(\xi_i) - A_i'(\xi_i) B_i(\xi_i)] = \dfrac{r_i}{\pi}$. As a result of (2.65)

$$M^{-1} = \frac{\pi}{r_i} \begin{bmatrix} r_i B_i'(\xi_i) & -B_i(\xi_i) \\ -r_i A_i'(\xi_i) & A_i(\xi_i) \end{bmatrix},$$

and (2.65) becomes

$$\begin{bmatrix} C_i^{(1)} \\ C_i^{(2)} \end{bmatrix} = \frac{\pi}{r_i} \begin{bmatrix} r_i B_i'(\xi_i) & -B_i(\xi_i) \\ -r_i A_i'(\xi_i) & A_i(\xi_i) \end{bmatrix} \begin{bmatrix} A_i(\xi_i) & B_i(\xi_i) \\ r_{i+1} A_i'(\xi_i) & r_{i+1} B_i'(\xi_i) \end{bmatrix} \begin{bmatrix} C_{i+1}^{(1)} \\ C_{i+1}^{(2)} \end{bmatrix} = M_i \begin{bmatrix} C_{i+1}^{(1)} \\ C_{i+1}^{(2)} \end{bmatrix}. \tag{2.66}$$

Now let us consider the case for initial boundary between region 0 and region 1. In region 0 the wave function is described as plane wave and in region 1 it is a combination of Airy functions. Then

$$\psi_0 = C_0^{(1)} e^{ik_o x} + C_0^{(2)} e^{-ik_o x},$$
$$\psi_1(\xi) = C_1^{(1)} A_i(\xi) + C_1^{(2)} B_i(\xi). \tag{2.67}$$

The continuity of the wave function and of the derivative of the wave function leads to

$$C_0^{(1)} + C_0^{(2)} = C_1^{(1)} A_i(\xi_0) + C_1^{(2)} B_i(\xi_0),$$
$$ik_0[C_0^{(1)} - C_0^{(2)}] = r_1 C_1^{(1)} A_i'(\xi_0) + r_1 C_1^{(2)} B_i'(\xi_0). \tag{2.68}$$

Dividing the second equation by ik_o one gets

$$C_0^{(1)} - C_0^{(2)} = \frac{r_1}{ik_0} C_1^{(1)} A_i'(\xi_0) + \frac{r_1}{ik_0} C_1^{(2)} B_i'(\xi_0). \tag{2.69}$$

Then

$$2\,C_0^{(1)} = \left[A_i(\xi_0) + \frac{r_1}{ik_0} A_i'(\xi_0)\right] C_1^{(1)} + \left[B_i(\xi_0) + \frac{r_1}{ik_0} B_i'(\xi_0)\right] C_1^{(2)},$$
$$2\,C_0^{(2)} = \left[A_i(\xi_0) - \frac{r_1}{ik_0} A_i'(\xi_0)\right] C_1^{(1)} + \left[B_i(\xi_0) + \frac{r_1}{ik_0} B_i'(\xi_0)\right] C_1^{(2)}. \tag{2.70}$$

In summary,

$$\begin{bmatrix} C_0^{(1)} \\ C_0^{(2)} \end{bmatrix} = \begin{bmatrix} \frac{1}{2}[A_i(\xi_0) + \frac{r_1}{ik_0} A_i'(\xi_0)] & \frac{1}{2}[B_i(\xi_0) + \frac{r_1}{ik_0} B_i'(\xi_0)] \\ \frac{1}{2}[A_i(\xi_0) - \frac{r_1}{ik_0} A_i'(\xi_0)] & \frac{1}{2}[B_i(\xi_0) + \frac{r_1}{ik_0} B_i'(\xi_0)] \end{bmatrix} \begin{bmatrix} C_1^{(1)} \\ C_1^{(2)} \end{bmatrix}. \tag{2.71}$$

We now consider the other boundary [N, N + 1]. In region N we have a combination of Airy functions and in region N + 1 we have plane waves. Hence, we have

$$\psi_N(\xi) = C_N^{(1)} A_i(\xi) + C_N^{(2)} B_i(\xi),$$
$$\psi_{N+1}(\xi) = C_{N+1}^{(1)} e^{ik_{N+1}x} + C_{N+1}^{(2)} e^{-ik_{N+1}x}. \tag{2.72}$$

The continuity of the wave function and of the derivative of the wave function then implies

$$\begin{aligned} C_N^{(1)} A_i(\xi_N) + C_N^{(2)} B_i(\xi_N) &= C_{N+1}^{(1)} e^{ik_{N+1}a_{N+1}} + C_{N+1}^{(2)} e^{-ik_{N+1}a_{N+1}}, r_N C_N^{(1)} A_i'(\xi_N) \\ &\quad + r_N C_N^{(2)} B_i'(\xi_N) \\ &= ik_{N+1}[C_{N+1}^{(1)} e^{ik_{N+1}a_N} - C_{N+1}^{(1)} e^{-ik_{N+1}a_N}. \end{aligned} \tag{2.73}$$

In matrix form this can be represented as,

$$\begin{bmatrix} C_N^{(1)} \\ C_N^{(2)} \end{bmatrix} = \frac{\pi}{r_n} \begin{bmatrix} r_N B_i'(\xi_N) + ik_{N+1} B_i(\xi_N) & r_N B_i'(\xi_N) - ik_{N+1} B_i(\xi_N) \\ -r_N A_i'(\xi_N) + ik_{N+1} A_i(\xi_N) & -r_N A_i'(\xi_N) - ik_{N+1} A_i(\xi_N) \end{bmatrix} M_1 \begin{bmatrix} C_{N+1}^{(1)} \\ C_{N+1}^{(2)} \end{bmatrix}. \tag{2.74}$$

Now, combining (2.66), (2.71), and (2.74), one finally arrives at the total transmission matrix of the system,

$$M_T = M_{FI}M_1M_2 \ldots\ldots\ldots M_{N-1}M_{BI}\begin{bmatrix} e^{ik_{N+1}a_N} & 0 \\ 0 & e^{-ik_{N+1}a_N} \end{bmatrix}$$
$$= \begin{bmatrix} m_{11}^T & m_{12}^T \\ m_{21}^T & m_{22}^T \end{bmatrix}\begin{bmatrix} e^{ik_{N+1}a_N} & 0 \\ 0 & e^{-ik_{N+1}a_N} \end{bmatrix}. \tag{2.75}$$

The transmission coefficient is then given by,

$$T = \frac{k_{N+1}}{k_0}\frac{1}{|m_{11}^T|^2}, \tag{2.76}$$

where m_{11}^T is the element of the matrix $M_T = M_{FI}M_1M_2 \ldots\ldots\ldots M_{N-1}M_{BI}$ and the various matrices that appear in (2.75) are defined as follows:

$$M_{FI} = \begin{bmatrix} \frac{1}{2}[A_i(\xi_0) + \frac{r_1}{ik_0}A_i'(\xi_0)] & \frac{1}{2}[B_i(\xi_0) + \frac{r_1}{ik_0}B_i'(\xi_0)] \\ \frac{1}{2}[A_i(\xi_0) - \frac{r_1}{ik_0}A_i'(\xi_0)] & \frac{1}{2}[B_i(\xi_0) + \frac{r_1}{ik_0}B_i'(\xi_0)] \end{bmatrix},$$

$$M_{BI} = \frac{\pi}{r_n}\begin{bmatrix} r_N B_i'(\xi_N) + ik_{N+1}B_i(\xi_N) & r_N B_i'(\xi_N) - ik_{N+1}B_i(\xi_N) \\ -r_N A_i'(\xi_N) + ik_{N+1}A_i(\xi_N) & -r_N A_i'(\xi_N) - ik_{N+1}A_i(\xi_N) \end{bmatrix},$$

$$M_i = \frac{\pi}{r_i}\begin{bmatrix} r_i B_i'(\xi_i) & -B_i(\xi_i) \\ -r_i A_i'(\xi_i) & A_i(\xi_i) \end{bmatrix}\begin{bmatrix} A_i(\xi_i) & B_i(\xi_i) \\ r_{i+1}A_i'(\xi_i) & r_{i+1}B_i'(\xi_i) \end{bmatrix}. \tag{2.77}$$

In the actual implementation of the method outlined above in the simulation of devices with Schottky barriers, we are considering the electrons between the gate and the buried oxide layer (in the active region) and we calculate the potential profile along the thickness of the device by solving Poisson's equation. Then, applying the Airy function transfer matrix method, we calculate the transmission probability for each particle in the MESFET device. On the basis of particle's position we calculate its potential energy. Then, we compare each particle's energy with the corresponding grid point potential energy. Now, using random number generation method, we evaluate whether each particle is going to tunnel through the Schottky barrier or not. If the transmission probability is greater than the random number then tunneling occurs. Once the particle tunnels, we use a rejection technique to make it inactive for the next iterative steps. For each time increment, we count the number of particles that tunnel through the barrier. After reaching a steady state condition, we calculate the tunneling current from the number of tunneled particles. We apply the piece-wise linear transfer matrix technique in a nonlinear potential barrier as shown in Fig. 2.47 to calculate the transmission probability. Following the technique, we have obtained the transmission probability which is shown in Fig. 2.48. From Fig. 2.48 it is observed that our result is properly matched with calculation previously performed by Lui et al. [78].

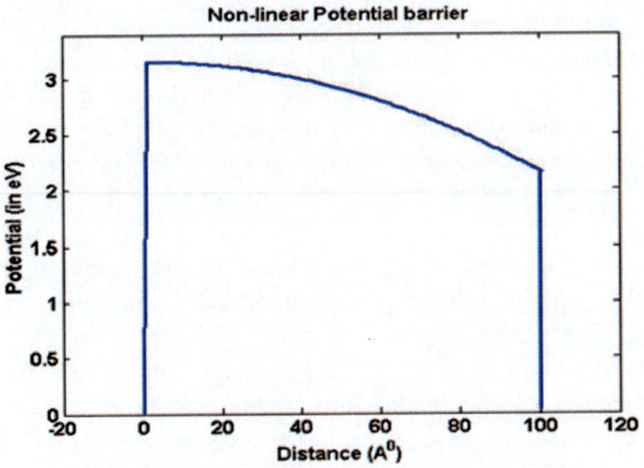

Fig. 2.47 Nonlinear potential barrier is used to calculate quantum mechanical transmission probability

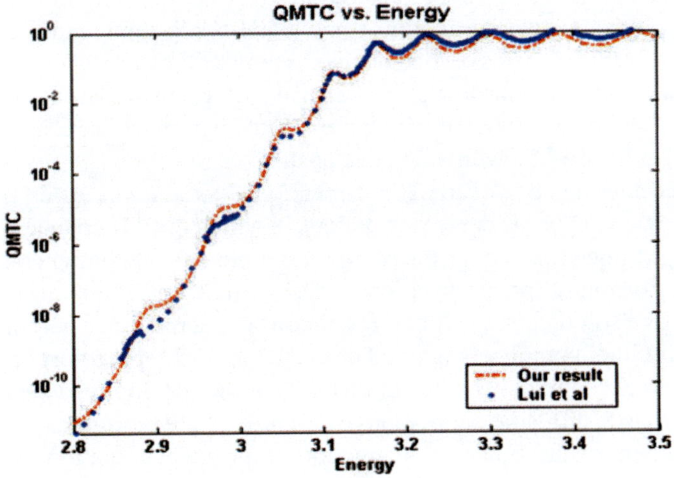

Fig. 2.48 Quantum mechanical transmission probability variation with respect to particle energy and validates our model's exactness

3 Discrete Impurity Effects

The pioneering experimental studies by Mizuno and co-workers [79] in the mid 1990s clearly demonstrated that threshold voltage fluctuations due to the discrete nature of the impurity atoms, are going to be a significant problem in future ultra-small devices. They had shown that the threshold voltage standard deviation is inversely proportional to the square root of the gate area, to the oxide thickness, and to the fourth root of the average doping in the device channel region. They also observed that the statistical variation of the channel dopant number accounts for about 60% of the experimentally derived threshold voltage fluctuation. In a later study, Mizuno [80] also found that the lateral and vertical arrangement of ions produces variations in the threshold voltage that depend upon the drain and substrate biases. Horstmann and co-workers [81] investigated global and local matching of sub-100 nm n-channel metal-oxide-semiconductor (NMOS) and p-channel metal-oxide-semiconductor (PMOS)-transistors and confirmed the area law proposed in [80]. The empirical analytical expression by Mizuno was generalized by Stolk et al. [82] by taking into account the finite thickness of the inversion layer, the depth-distribution of the charge in the depletion layer and the influence of the source and drain impurity distributions.

Numerical drift-diffusion and hydrodynamic simulations [83–86] have also confirmed the existence of the fluctuations in the threshold voltage in ultra-small devices. Two-dimensional (2D) [87] and three-dimensional (3D) [88–91] ensemble Monte Carlo (EMC) particle-based simulations have also been carried out. An important observation was made in [10], where it was shown that there is a significant correlation between the threshold voltage shift and the actual position of the impurity atoms. A rather systematic analysis of the random dopant induced threshold voltage fluctuations in ultra-small metal-oxide-semiconductor field-effect transistors (MOSFETs) was carried out by Asenov [92] using 3D drift-diffusion device simulations and confirming previous results. Recent simulation experiments by Asenov and Saini [93] have shown that discrete impurity effects are significantly suppressed in MOSFETs with a δ-doped channel.

However, the majority of the above-mentioned simulation experiments, except [10,91], utilized 2D or 3D device simulators, in which the "discreteness" of the ions was only accounted for through the charge assignment to the mesh nodes. There, the long-range portion of the electron-ion forces are inherent in the mesh force and is found from the solution of the Poisson equation. The short-range portion of these interactions is either completely ignored or treated in the $\mathbf{k}$-space portion of the EMC transport kernel (in particle based simulations) or via the doping dependence of the mobility (in drift-diffusion simulations). Because of the complexity and obscurity of the treatment of the Coulomb interaction in the MC simulations, a more direct approach has been introduced [10], in which the MC method is supplemented by a *molecular dynamics* (MD) routine. In this approach, the mutual Coulomb interaction among electrons and impurities is treated in the drift part of the MC transport kernel. Indeed, the various aspects associated with the Coulomb interaction, such as dynamical screening and multiple scatterings, are automatically taken into account. Very recently, the MC/MD method has been extended for spatially inhomogeneous systems. Since a part of the Coulomb interaction is already taken into account by the solution of the Poisson equation, the MD treatment of the Coulomb interaction is restricted only to the limited area near the charged particles. It is claimed that the full incorporation of the Coulomb interaction is indispensable to reproduce the correct electron mobility in highly doped silicon samples.

Although real space treatments eliminate the problem of double counting of the force, a drawback is that the 3D Poisson equation must be solved repeatedly to properly describe the self-consistent fields which consumes over 80% of the total simulation time. To further speed up simulations, in this work a new idea has been proposed: to use a 3D Fast Multi-Pole Method (FMM) [94–97] instead. The FMM allows calculation of the field and the potential in a system of n particles connected by a central force within $O(n)$ operations given certain prescribed accuracy. The FMM is based on the idea of condensing the information of the potential generated by point sources in truncated series expansions. After calculating suitable expansions, the long range part of the potential is obtained by evaluating the truncated series at the point in question and the short range part is calculated by direct summation. The field due to the applied boundary biases is obtained at the beginning of the simulation by solving the Poisson equation. Hence the total field acting on each electron is the sum of this constant field and the contribution from the electron–electron and electron–impurity interactions handled by the FMM calculations. *The image charges, which arrise because of the dielectric discontinuity, are handled by the method of images.*

Quite recently, several groups, including ours [39], have shown that the Coulomb effects become even more prominent when the device size scales into the nm range. Even in undoped samples, a single unintentional dopant atom can cause significant fluctuations in the threshold voltage and therefore in the device on-state current due to the randomness of its position within the device active area. Thus, *proper inclusion of the short – range Coulomb interactions is a MUST when considering state of the art SOI FD-MOSFETs and alternate device structures, such as dual gate and FinFET devices.*

3.1 The P^{3}M *Method*

The particle-particle-particle-mesh (P^{3}M) algorithms are a class of hybrid algorithms developed by Hockney and Eastwood [98]. These algorithms enable correlated systems with long-range forces to be simulated for a large ensemble of particles. The essence of P^{3}M algorithms is to express the inter-particle force as a sum of a short-range part calculated by a direct particle–particle force summation and a long-range part approximated by the particle-mesh (PM) force calculation. Using the notation of Hockney, the total force on a particle i may be written as

$$F_i = \sum_{j \neq i} F_{ij}^{coul} + F_i^{ext}. \tag{2.78}$$

F_i^{ext} represents the external field or boundary effects of the global Poisson solution. F_{ij}^{coul}, is the force of particle j on particle i given by Coulomb's law as

$$F_{ij}^{coul} = \frac{q_i q_j}{4\pi\varepsilon} \frac{(r_i - r_j)}{|r_i - r_j|^3}, \tag{2.79}$$

where q_i and q_j are particle charges and r_i and r_j are particle positions. In a P³M algorithm, the total force on particle i is split into two sums

$$F_i = \sum_{\substack{j \neq i \\ SRD}} F_{ij}^{sr} + \sum_{\substack{j \neq i \\ GD}} F_{ij}^{m}. \tag{2.80}$$

The first sum represents the direct forces of particles j on particle i within the short-range domain (SRD), while the second sum represents the mesh forces of particles j on particle i over the global problem domain (GD) that includes the effect of material boundaries and the boundary conditions on particle i. F_{ij}^{sr} is the short-range particle force of particle j on particle i, and F_{ij}^{m} is the long-range mesh force of particle j on particle i. The short-range Coulomb force can be further defined as,

$$F_{ij}^{sr} = F_{ij}^{coul} - R_{ij}, \tag{2.81}$$

where F_{ij}^{coul} is given by (2.79) and R_{ij} is called the reference force. The reference force in (2.81) is needed to avoid double counting of the short-range force due to the overlapping domains in (2.80). The reference force should correspond to the mesh force inside the short-range domain (SRD) and equal to the Coulomb force outside the short-range domain. In other words, a suitable form of the reference force for a Coulombic long-range force is one which follows the point particle force law beyond the cutoff radius r_{sr}, and goes smoothly to zero within that radius. Such smoothing procedure is equivalent to ascribing a finite size to the charged particle. As a result, a straightforward method of including smoothing is to ascribe some simple density profile $S(r)$ to the reference inter-particle force. Examples of shapes which are used in practice and give comparable total force accuracy are the uniformly charged sphere, the sphere with uniformly decreasing density

$$S(\mathrm{r}) = \begin{cases} \dfrac{48}{\pi r_{sr}^4}\left(\dfrac{r_{sr}}{2} - r\right), & r \leq r_{sr}/2 \\ 0, & \textit{otherwise,} \end{cases} \tag{2.82}$$

and the Gaussian distribution of density. The second scheme gives marginally better accuracies in 3D simulations. For this case the reference force can be obtained [99] as,

$$\begin{cases} R_{ij}(r) = \frac{q_i q_j}{4\pi\varepsilon} \times \frac{1}{35 r_{sr}^2}\left(224\xi - 224\xi^3 \right. & \xi = \frac{2r}{r_{sr}} \text{ and } 0 \leq r \leq r_{sr}/2 \\ \qquad\qquad \left. + 70\xi^4 + 48\xi^5 - 21\xi^6\right) & \\ R_{ij}(r) = \frac{q_i q_j}{4\pi\varepsilon} \times \frac{1}{35 r_{sr}^2}\left(\frac{12}{\xi^2} - 224 + 896\xi \right. & r_{sr}/2 \leq r \leq r_{sr} \\ \qquad\qquad -840\xi^2 - 224\xi^3 + 70\xi^4 & \\ \qquad\qquad \left. +48\xi^5 - 7\xi^6\right) & \\ R_{ij}(r) = \frac{q_i q_j}{4\pi\varepsilon} \times \frac{1}{r^2} & r > r_{sr} \end{cases} \tag{2.83}$$

Hockney advocates pre-calculating the short-range force, $F_{ij}^{sr}(r)$ in (2.81) including the reference force above for a fixed mesh. It is important to extend the P^{3}M algorithm to nonuniform meshes for the purpose of semiconductor device simulation since practical device applications involve rapidly varying doping profiles and narrow conducting channels which need to be adequately resolved. Since the mesh force from the solution to the Poisson equation is a good approximation within about two mesh spaces, r_{sr} is locally chosen as the shortest distance which spans two mesh cells in each direction of every dimension of the mesh at charge i.

In order to incorporate the effects of material boundaries and boundary conditions, the reference force would be found most precisely in the short-range domain by associating particle j with the particle-mesh and calculating the resulting force on particle i with $F_i^{ext} = 0$. Since such a procedure would be required for each particle, it is obviously too costly for reasonable ensemble sizes and defeats the purpose of the P^{3}M algorithm [100]. Instead, it is desirable to use an approximation for this force, which minimizes the effects of the transition error in going from the long-range domain to the short-range domain. One approach developed in [100] is to choose a particular orientation of approaching particles relative to the mesh and find a radial approximation to the reference force. This method is straightforward and computationally efficient per particle for a fixed uniform mesh, but it is not easily adaptable to nonuniform meshes where the mesh force is not isotropic.

3.2 *The Fast Multipole Method*

FMM was first introduced by Rokhlin [95] and was later refined by Greengard [96] for the application of two and three-dimensional N-body problems whose interactions are Coulombic or gravitational in nature. In a system of N particles, the decay of the Coulombic or gravitational potential is sufficiently slow so that all interactions must be accounted for, resulting in CPU time requirements on the order of $O(N^2)$. On the other hand, the FMM requires an amount of work proportional to N to evaluate all interactions to within a round off error, making it practical for large-scale problems encountered in plasma physics, fluid dynamics, molecular dynamics, and celestial mechanics.

There have been a number of previous efforts aimed at reducing the computational complexity of the N-body problem. Assuming the potential satisfies Poisson's equation, a regular mesh is laid out over the computational domain and the method proceeds by: (1) interpolating the source density at mesh points; (2) using a fast Poisson solver to obtain potential values on the mesh; (3) computing the force from the potential and interpolating to the particle positions. The complexity of these methods is of the order of $O(N + M \log M)$, where M is the number of mesh points. The number of mesh points is usually chosen to be proportional to the number of particles, but with a small constant of proportionality so that $M \langle\langle N$. Therefore, although the asymptotic complexity for the method is $O(N \log N)$ the computational cost in practical calculations is usually observed to be proportional to N. Unfortunately, the mesh provides limited resolution, and highly non-uniform source distributions cause a significant degradation of performance. Further errors are introduced in step (3) by the necessity for numerical differentiation to obtain the force. To improve the accuracy of particle-in-cell calculations, short-range interactions can be handled by direct computation, while far-field interactions are obtained from the mesh, giving rise to the so-called particle-particle-particle-mesh (P^3M) method described previously. While these algorithms still depend for their efficient performance on a reasonably uniform distribution of particles, in theory they do permit arbitrarily high accuracy to be obtained. As a rule, when the required precision is relatively low, and the particles are distributed more or less uniformly in a rectangular region, P^3M methods perform satisfactorily. However, when the required precision is high (for example in the modeling of highly correlated systems), the CPU time requirements of such algorithms tend to become excessive.

3.2.1 Multipole Moment

A multipole expansion is a series expansion which describes the effect produced by a given system in terms of an expansion parameter [95] that becomes smaller as the distance of the observation point from the source point increases. Therefore the leading order terms in a multipole expansion are generally the dominant. The first order behavior of the system at large distances can therefore be predicted from the first terms of the series, which is much easier to compute than the general solution.

Let r be the vector from the fixed reference point to a point in the system and r_1 be the vector from reference point to the observation point, and $d \equiv r_1 - r$ be the vector from a point in the system to the observation point. From the laws of cosines, d can be expressed as

$$d^2 = r_1^2 + r^2 - 2r_1 r\cos\varphi = r_1^2\left(1 + \frac{r^2}{r_1^2} - 2\frac{r}{r_1}\cos\varphi\right) \tag{2.84}$$

where $\cos\varphi \equiv \hat{r}.\hat{r}_1$. Therefore,

$$d = r_1\sqrt{1 + \frac{r^2}{r_1^2} - 2\frac{r}{r_1}\cos\varphi} \tag{2.85}$$

Let $\xi \equiv \frac{r}{r_1}$ and $y = \cos\varphi$. Then

$$\frac{1}{d} = \frac{1}{r_1}\left(1 - 2\xi y + \xi^2\right)^{-1/2} \tag{2.86}$$

But $(1 - 2\xi y + \xi^2)^{-1/2}$ is the generating function for Legendre Polynomials, i.e.

$$(1 - 2\xi y + \xi^2)^{-1/2} = \sum_{i=0}^{\infty} \xi^i P_i(y) \tag{2.87}$$

so,

$$\frac{1}{d} = \frac{1}{r_1}\sum_{i=0}^{\infty}\left(\frac{r}{r_1}\right)^i P_i(\cos\varphi) = \sum_{i=0}^{\infty}\frac{1}{r_1^{i+1}} r^i P_i(\cos\varphi). \tag{2.88}$$

Any physical potential that obeys a $1/d$ law can therefore be expressed as a multipole expansion,

$$V = \sum_{i=0}^{\infty}\frac{1}{r_1^{i+1}}\int r^i P_i(\cos\varphi)\rho(r)d^3r. \tag{2.89}$$

In MKS unit,

$$V = \frac{1}{4\pi\varepsilon_0\varepsilon_r}\sum_{i=0}^{\infty}\frac{1}{r_1^{i+1}}\int r^i P_i(\cos\varphi)\rho(r)d^3r, \tag{2.90}$$

where ε_0 is the permittivity of the free space, ε_r is the dielectric constant of the medium and $\rho(r)$ is the charge density.

3.2.2 How FMM Speeds Up the Computation?

In FMM *multipole moments* are used to represent distant particle groups and a *local expansion* is used to evaluate the contribution from distant particles in the form of a series. The multipole moment associated with a distant group can be *translated* into the coefficient of the local expansion associated with a local group. In FMM the computational domain is decomposed in a hierarchical manner with a quad-tree in two dimensions and an oct-tree in three dimensions to carry out efficient and systematic grouping of particles with tree structures. The hierarchical decomposition is used to cluster particles at various spatial lengths and compute interactions with other clusters that are sufficiently far away by means of the series expansions.

For a given input configuration of particles, the sequential FMM first decomposes the data-space in a hierarchy of blocks and computes local neighborhoods and *interaction-lists* involved in subsequent computations. Then, it performs two passes on the decomposition tree. The first pass starts at the leaves of the tree, computing *multipole expansion coefficients* for the Columbic field. It proceeds towards the root accumulating the multipole coefficients at intermediate tree-nodes. When the root is reached, the second pass starts. It moves towards the leaves of the tree, *exchanging* data between blocks belonging to the neighborhoods and interaction-lists calculated at tree-construction. At the end of the downward pass all long-range interactions have been computed. Subsequently, nearest-neighbor computations are performed directly to take into consideration interactions from nearby bodies. Finally, short- and long-range interactions are accumulated and the total forces exerted upon particles are computed. The algorithm repeats the above steps and simulates the evolution of the particle system for each successive time-step.

3.3 *The Role of Discrete Impurities as Observed by Simulations and with Comparisons to Experiments*

In the three subsequent subsections first the role of discrete impurities on the operation of conventional device designs is discussed, then unintentional dopants are being examined and finally the role of unintentional dopants on the FinFET transfer and output characteristics is being examined.

3.3.1 Previous Knowledge on Threshold Voltage and On-State Current Fluctuations in Sub-Micrometer MOSFET Devices

As already discussed in the introduction part of this book chapter, continued scaling of devices has led to a number of undesirable effects, including fluctuations in the threshold voltage that arise because of the discrete, or atomistic nature of the impurity atoms in the device active region. For better insight of the importance of this issue, we have considered a prototypical MOSFET with 0.07 μm channel length, 0.07 μm channel width and channel doping of $10^{18}\,\mathrm{cm}^{-3}$. The number of dopant atoms in the depletion region of this device is on the order of several hundreds, and well below 100 in the active region. In addition, there are regions where the impurity atoms cluster and other regions in which the impurity density is well below the average value expected from the doping level. With such a small number of the impurity atoms in the device active region, the local variations in the "doping concentration" across the channel become a significant factor in determining the threshold voltage, mobility and drain current characteristics. This in turn, causes considerable problems for circuit design, especially for circuits in which the devices must be well matched, such as operational amplifiers [101] and static random access memories [102]. The *SIA* roadmap technology requirements state that the variation in gate length should be less than 10% and the variation in threshold voltage should be less than 40 mV for devices in the 150 nm generation and beyond [103].

It is interesting to note that the existence of these surface potential fluctuations in MOS devices was postulated by Nicollian and Goetzberger [104] in order to explain the departures from the theoretical predictions in conductance vs. frequency measurements in MOS structures. In addition to their effect on the *ac*-conductance results, surface potential fluctuations were also found to have significant influence on a variety of other device characteristics, such as threshold voltage, transconductance, substrate current and off-state leakage currents. Experimental studies by Mizuno, Okamura, and Toriumi [6] have shown that the threshold voltage standard deviation is related to the average number of ionized impurities beneath the channel according to

$$\sigma_{vt} = \left(\frac{\sqrt[4]{q^3\varepsilon_s\phi_b}}{\sqrt{2}\varepsilon_{ox}}\right)\frac{T_{ox}\sqrt[4]{N}}{\sqrt{L_{eff}W_{eff}}}, \tag{2.91}$$

where N is the average channel doping density, ϕ_b is the built-in potential, T_{ox} is the oxide thickness, L_{eff} and W_{eff} are the effective channel length and width, and ε_s and ε_{ox} are the semiconductor and oxide permittivity, respectively. They found that the statistical variation of the channel dopant number accounts for about 60% of the experimentally derived threshold voltage fluctuations. In a later study, Mizuno [81] also found that the lateral and vertical arrangement of ions produces variations in the threshold voltage dependence upon the drain and substrate bias. Quite recently, Horstmann, Hilleringmann and Goser [105], who investigated the global and local matching of sub-100 nm NMOS- and PMOS-transistors, confirmed the law of area given in (2.91). Also, Stolk et al. [106] generalized the analytical result by Mizuno and his co-workers by taking into account the finite thickness of the inversion layer, depth-distribution of charges in the depletion layer and the influence of the source and drain dopant distributions and depletion regions. For a uniform channel dopant distribution, the analytical expression for the threshold voltage standard deviation given in [107] simplifies to

$$\sigma_{vt} = \left(\frac{\sqrt[4]{q^3\varepsilon_s\phi_b}}{\sqrt{3}}\right)\left[\frac{k_bT}{q}\cdot\frac{1}{\sqrt{4\varepsilon_s\phi_b N_a}}+\frac{T_{ox}}{\varepsilon_{ox}}\right]\frac{\sqrt[4]{N}}{\sqrt{L_{eff}W_{eff}}}. \tag{2.92}$$

In (2.92), the first term in the square brackets represents the surface potential fluctuations whereas the second term represents the fluctuations in the electric field.

The purpose of this section is twofold. First, we will clarify some issues related to the origin of the threshold voltage fluctuations in ultra-small devices. The second, and more important issue discussed here is how discrete impurities affect device high-field characteristics, such as carrier drift velocity and the on-state currents in conventional MOSFETs.

The Role of the Short-Range e–e and e–i Interactions

To be able to study the effect of the proper inclusion of the short-range Coulomb force to the mesh force, the energy and position of several electrons were monitored during a simulation run. The simulated device has channel length $L_G = 80$ nm, channel width $W_G = 80$ nm and oxide thickness $T_{ox} = 3$ nm. The lateral extension of the source and drain regions is 50 nm. The channel doping equals $3 \times 10^{18}\,\text{cm}^{-3}$. The applied bias is $V_G = V_D = 1$ V. Only those electrons that entered the channel region from the source side were "tagged" and their energy and position was monitored and used in the average energy calculation. The average velocity and the average energy of the electrons that reach the drain end of the device is shown in Fig. 2.49. From the average velocity simulation results, it follows that the short-range electron–electron (*e–e*) and electron–ion (*e–i*) interaction terms damp the velocity overshoot effect, thus increasing the transit time of the carriers through the device, in turn reducing its cut-off frequency (Fig. 2.49a). It is also quite clear that when we use the mesh force only, i.e. we skip the molecular dynamics (MD) loop that allows us to correct for the short-range *e–e* and *e–i* interactions, those electrons that enter the drain end of the device from the channel never reach equilibrium (Fig. 2.49b). Their average

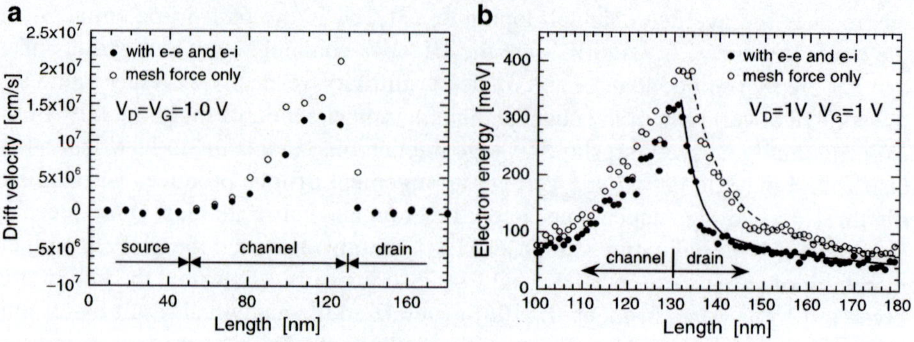

Fig. 2.49 (**a**) Average velocity of the electrons along the channel, with and without the inclusion of the *e–e* and *e–i* interactions. (**b**) Average energy of the electrons coming to the drain from the channel. The applied bias equals $VD = VG = 1$ V. *Filled (open) circles* correspond to the case when the short-range $e - e$ and $e - i$ interactions are included (omitted) in the simulations

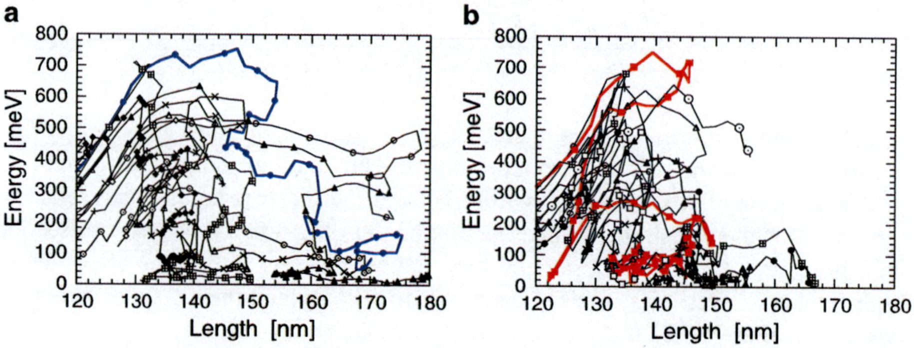

Fig. 2.50 (**a**) Phase-space trajectories of ten randomly chosen electrons for the case when the mesh force is only considered in the free-flight portion of the simulator. (**b**) Phase-space trajectories of ten randomly chosen electrons for the case when the short-range e–e and e–i interactions are included via our MD routine

energy is more than 60 meV far into the drain region. Also, the average energy peaks past the drain junction. The addition of the short-range Coulomb forces to the mesh force via the MD loop, leads to rapid thermalization of the carriers once they enter the drain region. The characteristic distance over which carriers thermalize is on the order of a few nm.

In Fig. 2.50, we show the phase-space trajectory of 10 randomly selected electrons that reach the drain region. We use $V_G = 0.5$ V, $V_D = 0.8$ V, $T_{ox} = 3$ nm, and $N_A = 3 \times 10^{17}\,\mathrm{cm}^{-3}$ in these simulations. Notice that some of the electrons reach the end of the device and are reflected back without losing much energy when we use the mesh force only (Fig. 2.50a). The addition of the short-range Coulomb force leads to very fast thermalization of the carrier energy once they enter the drain end (Fig. 2.50b). None of the randomly selected electrons reach the device boundary, as opposed to 3 out of 10 electrons reaching the boundary when the short-range Coulomb force is turned off.

Threshold Voltage Fluctuations

The threshold voltage fluctuations vs. device gate width, channel doping and oxide thickness, are shown in Fig. 2.51. Also shown in this figure are the analytical model predictions given by (2.91) and (2.92). The decrease of the threshold voltage

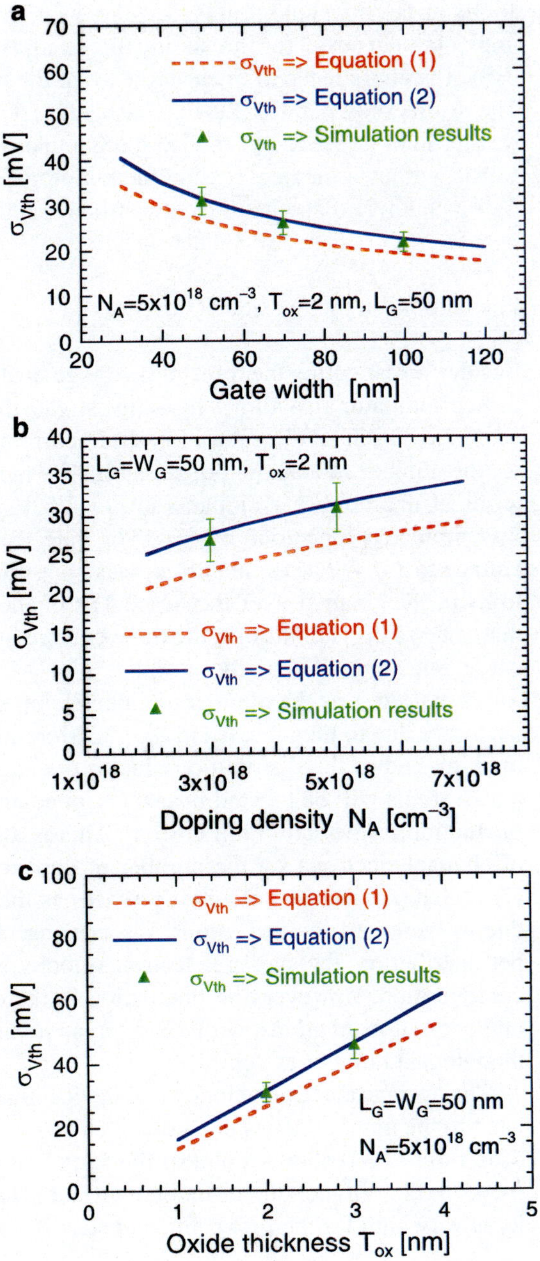

Fig. 2.51 Variation of the threshold voltage with (**a**) gate width, (**b**) channel doping, and (**c**) oxide thickness

fluctuations with increasing the width of the gate is due to the averaging effects, in agreement with the experimental findings by Horstmann et al. [82]. We want to point out that we still observed significant spread of the device transfer characteristics along the gate voltage axis even for devices with $W_G = 100$ nm. This is due to the nonuniformity of the potential barrier, which allows for early turn-on of some parts of the channel. As expected, the increase in the channel doping leads to larger threshold voltage standard deviation $\sigma_{V_{TH}}$. These results also imply that the fluctuations in the threshold voltage can be even larger in devices in which counter ion implantation is used for threshold voltage adjustments. Similarly, the increase in the oxide thickness leads to linear increase in the threshold voltage standard deviation. The results shown in Fig. 2.51a–c also suggest that reconstruction of the established scaling laws is needed to reduce the fluctuations in the threshold voltage. In other words, within some new scaling methodology, T_{ox} should become much thinner, or N_A much lower that what the conventional scaling laws give.

Fluctuations in the On-State Currents

Besides investigating the threshold voltage fluctuations, our 3D EMC particle-based device simulator also allows us to investigate the fluctuations in the high-field characteristics, such as the saturation drain current. The variation of the drain current vs. the number of channel dopant atoms for the 15 devices from [107] described in terms of the number of dopants in Fig. 2.52a, is shown in Fig. 2.52c. Each device was simulated for a total of 4 ps. The gate voltage was set to 1.5 V and the drain voltage to 1.0 V. The drain current was measured by averaging the velocity of electrons in the channel over the last 2.4 ps of the simulation. It is important to note that at these bias conditions, the devices were in the saturation region of the $I_D - V_G$ curve, but were not velocity saturated.

As expected, as the number of channel dopant atoms increases, the drain current decreases due to the increase in the V_T. More importantly, for the five devices from the high-end of the distribution, due to the larger probability that some of the impurity atoms will be located near the semiconductor/oxide interface, there is larger fluctuation in the saturation current. This is also reflected in the average velocity of channel electrons vs. the number of dopant atoms in the channel, as shown in Fig. 2.52d. Again, the velocity decreases as the number of dopant atoms increases due to increased ionized impurity scattering. At the low end of the dopant number distribution, the average electron velocity is roughly the same for each dopant configuration. However, the fluctuation in the electron velocity increases with the number of dopant atoms, with a $3\times$ spread in the velocity seen for the devices at the high dopant number extreme.

The average electron velocity and device drain current characteristics were correlated to the number of dopant atoms in a 10 nm range at various depths. Figure 2.52 (Top right panel) shows a plot of the square of the correlation coefficient vs. depth (beneath the semiconductor/oxide interface). The correlation to the electron velocity is very high for the first 6 nm, and steadily decreases up to 18 nm depth, beyond

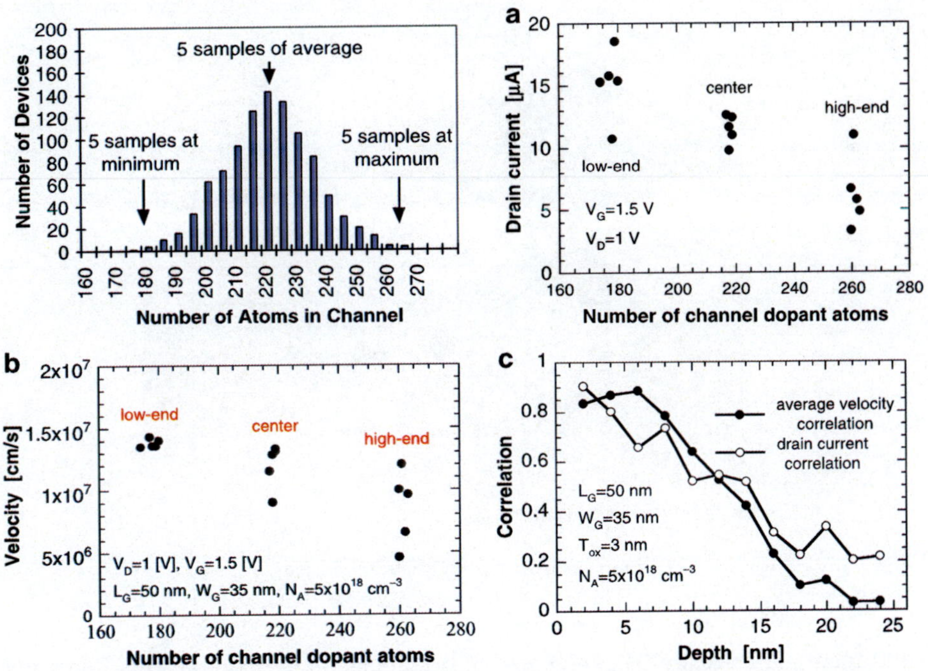

Fig. 2.52 *Top left*: Histogram of the number of dopant atoms in the channel for a population of 1,000 devices. *Top right*: Correlation of the drain current and average electron velocity to the number of dopant atoms within a 10 nm range at various depths beneath the channel. *Bottom left*: Drain current vs. the number of channel dopant atoms. *Bottom right*: Average velocity of channel electrons vs. the number of channel dopant atoms

which the correlation is nearly zero. It appears that only the dopant atoms in the first 6–10 nm from the semiconductor/oxide interface have significant effect on the velocity. This is reinforced by the fact that the correlation nearly goes to zero at a depth of 18 nm, as opposed to the threshold voltage correlation, which remains fairly high at a larger depth. The correlation of the drain current to the number of dopant atoms is also high near the surface, but the drop-off is not as steep as the velocity correlation. Beyond 18 nm depth, the correlation of the drain current is non-zero due to the correlation of the threshold voltage to the number of dopant atoms (see previous discussion).

3.3.2 Threshold Voltage Fluctuations Due to Unintentional Doping in Narrow-Width SOI Device Structures

The SOI device structure that has been simulated in this work to study comprehensively the effects of quantum mechanical size-quantization and discrete/unintentional dopant effects on the performance of nanoscale devices is shown in Fig. 2.53. It consists of a thick (600 nm) silicon substrate, on top of which is grown

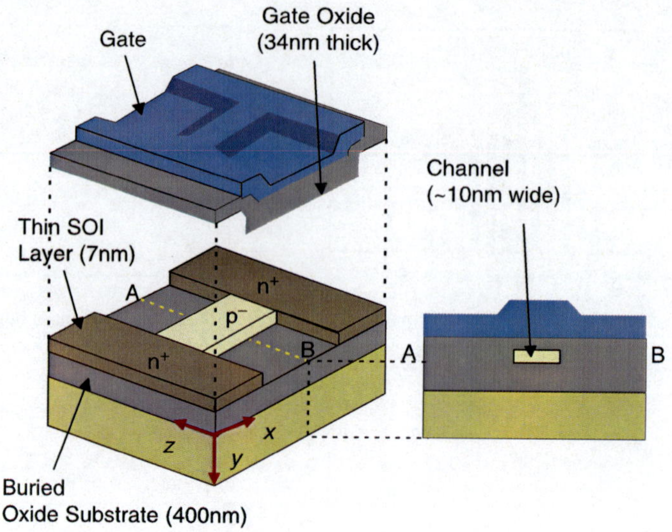

Fig. 2.53 Device structure of ultra-narrow channel FD-SOI device

400 nm of buried oxide. The thickness of the silicon on insulator layer is 7 nm, *with p^- region width of 10 nm* (if not stated otherwise) making it a fully-depleted device under normal operating conditions. The channel length is 50 nm and the doping of the p^- active layer is $10^{16}\,\mathrm{cm}^{-3}$ which corresponds to a nearly undoped channel region. The source/drain length is 15 nm, width being three times the channel width i.e. 30 nm. On top of the SOI layer sits the gate-oxide layer with a thickness of 34 nm. This is rather a thick gate oxide, but it is used to compare the simulation results with the experimental data of Majima et al. [108]. The doping of the source/drain junctions equals $10^{19}\,\mathrm{cm}^{-3}$ (if not stated otherwise), and the gate is assumed to be a metal gate with workfunction equal to the semiconductor affinity. The use of the low source-drain doping is justified by the fact that most of the carriers that are being simulated are residing in the source/drain regions and the reduction of the source/drain doping leads to a smaller ensemble of carriers. It has been found via Silvaco ATLAS Drift-Diffusion simulations of similar device structures that a reduction in the source/drain doping by one order of magnitude leads to approximately 20–30% decrease in the on-state current due to the additional source/drain series resistances.

In a 50 by 10 by 7 nm SOI device structure in Fig. 2.53, with a channel doping of $10^{16}\,\mathrm{cm}^{-3}$, one has merely a single dopant atom in the channel region. Even if the channel is undoped, the unavoidable background doping gives rise to at least one ionized dopant being present at a random location within the channel. Also, if an electron becomes trapped in a defect state at the interface, or in the active silicon body, it will introduce a fixed charge in the channel region. These potential sources of localized single charge will introduce a highly localized barrier to the carrier/current flow. Such a *localized barrier* is shown in Fig. 2.54. The device

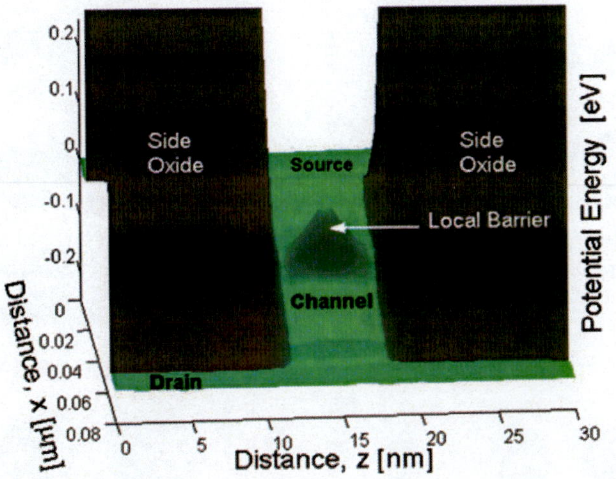

Fig. 2.54 Shape of the conduction band profile when a single impurity is localized in the center of the channel

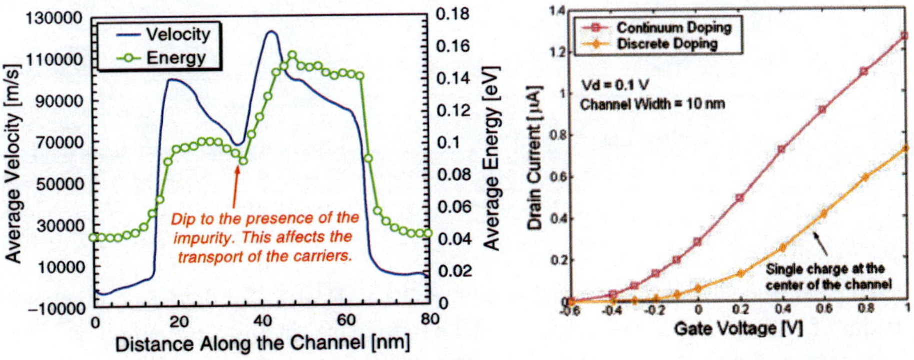

Fig. 2.55 *Left panel*: Velocity and energy plots for $V_G = 1.0$ and $V_D = 0.2$ V when a single impurity is present at the center of the channel. *Right panel*: Device transfer characteristics for the case of a continuum and discrete impurity model with a single charge at the center of the channel

operation is affected by this localized barrier from both electrostatics (effective increase in doping) and dynamics (transport) points of view. The transport is affected through modulation of carrier velocity and energy characteristics as shown in Fig. 2.55 (left panel) where the dip is due to the presence of a single impurity in the center of the channel region. In Fig. 2.55 (right panel), the device transfer characteristics are shown for a device with continuum doping and with an unintentional dopant present in the center of the channel. The channel width is 10 nm. One observes increase in the device threshold voltage V_{th} and degradation of the drain current due to the presence of a single charge.

In Fig. 2.56 shown are the fluctuations in the drain current as a function of the position of a single dopant ion in the channel region of the device. Simulations have

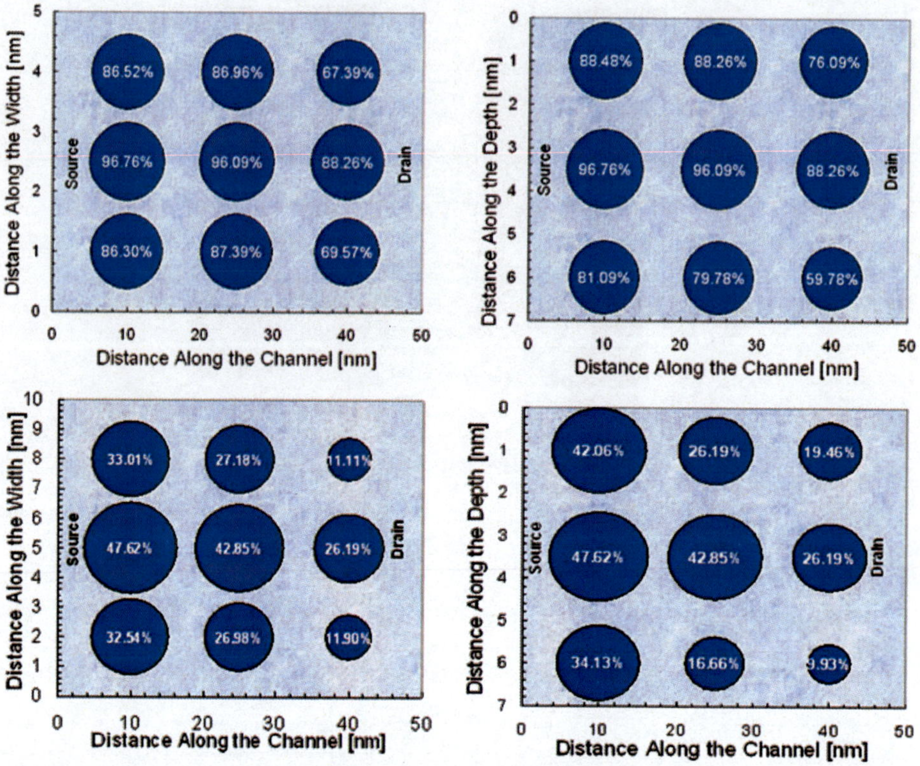

Fig. 2.56 Slicing of the region and corresponding variables in the slices

been performed using $V_G = 1.0$ V and $V_D = 0.1$ V. Results for devices with channel width of both 10 and 5 nm are shown. Due to the size-quantization effect, as a consequence of the charge set-back, results in the majority of current flowing through the middle portion of the channel. Thus a dopant ion trapped in the center region of the channel produces maximum fluctuations in the on-state current. The drain-end is less affected due to two reasons: (a) the presence of a weaker quantization effect therein due to the least vertical field experienced by the electrons and (b) the presence of the largest in-plane (x-component) electric field along the length of channel region which obviously minimizes the effect of the single dopant.

To investigate the impact of screening effect for the impurity positioned along the center of the channel region on the drain current detailed simulations were performed. The results are shown in Fig. 2.56. One can see that the impurity positioned in the very vicinity of the source-end has lower effect than when it is positioned a little away from the source-end. This is attributed to the fact that the very presence of a large number of electrons in the source region try to screen further the impurity and thereby its effect on the drain current.

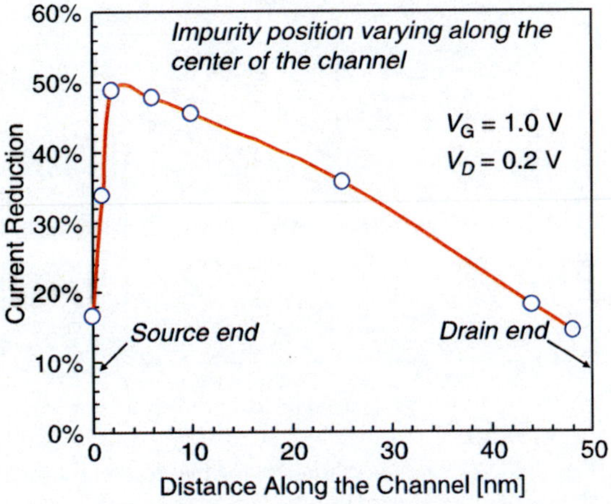

Fig. 2.57 Impact of screening on the drain current

The impurity position dependence of the drain current is shown in Fig. 2.57 (left panel) in the device output characteristics. There are several noteworthy conclusions that can be drawn from these simulations:

- Single impurity at the source-end of the channel affects the drain current the most.
- Impurities at the drain-end of the channel reduce the DIBL (drain-induced-barrier-lowering) in the output characteristics.
- Dopant atoms trapped in the center region of the channel produce the maximum fluctuations than the dopant atoms near the interface.

The observed impurity position dependence of the drain current may be attributed to both the inhomogeneities in the electrostatics and the non-uniform carrier quantization in the channel region. Another potential source arises from the modulation of the transport characteristics, which is reflected in the carrier velocity behavior as shown in the right panel of Fig. 2.58. Here, the velocity profiles for impurities at three different positions are shown. One can see that the impurity near the source end affects (reduces) the electron velocity most, throughout the channel region. Simulations have been performed using $V_G = 1.0\,\text{V}$ and $V_D = 0.2\,\text{V}$.

The results presented in Fig. 2.58 also suggest that there might be fluctuations in the device threshold voltage for devices fabricated on the same chip due to unintentional doping and random positioning of the impurity atoms. This can also be deduced from the scatter of the experimental data from [109]. The simulation results of the transfer characteristics with a single impurity present in different regions in the channel of the device, shown in the left panel of Fig. 2.59 clearly demonstrates the origin of the threshold voltage shifts for devices with 10 and 5 nm channel width. The width dependence of the threshold voltage for the case of a uniform (undoped) and a discrete impurity model is shown in the right panel of Fig. 2.59. This figure suggests that *both size-quantization effects and unintentional doping must be concurrently considered to explain threshold voltage variation in small devises.*

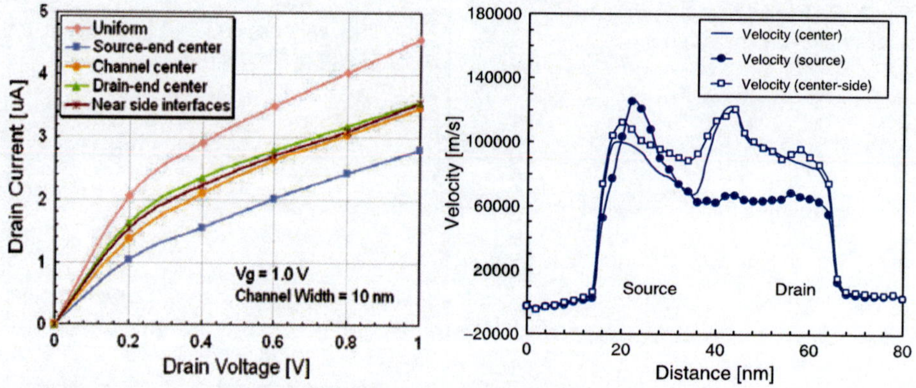

Fig. 2.58 *Left panel*: Variations of the device drain current as a function of the placement of a single impurity at various positions in the channel. We have used $V_G = 1.0\,\text{V}$ in these simulations. *Right panel*: Variations of the electron velocity as a function of the placement of a single impurity at various positions in the channel. We have used $V_G = 1.0\,\text{V}$ and $V_D = 0.2\,\text{V}$ in these simulations

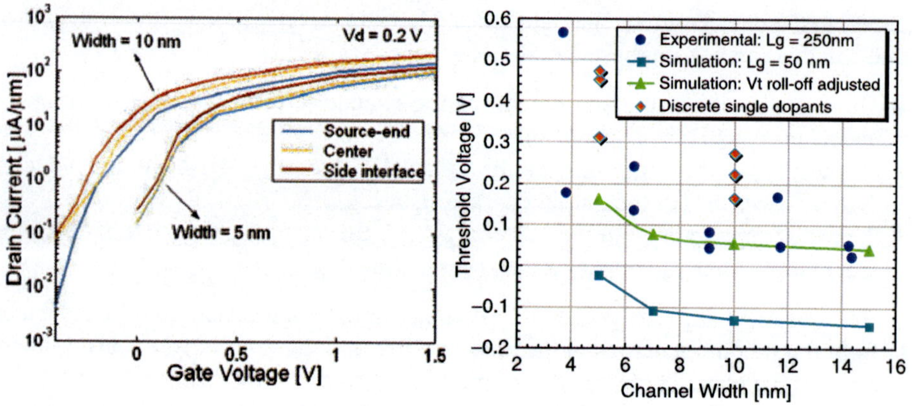

Fig. 2.59 *Left panel*: Transfer characteristics of the device with 10 and 5 nm channel widths and different location of the impurity atoms. We have used $V_G = 1.0\,\text{V}$ in these simulations. *Right panel*: Width dependence of the threshold voltage for the case of a uniform and a discrete impurity model. Clearly seen in this figure are two trends: **(a)** Threshold voltage increase with decreasing channel width due to quantum-mechanical size quantization effects, and **(b)** Scatter in the threshold voltage data due to unintentional doping

3.3.3 The Role of Unintentional Doping on FinFET Device Design Parameters

The FinFET device structure that has been simulated in this work is shown in Fig. 2.60 [109]. It consists of a thick (100 nm) buried oxide on top of which source/drain regions and a vertical fin are formed. The channel length is 40 nm with a gate length of 20 nm and a fin extension length of 10 nm on each side of the gate. The fin height and width are 30 and 10 nm, respectively. The source/drain length

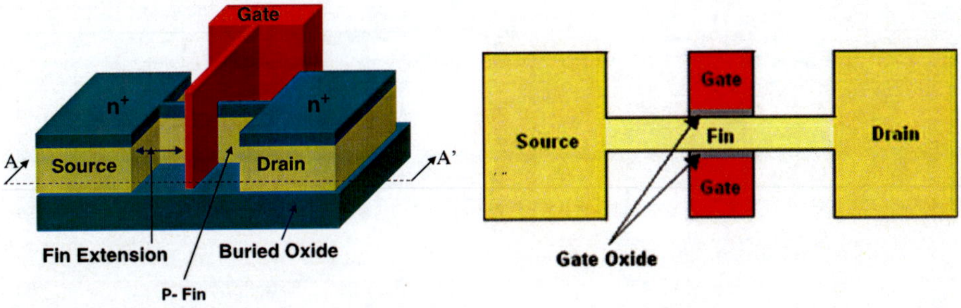

Fig. 2.60 *Left panel*: 3D schematic view of FinFET. *Right panel*: Top view of the FinFET shown in top panel along the cross section A-A'

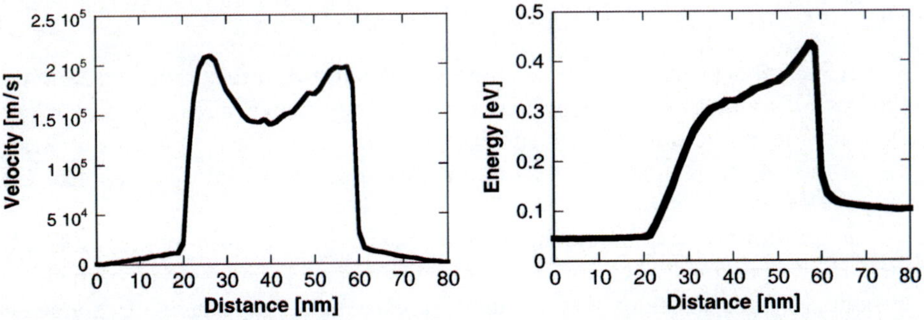

Fig. 2.61 *Left panel*: Average velocity (x-component) profile of carriers along the channel. *Right panel*: Average energy of carriers along the length of the device. $V_G = V_D = 0.8$ V and S/G = D/G = 10 nm

is 20 nm, the width being three times the channel width, i.e. 30 nm. The doping of the source/drain junctions equals $2 \times 10^{19}\,\text{cm}^{-3}$. The fin is assumed intrinsic. The gate is assumed to be n^+ polysilicon with work function equal to the semiconductor affinity. Gate oxide of 2.5 nm has been used for both side and top gates. To simulate this device structure, a convenient meshing scheme has been adopted. Meshing is uniform along the x (channel length) and z (width) directions and is non-uniform along the y (depth) direction, with the exception of the semiconductor region, where uniformity in meshing has been kept in order to facilitate the Monte Carlo transport simulations.

Significant velocity overshoot is observed in small geometry devices due to the presence of very high electric fields. Figure 2.61 (left panel) depicts the average velocity profile along the channel length of a FinFET device. Equal amount of velocity overshoot is observed near the source and the drain end of the channel when fin extension length on each side of the gate is equal. Note that the magnitude of the velocity overshoot also depends on the fin extension length on each side of the gate and this observation is discussed later in the text. Figure 2.61 (right panel) depicts the average energy profile along the device channel length. Near the source end the average carrier energy equals the thermal energy. Along the channel the average

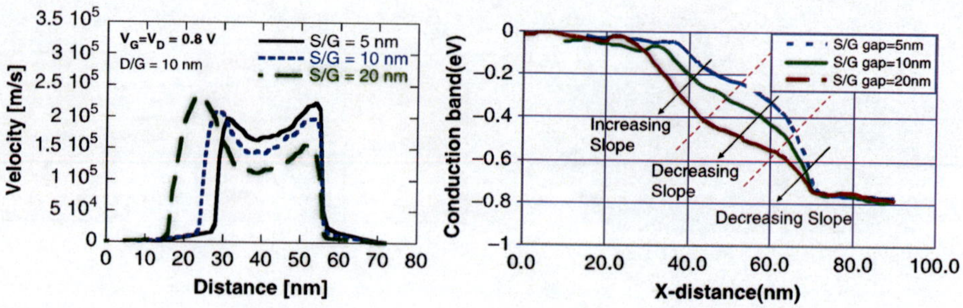

Fig. 2.62 *Left panel*: Average velocity (x-component) profile of carriers along the channel as a function of S/G gap. The applied bias equals $V_{\mathrm{G}} = V_{\mathrm{D}} = 0.8\,\mathrm{V}$. *Right panel*: Conduction band profile along x-direction

energy increases progressively reaching its peak value near the drain end. Note that carriers are not thermalized near the drain end of the channel due to the omission of the short-range electron–electron and electron–ion interactions in these simulations. Fin extension of 10 nm has been used on each side of the gate. The applied bias equals $V_D = V_G = 0.8\,\mathrm{V}$.

The amount of velocity overshoot the carriers experience within the FinFET devices shown previously heavily depends on the fin extension length on each side of the gate. Keeping D/G gap fixed, gradual increase in S/G gap causes the source end to experience more overshoot and the drain side overshoot to gradually diminish as shown in Fig. 2.62 (left panel). This is due to the fact that with an increase in extension length, source and drain lateral fields along the channel redistribute which changes the velocity profiles which can be seen from the 1-D conduction band profile along the x-direction as shown in Fig. 2.62 (right panel). Near the drain end and in the channel the slope of conduction band decreases with increase in S/G gap, resulting in lower electric field. Also note that near the source end the slope of conduction band increases giving higher electric field at that region. D/G gap is fixed at 10 nm and $V_D = V_G = 0.8\,\mathrm{V}$ is used in the simulation. The same phenomena happen for varying the D/G gap while keeping S/G gap constant at 10 nm.

From the transfer characteristics of the device as shown in Fig. 2.63 (left panel), it is evident that the threshold voltage is negative and is around $-0.1\,\mathrm{V}$. Negative threshold voltage results due to the use of n^+-polysilicon as a gate electrode. The metal work function equal to the electron affinity of Si is assumed in the simulation. Polysilicon gates also suffer from depletion and high gate resistance. A nominal threshold voltage of 0.2–0.4 V for n-channel FinFET can be achieved using metal gates with work function close to the mid band-gap of silicon ($\sim$4.6 eV). Achieving symmetric threshold voltages for both n-channel and p-channel FinFETs requires metals with different work functions [110]. The output characteristics of the device from Fig. 2.60 are presented in Fig. 2.63 (right panel). Equal fin extension of 10 nm is assumed on both sides of the gate. Gate voltage $V_G = 0.4\,\mathrm{V}$ is used. The inclusion of the electron–electron and electron–ion interaction results in lower drain current.

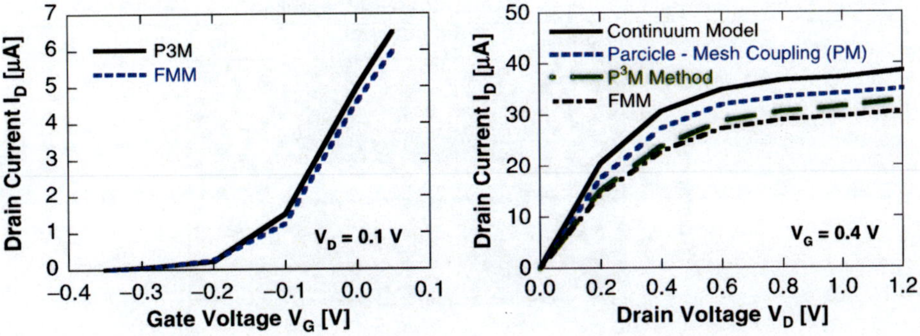

Fig. 2.63 *Left panel*: Transfer characteristics. *Right panel*: Output characteristics

Table 2.3 P^3M vs. FMM speed-up

Approach	CPU time per iteration (s)
P^3M	~24
FMM	<1

Also the Fast Multipole method (FMM) gives output characteristic which is in good agreement with that using the P^3M approach.

It is important to note that the CPU time requirement when using the FMM is much smaller compared to the traditional P^3M approach. Table 2.3, gives a comparison of the CPU time requirements for simulating FinFET device with a 3D mesh of $64 \times 24 \times 24$ node points. The number of particles simulated is around 1,500. The speedup due to using FMM depends on the number of particles, mesh size and computational resources. As the number of particles increases, FMM becomes slower but still much faster when compared to the P3M approach. Also for very small number of particles, it is better to calculate e–e and e–ion interaction directly than using FMM [111]. Correction for image charges is incorporated in our simulator to get the precise results.

FinFET devices use undoped or lightly doped fin. In a 40 by 10 by 30 nm channel region, with a channel doping of $10^{16}\,\text{cm}^{-3}$, one has merely 0.12 dopant atoms in the channel region. Even if the channel is undoped, the unavoidable background doping gives rise to at least one ionized dopant being present at a random location within the channel. Also, if an electron becomes trapped in a defect state at the interface or in the silicon body, it will introduce a fixed charge in the channel region. These potential sources of localized single charge will introduce a localized barrier to current flow. The position of a single dopant at the center of the channel along with the localized barrier it creates is shown in Fig. 2.64 (left and right panel). The device operation is affected by this localized barrier from both electrostatics (effective increase in doping) and dynamics (transport) points of view. The effective increase in doping in the channel region results in increase in the threshold voltage and consequently, the drain current reduces. The transport is affected through modulation of the carrier velocity and energy characteristics.

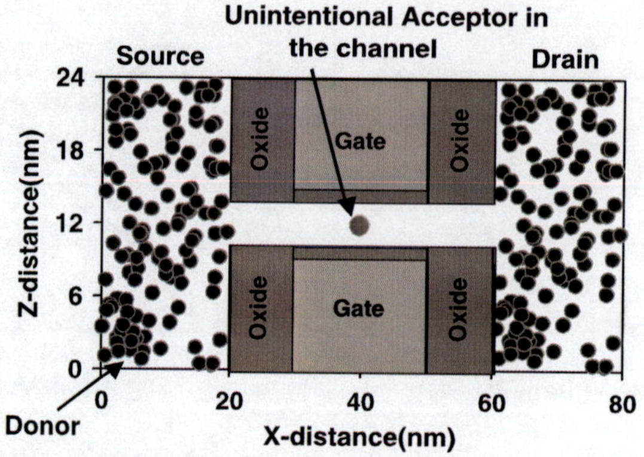

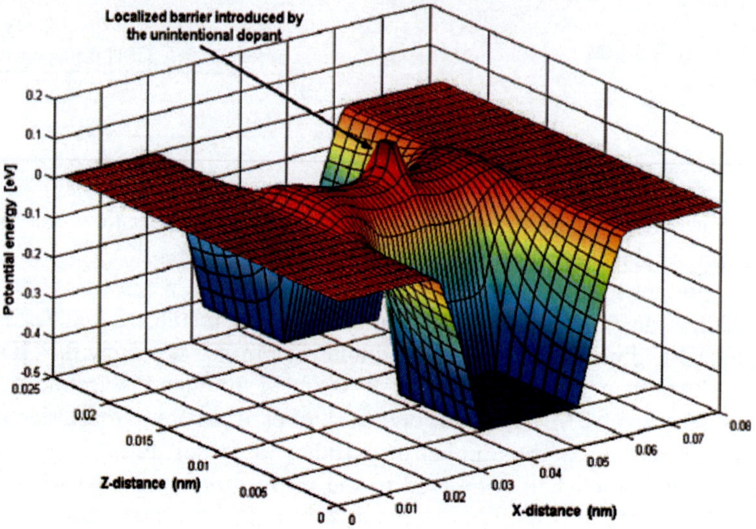

Fig. 2.64 *Left panel*: Top view of the FinFET device showing dopant position at the center region of the channel. *Right panel*: Potential profile showing the localized barrier introduced by the unintentional dopant

Due to the presence of multiple channels in the FinFET device, the effect of unintentional doping is not that much pronounced. The reduction in drain current heavily depends on the fin width. With decrease in fin width, the localized barrier has more pronounced effect on carrier motion through the channel, and the reduction in drain current is significant. This trend is schematically shown on the left panel of Fig. 2.65 . Fin extension length of 10 nm is used on each side of the gate. $V_D = 0.1\,\text{V}$, $V_G = 0.4\,\text{V}$ is used in the simulation. The unintentional dopant is placed near the source end close to the top interface. Fin extension length on each side also influences the reduction in drain current due to unintentional dopant as it is shown

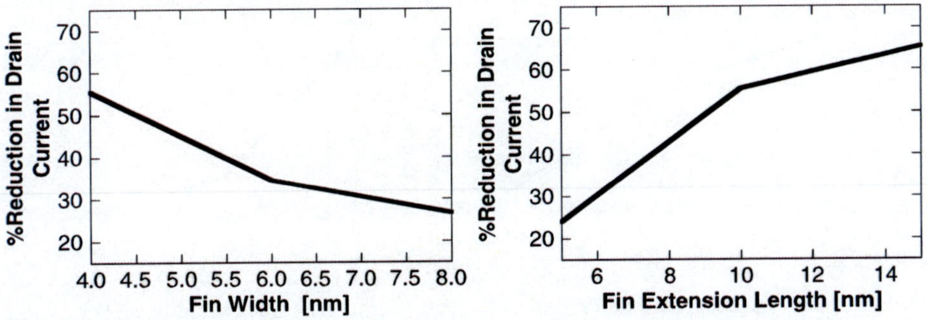

Fig. 2.65 *Left panel*: Reduction in drain current due to unintentional dopant as a function of fin width. $V_G = 0.4\,\text{V}$, $V_D = 0.1\,\text{V}$. *Right panel*: Reduction in drain current due to unintentional dopant as a function of fin extension length. $V_G = 0.4\,\text{V}$, $V_D = 0.1\,\text{V}$

in the right panel of Fig. 2.65. Longer fin extension results in more reduction in drain current than that due to smaller fin extension for any dopant position. With longer fin extension, lateral field from source and drain has less influence on the barrier produced by the unintentional dopant thereby, reducing the drain current more when compared to the case with smaller fin extension. Fin extension length can therefore, be optimized for suppressing unintentional doping effects while keeping the drive current within required range. $V_G = 0.4\,\text{V}$ and $V_D = 0.1\,\text{V}$ is used. The dopant atom is placed near the source end close to the top interface. Fin width of 4 nm is used. As noted in earlier device structures, the reduction in drain current due to unintentional dopant significantly depends on the position of the dopant atom in the channel. It is found that dopant placed near the source end has greater effect on the drain current. Near the drain end, the effect is less pronounced. Since in FinFET devices channels are formed symmetrically in vertical plane on each side of the fin, placing the unintentional dopant near the center along the width will reduce drain current more than that caused by dopant for any other position.

The effect of unintentional doping on device operation is relatively strong near sub threshold regime/weak inversion when few carriers are present in the channel. Thus the presence of unintentional dopant in the channel is expected to affect the switching behavior of the device. Increasing either the gate voltage or the drain bias will reduce the effect. As the gate voltage is increased, the number of carriers in the channel region increases and screens the localized potential produced by the unintentional dopant as shown in the left panel of Fig. 2.66. Drain bias of 0.1 V is applied in the simulation. Unintentional dopant is placed at the center of the channel near the top interface. Similarly with increase in drain voltage carriers are accelerated more along the channel and thus, can easily overcome the localized barrier. Therefore the reduction in drain current gradually decreases with increasing drain bias as shown in the right panel of Fig. 2.66. Gate bias of 0.4 V is applied in the simulation. Dopant is placed near the source end of the fin close to the top interface.

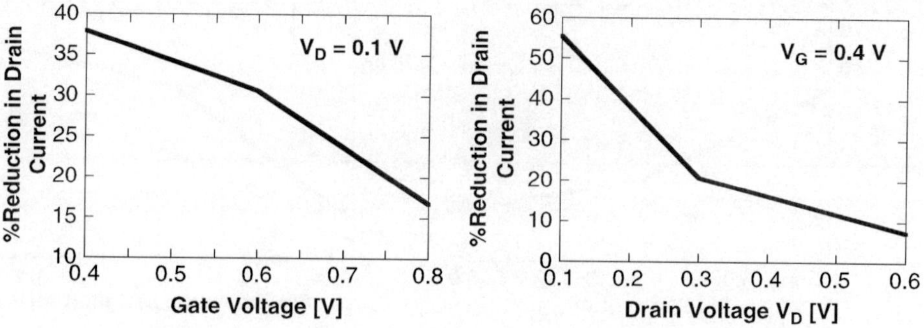

Fig. 2.66 *Left panel*: Screening behavior of the carriers on reduction of drain current due to unintentional dopant. *Right panel*: Reduction in drain current due to unintentional dopant as a function of drain voltage

4 Conclusions

A recently proposed *effective potential* approach has been utilized to successfully simulate two-dimensional space-quantization effects in a model of a narrow-channel SOI device structure. The incorporation of the *effective potential* approach into a full 3D Monte Carlo particle-based simulator allows one to investigate the device transfer and output characteristics with proper treatment of the size-quantization effects, velocity overshoot and carrier heating on an equal footing. The *effective potential* provides a set-back of the charge from the interface proper and quantization energy within the channel. Both of these effects lead to an increase in the threshold voltage. A threshold voltage increase of about 180 mV has been observed when the effective potential is included in the SOI device with 10 nm channel width. Also, observed is a pronounced channel width dependency of the threshold voltage which is termed as the *quantum mechanical narrow channel effect*. The width dependence of the threshold voltage is in close agreement with the experimental results. The increase in the threshold voltage is found to give rise to a significant on-state current reduction (20–30%), which depends upon the gate bias. Larger degradation is observed for larger gate voltages. The energy characteristics along the channel do not change with the inclusion of quantum mechanical size-quantization effects. The average drift velocity shows a small decrease due to the smearing of the potential.

A novel effective potential approach has been proposed and tested in the simulations of quantization effects in 25 nm nano-MOSFET device. The approach is parameter free as the size of the electron depends upon its energy. We have justified the correctness of the approach with simulations of the gate voltage dependence of the sheet electron density. The excellent agreement between the simulations and SCHRED results suggests that one is able to correctly predict the effective oxide thickness increase due to quantum-mechanical size-quantization effects that leads to a reduction of the sheet electron density. The nano-MOSFET simulation results also confirm this charge displacement effect near the source end of the channel where quantization effects play significant role. Due to the larger smearing of the potential for high energy electrons, we see a decrease in the carrier velocity when quantization effects are included in the model. This leads to a smaller drain current in both the device transfer and output characteristics. The charge displacement from the interface, and the effective increase of the oxide thickness, gives rise to a threshold voltage shift of $\sim$220 mV which is consistent with earlier observations. The shift in the threshold voltage leads in turn, to a drain current degradation of about 30%. Hence, the observations presented here that utilize the new effective potential approach, confirm that quantum-mechanical space-quantization effects must be included in the theoretical model to correctly predict the device behavior. In some cases, this can be achieved with the incorporation of the barrier field that is precomputed in the initial stages of the simulation and does not require additional CPU time during the simulation sequence. We believe that this new effective potential approach is more reliable in simulation of quantization effects in nano-scale devices with barriers that have different size and shape.

To treat the short-range Coulomb (electron–ion and electron–electron) interactions properly, *three* different but consistent real-space *molecular dynamics* (MD) schemes have been implemented in the simulator: the particle-particle-particle-mesh (P^3M) method, the corrected Coulomb approach and the Fast Multipole Method (FMM). *It is believed that the FMM algorithm has been used for the first time in the simulations of semiconductor devices.* The correctness of the approaches is verified via the simulations of the doping dependence of the low-field electron mobility in a 3D resistor and through its comparison with available experimental data. These approaches are then applied in the investigations of the role of unintentional doping on the operation of narrow-width SOI devices. We find significant correlation between the location of the impurity atom and the magnitude of the drain current. Namely, impurities near the source end of the channel have maximum influence on the drain current. This observation suggests that one has to take into account transistor mismatches due to unintentional doping when performing circuit designs. We have also investigated in depth the fluctuations in the threshold voltage due to discrete distribution of the impurity atoms in narrow width SOI devices with 10 and 5 nm channel width. The simulated data for the threshold voltages are in perfect agreement with the experimental values and they explain the fluctuations in the experimentally derived threshold voltage data.

Another device structure that has been investigated regarding the influence of the discrete impurities is the FinFET. Among different double gate structures FinFET attracts the researchers due to its inherent immunity to short channel effects and ease of fabrication using the existing planar fabrication process flow. Single fin FinFET can easily be extended to multiple fin structure for higher drive current. Again, in this structure as well, we find significant correlation between the magnitude of the drain current and the position of the discrete dopant for the case when screening effects do not play considerable role.

Acknowledgements We would like to thank the financial support from the National Science Foundation under Contract Number ECCS 0901251: Modeling Heating Effects in Low-Power Multi-Gate SOI Devices and High-Power GaN HEMTs. Program Director: Paul Werbos.

References

1. W. Hansch, Th. Vogelsang, R. Kirchner and M. Orlowski., "Carrier Transport Near the Si/SiO2 Interface of a MOSFET", *Solid State Elec.*, vol 32, no. 10, pp. 839–849, Oct. 1989.
2. M.J. Van Dort, PH. Woerlee and AJ. Walker, "A Simple Model for Quantization Effects in Heavily-Doped Silicon MOSFETs at Inversion Conditions.", *Solid State Elec.*, vol. 37, no. 3, pp. 411–415, Mar. 1994.
3. F. F. Fang and W. E. Howard, "Negative Field-Effect Mobility on (100) Si Surfaces", *Phys. Rev. Lett.*, vol. 16, no. 18, pp. 797–799, May. 1966.
4. B. Winstead and U. Ravaioli, "Simulation of Schottky barrier MOSFET's with a coupled quantum injection/Monte Carlo technique," *IEEE Trans. Electron Devices*, vol. 47, no. 6, pp. 1241–1246, Jun. 2000.
5. R. W. Keyes, "The effect of randomness in the distribution of impurity atoms on FET thresholds," *Appl. Phys.*, vol. 8, no. 3, pp. 251–259, Jun. 1975.
6. T. Mizuno, J. Okamura, and A. Toriumi, "Experimental study of threshold voltage fuctuation due to statistical variation of channel dopant number in MOSFET's," *IEEE Trans. Electron Devices*, vol. 41, pp. 2216–2221, Nov. 1994.
7. H. S. Wong and Y. Taur, "Three dimensional 'atomistic' simulation of discrete random dopant distribution effects in sub-0.1 mm MOSFET's, in *IEDM Tech. Dig.*, pp. 705–708, Dec. 1993.
8. W. J. Gross, D. Vasileska, and D. K. Ferry, "3-D Simulations of ultrasmall MOSFET's with real-space treatment of the electron–electron and electron–ion interactions," *VLSI Design*, vol. 10, pp. 437–452, no. 4, 2000.
9. A. Asenov, "Random dopant induced threshold voltage lowering and fluctuations in sub 0.1 micron MOSFETs: A 3D 'atomistic' simulation," *IEEE Trans. Electron Devices*, vol. 45, no. 12, pp. 2505–2513, Dec. 1988.
10. William J. Gross, *Ph. D. Dissertation*, Arizona State University, Dec. 2000.
11. N. Sano, K. Matsuzawa, M. Mukai, and N. Nakayama, "Role of longrange and short-range Coulomb potentials in threshold characteristics under discrete dopants in sub-0.1um Si-MOSFETs," *IEDM Tech., Dig.*, pp. 275–283, Dec. 2000.
12. D. K. Ferry, A. M. Kriman, M. J. Kann, and R. P. Joshi, "Molecular dynamics extensions of Monte Carlo simulation in semiconductor device modeling", *Comp. Phys. Comm.*, vol. 67, no. 1, pp. 119–134, Aug. 1991.
13. L. R. Logan and J. L. Egley, "Dielectric response in p-type silicon: Screening and band-gap narrowing", *Phys. Rev. B*, vol. 47, no. 19, pp. 12532–12539, May. 1993.
14. C. Jacoboni and P. Lugli, *"The Monte Carlo Method for Semiconductor Device Simulation."*, Vienna, Austria: Springer-Verlag, 1989.
15. R. H. Dennard, F. H. Gaensslen, H.-N. Yu, V. L. Rideout, E. Bassous and A. R. leBlanc, "Design of ion-implanted MOSFET's with very small physical dimensions", *IEEE J. Solid-State Circuits*, vol. 9, pp. 256, 1974.
16. J. R. Brews, W. Fichtner, E. H. Nicollian and S. M. Sze, "Generalized guide for MOSFET miniaturization", *IEEE Electron Dev. Lett.*, vol 1, no. 2, pp. 2, Jan. 1980.
17. G. Bacarani and M. R. Worderman, "Transconductance degradation in thin-Oxide MOS-FET's", *Electron Devices Meeting*, pp. 278–281, (1982).
18. M.-S. Liang, J. Y. Choi, P.-K. Ko and C. Hu, "Inversion-Layer Capacitance and Mobility of Very Thin Gate-Oxide MOSFET's", *IEEE Trans. Electron Devices*, vol. 33, no. 3, pp. 409–413, Mar. 1986.
19. A. Hartstein and N. F. Albert, "Determination of the inversion-layer thickness from capacitance measurements of metal-oxide-semiconductor field-effect transistors with ultrathin oxide layers", *Phys. Rev. B*, vol. 38, no. 2, pp. 1235–1240, Jul. 1988.
20. M. J. van Dort, P. H. Woerlee, A. J. Walker, C. A. H. Juffermans and H. Lifka, "Influence of high substrate doping levels on the threshold voltage and the mobility of deep-submicrometer MOSFETs", *IEEE Trans. Electron Dev.*, vol. 39, no. 4, pp. 932–938, 1992.
21. M. J. van Dort, P. H. Woerlee and A. J. Walker, "A simple model for quantisation effects in heavily-doped silicon MOSFETs at inversion conditions", *Solid-State Electronics* **37**, 411 (1994).

22. D. Vasileska, and D.K. Ferry, "The influence of space quantization effects on the threshold voltage, inversion layer and total gate capacitance in scaled Si-MOSFETs," *Technical Proceedings of the First International Conference on Modeling and Simulation of Microsystems, Semiconductors, Sensors and Actuators, Santa Clara, California*, pp. 408–413, Apr. 1998.
23. S. Takagi and A. Toriumi, "Quantitative understanding of inversion-layer capacitance in Si MOSFET's", *IEEE Trans. Electron Devices*, vol. 42, no. 12, pp. 2125–2130, Dec. 1995.
24. S. A. Hareland, S. Krishnamurthy, S. Jallepali, C. F. Yeap, K. Hasnat, A. F. Tasch Jr. and C. M. Maziar, "A computationally efficient model for inversion layer quantization effects in deep submicron N-channel MOSFETs", *IEEE Trans. Electron Devices*, vol. 43, no. 1, pp. 90–96, Jan. 1996.
25. D. Vasileska, D. K. Schroder and D. K. Ferry, "Scaled silicon MOSFETs: degradation of the total gate capacitance", *IEEE Trans. Electron Devices*, vol. 44, no. 4, pp. 584–587, 1997.
26. K. S. Krisch, J. D. Bude and L. Manchanda, "Gate capacitance attenuation in MOS devices with thin gate dielectrics", *IEEE Electron Dev. Lett*, vol. 17, no. 11, pp. 521–524, Nov. 1996.
27. L. de Broglie, *C. R. Acad. Sci. Paris*, vol. 183, 447, 1926.
28. L. de Broglie, *C. R. Acad. Sci. Paris*, vol. 184, 273, 1927.
29. E. Madelung, "Quantum theory in hydrodynamical form", *Z. Phys.*, 40, 322, 1926.
30. D. Bohm, "A Suggested Interpretation of the Quantum Theory in Terms of "Hidden" Variables. I", *Phys. Rev.*, 85, no. 2, 166–179, Jan. 1952.
31. D. Bohm, "A suggested interpretation of the quantum theory in terms of hidden variables. II", *Phys. Rev.*, Vol. 85, 180 (1952).
32. C. Dewdney and B. J. Hiley, "A Quantum Potential Description of One-Dimensional Time-Dependant Scattering From Square Barriers and Square Wells", *Found. Phys.*, vol. 12, no. 1, pp. 27–48, Jan. 1982.
33. G. J. Iafrate, H. L. Grubin, and D.K Ferry, "Utilization of Quantum Distribution Functions for Ultra-Submicron Device Transport", *Journal de Physique.*, vol. 42 (Colloq. 7), 10, 307–312, Oct. 1981.
34. E. Wigner, "On the Quantum Correction For Thermodynamic Equilibrium", *Phys. Rev.*, vol. 40, no. 5, pp. 749–759, Jun. 1932.
35. D. K. Ferry and J.-R. Zhou, "Form of the quantum potential for use in hydrodynamic equations for semiconductor device modeling", *Phys. Rev. B.*, vol. 48, no. 11, pp. 7944–7950, Sep. 1993.
36. P. Feynman and H. Kleinert, "Effective classical partition functions", *Phys. Rev. A*, 34, no. 6, pp. 5080–5084, Dec. 1986.
37. C. L. Gardner and C. Ringhofer, "Smooth quantum potential for the hydrodynamic model", *Phys. Rev. E*, vol. 53, no. 1, pp. 157–166, Jan. 1996.
38. C. Ringhofer and C. L. Gardner, "Smooth quantum hydrodynamic model simulation of the resonant tunneling diode", *VLSI Design*, vol. 8, 1–4, 143–146, 1998.
39. D. Vasileska and S. S. Ahmed, "Narrow-Width SOI Devices: The Role of Quantum–Mechanical Size Quantization Effect and Unintentional Doping on the Device Operation", *IEEE Trans. Electron Devices*, vol. 52, no. 2, pp. 227–236, Feb. 2005.
40. D. K. Ferry, "The onset of quantization in ultra-submicron semiconductor devices", *Superlattices and Microstructures*, vol. 27, no. 2–3, pp. 61–66, Jan. 2000.
41. C. Ringhofer, S. Ahmed and D. Vasileska, "An effective potential approach to modeling 25 nm MOSFET devices", *Journal of Computational Electronics*, vol. 2, pp. 113–117, 2003.
42. C. Ringhofer, C. Gardner and D. Vasileska, "Effective potentials and quantum fluid models: a thermodynamic approach", *Inter. J. on High Speed Electronics and Systems*, vol. 13, no. 3, pp. 771–804, Jan. 2003.
43. Shaikh Shahid Ahmed, "Quantum and Coulomb Effects in Nanoscale Devices", Ph. D. Dissertation, *Arizona State University*, Dec. 2004.
44. R. Akis, S. Milicic, D. K. Ferry, D. Vasileska, "An Effective Potential Method for Including Quantum Effects Into the Simulation of Ultra-Short and Ultra-Narrow Channel MOSFETs", *Proceedings of the 4th International Conference on Modeling and Simulation of Microsystems, Hilton Head Island, SC*, pp. 550–3, Mar. 2001.

45. C. Ringhofer, S. S. Ahmed and D. Vasileska, "Effective potential approach to modeling of 25 nm MOSFET devices", *Superlattices and Microstructures*, vol. 34, no. 3–6, pp. 311–317, 2003.
46. http://www.intel.com
47. Y. Omura, S. Horiguchi, M. Tabe, and K. Kishi, "Quantum-mechanical effects on the threshold voltage of ultrathin-SOI nMOSFETs", *IEEE Elec. Device Lett.*, vol. 14, no. 12, pp. 569–571, Dec. 1993.
48. S. M. Ramey and D. K. Ferry, "Implementation of surface roughness scattering in Monte Carlo modeling of thin SOI MOSFETs using the effective potential", *IEEE Transactions on Nanotechnology*, vol. 2, no. 2, pp. 110–114, Jun. 2003.
49. S. Hasan, J. Wang, and M. Lundstrom, "Device design and manufacturing issues for 10 nm-scale MOSFETs: a computational study", *Solid–State Elect*, vol. 48, no. 6, pp. 867–875, 2004.
50. S. Datta, *"Electronic Transport in Mesoscopic Systems"*, Cambridge Studies in Semiconductor Physics Series, ISBN 0-521-59943-1, paperback, 1998.
51. D. Vasileska, S. M. Goodnick and Gerhard Klimeck, *Computational Electronics: Semiclassical and Quantum Transport Modeling*, CRC Press, June 2010.
52. P. Hohenberg and W. Kohn, "Inhomogeneous Electron Gas", *Phys. Rev.*, vol. 136, no. 3b, B864-B871, Nov. 1964.
53. Kohn, and L. J. Sham, "Self-Consistent Equations Including Exchange and Correlation Effects", *Phys. Rev.*, vol. 140, no. 4a, pp. A1133–A1138, Nov. 1965.
54. L. Hedin and B. I. Lundqvist, "Explicit local exchange-correlation potentials", *J. Phys. C*, vol. 4, no. 14, pp. 2064–2082, Mar. 1971.
55. C. Hu, S. Banerjee, k. Sadra, B.G. Streetman and R. Sivan, "Quantization Effects in Inversion Layers of PMOSFET's on Si (100) Substrates", *IEEE Electron Dev. Lett.*, vol. 17, no. 6, pp. 276–278, Jun. 1996
56. S. Takagi, M. Takayanagi, and A. Toriumi, "Characterization of Inversion-Layer Capacitance of Holes in Si MOSFET's", *IEEE Trans. Electron Devices*, vol. 46, no. 7, pp. 1446–1450, Jul. 1999.
57. D. Vasileska, D. K. Schroder and D.K. Ferry, "Scaled silicon MOSFET's: Part II-Degradation of the total gate capacitance", *IEEE Trans. Electron Devices*, vol. 44, no. 4, pp. 584–587, Apr. 1997.
58. D. Vasileska, and D.K. Ferry, "The influence of space quantization effects on the threshold voltage, inversion layer and total gate capacitance in scaled Si-MOSFETs", *Technical Proceedings of the First International Conference on Modeling and Simulation of Microsystems, Semiconductors, Sensors and Actuators, Santa Clara, California*, vol. 10, no. 2, pp. 408–413, Apr. 1998.
59. J. Fossum, Z. Ren, K. Kim and M. Lundstrom "Extraordinarily High Drive Currents in Asymmetrical Double-Gate MOSFETs", *Superlattices and Microstructures*, vol. 28, no. 5–6, pp. 525–530, Jun. 2000.
60. J. P. Colinge, X. Baie, V. Bayot, and E. Grivei, "A silicon-on-insulator quantum wire," *Solid-State Electron.*, vol. 39, no. 1, pp. 49–51, Jan. 1996.
61. X. Huang, W. C. Lee, C. Kuo, D. Hisamoto, L. Chang, J. Kedzierski, E. Anderson, H. Takeuchi, Y. K. Choi, K. Asano, V. Subramanian, T. J. King, J. Bokor, and C. Hu, "Sub 50-nm FinFET: PMOS," in *IEDM Tech. Dig.*, pp. 67–70, Dec. 1999.
62. Z. Jiao and C. A. T. Salama, "A fully depleted _-channel SOI nMOSFET," *Electrochem. Soc. Proc.*, vol. 3, pp. 403–408, 2001.
63. J. P. Coolinge, M. H. Gao, A. Romano, H. Maes, and C. Claeys, "Silicon- on-insulator "gate-all-around" MOS device," *SOI Conf. Dig.*, pp. 137–138, 1990.
64. D. Hisamoto, T. Kaga, Y. Kawamoto, and E. Takeda, "A fully depleted lean-channel transistor (DELTA)—A novel vertical ultra-thin SOI MOSFET," in *IEDM Tech. Dig.*, pp. 833–836, Dec. 1989.
65. C. P. Auth and J. D. Plummer, "A simple model for threshold voltage of surrounding-gate MOSFETs,", *IEEE Trans. Electron Devices*, vol. 45, no.11, pp. 2381–2383, Nov. 1998.

66. T. Sekigawa and Y. Hayashi, "Calculated threshold voltage characteristics of an XMOS transistor having an additional bottom gate," *Solid- State Electron.*, vol. 27, no. 8–9, pp. 827–828, Jan. 1984.
67. A. Rahman, M. S. Lundstrom, and A. W. Ghosh, "Generalized effective-mass approach for n-type metal-oxide-semiconductor field-effect transistors on arbitrarily oriented wafers", *Journal of applied physics*, vol. 97, no. 5, pp. 053702–053714, Feb. 2005.
68. A. Rahman, "Exploring new channel materials for nanoscale CMOS devices: A simulation approach", Ph.D. Dissertation, Purdue University.
69. I. H. Tan, G.L. Snider, L. D. Chang and E. L. Hu, "A self-consistent Solution of Schrödinger–Poisson Equations using a Non-uniform Mesh," *J. Appl. Phys.*, vol. 68, pp. 4071–4076, Oct. 1990.
70. T. Yang, Y. Liu, P.D. Ye, Y. Xuan, H. Pal, M. S. Lundstrom, "Inversion Capacitance-Voltage Studies on GaAs Metal-Oxide-Semiconductor Structure using Transparent Conducting Oxide as Metal Gate", *Applied Physics Letters*, vol. 92, pp. 252105–252108, Jun. 2008.
71. F. Gilibert, D. Rideau, F. Payet, F. Boeuf, E. Batail, M. Minondo, R. Bouchakour, T. Skotnicki, H. Jaouen, "Strained Si/SiGe MOSFET capacitance modeling based on band structure analysis", *Proceedings of the 35th European Solid State Device Research Conference (ESSDERC'2005), Grenoble*, no. 12–16, pp. 281–284, Sep. 2005.
72. L. Rayleigh, "On the propagation of waves through a stratified medium, with special reference to the question of reflection", *Proc. Roy. Soc. A*, vol. 86, no. 586, pp. 207, 1912.
73. ET. Jaynes. Probability Theory: The Logic of Science, Cambridge University Press, (2003).
74. Usuki T., Saito M., Takatsu M., Kiehl R.A., Yokoyama N.:*Numerical analysis of electron wave detection by a wedge shaped point contact*. Phys. Rev. B 520, 7615–7625 (1994).
75. Datta S.: *Nanoscale device modeling: the Green's function method*. Superlattices and Microstuctures 28, 253–278 (2000).
76. D. K. Ferry, *Quantum Mechanics for Electrical Engineers*, IOP Press (2000).
77. C. B. Duke, in Solid State Physics, edited by F. Seitz, D. Turnbull, and H. Ehrenreich ~ Academic, New York, 1969!
78. W. W. Lui and M. Fukuma "Exact solution of the Schrödinger equation across an arbitrary one-dimensional piecewise-linear potential barrier" *J. Appl. Phys.* 60, 1555–1559 (1986).
79. T. Mizuno, J. Okamura and A. Toriumi, "Experimental study of threshold voltage fluctuation due to statistical variation of channel dopant number in MOSFET's", *IEEE Trans. Electron Devices*, vol. 41, no. 11, pp. 2216–2221, Nov. 1994.
80. T. Mizuno, "Influence of Statistical Spatial-Nonuniformity of Dopant Atoms on Threshold Voltage in a System of Many MOSFETs", *Jpn. J. Appl. Phys*, vol. 35, pp. 842–848, Jan. 1996.
81. J. T. Horstmann, U. Hilleringmann and K. F. Goser, "Matching analysis of deposition defined 50-nm MOSFET's", *IEEE Trans. Electron Devices*, vol. 45, no. 1, pp. 299–306, Jan. 1998.
82. P. A. Stolk, F. P. Widdershoven and D. B. M. Klaassen, "Modeling statistical dopant fluctuations in MOS transistors", *IEEE Trans. Electron Devices*, vol. 45, pp. 1960–1971, Sep. 1998.
83. K. Nishinohara, N. Shigyo and T. Wada, "Effects of microscopic fluctuations in dopant distributions on MOSFET threshold voltage", *IEEE Trans. Electron Devices*, vol. 39, no. 3, pp. 634–639, Mar. 1992.
84. J.-R. Zhou and D. K. Ferry, "3D simulation of deep-submicron devices. How impurity atoms affect conductance", *IEEE Comput. Science and Eng.*, vol. 2, no. 2, pp. 30–36, May. 1995.
85. D. Vasileska, W. J. Gross, V. Kafedziski and D. K. Ferry, "Continuity Equations for Scaled Si MOSFETs", *VLSI Design*, vol. 8, no. 1–4, pp. 301, 1998.
86. D. Vasileska, W. J. Gross and D. K. Ferry, "Modeling of deep-submicrometer MOSFETs: random impurity effects, threshold voltage shifts and gate capacitance attenuation", *Extended Abstracts IWCE-6, Osaka, IEEE Cat.* No. 98EX116, pp. 259–262, 1998.
87. X. Tang, V. K. De and J. D. Meindl, "Intrinsic MOSFET parameter fluctuations due to random dopant placement", *IEEE Trans. on VLSI Systems*, vol. 5, no. 4, pp. 369–376, Dec. 1997.

88. P. Lugli and D. K. Ferry, "Degeneracy in the ensemble Monte Carlo method for high-field transport in semiconductors", *IEEE Trans. Electron Dev.*, vol. 32, no. 11, pp. 2431–2437, Nov. 1985.
89. A. M. Kriman, M. J. Kann, D. K. Ferry and R. Joshi, "Role of the exchange interaction in the short-time relaxation of a high-density electron plasma", *Phys. Rev. Lett.*, vol. 65, no. 13, pp. 1619–1622, Sep. 1990.
90. W. J. Gross, D. Vasileska, and D. K. Ferry, "3 D simulations of ultra-small MOSFETs with real-space treatment of the electron-electron and electron-ion interactions", *VLSI Design*, vol. 10, no. 4, pp. 437-, 2000.
91. D. Vasileska, W. J. Gross, and D. K. Ferry, "Monte Carlo particle-based simulations of deep-submicron n-MOSFETs with real-space treatment of electron-electron and electron-impurity interactions", *Superlattices and Microstructures*, vol. 27, no. 2–3, pp. 147–157, Feb. 2000.
92. A. Asenov, "Random dopant induced threshold voltage lowering and fuctuations in sub 0.1μm MOSFETs: A 3-D 'atomistic' simulation study", *IEEE Trans. Electron Dev.*, vol. 45, no. 12, pp. 2505–2513, Dec. 1998.
93. A. Asenov and S. Saini, "Suppression of random dopant-induced threshold voltage fluctuations in sub-0.1-μm MOSFET's with epitaxial and δ-doped channels", *IEEE Trans. Electron Dev.*, vol. 46, no. 8, pp. 1718–1724, Aug. 1999.
94. L. Greengard and V. Rokhlin, "A Fast Algorithm for Particle Simulations*1,*2", *J. Comput. Phys.*, vol. 135, no. 2, pp. 280–292, Aug 1997.
95. R. Beatson and L. Greengard, "A short course on fast multipole methods,", *Wavelets, Multi-level Methods and Elliptic PDEs (Leicester, 1996), ser. Numer. Math. Sci. Comput. New York: Oxford Univ. Press*, pp. 1–37, 1997.
96. H. Cheng, L. Greengard, and V. Rokhlin, "A fast adaptive multipole algorithm in three dimensions", *J. Comput. Phys.*, vol. 155, no. 2, pp. 468–498, Aug. 1999.
97. FMMPART3D user's guide, version 1.0 ed., *MadMax Optics*, Hamden, CT, USA.
98. R. W. Hockney and J. W. Eastwood, "Computer Simulation Using Particles", *New York, McGraw-Hill*, 1981.
99. C. J. Wordelman and U. Ravaioli, "Integration of a particle-particle-particle-mesh algorithm with the ensemble Monte Carlo method for the simulation of ultra-small semiconductor devices", *IEEE Tran. Electron Devices*, vol. 47, no. 2, pp. 410–416, Feb. 2000.
100. W. J. Gross, D. Vasileska, and D. K. Ferry, "Ultrasmall MOSFETs: the importance of the full Coulomb interaction on device characteristics", *IEEE Electron Devices*, vol. 47, no. 10, pp. 1831–1837, Oct. 2000.
101. Allen, D. Holberg, "CMOS Analog Circuit Design", *Saunders College Publishing, New York*, 1987.
102. Bohr, Y., A. El-Mansy, "Technology for advanced high-performance microprocessors",*IEEE Trans. Electron Dev.*, vol. 45, no. 3, pp. 620–625, Mar. 1998.
103. SIA Technology Roadmap of Semiconductors: http://www.itrs.net/
104. E. H. Nicollian and A. Goetzberger, "The Si-SiO2 interface-electrical properties as determined by the metal-insulator-silicon conductance technique", *Bell Syst. Techn. J.*, vol. 46, no. 6, pp. 1055–1133, 1967.
105. J. T. Horstmann, U. Hilleringmann and K. F. Goser, "Matching analysis of deposition defined 50-nm MOSFET's", *IEEE Trans. Electron Devices*, vol. 45, no. 1, pp. 299–306, Jan. 1998.
106. P. A. Stolk, F. P. Widdershoven and D. B. M. Klaassen, "Modeling statistical dopant fluctuations in MOS transistors", *IEEE Trans. Electron Devices*, vol. 45, no. 1, pp. 1960–1971, Sep. 1998.
107. W. J. Gross, D. Vasileska and D. K. Ferry, "Three-dimensional simulations of ultrasmall metal–oxide–semiconductor field-effect transistors: The role of the discrete impurities on the device terminal characteristics", Journal of Applied Physics, vol. 91, no. 6, pp. 3737–3740, Mar. 2002.
108. H. Majima, H. Ishikuro, and T. Hiramoto, "Experimental evidence for quantum mechanical narrow channel effect in ultra-narrow MOSFET's", *IEEE Electron Dev. Lett.*, vol. 21, no. 8, pp. 396–398, Aug. 2000.

109. H.R. Khan, D. Vasileska, S.S. Ahmed, C. Ringhofer and C. Heitzinger, "Modeling of FinFETs: 3D MC Simulation Using FMM and Unintentional Doping Effects on Device Operation", *Journal of Computational Electronics*, Vol. 3, Nos. 3–4, pp. 337–340 (2005).
110. L. Chang, S. Tang, T.-J. King, J. Bokor, and C. Hu,"Gate length scaling and threshold voltage control of double-gate MOSFETs," *IEDM Tech. Dig.*, vol. 6, pp. 719–722, Dec. 2000.
111. Clemens Heitzinger, Christian Ringhofer, Shaikh Ahmed and Dragica Vasileska, "3D Monte-Carlo device simulations using an effective quantum potential including electron-electron interactions", DOI 10.1007/s10825-006-0058-x, *Journal of Computational Electronics*, Volume 6, Numbers 1—3/September, pp. 15—18, 2007.

Chapter 3
Semiclassical and Quantum Electronic Transport in Nanometer-Scale Structures: Empirical Pseudopotential Band Structure, Monte Carlo Simulations and Pauli Master Equation

Massimo V. Fischetti, Bo Fu, Sudarshan Narayanan, and Jiseok Kim

Abstract The study of electronic transport in nanometer-scale devices requires an accurate knowledge of the excitation spectrum (i.e., the band structure) of the systems and, for short devices, a formulation of transport which transcends the semiclassical Boltzmann formulation. Here we show that the use of 'judiciously' chosen empirical pseudopotentials, coupled to the supercell method, can provide a sufficiently accurate description of the band structure of thin semiconductor films, hetero-structures, nanowires, and carbon-based structures such as graphene, graphene nanoribbons, and nanotubes. We discuss semiclassical Monte Carlo simulations employing the supercell-pseudopotential band structure, considering transport in thin Si bodies. This example illustrates the importance of the full-band approach since in this case it yields the low value of the saturated high-field electron drift velocity, observed experimentally but never predicted when employing effective-mass band structures. Finally, we discuss a mixed envelope-supercell approach to deal with open systems within the full-band supercell scheme and review the Master-equation approach to quantum transport. Finally, we present some results of fully dissipative quantum transport using, for now, the effective mass approximation, emphasizing the role of impurity scattering in determining the 'quantum access resistance' in thin-body devices.

Keywords Pseudopotentials · Electron transport · Nanostructures

1 Introduction

The study of electronic transport in semiconductor devices at the nanometer scale involves several additional complications when compared to what is needed to handle transport in larger structures: The band structure is not necessarily well

M.V. Fischetti (✉)
Department of Materials Science and Engineering, University of Texas at Dallas, 800 W. Campbell Rd., Richardson, TX 75080, USA
e-mail: max.fischetti@utdallas.edu

D. Vasileska and S.M. Goodnick (eds.), *Nano-Electronic Devices: Semiclassical and Quantum Transport Modeling*, DOI 10.1007/978-1-4419-8840-9_3,

approximated by the bulk band structure, but it may depend on the geometry and size of the device itself. In addition, one may need a formulation of charge transport which goes beyond the semiclassical picture on which the Boltzmann transport equation rests. Finally, the physics of the collision processes may depend strongly on the structure itself because of the presence of localized phonons, of interfacial excitations, of non-ideal surfaces and interfaces, etc.

Here we outline a possible scheme to tackle the problems just mentioned. Regarding the band structure of the system, in Sect. 2 we will consider empirical local pseudopotentials and show that qualitatively (and, often, quantitative) accurate results can be obtained in a variety of systems of interest, such as thin semiconducting bodies, hetero-layers, nanowires, and also carbon-base structures (graphene, graphene nanoribbons, carbon nanotubes). The reasons for focusing on 'empirical' pseudopotentials is twofold: First, we wish to develop a general scheme based on accurate **k**-space band-structure methods leading to the solution of quantum transport in small devices. From this perspective, empirical or ab-initio pseudopotentials are equivalent algorithmically and numerically. Second, while clearly ab-initio methods potentially can provide the most accurate picture since ionic and charge redistribution can be accounted for, on the other hand their significant computational cost and questionable quantitative accuracy (remember that charge transport is often sensitive to energy changes of the order of the thermal energy, $k_B T$) may render them unsuitable in some cases. 'Empiricism' may allow us calibrate input parameters in order to fit experimental data, when available. Our work is obviously based on the standard 'supercell' idea.

Regarding open systems, in Sect. 3.1 we outline a Monte Carlo scheme to deal with semiclassical transport – that is, incoherent along the transport direction – employing the band structure obtained from the **k**-space approach described before, presenting the simple example of two-dimensional transport in thin Si films in the presence of an additional confining potential computed self-consistently with the band-structure calculations. Despite the simplicity of the example, it is worth to note that already in this simple case the accurate band-structure yields a result – an electron saturated velocity lower than in bulk Si – which cannot be obtained using the effective-mass approximation, even when corrected for nonparabolic effects.

Moving to quantum transport – i.e., coherent along the transport direction – in Sect. 3.2 we present a scheme to deal with open systems within the pseudopotential framework, paying attention to the rather complicated boundary conditions (explicitly dealt with only in the one dimensional case) already required even in the simpler case of ballistic transport. We then review the use of the Pauli Master equation to handle dissipative transport in short devices. Lacking at present quantum-transport results in the full-band case, we review some results regarding the effect of phonon and impurity scattering in *n-i-n* diodes, resonant tunnel diodes, and double-gate field-effect transistors (FETs).

2 A k-Space Full-Band (Supercell) Approach for Closed Systems

In this section we briefly discuss the use of local empirical pseudopotentials to calculate the band structure of nanoscale systems and present results regarding structures of current technological interest.

2.1 The Method

The use of plane waves and empirical pseudopotentials to calculate the band structure of semiconductors (and other crystals) has proven useful to gain insight into the electronic excitation spectrum of solids. The 'empirical' nature implies loss of strong predictive power and 'portability' of ionic (pseudo)potentials, but it results in a vast simplification of the numerical problem (compared to ab-initio methods) and the small degree of fitting allowed by the technique affords – by definition – excellent agreement with experimental data. Since our focus is on electronic transport, not on structure calculations, it represents the best choice to perform accurate calculations, much as it has been in the case of semiclassical transport in the device simulation code DAMOCLES [16].

The 'supercell' method is conceptually a trivial extension of the standard 'bulk solid' plane wave method. In the latter, the primitive cell of the crystal (containing only two atoms in the diamond- or zincblende-structure face-centered cubic, fcc, semiconductors of interest here) is considered. If, instead of bulk homogenous solids, we are interested in studying more complex structures, instead of employing the primitive lattice cell, a larger cell is considered, such as many Si cells replicated N times along the z axis to form a Si layer of thickness Na_0 (where a_0 is the Si lattice constant) and M empty cells ('vacuum'), resulting in a cell of total extension $(N+M)a_0$ along the z axis. This is the supercell required to deal with inversion layers, thin semiconductor bodies, or quantum wells. Similarly, the cell may be extended along two directions, mimicking a quantum wire. A sufficiently large number of vacuum or insulating cells between adjacent layers or wires will guarantee avoiding artifacts due to the possible coupling between them (as in a superlattice). An external potential can be added by considering it extended periodically with the period of the supercell. Thus its non-vanishing Fourier components will be of the form:

$$\Phi_{\mathbf{G}} = \frac{1}{\Omega_{sc}} \int_{\Omega_{sc}} d\mathbf{r}\, \Phi(\mathbf{r})\, e^{-i\mathbf{G}\cdot\mathbf{r}}, \tag{3.1}$$

where Ω_{sc} is the volume of the supercell and $\mathbf{G}$ are the vectors of the reciprocal lattice. In the case of inversion layers, quantum wells, and thin bodies, $\Phi(\mathbf{r})$ depends only on z, so that:

$$\Phi_{\mathbf{G}} = \delta_{\mathbf{G}_\parallel}\, \Phi_{G_z} = \delta_{\mathbf{G}_\parallel} \frac{1}{L_z} \int_0^{L_z} dz\, \Phi(z)\, e^{-iG_z z}, \tag{3.2}$$

where L_z is the extension of the 1D supercell in the z direction and $\mathbf{G}_{\|}$ indicates the component of the reciprocal-lattice vector $\mathbf{G}$ on the plane parallel to the plane of the layer or film. In the case of graphene nanoribbons, nanowires, or carbon nanotubes $\Phi(\mathbf{r})$ will depend only on the in-plane coordinates $\mathbf{R}$, so that:

$$\Phi_{\mathbf{G}} = \delta_{G_z} \Phi_{\mathbf{G}_{\|}} = \delta_{G_z} \frac{1}{A} \int_A d\mathbf{R}\, \Phi(\mathbf{R})\, e^{-i\mathbf{G}_{\|}\cdot\mathbf{R}}, \tag{3.3}$$

where A is the cross-sectional area of the 2D supercell. The electronic structure of the system will be obtained by solving the eigenvalue problem:

$$\sum_{\mathbf{G}'} \left[\frac{\hbar^2}{2m} |\mathbf{k}+\mathbf{G}|^2 \delta_{\mathbf{G}\mathbf{G}'} + V^{(lat)}_{\mathbf{G}-\mathbf{G}'} + \Phi_{\mathbf{G}-\mathbf{G}'} \right] \phi^{(n)}_{\mathbf{k}\mathbf{G}'} = E_n(\mathbf{k})\, \phi^{(n)}_{\mathbf{k}\mathbf{G}}. \tag{3.4}$$

For perfect confinement the solution does not depend on the component(s) of the wavevector $\mathbf{k}$ along the confinement direction(s), so for 1D supercells we may think of $\mathbf{k}$ as given by $\mathbf{k} = (\mathbf{K}, 0)$, $\mathbf{K}$ representing the 2D wavevector on the plane perpendicular to the confinement direction, while for 2D supercells $\mathbf{k} = (\mathbf{0}, k_z)$, where k_z is the wavenumber along the direction perpendicular to the confinement (e.g., the axial direction of a nanowire). (We employ bold uppercase symbols for 2D vectors). Also, $V^{(lat)}$ is the lattice (pseudo)potential resulting from the sum over all ions α in the supercell of the ionic pseudopotentials $V^{(\alpha)}_{\mathbf{G}}$, normalized to each atomic volume Ω_α multiplied by the 'structure factor' $e^{i\mathbf{G}\cdot\boldsymbol{\tau}_\alpha}$:

$$V^{(lat)}_{\mathbf{G}} = \frac{1}{\Omega_{sc}} \sum_{\alpha} e^{-i\mathbf{G}\cdot\boldsymbol{\tau}_\alpha}\, \Omega_\alpha\, V^{(\alpha)}_{\mathbf{G}}, \tag{3.5}$$

where $\boldsymbol{\tau}_\alpha$ is the position of ion α in the supercell. The wavefunction corresponding to the eigenvalue $E_n(\mathbf{k})$ is given by the Bloch expression:

$$\psi^{(n)}_{\mathbf{k}}(\mathbf{r}) = e^{i\mathbf{k}\cdot\mathbf{r}} \sum_{\mathbf{G}} \phi^{(n)}_{\mathbf{k}\mathbf{G}}\, e^{i\mathbf{G}\cdot\mathbf{r}}. \tag{3.6}$$

This method is 'exact' (within the single-electron EP framework), but it can handle only closed systems. Thus, it is ideally suited to treat low-dimensionality confined cases, but it must be augmented by other techniques (such as the 'envelope' approximation considered in Sect. 3.2 below) when we must deal with transport (i.e., open boundary conditions) problems.

2.2 Local Empirical Pseudopotential Parameters

In the following subsections examples will be given regarding the use of the supercell method to study Si inversion layers and nanowires, III–V hetero-channels, graphene sheets and nanoribbons, and C nanotubes (CNTs). These structures have been treated by employing several forms of local empirical pseudopotentials presented in the literature. Since lack of 'portability' is one of the major drawbacks of the empirical nature of the technique, care must be taken in employing the model pseudopotential best suited to the system of interest.

For Si, Ge, and H we have employed the empirical pseudopotentials proposed by Zhang et al. [65]. These yield the correct workfunction (and so, band alignments), and they can be used to study free-standing Si layers bounded by vacuum. The surface states which may arise from the dangling bonds can be removed by terminating them with H, whose local empirical pseudopotential is given by the expression given in Ref. [63].

For C (and H) Kurokawa et al. [36] have studied the band structure of bulk C (diamond) as well amorphous C-H-based crystallites, but the C pseudopotential yields – quite surprisingly – reasonable results also for graphene and CNTs, as shown below. As for Si and III–V compound semiconductors, several other empirical pseudopotentials have been given for C in the diamond structure, for both nonlocal [29] and local [56] models. However, Kurokawa et al. have provided an expression which is suitable for interpolation while also yielding the correct workfunction for C in the diamond structure. In addition, in combination with the associated local empirical pseudopotential for H, also given by Kurokawa and co-workers, the electronic structure of *trans*-polyacytelene is correctly reproduced. This is significant, since these C pseudopotentials appear to reproduce both the hybridized sp^3 tetragonal diamond bonding coordination as well as the planar sp^2 (with p_z out-of-plane orbitals hybridizing into π^* orbitals) coordination found in graphene, nanoribbons and, with the addition of curvature effects we shall discuss, also nanotubes. Using a cutoff energy of 25 Ry the plane-wave method used here yields an indirect Γ_{25}'-Δ_1 gap of about 5.35 eV and values consistent with Kurokawa's results for the direct Γ_{25}'-Γ_2, X_4-X_1, and L_3'-L_3 gaps. For numerical convenience we have employed a cutoff energy of 15 Ry which yields similar values for the direct gap, but a slightly smaller value for the indirect Γ_{25}'-Δ_1 gap of 4.57 eV. We have spot-checked our results for graphene, nanoribbons, and CNTs employing a larger cutoff of 25 Ry without finding any significant difference. One such example will be shown below (see Fig. 3.25). One-electron empirical local pseudopotentials for C fitted to the band structure of graphene have also been proposed by Mayer [41] in terms of their real-space form given by the sum of three Gaussians. Note that these are *one-electron* empirical forms which would provide only one valence band per diamond primitive cell, so they may be suitable for transport, but not to handle the physics of the nanostructures of interest.

Regarding III–V compound semiconductors, pseudopotentials have been provided by Zunger's group for As, Al, and Ga in [4] and [40], P in [4], and In in [5]. The form chosen for the q-dependence of the ion pseudopotential is the sum of four Gaussians with additional parameters required to provide the correct band-alignment among the various elements, thus making it possible to study the electronic states of hetero-structures.

2.3 Examples

In this section we present results regarding the band structure (and, occasionally, properties of the wavefunctions) of several nanometer-scales structures: Thin Si layers, InGaAs/InP/AlInAs hetero-channels, square cross-section Si nanowires, infinite graphene sheets and nanoribbons, and carbon nanotubes. We consider this selection of examples sufficiently wide and of potential technological interest to assess the potential usefulness of the supercell technique to yield the correct excitation spectrum of systems of interest.

2.3.1 Thin Si Layers

The band structure of these systems – with and without an external confining field – is obtained by solving (3.4) conventionally using the Si and H pseudopotentials from Zunger's group [63, 65]. The density of states is computed by first employing a discretization of the 2D Brillouin Zone into a square mesh of size ΔK of points $\mathbf{K}_j$ labeled by an index j, computing the energy $E_n(\mathbf{K}_j) = E_{n,j}$ and energy gradient $\nabla_{2D}E_n(\mathbf{K}_j) = \nabla_{2D}E_{nj}$ for all relevant band (or subbands) labeled by the index n, and finally using the two-dimensional version of the Gilat–Raubenheimer algorithm [25]:

$$\rho_{2D}(E) = 2\sum_n \int \frac{d\mathbf{K}}{(2\pi)^2}\,\delta[E_n(\mathbf{K}) - E] = \frac{1}{2\pi^2}\sum_{jn}^{*} \frac{L_n(w_j)}{\nabla_{2D}E_{nj}}. \tag{3.7}$$

Here $L_n(w_j)/\nabla_{2D}E_{nj}$ is the density of states in the jth square, $L_n(w_j)$ being the length of the segment of the equienergy surface intersecting the square mesh element j:

$$L(w) = \begin{cases} \frac{\Delta K}{\cos\alpha} & (w \le w_0) \\ \frac{w_1 - w}{\cos\alpha \sin\alpha} & (w_0 \le w \le w_1) \end{cases}, \tag{3.8}$$

where $w_0 = (\Delta K/2)(\cos\alpha - \sin\alpha)$, $w_1 = (\Delta K/2)(\cos\alpha + \sin\alpha)$, α is the angle between the K_y-axis and $\nabla_{2D}E_j$, $w = (E - E_j)/|\nabla_{2D}E_j|$. Each square spans the energy range $(E_j - |\nabla_{2D}E_j|w_1, E_j + |\nabla_{2D}E_j|w_1)$ and the 'star' over the symbol of sum in (3.7) implies that only squares containing the final energy E are considered. The ballistic conductance along the direction characterized by the unit vector $\hat{\mathbf{n}}$ can be computed in a similar way:

$$G_{2D}(E) = 2\sum_n \int \frac{d\mathbf{K}}{(2\pi)^2}\, \upsilon_n(\mathbf{K})\cdot\hat{\mathbf{n}}\,\delta[E_n(\mathbf{K}) - E] = \frac{e^2}{\pi h}\sum_{jn}^{*} \frac{\nabla_{2D}E_{nj}\cdot\hat{\mathbf{n}}}{|\nabla_{2D}E_{nj}|} L_n(w_j), \tag{3.9}$$

where $\upsilon_n(\mathbf{K})$ is the group velocity in band n at the point $\mathbf{K}$ and the integration must be extended only over states whose group velocity along the direction $\hat{\mathbf{n}}$ is positive. Figure 3.1 shows the band structure of (100) Si layers of thickness equal to $9a_{Si}$,

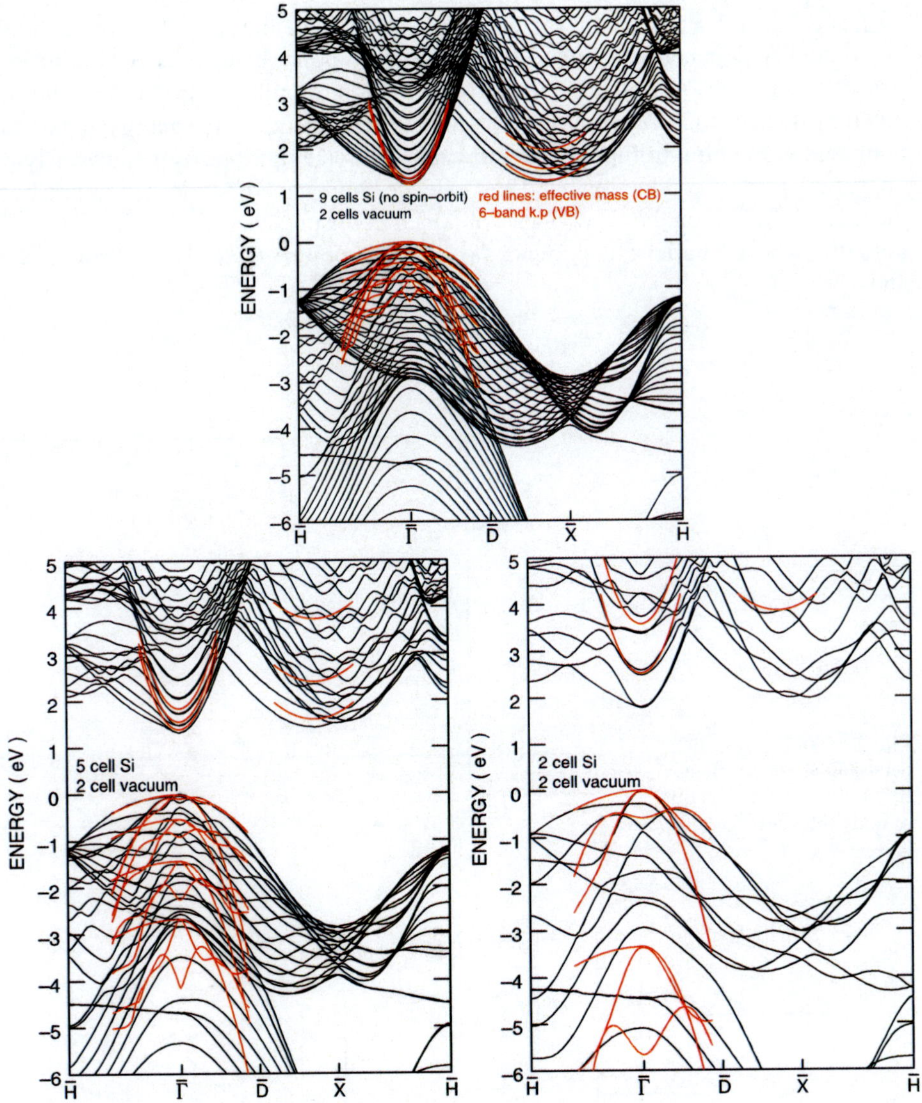

Fig. 3.1 Band structure of a (100) Si slab in vacuum terminated by H atoms. The slab thickness is 9-Si cells (*top*), 5-cells (*bottom left*), and 2-cells (*bottom right*). Note the quantized subbands in the conduction and valence bands, the widening of the gap caused by the confinement and, barely visible, the lifting of the twofold degeneracy of the unprimed states (known as 'valley splitting') caused by the symmetry breaking due to the external potential. Note also at the $\bar{X}$-point the presence of two additional 2D valleys. The *dashed lines* are parabolic-bands (conduction) and $\mathbf{k} \cdot \mathbf{p}$ approximations to the problem (calculated assuming vanishing wavefunction at the Si-vacuum interface) illustrating the significant effect of the EP band-structure. The spin-orbit interaction has been neglected and the zero of the energy has been set arbitrarily at the top of the valence bands

$5a_{Si}$, and $2a_{Si}$, all separated by vacuum $2a_{Si}$ 'thick'. Note the appearance of an additional doubly-degenerate conduction band at the $\bar{X}$ point, band already obtained by Esseni and Palestri [13] using a linear combination of bulk bands (LCBB) and denoted by them as M3, M4. Figure 3.2 shows the dependence of the energy gap on the thickness of the film. In Fig. 3.3 we show the squared amplitude of the lowest-lying unprimed, primed conduction band and valence band wavefunctions (averaged over the area of the cell on the (x,y) plane) obtained in the presence of a 'triangular well' potential whose Fourier components Φ_{G_z} are given by $a_{Si}F_s(N+N_v)/2$ for $G_z = 0$ and

$$\Phi_{G_z} = \frac{iF_s}{G_z}, \tag{3.10}$$

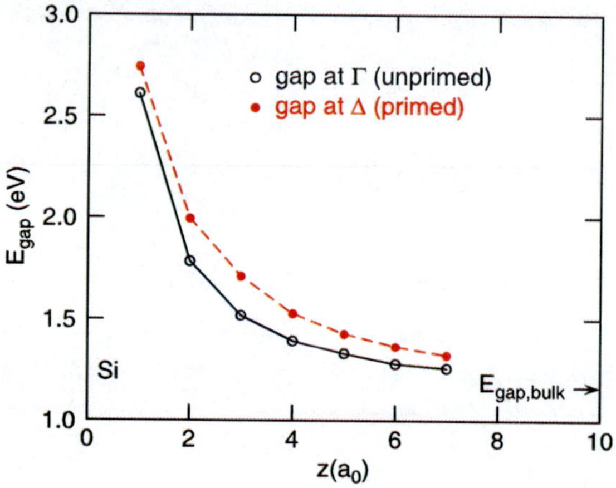

Fig. 3.2 Variation of the band-gap at the (100) and (001) minima as a function of body thickness for (100) Si layers

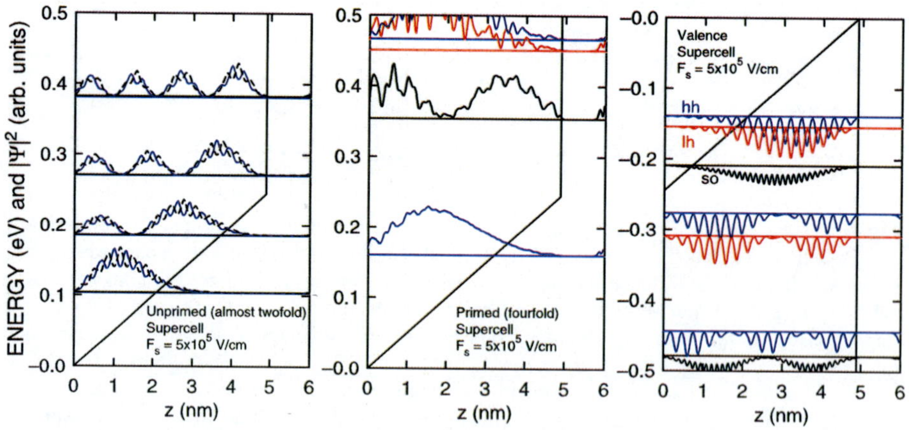

Fig. 3.3 Squared amplitude – averaged over a cell on the plane of the slab – of the wavefunctions in a 9-cell-thick Si layer in vacuum with H termination and a triangular-well potential with a field of 5×10^5 V cm^{-1}. At *left* are shown the wavefunctions of the unprimed states, at *center* those of the primed electron states, at *right* the hole states

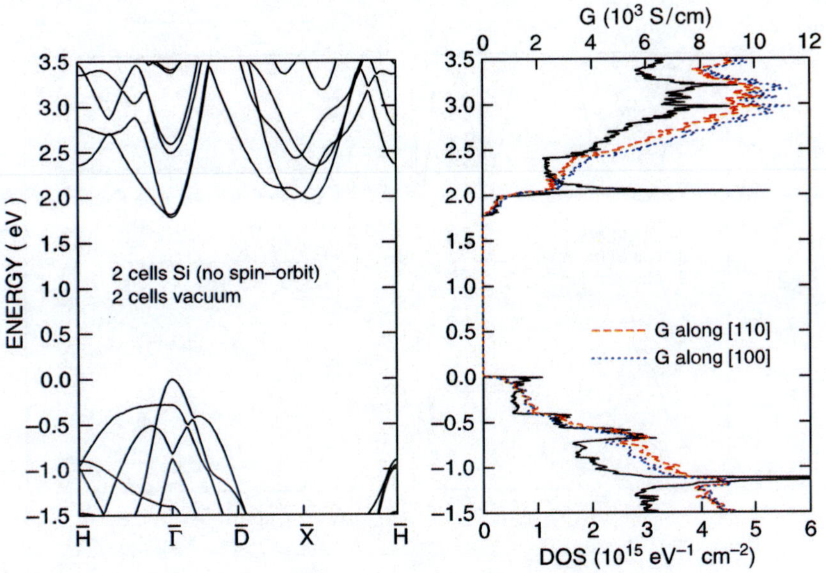

Fig. 3.4 Band structure (*left*), density of states, and ballistic conductance along the [100] and [110] directions (*right*) of a (100) Si slab as in the right frame of Fig. 3.1

for $G_z \neq 0$, where F_s is the surface field ($=5 \times 10^5$ V cm^{-1} in Fig. 3.3). Note the oscillations due to the Bloch components and the expected shape of the envelope. Figures 3.4–3.8 show details of the band structure and DOS near the gap for 2-cell, 3-cell, and 9-cell thin Si layers (the latter ones in the presence of a constant confining field) with surfaces of different orientations, while Fig. 3.9 compares the DOS obtained using the supercell method with what is obtained for parabolic bands for various surface orientations.

2.3.2 III–V Hetero-Channels

A similar scheme can be employed to calculate the band structure of channels consisting of different layers of III–V compound semiconductors. The pseudopotentials from Zunger's group [4, 5, 40] can be used in order to obtain the 'correct' band alignment and so the correct barriers of the confining wells. Figure 3.10 shows the band alignment and band structure of a (100) lattice-matched (to InP) $In_{0.53}Ga_{0.47}As$/InP/$Al_{48}In_{0.52}As$ hetero-channel mimicking a typical III–V MOSFET channel. The supercell consists of a composite $In_{0.53}Ga_{0.47}As$/InP channel, with 4-cell-thick $In_{0.53}Ga_{0.47}As$ layer and an equally thick InP layer, and of an 'insulating' 3-cell-thick $Al_{48}In_{0.52}As$ back layer. Figure 3.11 shows the variation of the energy gap as a function of the thickness of the composite channel, while Fig. 3.12

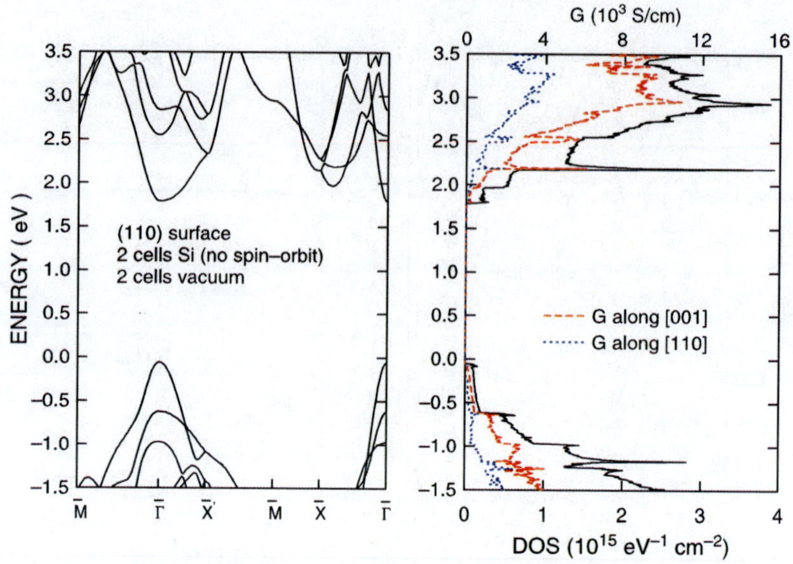

Fig. 3.5 Band structure (*left*), density of states, and ballistic conductance along the [001] and [1$\bar{1}$0] directions (*right*) of a (110) Si slab. The slab is 2-cell ($2a_0/\sqrt{2}$) thick and H terminated and a similar thickness of vacuum padding has been employed. The twofold-degenerate absolute minimum of the conduction band is at $\mathbf{K} = (0, 0.15)(2\pi/a_0)$, while a fourfold degenerate minimum is at $\mathbf{K} = (0.85/\sqrt{2}, 0)(2\pi/a_0)$. Note that the energetic ordering of these minima is opposite to what shown in Fig. 3.8 because the large nonparabolicity of the dispersion around the twofold minimum along the [110] direction weighs heavily at the high energies shown here for this very thin film

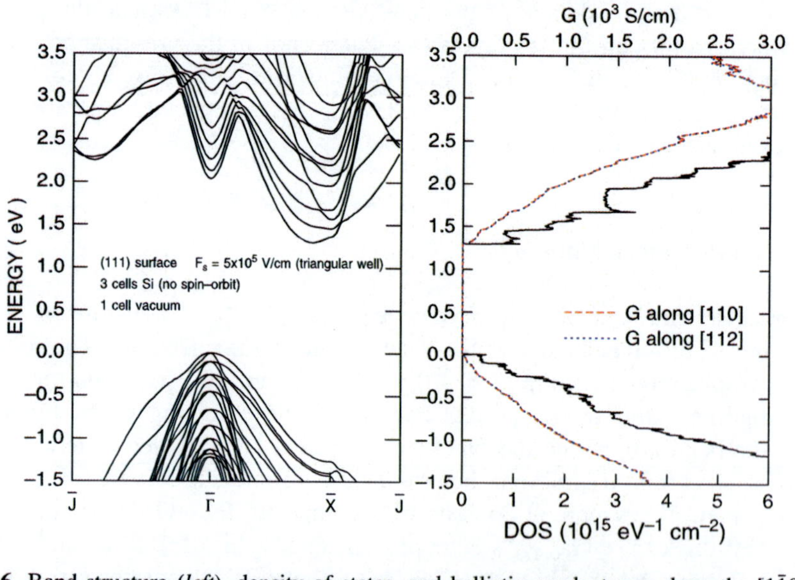

Fig. 3.6 Band structure (*left*), density of states, and ballistic conductance along the [1$\bar{1}$0] and [11$\bar{2}$] directions (*right*) of a (111) Si slab. The film is 3-cell ($3\sqrt{3}a_0$) thick, H terminated and 1-cell of vacuum separates the periodically repeated films. A constant field of 5×10^5 V cm^{-1} is applied perpendicularly to the slab to mimic a triangular-well confining potential

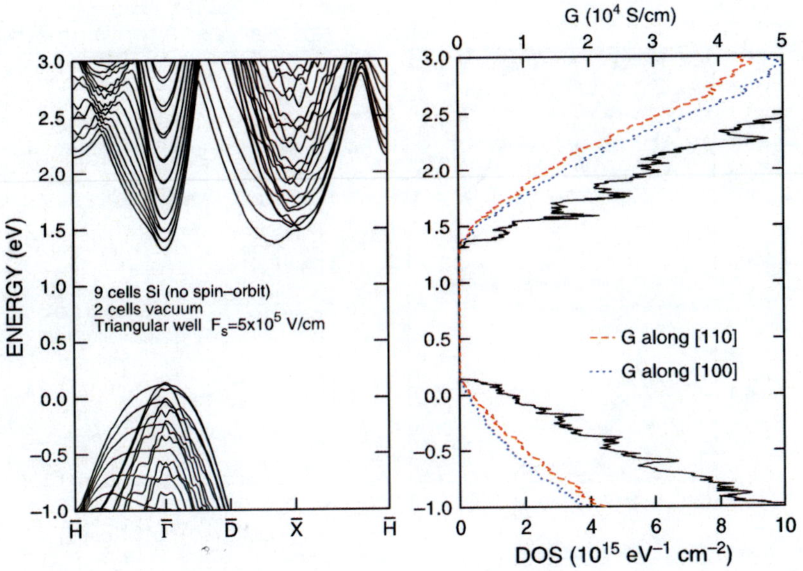

Fig. 3.7 Band structure (*left*), density of states, and ballistic conductance along the [100] and [110] directions (*right*) of a (100) Si slab as in the *left* frame of Fig. 3.1, but with a triangular-well potential with a field of $5\times10^5\,\mathrm{V\,cm^{-1}}$. The zero for the energy has been set at the top of the valence bands in the absence of the applied field, as in the *left* frame of Fig. 3.1, to judge the shift of the subband energies in the presence of the external field

shows how the band structure is modified by the application of a parabolic potential (mimicking the potential of an inversion layer) of the form:

$$\Phi(z) = V_0\left(1 - \frac{2z}{L} + \frac{z^2}{L^2}\right), \tag{3.11}$$

(where L is the extension of the supercell in the z direction, $L = a_{InP}N$, with N the total number of cells employed and $V_0 = F_sL/2$ is the total voltage drop in the cell expressed in terms of the F_s) with Fourier components $F_sL/3$ for $G_z = 0$ and

$$\Phi_{G_z} = -\frac{F_s}{2}\left(\frac{2}{LG_z^2} - \frac{i}{G_z}\right) \tag{3.12}$$

for $G_z \neq 0$. The figure presents results obtained for $F_s = \pm 5\times10^5\,\mathrm{V\,cm^{-1}}$, for electron and hole confinement, respectively. Finally, the electron and hole wavefunctions in the $In_{0.53}Ga_{0.47}As/InP/Al_{48}In_{0.52}As$ hetero-layer for the cases of flat-band, electron and hole confinement are shown in Fig. 3.13. Note how potentially intricate issues (such as matching the wavefunctions at interfaces, determining the effective mass to be used when the wavefunction extends over two materials, etc.) are bypassed by the supercell method. The shape of envelope of the wavefunctions agrees with our naïve expectations based on the band discontinuities shown and on the 'envelope' idea.

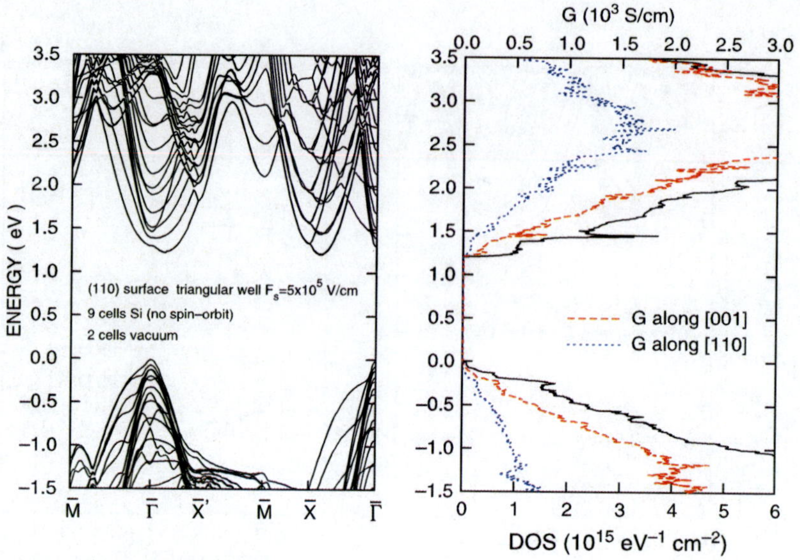

Fig. 3.8 Band structure (*left*), density of states, and ballistic conductance along the [001] and $[1\bar{1}0]$ directions (*right*) of a (110) Si slab 9-cell ($9a_0/\sqrt{2}$) thick with a triangular-well potential with a field of $5\times10^5\,\mathrm{V\,cm^{-1}}$. Comparing with Fig. 3.5, note that the twofold minimum $\mathbf{K} = (0, 0.15)(2\pi/a_0)$ is now at an energy about 12 meV higher than the fourfold minimum at $\mathbf{K} = (0.85/\sqrt{2}, 0)(2\pi/a_0)$. From estimates based on conventional effective masses [2] we expect this same ordering, but an energy difference of about 28 meV. The difference is likely due to non-parabolic effects

2.3.3 Si Nanowires

We have considered Si nanowires (NWs) with axis along the [100], [110], and [111] directions. Published results used to validate our results are from Nehari [44] and Neophytou [45] – who have employed a tight-binding model –, by Sacconi [54], whose results have been obtained using linear combination of bulk bands (LCBB) and empirical tight-binding (ETB), and by Scheel [57], Lee [38] who have employed first-principles DFT methods.

Figure 3.14 shows the band structure, density of states and ballistic conductance for square-section Si NWs with sides 2 to 5-cells long separated by 1-cell of 'vacuum'. Note the energy gap increasing with decreasing wire size (as expected). The DOS for all 'subbands' n has been computed from the expression:

$$\rho_{1D}(E) = 2\sum_{n,i} \int \frac{dE_n}{2\pi} \left| \frac{dE_{n,i}}{dk_z} \right|^{-1} \delta(E_{n,i} - E), \tag{3.13}$$

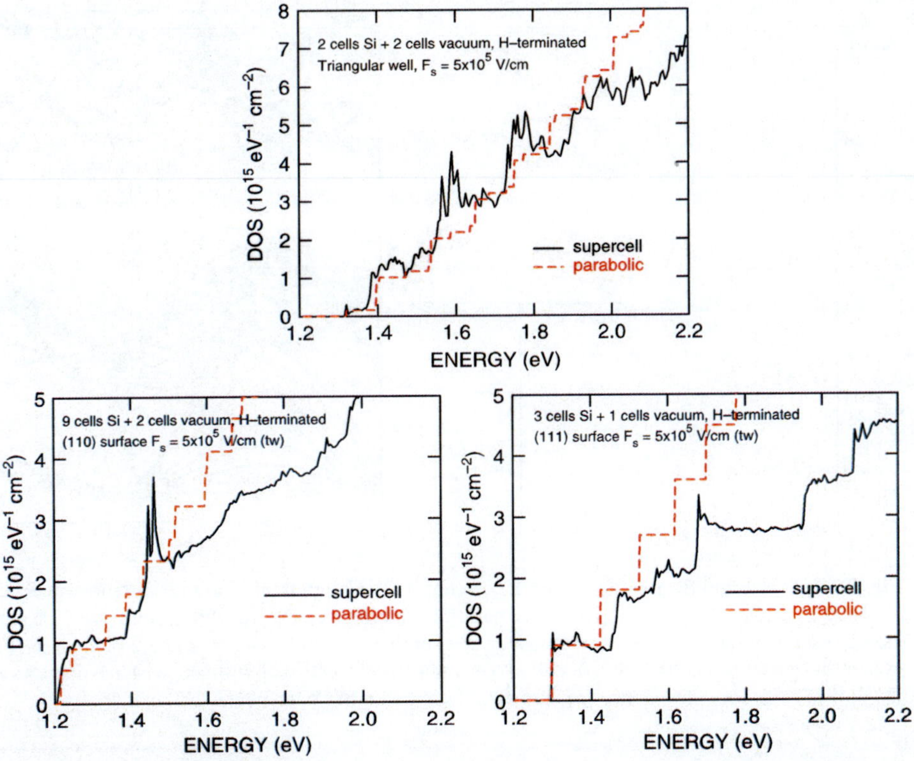

Fig. 3.9 Density of states of the (100) (*top*), (110) (*bottom left*), and (111) (*bottom right*) Si slabs of the previous figures (*solid black lines*) compared to the DOS calculated using a two-ladder (for the (100) and (110) surfaces) or one-ladder (for the (111) surface) parabolic band structure with longitudinal and transverse effective masses of 0.19 and 0.91 *m* (*dashed lines*). The energies are measured from the top of the valence band obtained in the zero-field case and the effective-mass ground state energy is has been shifted to coincide with the pseudopotential result

where the index i labels the solutions $k_{z,n,i}$ such that $E_{n,i} = E(k_{z,n,i}) = E$, and the ballistic conductance as:

$$G_{1D}(E) = 2e^2 \frac{1}{2} \sum_{n,i} \int \frac{dE_{n,i}}{2\pi} \, \upsilon_{n,i}(E) \left| \frac{dE_{n,i}}{dk_z} \right|^{-1} \delta(E_{n,i} - E)$$

$$= \frac{2e^2}{h} \frac{1}{2} \sum_{n,i} \int dE_n \, \delta(E_{n,i} - E), \tag{3.14}$$

where $\upsilon_{n,i}(E)$ is the group velocity $(1/\hbar)dE_{n,i}/dk_z$ at energy E is subband n at the point $k_{z,n,i}$ and the factor of 1/2 in the expression for $G_{1D}(E)$ comes from having to sum only over positive $\upsilon_n(E)$, and so over 1/2 of the 1D BZ. The effect of the vacuum 'padding' decoupling the nanowires can be judged by comparing the bottom

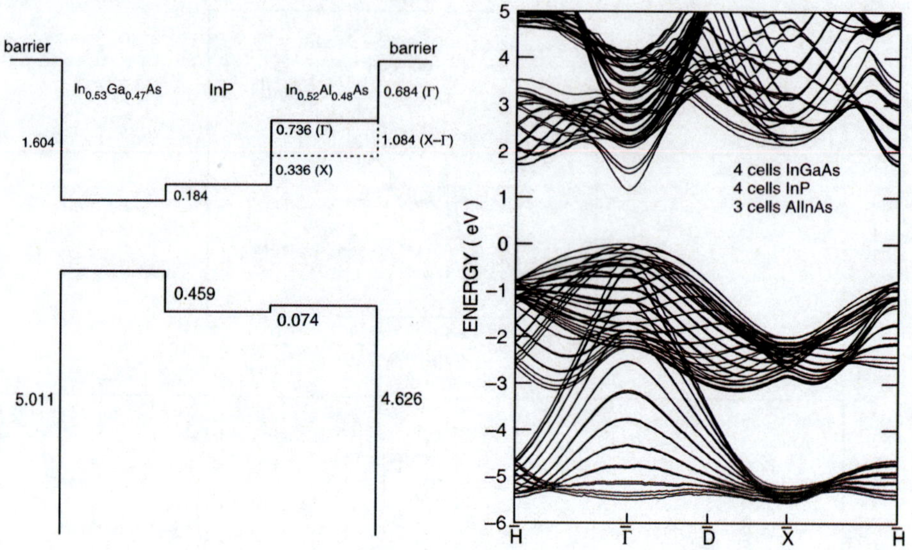

Fig. 3.10 *Left*: Band diagram showing the discontinuities/alignments for a (100) lattice-matched (to InP) $In_{0.53}Ga_{0.47}As/InP/Al_{48}In_{0.52}As$ periodic hetero-structure resulting from the Zunger's atomic pseudopotentials and accounting for spin-orbit interaction. *Right*: Band structure for the system at left with 4-cells/4-cells/3-cells layer thickness (1-cell = 1-InP cell = 0.586 nm). As in the thin-Si case, for convenience the spin-orbit interaction has been neglected here

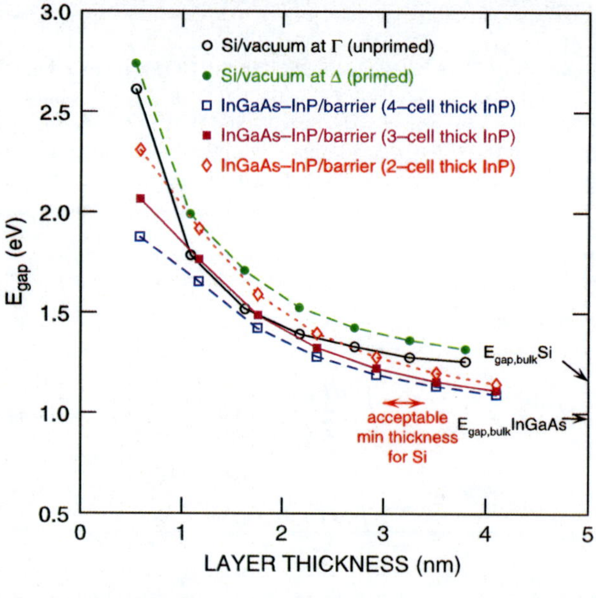

Fig. 3.11 Variation of the band gap of the AlInAs/InGaAs/InP quantum-well structure as a function of InGaAs thickness for 3 different values of the InP thickness and a 3-cell-thick insulating barrier layer compared with the Si (100) and (001) band-gaps. The Si data shown in Fig. 3.2 are also shown here for comparison. This plot may be used to estimate the sensitivity of the threshold voltage to thickness variations in ultra-thin-body Si and InP-InGaAs channel FETs

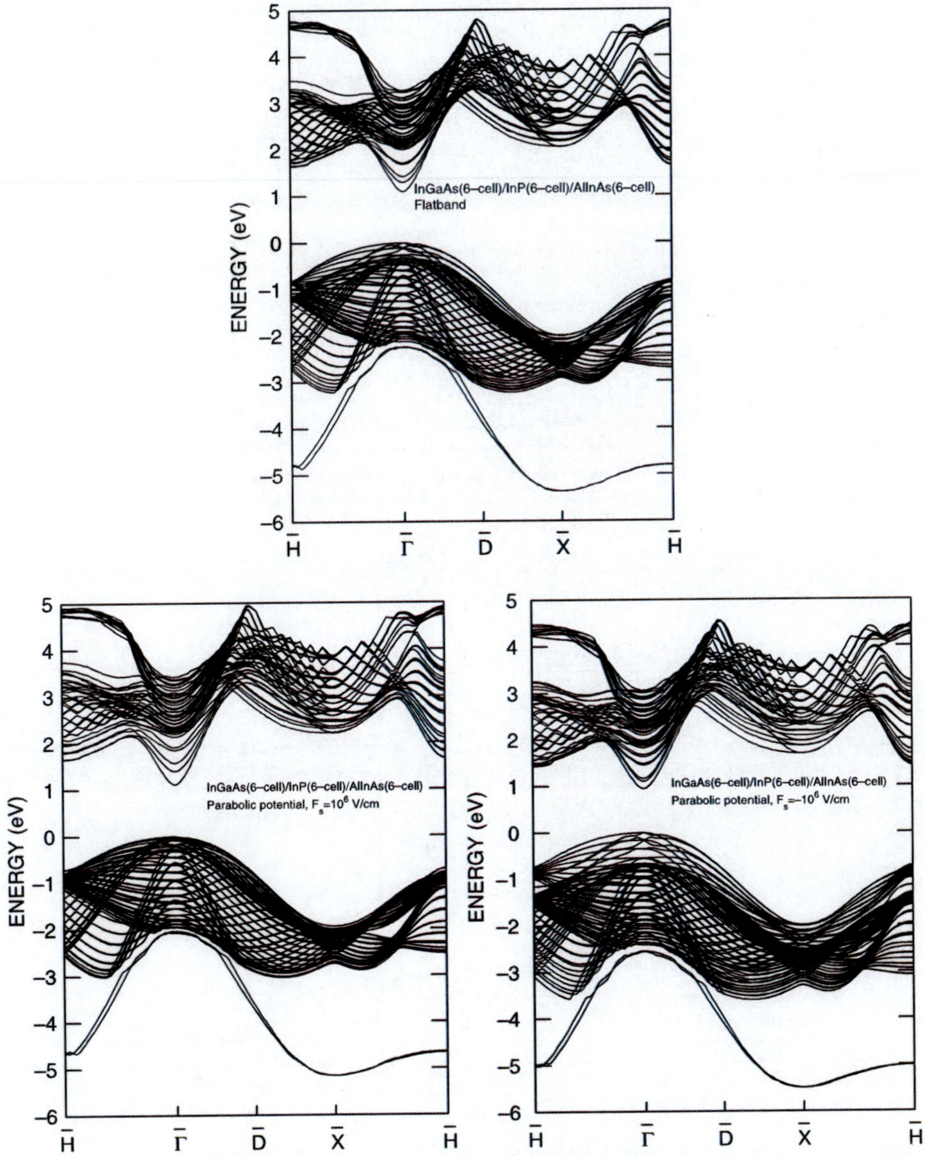

Fig. 3.12 In-plane dispersion for an InGaAs/InP/AlInAs hetero-channel under flatband conditions (*top*), with a parabolic potential with a surface electric field of 10^6 V cm^{-1} (*bottom left*, confinement for electrons) and -10^6 V cm^{-1} (*bottom right*, confinement for holes)

left frame of Fig. 3.14 with the top frame of Fig. 3.15, as these two plots show the band structure of identical nanowires but using 1-cell or 2-cells of vacuum padding, respectively. The band structure, DOS, and ballistic conductance of free-standing [100], [110] and [111] nanowires of similar square cross-sections ('approximately' square for the [110] and [111] wires) are shown in Fig. 3.15: As expected, only

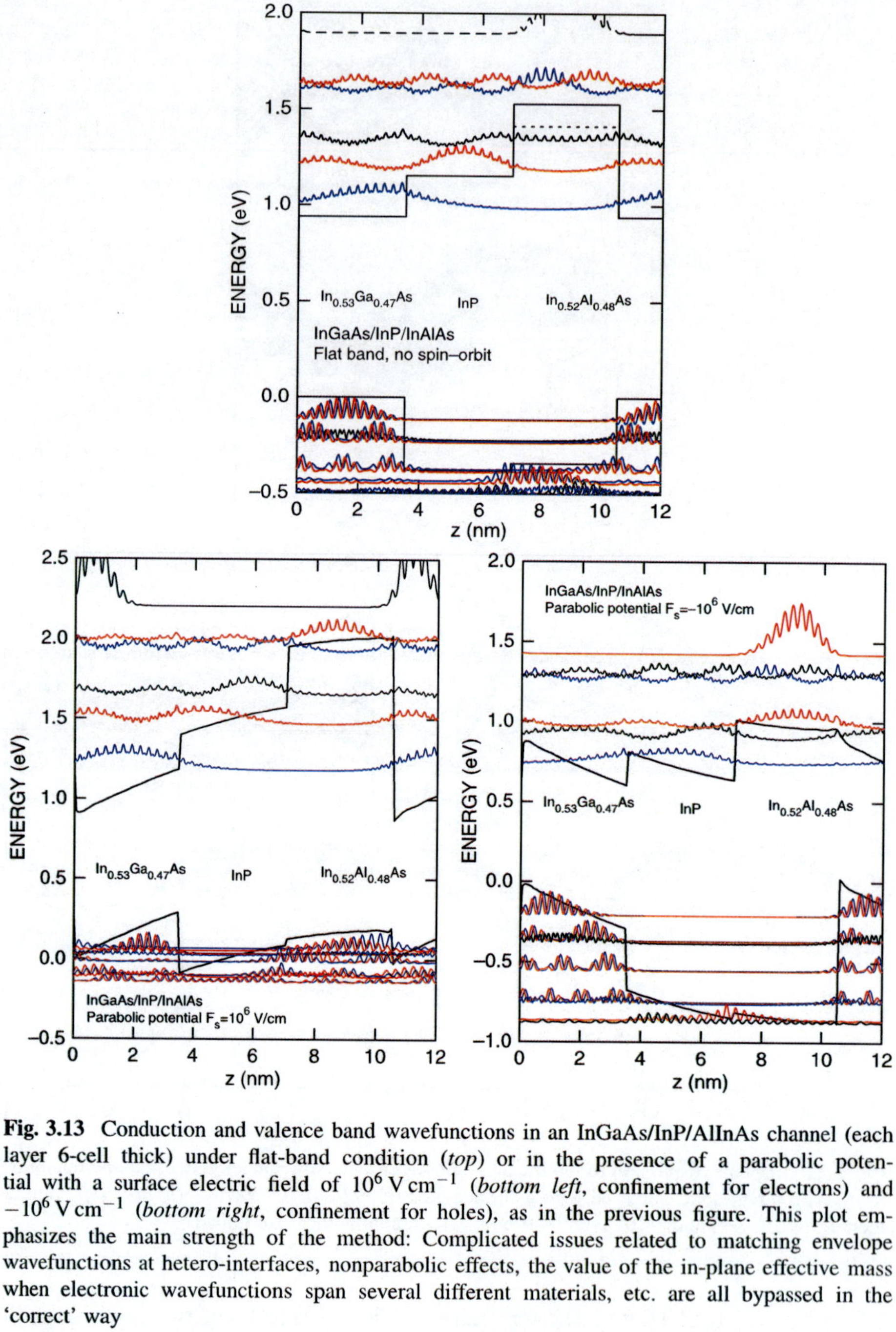

Fig. 3.13 Conduction and valence band wavefunctions in an InGaAs/InP/AlInAs channel (each layer 6-cell thick) under flat-band condition (*top*) or in the presence of a parabolic potential with a surface electric field of $10^6\,\text{V}\,\text{cm}^{-1}$ (*bottom left*, confinement for electrons) and $-10^6\,\text{V}\,\text{cm}^{-1}$ (*bottom right*, confinement for holes), as in the previous figure. This plot emphasizes the main strength of the method: Complicated issues related to matching envelope wavefunctions at hetero-interfaces, nonparabolic effects, the value of the in-plane effective mass when electronic wavefunctions span several different materials, etc. are all bypassed in the 'correct' way

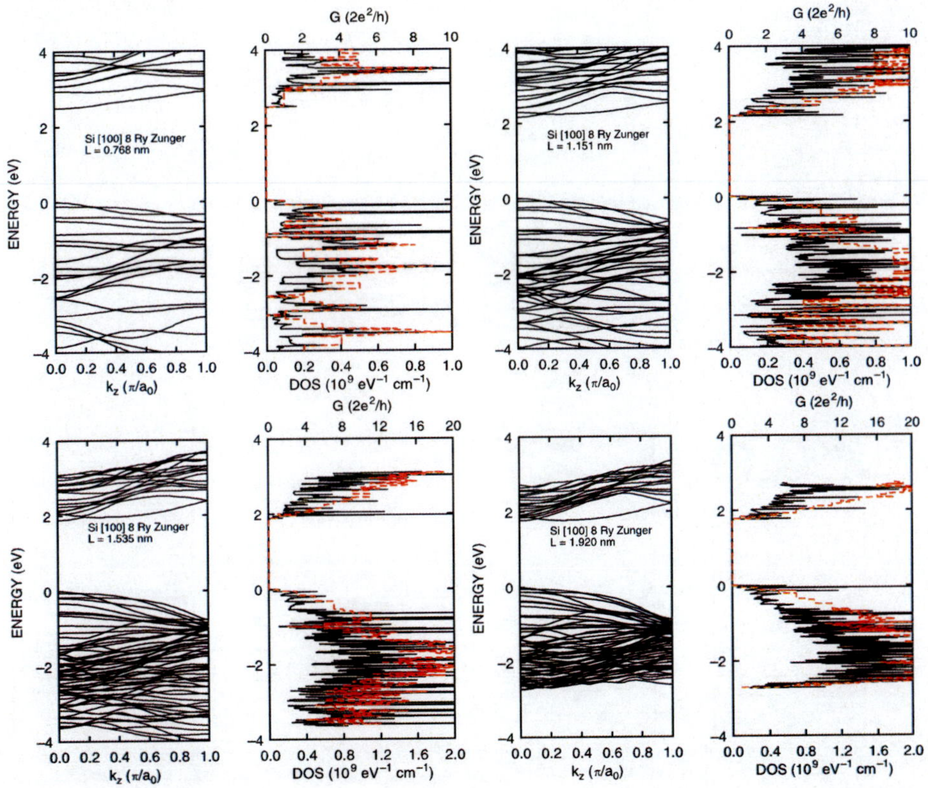

Fig. 3.14 *Top-left to bottom-right*: Band structure (*left*), density of states (*right, black solid line, bottom axis*) and conductance (*right, red dashed line, top axis*) of four square cross-section H-terminated [100] Si nanowires with sides of four different lengths. The wires are separated by a 1-cell thickness of vacuum. The results have been obtained using a cutoff energy of 8 Ry using the empirical pseudopotentials proposed by Zunger

[111]-oriented wires exhibit an indirect gap. Finally, the ballistic conductance of these wires is shown in Fig. 3.16: Note the larger conductance for both electrons and holes in the [100] wire and the smaller conductance of the [111] nanowire whose many bands are 'flat' and exhibit few crossings.

2.3.4 Graphene

The band structure of an infinite graphene sheet can be calculated assuming the sheet is a supercell layer separated periodically from the neighbor sheets by a distance $N\frac{a_0\sqrt{3}}{2}$, thus using 1D supercells as in the case of thin Si layers or hetero-layers discussed above. Examples of results available in the literature with which we can compare the quality of the supercell method and, more important, of the empirical pseudopotentials we have used are the qualitative results by Ajiki and Ando [1], by

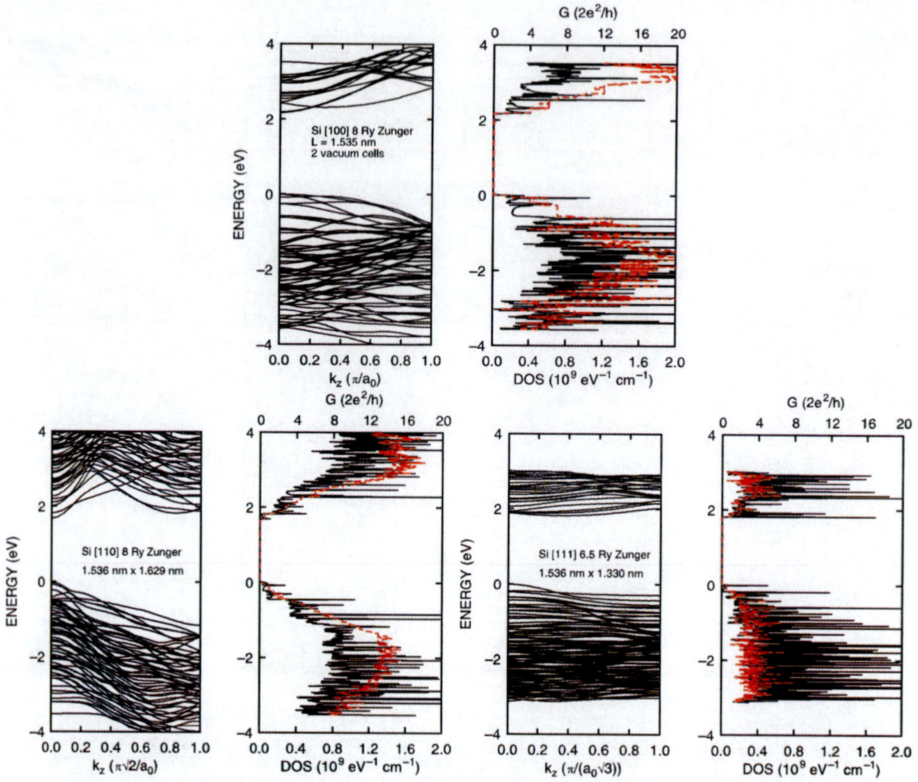

Fig. 3.15 Band structure, density of states, and conductance of free-standing H-terminated [100] (*top*), [110] (*bottom left*) and [111] (*bottom right*) Si nanowires with an ('almost' for the [110] and [111] wires) square cross-section of the indicated dimensions with 2-cells of vacuum padding. Note that for the [111] wire the cut-off energy has been set to 6.5 Ry, as indicated

Reich [53] for ab-initio results, by Khoshnevisan [33] for graphene (and also (5,5) CNT, see below) using the "Quantum Espresso" ab-initio DFT/LDA method [24].

Figure 3.17 shows the band structure obtained using the Kurokawa and Mayer pseudopotentials as well as the density of states. Compared to ab-initio results [33, 53], the Kurokawa pseudopotentials exhibits the 'correct' behavior of the $\pi - \pi^*$ band at energies close to the Fermi level and the correct band-crossing ('Dirac' point) at the symmetry point K, but exhibit a set of bands at Γ which have lower and compress the $\pi - \pi^*$-band energetic separation near $k = 0$. The Mayer pseudopotentials by definition fail to account for the $2s$ valence states (having been designed as 'one-electron' potentials) and also miss many higher energy states, while reproducing satisfactorily the $\pi - \pi^*$-band energetic separation near the Fermi level. The Fermi velocity at the Dirac point is about $\upsilon_F \approx 9.5 \times 10^7\,\mathrm{cm\,s^{-1}}$ from the Kurokawa pseudopotentials and $\approx 8.8 \times 10^7\,\mathrm{cm\,s^{-1}}$ using the Mayer pseudopotentials, both values in good agreement with DFT results, but about 15% smaller than experimental data and GW-corrected DFT values [61].

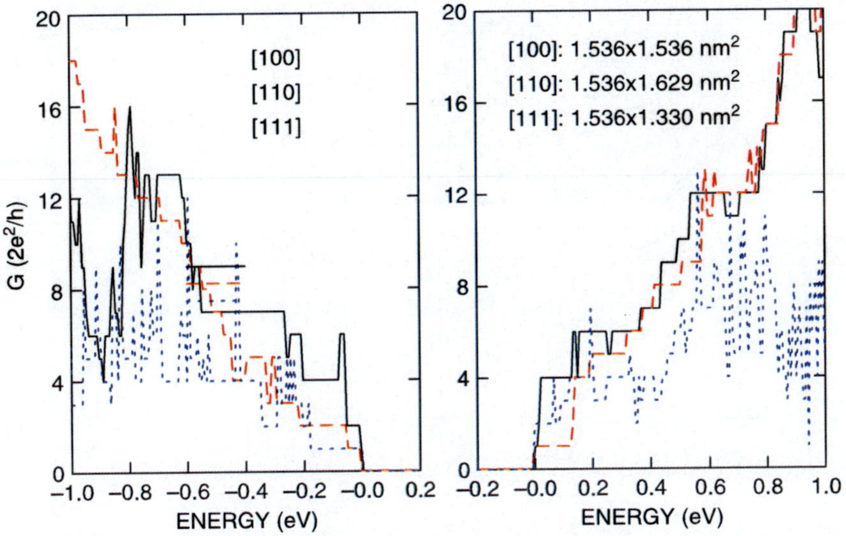

Fig. 3.16 Valence (*left*) and conduction band (*right*) ballistic conductance of the [100], [110], and [111] Si nanowires of the previous figures. The zero-energy has been set at the band-edge in both plots

2.3.5 Graphene Nanoribbons

Graphene nanoribbons (GNRs) can be described by the their chirality and their width characterized by the number N_a of atomic lines. For nanoribbons with armchair-type edges (AGNR) the dependence of the energy gap on their width follows three types of trends depending on whether $N_a = 3p$, $3p+1$, or $3p+2$, Tight-binding calculations [14, 15, 23] and calculations based on the massless Dirac equation [8, 55] yield the ordering $E_{g,3p} \geq E_{g,3p+1} > E_{g,3p+2} = 0$ of the gap, so $3p+2$-type GNRs are predicted to be semimetallic. On the contrary, ab-initio DFT (LDA and GW) calculations [3, 60, 64] predict $E_{g,3p+1} \geq E_{g,3p} > E_{g,3p+2} \neq 0$. Thus, all GNRs should be semiconducting, the difference between tight-binding and ab-initio results originating mainly, according to Son et al. [60], from the change of the C–C bond length along the edges (however, as we shall see, we find the same ab-initio behavior without accounting for this effect, hinting, instead, at some inherent inadequacy of the tight-binding method). Note also that the value of the calculated bandgap increases dramatically when accounting for GW corrections (compare the results of [60] with those of [64], for example). We have found that empirical pseudopotential approaches also reproduce the behavior found using first-principles calculations even without accounting for the edge-bonds distortion, as just stated. The use of Kurokawa pseudopotential also accounts for the correct behavior of bare edge states. The main problems with the use of these local pseudopotentials stem from their empirical non-self-consistent nature leading to their inability to predict the correct semiconducting behavior of zigzag-edge nanoribbons, as we shall

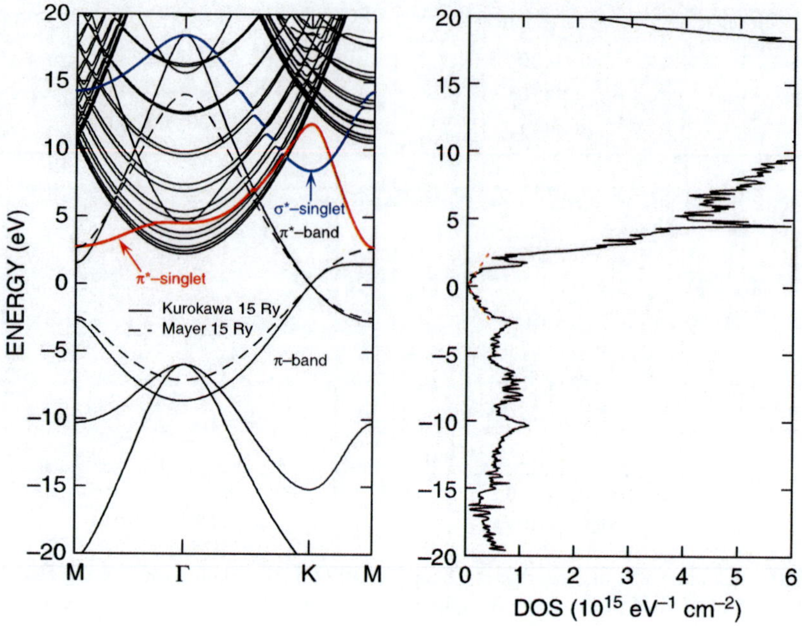

Fig. 3.17 *Left*: Band structure for graphene obtained using the Kurokawa (*solid lines*) and Mayer (*dashed lines*) pseudopotentials. A separation of $10\sqrt{3}a_0/2$ along z has been assumed between adjacent sheets. The bands indicated by *arrows* are the σ^* (*blue*) and π^* (*red*) singlet bands whose interaction and hybridization result in an interesting and unexpected behavior of the band-gap in single-wall zigzag $(n,0)$ carbon nanotubes of small diameter. The use of Kurokawa pseudopotentials, in particular, yields relatively small energies for the π^*-singlet band along the M-Γ line ($\approx$2.5 eV above the Fermi level, compared to energies three times as large obtained using the self-consistent LDA [33, 53]). *Right*: Density of states of graphene. Note how the DOS in the 'gap' region is accurately described, as it approaches quite closely the analytic expression $2E/[\pi(\hbar v_F)^2]$ around the Dirac point (*dotted red lines*)

discuss below, since dealing with spin-polarization effects requires self-consistent methods including exchange-correlation (actually, mainly exchange). In addition, and possibly unrelated to this, is the problem that, when applied to carbon nanotubes, they predict an excessively low energy of the π^* singlet in CNTs of some chirality (as in the $(n,0)$ CNTs with $n \leq 10$ discussed below), in disagreement with first-principles results.

Figure 3.18 shows the band-structure of a 'bare' (as opposite to H-terminated) 9-AGNR. Note that the use of the Mayer pseudopotentials yields a reasonable energy gap (when compared to first-principles results [60]) and also a reasonable dispersion for the topmost valence band and lowest-energy conduction bands, while missing by definition other valence bands, several highest-energy conduction bands, and also the well-known edge-states which enter the $\pi - \pi^*$ gap. These states are clearly noticeable in the results obtained by using the Kurokawa pseudopotentials: These pseudopotentials account for the existence of all bands found by ab-initio calculations, yield a reasonable gap at $k = 0$, as well as the edge states which

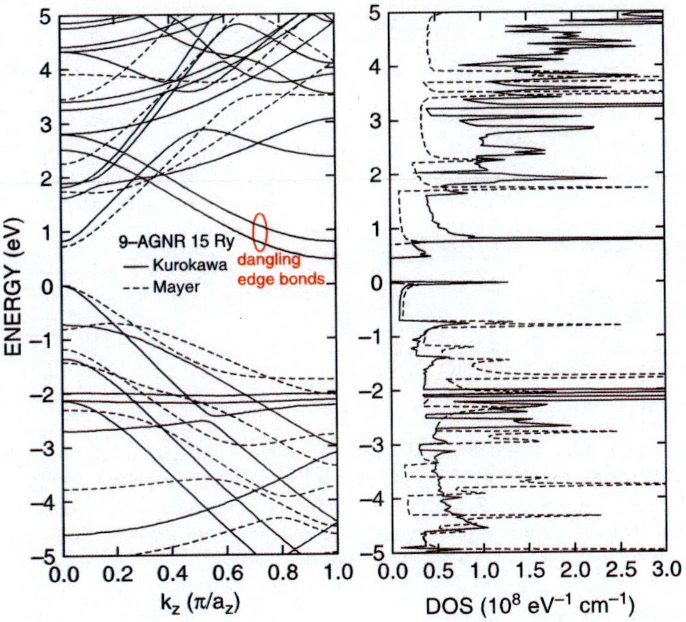

Fig. 3.18 Band structure and density of states for the bare-edge 9-AGNR illustrated at *right*. Results obtained using Kurokawa (*solid lines*) and Mayer (*dashed lines*) pseudopotentials are shown. Note the bands associated with edge states resulting from the edge-C dangling bonds. Here and in the following graphene ribbons are separated by $N_v\sqrt{3}a_0$, with $N_v = 4$ (unlike the choice of $N_v = 3$ made to sketch the ribbon in the *left panel* of this figure), along the plane of the ribbon and by $N_y\sqrt{3}a_0$, with $N_y = 3$, along the direction perpendicular to the sheets. The energy has been set to zero at the top of the valence band

can be removed by H termination [58]. In Fig. 3.19 we show similar results for a bare 5-AGNR and a bare 7-AGNRs. Note that the $k = 0$ energy gap obtained for the 5-AGNR (0.266 and 0.290 eV using the Mayer and Kurokawa pseudopotentials, respectively, in either case nonzero, unlike the tight-binding predictions) is much smaller than the gap obtained for the 7-AGNR (1.730 and 1.586 eV using the Mayer and Kurokawa pseudopotentials, respectively), as expected from the $E_{g,3p+1} \geq E_{g,3p} > E_{g,3p+2} \neq 0$ ordering predicted by first-principles approaches. Also, edge-states lower the gap [58], but disappear when terminating the edge-bonds with H, as shown in Fig. 3.20. In this case using the Kurokawa pseudopotentials the bandgap for the 5-AGNR drops from 0.290 eV (bare edges) to 0.197 eV (H-terminated edges), while for the 7-AGNR it increases from 1.586 eV (bare) to 1.612 eV (H-terminated). The 9-AGNR exhibits a gap of 0.820 eV (bare) and 0.868 (H-terminated, see Fig. 3.21). In all cases these gaps are in agreement with the first-principles, non-GW-corrected results of [60], as shown in Fig. 3.22. Only for the smallest-width ribbon (3-AGNR) the gap is noticeably smaller than what obtained from first-principles calculations, presumably because of the growing importance of edge-bond distortion noticed by Son et al. [60]. Finally, in Fig. 3.21 we show

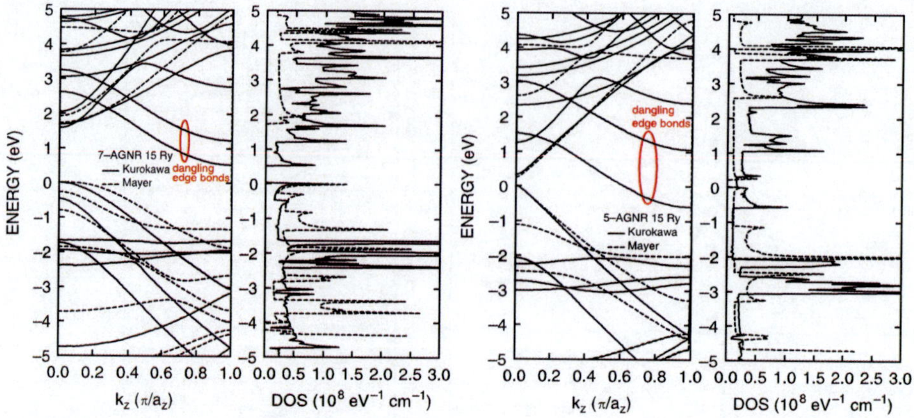

Fig. 3.19 Band structure and density of states for an $N_a = 7$ (*left*) and $N_a = 5$ (*right*) bare-edge graphene nanoribbon as in Fig. 3.18. Note that while tight-binding models predict a semimetallic (no gap) behavior for the 5-AGNR, empirical pseudopotentials yield semiconducting behavior even in the absence of the distortion of the edge C–C bonds found by first-principles calculations [60]. As in the previous figure, note the edge-states bands

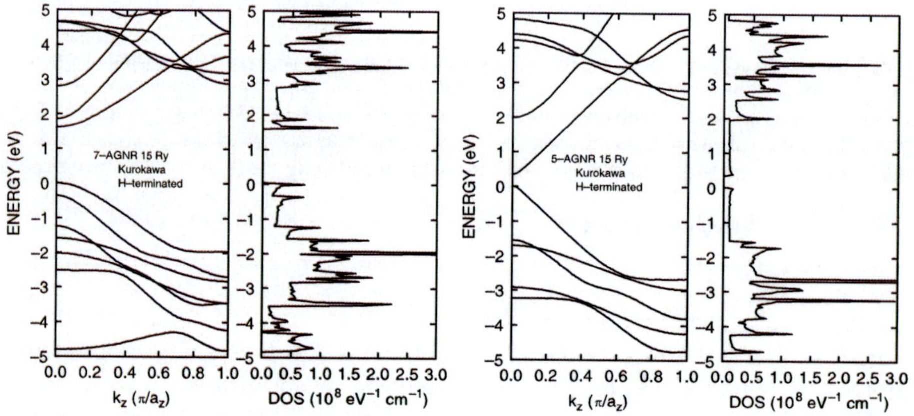

Fig. 3.20 Band structure and density of states for an $N_a = 7$ (*left*) and $N_a = 5$ (*right*) graphene nanoribbon with armchair edges, as in Fig. 3.19, but with H termination of the edge C atoms

the band structure of a 9-AGNR ($N_a = 3p$), a 10-AGNR ($N_a = 3p + 1$), and an 11-AGNR ($N_a = 3p + 2$), to show directly the energy gap for the three ‘ladders’ considered in Fig. 3.22.

In Fig. 3.23 we show the band structure of an $N_a = 4$ zigzag-edge graphene nanoribbon (4-ZGNR), of an 8-ZGNR, and of a 12-ZGNR obtained using the Mayer pseudopotentials with dangling bonds for the edge C atoms and using the Kurokawa pseudopotentials with H-terminated C-edge-bonds. Note that in this case both choices of pseudopotentials result in a semimetallic behavior, the π and π^* bands overlapping slightly, in agreement with the results obtained by Ezawa [15] (who,

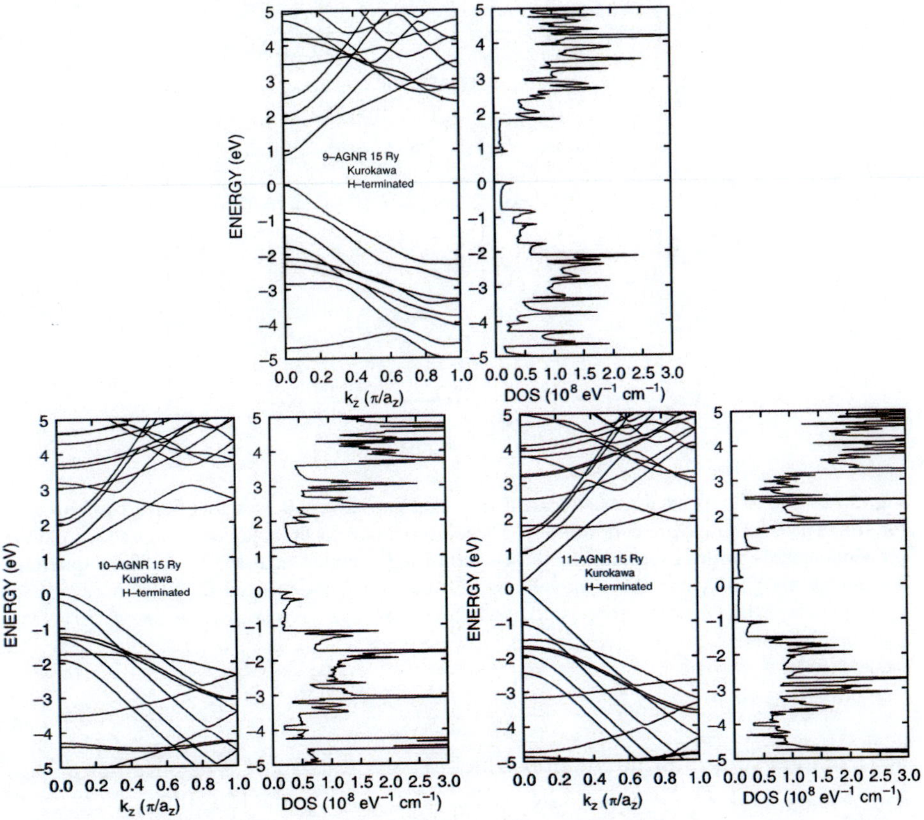

Fig. 3.21 *Top*: Band structure and density of states for an $N_a = 9$ graphene nanoribbon with armchair edges, as in the left frame of Fig. 3.18, but with H termination of the edge C atoms. *Bottom*: Band structure of an H-terminated 10-AGNR (*left*) and an 11-AGNR (*right*), to show the values of the energy gap for $N_a = 3p+1$ and $N_a = 3p+2$ compared to the $N_a = 3p$ case at *left*

however, also predicts metallic behavior for $N_a = 2p+2$ armchair nanoribbons) and of the LDA results by Pisani et al. [51] for monohydrogenated non-magnetic nanoribbons. Although the shape of the bands appear qualitatively in agreement with the LDA results of [64], the semimetallic behavior of this ZGNR emerges from the fact that we have not accounted for spin polarization effects [28].

One can speculate about other possible shortcomings of our empirical pseudopotential calculations. The obvious first concern stems from a possible inaccuracy of the empirical C pseudopotentials. However, while Kurokawa's C pseudopotentials were calibrated to the diamond structure, the H pseudopotentials had been fit to the electronic structure of trans-polyacetylene, which resembles very closely the hydrogenated edges of ZGNRs. The fact that spin-orbit interaction has been ignored may constitute another possible cause of concern. On the one hand, Kan and coworkers [31] have shown that the ZGRN bands are not spin-degenerate. However, spin polarization of edge states is known to emerge from a Hubbard-like interaction,

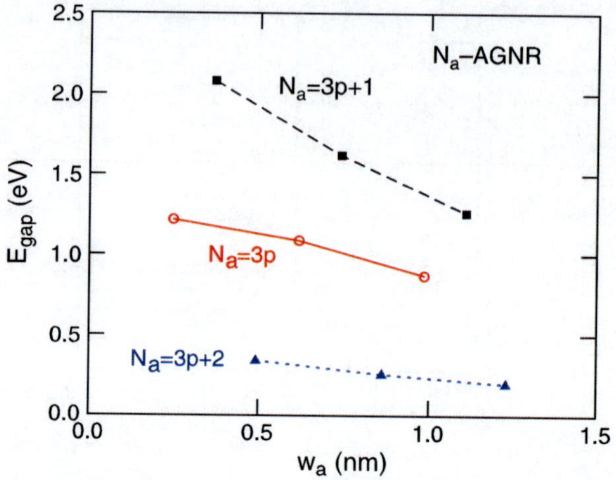

Fig. 3.22 The calculated three ladders of the band-gap at $k = 0$ as a function of ribbon-width for armchair-edge graphene nanoribbons. These results have been obtained using the Kurokawa's pseudopotentials with H-terminated sp^2 σ carbon edge-bonds and are in excellent qualitative agreement and good quantitative agreement with the non-GW-corrected DFT+LDA calculations by Son et al. [60]. GW corrections yield much larger gaps for the quasiparticle energy [64], while tight-binding models predict qualitatively incorrect gaps and an incorrect ordering of the three ladders [60]

not from the spin-orbit interaction which is very small in C [12]. Kan et al. have also shown that ZGNRs may be metallic or semiconducting depending on the functional groups (H, NH_2, CH_3, NO_2) used to terminate the sp^2 σ orbitals of the edge C atoms. This would point at some possibly wrong assumptions we might have made in terminating these orbitals with the Kurokawa's H pseudopotential and assuming the CH_4 C–H bond length. A final possible source of concerns may be the distortion of the edge C–C bonds emphasized by Son et al. [60] or of the C–H edge bonds.

2.3.6 Carbon Nanotubes

The band structure of carbon nanotubes has been obtained using the supercell method with atomic positions calculated using the on-line Java tool TubGen v3.3 [21]. Comparison can be made with the qualitative analysis by Ajiki and Ando [1], with by Reich [53], Gulseren [27], Sharma [59] and Miyake and Saito [42, 43] for the diameter dependence of zigzag semiconducting nanotubes, and by Mayer [41] and Khoshnevisan [33] for the band-structure of (5,5) and (10,0) CNTs. In Fig. 3.24 we show the band-structure and density of states for these CNTs. These data have been obtained using a supercell with square cross-section of sides 1 and 1.4 nm long for the (5,5) and (13,0) CNTs and both the Kurokawa and Mayer pseudopotentials with a cutoff energy of 15 Ry. (Figure 3.25 shows the

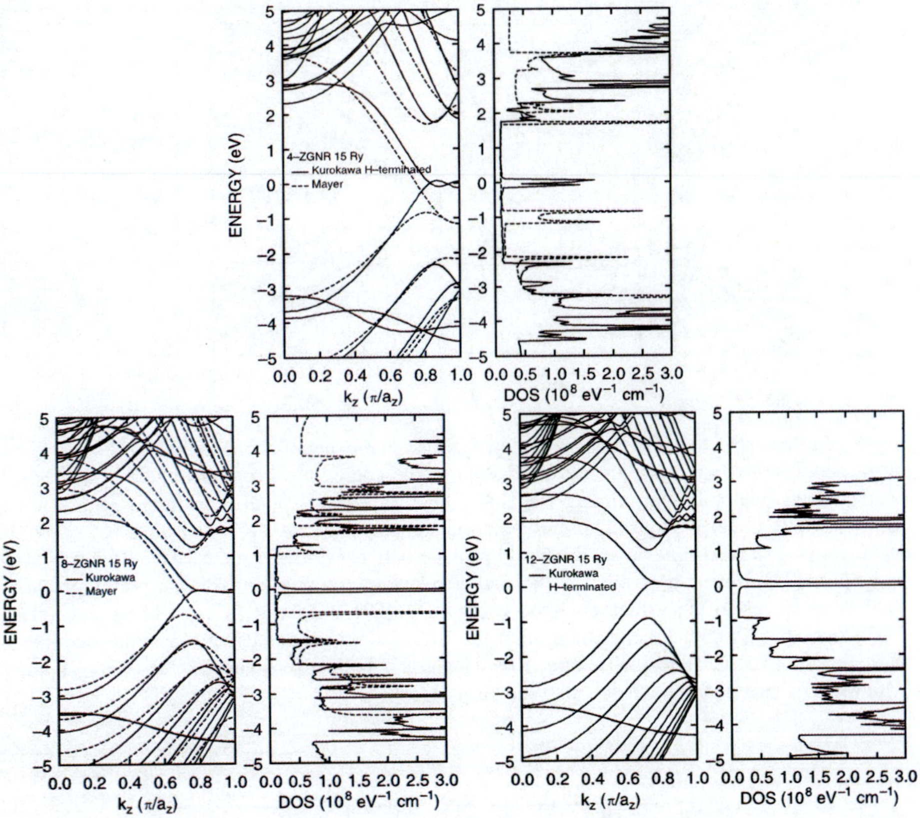

Fig. 3.23 Band structure and density of states of an $N_a = 4$ graphene nanoribbon (4-ZGNR, *left*), an 8-ZGNR (*center*), and a 12-ZGNR (*right*) with zigzag edges obtained using the Mayer pseudopotentials with bare edges (*dashed lines* for the 4-ZGNR and 8-ZGNR only) and the Kurokawa pseudopotentials with H termination of the edge C atoms (*solid lines*). In both cases the ZGNRs exhibit semimetallic behavior, in agreement with Ezawa's tight-binding results [15] and Pisani's 'non-magnetic' LDA calculations [51]. A gap is expected to open when accounting for spin polarization effects, as shown by first-principles self-consistent LSDA calculations [51,60]

negligible effect that the cutoff energy has as far as the few bands close to the Fermi level are concerned.) Results using a real-space approach [66] are also shown in the case of the (5,5) nanotube. The small difference obtained by using real-space or **k**-space methods can be attributed to the truncation at high spatial frequencies (large **G**-vectors) by the latter more than by the proximity of the 'neighbor' nanotube implied by the supercell periodicity. Indeed, increasing the size of the supercell beyond twice the diameter of the nanotube does not cause any appreciable difference in the results of the supercell **k**-space method (see the left and right panels of Fig. 3.26, for example). In all cases the Mayer pseudopotentials yield the correct behavior of the π and π^* bands near the Fermi level, but miss by construction the deeper $2s$ and $2p$ valence states as well as many additional higher-energy states. By contrast, the pseudopotentials proposed by Kurokawa yield results much closer

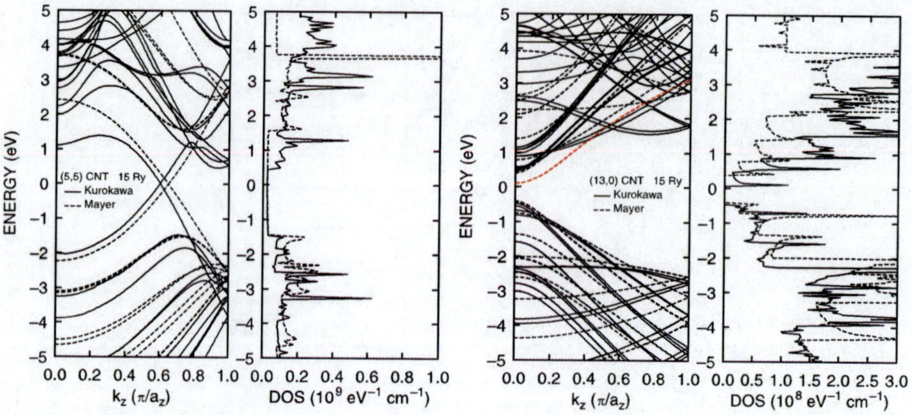

Fig. 3.24 Band structure and density-of-states (DOS) of the metallic armchair (5,5) CNT (*left*) and of the semiconducting (13,0) zigzag CNT (*right*). The energy has been set to zero at mid-gap or band crossing, which is approximately equal to the Fermi level. The dispersion has been obtained using the 'bulk C' local empirical pseudopotentials of Kurokawa et al. [36] (*solid lines*), and those of Mayer [41] (*dashed lines*), which should provide better results since they have been calibrated to graphene. However, the value of the bandgap obtained using Kurokawa's pseudopotentials (0.574 eV) agrees to the values obtained using LDA [59] and CGA [27] (yielding respectively 0.669 and 0.625 eV) much better than the result (0.817 eV) obtained using Mayer's pseudopotentials. For the (5,5) CNT results obtained by Zhang and Polizzi [66] using a real-space approach with Mayer's pseudopotentials are also shown (*circles*)

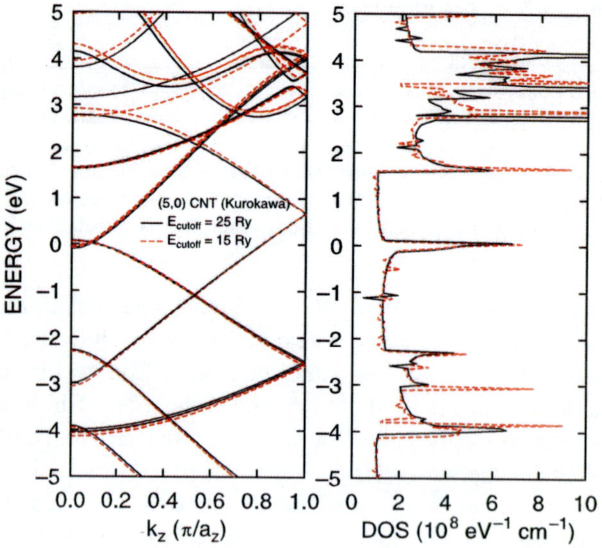

Fig. 3.25 Band structure and density-of-states (DOS) of the metallic zigzag (5,0) CNT (*left*) obtained using a cutoff energy of 25 Ry (*black solid lines*) and 15 Ry (*dashed red lines*)

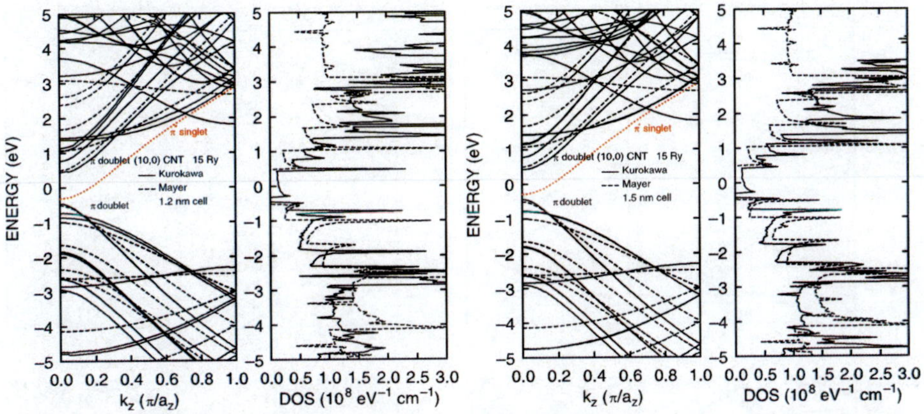

Fig. 3.26 Band structure and density-of-states (DOS) of the semiconducting (10,0) zigzag CNT obtained, as in Fig. 3.24, using the 'bulk C' local empirical pseudopotentials of Kurokawa et al. [36] (*solid lines*), and those of Mayer [41] (*dashed lines*). The energy has been set to zero at mid-gap or band crossing, which is approximately equal to the Fermi level. The *left panels* show results obtained using a supercell with square cross-section of 1.2 nm, at *right* results obtained by enlarging the cell side to a length of 1.5 nm. The bands 'move' by less than 20 meV when increasing the cell size. Note that the value of the bandgap obtained using Kurokawa's pseudopotentials (0.1443 eV) is significantly smaller than the values obtained using LDA [59] and CGA [27], 0.764 eV in both cases) because of the presence of the π^* singlet band (shown by a *red dotted line*) which the choice of Kurokawa pseudopotentials pushes to low energies inside the $\pi - \pi^*$ doublet gap. However, the magnitude of the $\pi - \pi^*$ doublet gap (0.8695 eV) is much closer to the expected value. Results obtained using Mayer's pseudopotentials yield a much larger gap, 1.0141 eV

to *ab-initio* results in the cases of (5,5) and large-diameter $(n,0)$ CNTs. However, in the latter cases (see for example the case of (10,0) nanotubes shown in Fig. 3.26) the π^* singlet is pushed within the $\pi - \pi^*$ doublet gap resulting in an energy gap at Γ much lower than expected trends and ab-initio calculations [27, 53]. Blase [6] and later Gulseren et al. [27] have attributed this to $\sigma^* - \pi^*$ hybridization due to the high curvature of small-radius CNTs, effect which seems to be overestimated by the Kurokawa pseudopotentials (see the σ^* and π^* singlet bands for graphene already mentioned as responsible for this effect in the caption of Fig. 3.17). This is emphasized in Fig. 3.27 – showing the band-structure of several $(n,0)$ zigzag CNTs – and especially in Fig. 3.28 showing the energy of the π and π^* (almost) doubly degenerate bands and of the π-singlet band as a function of diameter for $(n,0)$ CNTs with n spanning the range 4–15. It can be seen that first-principles calculations predict metallic behavior for *all* $(n,0)$ CNTs for $n \leq 6$, the Kurokawa empirical pseudopotentials predict this behavior for $n \leq 9$, while employing the Mayer one-electron pseudopotentials results in metallic behavior only for $n = 3p$ (p =integer) for any n, since these pseudopotentials do not account for the σ^* and π^* singlet bands.

Looking at Figs. 3.27 and 3.28 we should also note that for zigzag nanotubes with diameter smaller than about 2 nm (that is, for $(n,0)$ CNTs with $n \leq 5$), the Kurokawa pseudopotentials fail quite dramatically, always yielding semimetallic behavior and

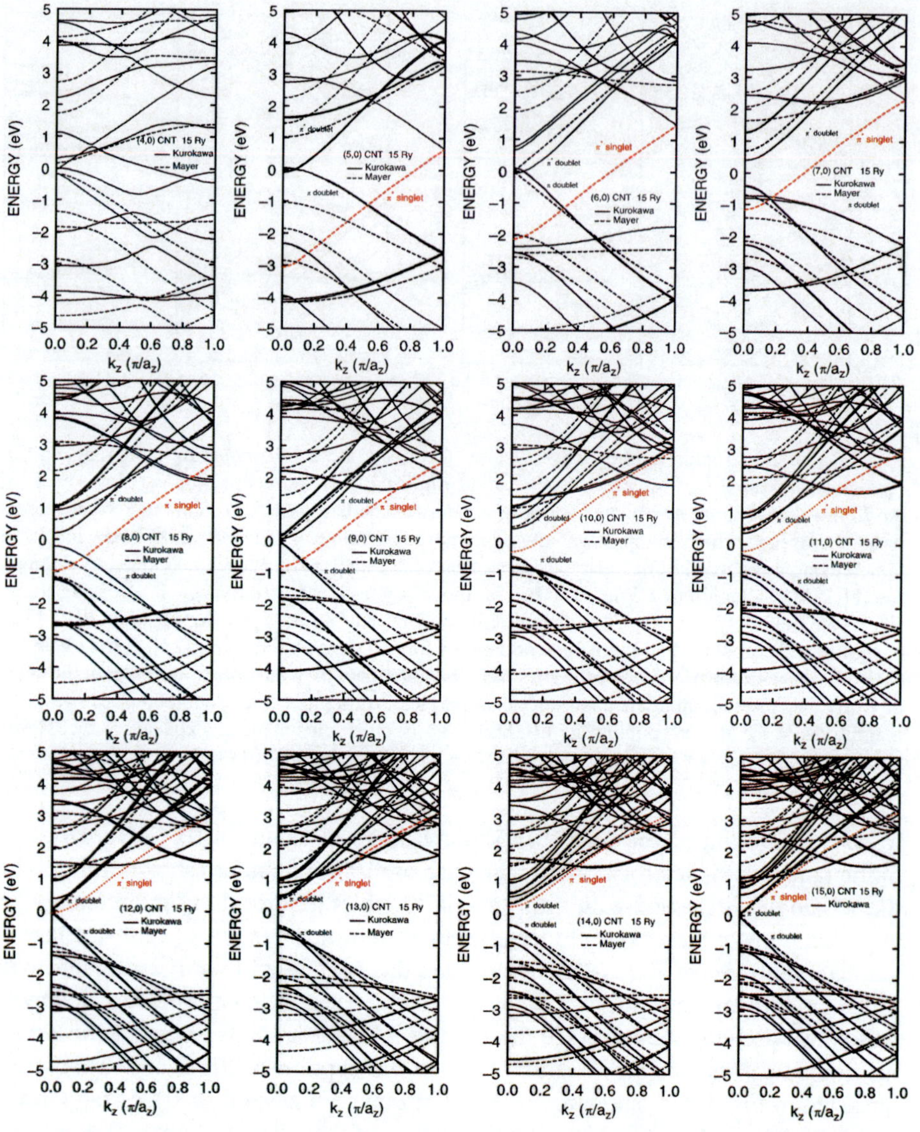

Fig. 3.27 Band structure of several $(n, 0)$ CNTs illustrating the opening and closing of the $\pi - \pi^*$ doublet gap as well as the 'intrusion' of the π^* singlet band within this gap as the tube diameter shrinks. These results have been obtained using the Kurokawa (*solid lines*) and Mayer (*dashed lines*) C pseudopotentials. The energy has been set to zero at $\pi - \pi^*$ doublets mid-gap, roughly equal to the Fermi level. Note how the use of the Kurokawa pseudopotentials accounts for curvature effects (the presence of the π^*-singlet band, the slight opening of the gap for the 'semimetallic' $(3p, 0)$ tubes), even in the absence of additional atomic displacement with the corresponding changes in bond distance and angles which have been deemed necessary to obtain these effects by some of those employing first-principles approaches [27]

unexpected dispersion. In Fig. 3.27 we summarize our results for all of the $(n,0)$ zigzag nanotubes we have considered.

A word of caution regarding the 'correctness' (or lack thereof) of our results using the Mayer or Kurokawa pseudopotentials. We have compared our results with 'first-principles' calculations (usually DFT+LDA and the occasional GGA or GW correction [27,33,53]). The energy gaps and dispersion found in these papers are not always consistent among themselves. For example the MP3 (Quantum Chemistry) approach followed by Bulusheva et al. [9] yields significantly different gaps, while the quality of experimental data on the gap dependence on tube diameter, all of them from the same Harvard group [46–48], is hard to assess given the daunting practical difficulty of isolating CNTs of the same chirality (and, so, diameter). Also, the small density of states associated with the π^*-singlet band could render it hard to detect optically and electrically.

3 Electronic Transport

The Schrödinger equation we have solved so far, (3.4), is limited in two important aspects: First, it assumes full periodicity of the external potential, so that **k**-space techniques may be employed. Second, it also assumes full periodicity of the boundary conditions for the wavefunctions themselves. In many cases we have made use of the latter property in an even stronger form, albeit implicitly, by essentially asking the wavefunctions to vanish at the 'edges' of the structures, thus assuming the systems to be 'closed' in the confinement direction (perpendicular to the plane of the surfaces or interfaces for layers and films) or directions (perpendicular to the axial direction of wires and nanotubes) and periodic along the transport direction(s). Dealing with electronic transport in these systems we may still retain the former assumption, but we must obviously relax the latter requirement.

Depending on the expected degree of coherence exhibited by the wavefunctions within the system, we may treat transport semiclassical or quantum mechanically. We shall always assume the wavefunctions as coherent along the cross section of the device (i.e., along the thickness of a layer, width of a nanoribbon, or over the cross section of a nanowire or nanotube). (If that were not the case, we would most likely return to the case of large devices in which a bulk, 3D band structure and bulk, 3D semiclassical Boltzmann equation would handle transport in a sufficiently accurate way.) But we shall also assume either devices large enough along the transport direction as to lead to incoherent semiclassical transport, or short enough to be amenable to a quantum description based on the Pauli Master equation [18,19]. This case will be reviewed in Sect. 3.2 below. Here we shall consider the opposite case of long structures.

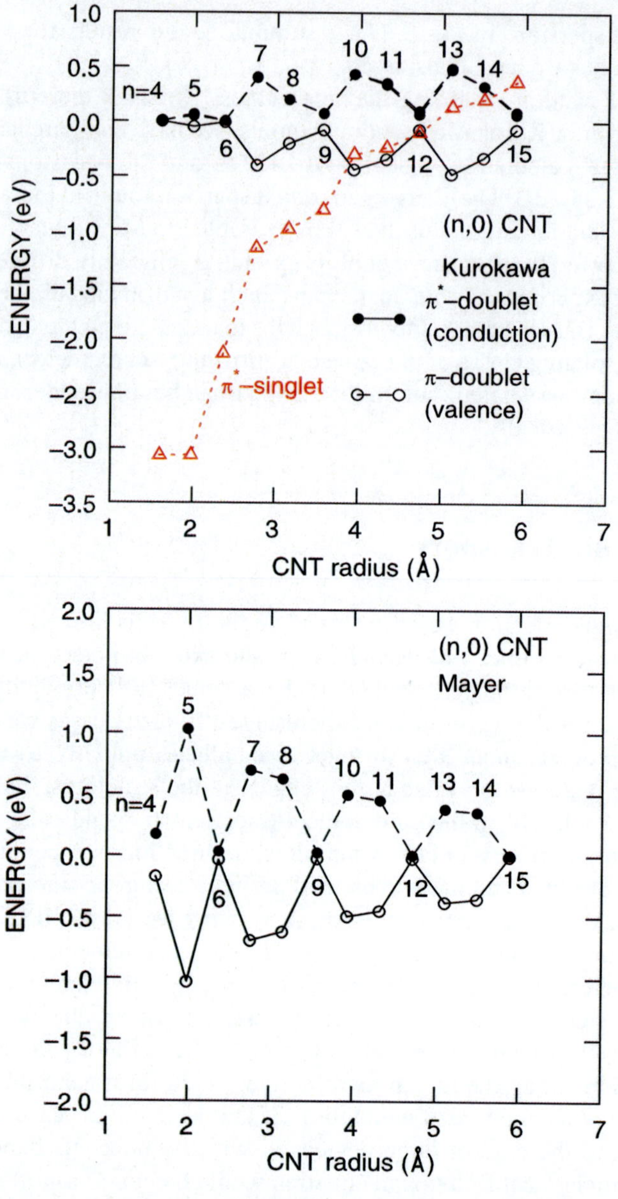

Fig. 3.28 *Top*: Maximum (minimum) energy of the conduction π^*-bands (valence π-bands) as a function of diameter of $(n,0)$ CNTs obtained using the Kurokawa pseudopotentials. The energies are measured from the $\pi-\pi^*$-doublet mid-gap (approximately equal to the Fermi level in the absence of the π^* singlet) to emphasize the periodic variation of the band gap with the chiral number n. Note the quasi-periodic oscillations of the $\pi-\pi^*$ gap as n varies between $3p$ (with p an integer), corresponding to a very small gap which would vanish in absence of curvature effects, $3p+2$ and $3p+1$, the latter case yielding the largest gap in analogy with the situation observed for graphene nanoribbons in Fig. 3.22. Note, however, that the hybridization of the σ^* and π^* orbitals

3.1 Semiclassical Transport

In dealing with transport in structures which exhibit electronic coherence over their cross-section, but incoherent along the transport direction, we are brought back to semiclassical transport in reduced dimensionality. To fix the ideas, let's consider a free-standing thin Si layers (with dangling bonds terminated by H), in the presence of a non-uniform field along the direction perpendicular to its surfaces (the 'confinement' direction along the z axis) and also along the transport directions on the (x,y) plane. For simplicity, let's assume uniformity along the y direction. Then we can imagine of 'slicing' the structure along $x = constant$-planes. For a potential varying along the x direction slowly enough so that this variation may be ignored as far as the band structure is concerned, each slice can be viewed as a 2D periodic structure, with ionic and external periodic potentials, amenable to the description employed before. If the 'thickness' of each slice is larger than the electronic coherence length, then each slice can be viewed as infinitely long along the z-axis and we may describe electronic transport by means of a two-dimensional Boltzmann equation. In essence, the dependence of the external potential $V(x,z)$ on the variable z can be viewed as a parametric dependence in the Schrödinger equation, decoherence effective killing memory of the phases of the wavefunctions as we move from one slice to the next (see the discussion in [18]). This is quite similar to the case of transport in Si inversion layers considered in [17], the main difference being the more complicated band structure considered here.

3.1.1 Electron Transport in Thin Si Inversion Layers as an Example

Let's consider the specific example of a free standing, H-terminated 12-cell-thick ($\approx$6.52 nm) Si layer. A potential difference between the two (100) surfaces is assumed to be applied externally – mimicking the effect of a gate contact – inducing a sheet electronic charge of 10^{13} electrons cm^{-2}. Equation (3.4) is solved self-consistently with Poisson equation, resulting in the band structure illustrated in Fig. 3.29. A uniform field is assumed to be applied along the (transport) x-axis, but assumed to be weak enough as to leave the band structure unaltered.

caused by the increasing curvature of the CNTs at small diameters pushes the energy of the π^*-singlet states within the $\pi - \pi^*$ gap for n smaller than about 15 and ultimately closes the gap for $n < 10$. First-principle results predict this 'gap closing' for $n < 7$, instead [27]. Results for $n < 5$ are questionable because of the expected extremely strong curvature effects which have to be treated using first-principles methods. *Bottom*: As in the *top* frame, but showing results obtained using the Mayer pseudopotentials. Note the correct periodic oscillations of the gap with chiral number n with period 3. However, the inability of the Mayer pseudopotentials to yield the singlet π^* and σ^* bands result in the prediction of semiconducting behavior also for small-diameter nanotubes, notably, the (4,0) and (5,0) CNTs

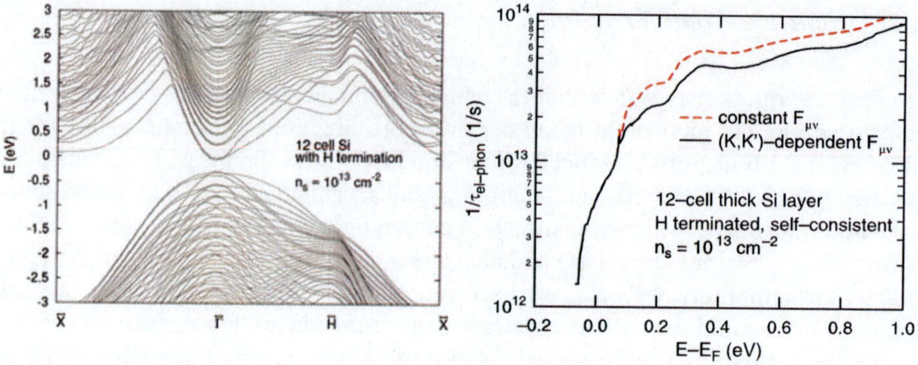

Fig. 3.29 *Left*: Band structure of a 12-cell-thick H-terminated Si layer in the presence of a self-consistent potential inducing an electron sheet density of 10^{13} cm^{-2}. *Right*: Nonpolar electron–phonon scattering rates at 300 K in the layer at left calculated using either the full $(\mathbf{K},\mathbf{K}')$-dependent overlap factor, (3.44) (*black solid line and symbols*), or the approximated expression from (3.45) (*red dashed line, open symbols*)

The solution of (3.4) yields the wavefunctions:

$$\psi_{\mathbf{K}n}(\mathbf{r}) = \frac{e^{i\mathbf{K}\cdot\mathbf{R}}}{A^{1/2}} \sum_{\mathbf{G}} \phi_{\mathbf{G}}^{(n)} \, e^{i\mathbf{G}\cdot\mathbf{r}}. \tag{3.15}$$

where A is the normalization area in the (x,y) plane. The $\mathbf{G}$ vectors span the reciprocal space of the supercell, the index n labels bands or subbands, $\mathbf{K}$ is a two-dimensional wavevector and $\mathbf{R}$ is the real-space coordinate on the transport plane, the (x,y)-plane. Of course the Bloch coefficients $\phi_{\mathbf{G}}^{(n)}$ are different from the bulk-Si coefficients.

We shall now consider electron transport in this layer using a Monte Carlo scheme and obtaining the dependence on the longitudinal electric field of the average electron energy and drift velocity. In addition to the self-consistent Poisson/supercell iteration already mentioned, the major numerical task is the calculation of the electron scattering rates.

3.1.2 Full-Band Scattering Rates in Reduced Dimensionality

We present here a general description of how to employ Fermi golden rule to calculate the first-order scattering rate of carriers with perturbing potentials due to phonons – polar and nonpolar – and ionized impurities.

Scattering Potentials

In general the scattering potential will be of the form:

$$V_{\mathbf{q}}^{(\eta)}(\mathbf{r}) = V_{\mathbf{q}}^{(\eta)} \, e^{i\mathbf{q}\cdot\mathbf{r}}, \tag{3.16}$$

where η may represent, for instance, the type of ionized impurity or the branch of the phonon dispersion. In particular:

$$V_{\mathbf{q}}^{(\eta)} = \left[\frac{\hbar\Delta_\eta^2(\mathbf{q})q^2}{2\rho\omega_{\mathbf{q}}^{(\eta)}}\right]^{1/2} \left\{\begin{array}{c} n_{\mathbf{q}}^{(\eta)1/2} \\ (n_{\mathbf{q}}^{(\eta)}+1)^{1/2} \end{array}\right\}, \tag{3.17}$$

for nonpolar scattering with acoustic phonons of branch η in a crystal of mass-density ρ, with wavevector $\mathbf{q}$, frequency $\omega_{\mathbf{q}}^{(\eta)}$, Bose occupation number $n_{\mathbf{q}}^{(\eta)}$, deformation potential (possibly $\mathbf{q}$-dependent) $\Delta_\eta(\mathbf{q})$:

$$V_{\mathbf{q}}^{(\eta)} = \left[\frac{\hbar(DK)_{op}^2}{2\rho\omega_{\mathbf{q}}^{(\eta)}}\right]^{1/2} \left\{\begin{array}{c} n_{\mathbf{q}}^{(\eta)1/2} \\ (n_{\mathbf{q}}^{(\eta)}+1)^{1/2} \end{array}\right\}, \tag{3.18}$$

for nonpolar scattering with optical acoustic phonons of branch η with optical deformation potential $(DK)_{op}$:

$$V_{\mathbf{q}}^{(\eta)} = \frac{e\mathscr{F}}{q\varepsilon_r(\mathbf{q},\omega_{\mathbf{q}}^{(\eta)})} \left\{\begin{array}{c} n_{\mathbf{q}}^{(\eta 1/2)} \\ (n_{\mathbf{q}}^{(\eta)}+1)^{1/2} \end{array}\right\}, \tag{3.19}$$

with $\mathscr{F}^2 = (\hbar\omega_{\mathbf{q}}^{(\eta)}/2)(1/\varepsilon_\infty - 1/\varepsilon_0)$ for polar scattering with optical phonons ($\varepsilon_r(\mathbf{q},\omega)$ being the relative dielectric function and ε_∞ and ε_0 the optical and static dielectric constants):

$$V_{\mathbf{q}}^{(\eta)} = \frac{e^2}{\varepsilon_{val}(\mathbf{q})} \frac{1}{q^2+\varepsilon_s\beta^2(\mathbf{q},\omega)/\varepsilon_{val}(\mathbf{q})}, \tag{3.20}$$

for scattering with ionized impurities. In the last equation the free-carrier screening wavevector is, as usual, $\beta^2(\mathbf{q}) = [(e^2n)/(\varepsilon_{sc}k_BT)][\mathscr{F}_{-1/2}(\eta)/\mathscr{F}_{1/2}(\eta)]G(\xi,\eta)$, and $\varepsilon_{val}(\mathbf{q})$ is the valence band dielectric function, usually taken simply as the static dielectric constant. (Note that the dimensions of $|V_{\mathbf{q}}^{(\eta)}|^2$ are joules$^2\times$ m^3 in (3.17)–(3.19) and joules$^2\times$ m^6 in (3.20). The latter should be multiplied by the density of ionized impurities, so also in this case the 'effective' dimension will be measured in joules$^2\times$ m^3).

General Formulation

It is worth recalling that we are describing the system within the supercell approach. This means that we could use a full 3D description of transport and express the scattering rates with the bulk expression employed in 3D full-band Monte Carlo simulations [16]. This would be a rigorous way to describe transport in mixed bulk/confined situation. Two important such cases are those of (1) electrons in an inversion layer which are confined in two-dimensions at low energy, but become fully delocalized bulk electrons at energies larger than the potential energy deep in the substrate and (2) electrons undergoing 'vertical transport' in a superlattice and undergoing 'trapping' into 2D quantum-well states. This mixed 'bulk-/2D' problem has been investigated treated phenomenologically in the past [17] and we see that the supercell approach can provide – at least in principle – a viable and correct solution.

Yet, it is convenient to note that whenever we deal only with fully confined states whose energy $E_n(\mathbf{k}) = E_n(\mathbf{K}, k_z)$ and Bloch coefficients $\phi^{(n)}_{\mathbf{G},\mathbf{k}} = \phi^{(n)}_{\mathbf{G},\mathbf{K},k_z}$ do not depend on k_z (or on $\mathbf{K}$ in the case of 1D confinement) it is convenient to make use of this fact and reduce the numerical complexity of the problem by writing the scattering rate of an electron in band (or subband) n and in-plane wavevector $\mathbf{K}$ due to a perturbation potential $V^{(\eta)}_{\mathbf{q}}$ as a sum only over 2D states as follows:

$$\frac{1}{\tau^{(\eta)}_n(\mathbf{K})} = \frac{2\pi}{\hbar} \sum_{\mathbf{K}'n'\mathbf{q}} |\langle \mathbf{K}'n'|V^{(\eta)}_{\mathbf{q}}|\mathbf{K}n\rangle|^2 \; \delta[E_n(\mathbf{K}) - E_{n'}(\mathbf{K}') \pm \hbar\omega^{(\eta)}_{\mathbf{q}}]. \tag{3.21}$$

The matrix element $\langle \mathbf{K}'n'|V^{(\eta)}_{\mathbf{q}}|\mathbf{K}n\rangle$ is given by:

$$\langle \mathbf{K}'n'|V^{(\eta)}_{\mathbf{q}}|\mathbf{K}n\rangle = \frac{1}{A}\int dz \int d\mathbf{R}\; \psi^*_{\mathbf{K}'n'}(\mathbf{R},z)\; V^{(\eta)}_{\mathbf{q}} e^{iq_z z}\, e^{i\mathbf{Q}\cdot\mathbf{R}}\; \psi_{\mathbf{K}n}(\mathbf{R},z). \tag{3.22}$$

Using (3.15) we can write the matrix element above as:

$$\begin{aligned}\langle \mathbf{K}'n'|V^{(\eta)}_{\mathbf{q}}|\mathbf{K}n\rangle = \frac{1}{A}V^{(\eta)}_{\mathbf{q}} \sum_{\mathbf{G}\mathbf{G}'} \int d\mathbf{R}\; e^{i(\mathbf{K}-\mathbf{K}'+\mathbf{G}_\parallel-\mathbf{G}'_\parallel+\mathbf{Q})\cdot\mathbf{R}} \\ \times \int dz\; \phi^{(n')*}_{\mathbf{G}'\mathbf{K}'}\; e^{i(q_z+G_z-G'_z)z} \phi^{(n)}_{\mathbf{G}\mathbf{K}}.\end{aligned} \tag{3.23}$$

We have employed a rather strange notation by leaving the 'constants' $\phi^{(n')*}_{\mathbf{G}'\mathbf{K}'}$ and $\phi^{(n)}_{\mathbf{G}\mathbf{K}}$ inside the integration over z. This will be convenient when attempting to recover from (3.23) the 'usual' expressions for the scattering rates of a two-dimensional electron gas (2DEG). Even more notably, when using the mixed supercell/envelope scheme described below (see Sect. 3.2.1 and in particular (3.61)), the same expressions, (3.21) and (3.23), will describe scattering in an open boundary-conditions 1D quantum transport situation inhomogeneous along z (such as in nanowires, nanoribbons, and nanotubes with a driving external field along the z-direction, this direction now being the transport direction) the only difference being that in this case the Bloch coefficients $\phi^{(n)}_{\mathbf{G}\mathbf{K}}$ will acquire a dependence on z as solutions of (3.61).

Now, proceeding in a conventional way, let us write $\mathbf{R} = \mathbf{R}_l + \rho$, where $\mathbf{R}_l$ is a 2D lattice site and ρ the 2D-vector spanning the 2D cell. Thus:

$$\begin{aligned}&\frac{1}{A}\sum_{\mathbf{G}\mathbf{G}'} \int d\mathbf{R}\; e^{i(\mathbf{K}-\mathbf{K}'+\mathbf{G}_\parallel-\mathbf{G}'_\parallel+\mathbf{Q})\cdot\mathbf{R}} \quad f_{\mathbf{G},\mathbf{G}',\mathbf{K},\mathbf{K}'} \\ &= \frac{1}{A}\sum_{\mathbf{G}\mathbf{G}'}\sum_{l} e^{i(\mathbf{K}-\mathbf{K}'+\mathbf{G}_\parallel-\mathbf{G}'_\parallel+\mathbf{Q})\cdot\mathbf{R}_l} \int_{\Omega_{2D}} d\rho\; e^{i(\mathbf{K}-\mathbf{K}'+\mathbf{G}_\parallel-\mathbf{G}'_\parallel+\mathbf{Q})\cdot\rho} \quad f_{\mathbf{G},\mathbf{G}',\mathbf{K},\mathbf{K}'},\end{aligned} \tag{3.24}$$

where Ω_{2D} is the area of the 2D cell and we have indicated with $f_{\mathbf{G},\mathbf{G}',\mathbf{K},\mathbf{K}'}$ the z-integral in (3.23). Now, the sum over 2D-lattice sites yields a non-vanishing contribution only when $\mathbf{K}-\mathbf{K}'+\mathbf{Q}$ is equal to some vector of the 2D reciprocal lattice, $\mathbf{G}''_{\parallel}$. Then, relabeling the dummy variables $\mathbf{G}''$ as $\mathbf{G}$, $\mathbf{G}$ as $\mathbf{G}'$, and $\mathbf{G}'$ as $\mathbf{G}''$, (3.24) becomes:

$$\sum_{\mathbf{G}_{\parallel}} \delta(\mathbf{K}-\mathbf{K}'+\mathbf{Q}+\mathbf{G}_{\parallel}) \sum_{\mathbf{G}'\mathbf{G}''} \int_{\Omega_{2D}} \frac{1}{\Omega_{2D}} d\rho\, e^{i(\mathbf{G}_{\parallel}+\mathbf{G}'_{\parallel}-\mathbf{G}''_{\parallel})\cdot\rho}\, f_{\mathbf{G}',\mathbf{G}'',\mathbf{K},\mathbf{K}'}. \tag{3.25}$$

Now, using the result of (3.25) into (3.23), we obtain:

$$\langle \mathbf{K}'n'|V_{\mathbf{q}}^{(\eta)}|\mathbf{K}n\rangle = \sum_{\mathbf{G}_{\parallel}} \delta(\mathbf{K}-\mathbf{K}'+\mathbf{Q}+\mathbf{G}_{\parallel})\, V_{\mathbf{Q},q_z}^{(\eta)} \sum_{\mathbf{G}'\mathbf{G}''} \frac{1}{\Omega_{2D}} \int_{\Omega_{2D}} \rho\, e^{i(\mathbf{G}_{\parallel}+\mathbf{G}'_{\parallel}-\mathbf{G}''_{\parallel})\cdot\rho}$$
$$\times \int dz\, \phi_{\mathbf{G}''\mathbf{K}'}^{(n')*}\, e^{i(q_z+G'_z-G''_z)z}\, \phi_{\mathbf{G}'\mathbf{K}}^{(n)}. \tag{3.26}$$

Thus the scattering rate is given by:

$$\frac{1}{\tau_n^{(\eta)}(\mathbf{K})} = \frac{2\pi}{\hbar} \sum_{\mathbf{K}'n'} \int \frac{dq_z}{2\pi} \sum_{\mathbf{G}_{\parallel}} \left| V_{\mathbf{K}-\mathbf{K}'+\mathbf{G}_{\parallel},q_z}^{(\eta)} \sum_{\mathbf{G}'\mathbf{G}''} \frac{1}{\Omega_{2D}} \int_{\Omega_{2D}} d\rho\, e^{i(\mathbf{G}_{\parallel}+\mathbf{G}'_{\parallel}-\mathbf{G}''_{\parallel})\cdot\rho} \right.$$
$$\left. \times \int dz\, \phi_{\mathbf{G}''\mathbf{K}'}^{(n')*}\, e^{i(q_z+G'_z-G''_z)z}\, \phi_{\mathbf{G}'\mathbf{K}}^{(n)} \right|^2 \delta\left[E_n(\mathbf{K})-E_{n'}(\mathbf{K}')\pm\hbar\omega_{\mathbf{K}-\mathbf{K}'+\mathbf{G}_{\parallel},q_z}^{(\eta)}\right]. \tag{3.27}$$

Finally, recalling that $\int_{\Omega_{2D}} d\rho\, e^{i\mathbf{G}_{\parallel}\cdot\rho} = \Omega_{2D}\,\delta_{\mathbf{G}\mathbf{0}}$, we can re-write this expression as:

$$\frac{1}{\tau_n^{(\eta)}(\mathbf{K})}$$
$$= \frac{2\pi}{\hbar} \sum_{\mathbf{K}'n'} \int \frac{dq_z}{2\pi} \sum_{\mathbf{G}_{\parallel}} \left| V_{\mathbf{K}-\mathbf{K}'+\mathbf{G}_{\parallel},q_z}^{(\eta)} \sum_{\mathbf{G}'_{\parallel}G'_zG''_z} \int dz\, \phi_{\mathbf{G}_{\parallel}+\mathbf{G}'_{\parallel},G''_z,\mathbf{K}'}^{(n')*}\, e^{i(q_z+G'_z-G''_z)z}\, \phi_{\mathbf{G}'_{\parallel},G'_z,\mathbf{K}}^{(n)} \right|^2$$
$$\times \delta\left[E_n(\mathbf{K})-E_{n'}(\mathbf{K}')\pm\hbar\omega_{\mathbf{K}-\mathbf{K}'+\mathbf{G}_{\parallel},q_z}^{(\eta)}\right]. \tag{3.28}$$

Considering only normal (N) processes (i.e., collapsing the sum over $\mathbf{G}_{\parallel}$ to the lone term $\mathbf{G}_{\parallel}$ needed to map $\mathbf{K}-\mathbf{K}'$ inside the first 2D BZ), we have:

$$\frac{1}{\tau_n^{(\eta)}(\mathbf{K})} \approx \frac{2\pi}{\hbar} \sum_{\mathbf{K}'n'} \int \frac{dq_z}{2\pi} \left| V_{\mathbf{K}-\mathbf{K}',q_z}^{(\eta)} \sum_{\mathbf{G}_{\parallel}G_zG'_z} \int dz\, \phi_{\mathbf{G}_{\parallel},G'_z,\mathbf{K}'}^{(n')*}\, e^{i(q_z+G_z-G'_z)z}\, \phi_{\mathbf{G}_{\parallel},G_z,\mathbf{K}}^{(n')} \right|^2$$
$$\times \delta\left[E_n(\mathbf{K})-E_{n'}(\mathbf{K}')\pm\hbar\omega_{\mathbf{K}-\mathbf{K}',q_z}^{(\eta)}\right]. \tag{3.29}$$

To express this result in a more compact notation better suited to simplifications, we can define the functions:

$$\xi^{(n)}_{\mathbf{G}_{\|}\mathbf{K}}(z) = \sum_{G_z} \phi^{(n)}_{\mathbf{GK}} e^{iG_z z}. \tag{3.30}$$

Then from (3.29) we have:

$$\begin{aligned}\mathscr{I}^{(2D)}_{\mathbf{K},\mathbf{K}',n,n',\mathbf{G}_{\|}}(q_z) &= \sum_{\mathbf{G}'_{\|}} \sum_{G'_z G''_z} \int dz\, \phi^{(n')*}_{\mathbf{G}_{\|}+\mathbf{G}'_{\|},G''_z,\mathbf{K}'}\, e^{i(q_z+G'_z-G''_z)z}\, \phi^{(n)}_{\mathbf{G}'_{\|},G'_z,\mathbf{K}} \\ &= \int dz \sum_{\mathbf{G}'_{\|}} \xi^{(n')*}_{\mathbf{G}_{\|}+\mathbf{G}'_{\|},\mathbf{K}'}(z)\, e^{iq_z z}\, \xi^{(n)}_{\mathbf{G}'_{\|},\mathbf{K}}(z).\end{aligned} \tag{3.31}$$

For N processes this becomes simply:

$$\mathscr{I}^{(2D)}_{\mathbf{K},\mathbf{K}',n,n'}(q_z) = \int dz \sum_{\mathbf{G}_{\|}} \xi^{(n')*}_{\mathbf{G}_{\|},\mathbf{K}'}(z)\, e^{iq_z z}\, \xi^{(n)}_{\mathbf{G}_{\|},\mathbf{K}}(z). \tag{3.32}$$

Using this expression for the overlap factor, (3.28) becomes:

$$\begin{aligned}\frac{1}{\tau^{(\eta)}_n(\mathbf{K})} &= \frac{2\pi}{\hbar} \sum_{\mathbf{K}'n'} \int \frac{dq_z}{2\pi} \sum_{\mathbf{G}_{\|}} \left| V^{(\eta)}_{\mathbf{K}-\mathbf{K}'+\mathbf{G}_{\|},q_z}\, \mathscr{I}^{(2D)}_{\mathbf{K},\mathbf{K}',n,n',\mathbf{G}_{\|}}(q_z)\right|^2 \\ &\times \delta\left[E_n(\mathbf{K}) - E_{n'}(\mathbf{K}') \pm \hbar\omega^{(\eta)}_{\mathbf{K}-\mathbf{K}'+\mathbf{G}_{\|},q_z}\right].\end{aligned} \tag{3.33}$$

With this notation the scattering rate obtained by accounting for N processes only reduces to:

$$\frac{1}{\tau^{(\eta)}_n(\mathbf{K})} = \frac{2\pi}{\hbar} \sum_{\mathbf{K}'n'} \int \frac{dq_z}{2\pi} \left| V^{(\eta)}_{\mathbf{K}-\mathbf{K}',q_z}\, \mathscr{I}^{(2D)}_{\mathbf{K},\mathbf{K}',n,n'}(q_z)\right|^2 \delta\left[E_n(\mathbf{K}) - E_{n'}(\mathbf{K}') \pm \hbar\omega^{(\eta)}_{\mathbf{K}-\mathbf{K}',q_z}\right]. \tag{3.34}$$

No additional simplification is possible when the full wavefunctions $\xi^{(n)}_{\mathbf{G}_{\|}\mathbf{K}}(z)$ (see (3.30) and (3.31)) must be used. This case presents the obvious numerical difficulty caused by the size of the array $\xi^{(n)}_{\mathbf{GK}}(z)$: Storing these (complex) wavefunctions for $\sim 10^4$ **G**-vectors at each of the $\sim 10^3$ **K**-points used to tabulate the band-structure over the wedge of the BZ and for each of the ~ 10 bands n requires ~ 1 GB of storage. However a customary simplification can be obtained by embracing the pure 'envelope' approximation by ignoring the Bloch components $e^{i\mathbf{G}_{\|}\cdot\mathbf{R}}$ in (3.23). This is fully equivalent to ignoring overlap-factor effects in bulk calculations. Then from the full wavefunction $\psi_{\mathbf{K}n}(\mathbf{R},z)$ given by (3.15) we can factor a z-only-dependent envelope:

$$\zeta^{(n)}_{\mathbf{K}}(z) = \sum_{\mathbf{G}} \phi^{(n)}_{\mathbf{GK}} e^{iG_z z}. \tag{3.35}$$

Inserting this into (3.23) we obtain:

$$\langle \mathbf{K}'n'|V_{\mathbf{q}}^{(\eta)}|\mathbf{K}n\rangle \approx \frac{1}{A}V_{\mathbf{q}}^{(\eta)} \sum_{\mathbf{GG}'}\int d\mathbf{R}\, e^{i(\mathbf{K}-\mathbf{K}'+\mathbf{Q})\cdot\mathbf{R}} \int dz\, \phi_{\mathbf{G}'\mathbf{K}'}^{(n')*}(z)\, e^{i(q_z+G_z-G'_z)z}\phi_{\mathbf{GK}}^{(n)}(z)$$

$$= V_{\mathbf{q}}^{(\eta)}\, \delta_{\mathbf{K}-\mathbf{K}+\mathbf{Q},\mathbf{0}} \int dz\, \zeta_{\mathbf{K}'}^{(n')*}(z)\, e^{iq_z z}\, \zeta_{\mathbf{K}}^{(n)}(z) \tag{3.36}$$

so that the 2D overlap factor $\mathscr{I}^{(2D)}_{\mathbf{K},\mathbf{K}',n,n',\mathbf{G}_\parallel}(q_z)$ becomes simply:

$$\tilde{\mathscr{I}}^{(2D)}_{\mathbf{K},\mathbf{K}',n,n'}(q_z) \approx \int dz\, \zeta_{\mathbf{K}'}^{(n')*}(z)\, e^{iq_z z}\, \zeta_{\mathbf{K}}^{(n)}(z), \tag{3.37}$$

which, except for the more complicated subband dispersion, allows us to formulate the scattering rate (3.34) in terms of its 'usual' expression in 2D.

Numerical Evaluation

In order to obtain a numerically computable expression, let's consider again the more generally valid expression, (3.33). Having discretized the 2D BZ with squares centered at points $\mathbf{K}_j$, we can write:

$$\frac{1}{\tau_n^{(\eta)}(\mathbf{K})} \approx \frac{2\pi}{\hbar} \sum_{jn'}^{*} \int \frac{dq_z}{2\pi} \sum_{\mathbf{G}_\parallel} \left| V^{(\eta)}_{\mathbf{K}-\mathbf{K}_j+\mathbf{G}_\parallel,q_z}\, \mathscr{I}^{(2D)}_{\mathbf{K},\mathbf{K}_j,n,n',\mathbf{G}_\parallel}(q_z)\right|^2 \frac{L_{n'}(w_j)}{|\nabla_{2D}E_{n'j}|}, \tag{3.38}$$

where the meaning of the symbols can be found following (3.8) above, $w = (E_{final} - E_{n'j})/|\nabla_{2D}E_{n'j}|$, and E_{final} is the final energy for a particular process and square j.

Special Cases

We are interested here in the specific case of a nonpolar semiconductor, Si, and we will present results obtained by considering only nonpolar scattering with phonons. However, having come so far, it is worth carrying on the discussion in full generality and consider several scattering processes and polar semiconductors as well.

Equation (3.38) still requires the numerical evaluation of a double integral: First over the z-component of the momentum transfer, q_z, then over the z-coordinate, hidden within the overlap factor $\mathscr{I}^{(2D)}_{\mathbf{K},\mathbf{K}_j,n,n',\mathbf{G}_\parallel}(q_z)$. The latter integration cannot be reduced to any closed-form expression since the wavefunctions are known only numerically. However, in some special cases the integration over q_z can be performed analytically.

First, more practical from a numerical perspective is the N-process-only approximation. This approximation is justified in many cases, since often the matrix element of the scattering potential decreases with increasing momentum transfer (and so with G). In this case (3.34) can be evaluated numerically as:

$$\frac{1}{\tau_n^{(\eta)}(\mathbf{K})} \approx \frac{2\pi}{\hbar} \sum_{jn'}^{*} \int \frac{dq_z}{2\pi} \left| V^{(\eta)}_{\mathbf{K}-\mathbf{K}_j,q_z} \mathscr{I}^{(2D)}_{\mathbf{K},\mathbf{K}_j,n,n'}(q_z) \right|^2 \frac{L_{n'}(w_j)}{|\nabla_{2D} E_{n'j}|}. \tag{3.39}$$

Specific examples which can be evaluated in general are the cases of non-polar electron–phonon scattering, of Fröhlich scattering with potential screened using a Debye-Hückel wavevector q_{DH} – approximated either by the 2D expression $(e^2/\varepsilon_s)n_s/(k_BT)$ (we ignore for now multisubband screening, dynamic effects, and such) or by the 3D expression, $[(e^2n)/(k_BT)]^{1/2}$ –

$$V_{\mathbf{q}}^{(pop,DH)} = \frac{e\mathscr{F}}{(q^2+q_{DH}^2)^{1/2}} \left\{ \begin{array}{c} n_{LO}^{1/2} \\ (n_{LO}+1)^{1/2} \end{array} \right\} = \frac{C_{pop}}{(q^2+q_{DH}^2)^{1/2}}, \tag{3.40}$$

and the case of similarly screened Coulomb interaction with a single ionized impurity:

$$V_{\mathbf{q}}^{(imp,DH)} = \frac{e^2}{\varepsilon_s(q^2+q_{DH}^2)} = \frac{C_{imp}}{q^2+q_{DH}^2}. \tag{3.41}$$

Since the overlap factors are multi-dimensional integrals, it is important to make any possible attempt to perform analytically as many integrations as possible to reduce the computational burden.

Let's consider the term

$$\int \frac{dq_z}{2\pi} \left| V^{(\eta)}_{\mathbf{K}-\mathbf{K}_j,q_z} \mathscr{I}^{(2D)}_{\mathbf{K},\mathbf{K}_j,n,n'}(q_z) \right|^2 \tag{3.42}$$

in (3.39) and see how we can handle it analytically in the special cases mentioned above.

Nonpolar Phonon Scattering (with momentum-independent matrix element). Non-polar scattering with acoustic and optical phonons can be simplified dramatically when the matrix element is assumed to be momentum-independent and the phonon energy is considered constant (zero for acoustic phonons, dispersionless for optical phonons). Thus neither $\hbar\omega_{\mathbf{q}}^{(\eta)}$ nor $V_{\mathbf{q}}^{(\eta)}$ depend on q_z and the only q_z dependence occurs within the overlap factor. Thus the term (3.42) can be handled as follows:

$$\begin{aligned} |V^{(\eta)}|^2 & \int \frac{dq_z}{2\pi} \left| \int dz\, \zeta_{\mathbf{K}'}^{(n')*}(z)\, e^{iq_z z}\, \zeta_{\mathbf{K}}^{(n)}(z) \right|^2 \\ &= |V^{(\eta)}|^2 \int dz \int dz'\, \zeta_{\mathbf{K}'}^{(n')*}(z)\, \zeta_{\mathbf{K}}^{(n)}(z)\, \zeta_{\mathbf{K}'}^{(n')}(z')\, \zeta_{\mathbf{K}}^{(n)*}(z') \int \frac{dq_z}{2\pi} e^{iq_z(z-z')} \\ &= |V^{(\eta)}|^2 \int dz\, \zeta_{\mathbf{K}'}^{(n')*}(z)\, \zeta_{\mathbf{K}}^{(n)}(z) \zeta_{\mathbf{K}'}^{(n')}(z)\, \zeta_{\mathbf{K}}^{(n)*}(z) \\ &= |V^{(\eta)}|^2 \int dz \left|\zeta_{\mathbf{K}'}^{(n')*}(z)\right|^2 \left|\zeta_{\mathbf{K}}^{(n)}(z)\right|^2 = |V^{(\eta)}|^2 \mathscr{F}^{(2D)}_{\mathbf{K}'\mathbf{K}nn'}. \end{aligned} \tag{3.43}$$

In this case (3.34) simplifies to:

$$\frac{1}{\tau_n^{(\eta)}(\mathbf{K})} \approx \frac{2\pi}{\hbar} |V^{(\eta)}|^2 \sum_{\mathbf{K}'n'} \mathscr{F}^{(2D)}_{\mathbf{K}'\mathbf{K}nn'} \delta[E_n(\mathbf{K}) - E_{n'}(\mathbf{K}') \pm \hbar\omega^{(\eta)}]. \tag{3.44}$$

Whenever we are interested in transport not too far from equilibrium (as in mobility calculations) we can assume that the carriers populate only regions of the first BZ not too far from a band extremum $\mathbf{K}_0$. Then one can ignore the $\mathbf{K}$-dependence of the wavefunctions and reduce the computational burden by having to calculate only a single overlap factor for each pair of (sub)bands (n, n'). So, the simplest possible expression for the scattering rate can be derived from (3.44):

$$\begin{aligned} \frac{1}{\tau_n^{(\eta)}(\mathbf{K})} &\approx \frac{2\pi}{\hbar} |V^{(\eta)}|^2 \sum_{\mathbf{K}'n'} \mathscr{F}^{(2D)}_{\mathbf{K}_0\mathbf{K}_0nn'} \delta[E_n(\mathbf{K}) - E_{n'}(\mathbf{K}') \pm \hbar\omega^{(\eta)}] \\ &= \frac{2\pi}{\hbar} |V^{(\eta)}|^2 \sum_{n'} \mathscr{F}^{(2D)}_{\mathbf{K}_0\mathbf{K}_0nn'} \rho^{(n')}[E_n(\mathbf{K}) \pm \hbar\omega^{(\eta)}], \end{aligned} \tag{3.45}$$

where $\rho^{(n)}(E)$ is the density of states in (sub)band n at energy E. Clearly, (3.33) can be simplified by using any (or any combination) of the approximations we have considered here (N-process only, $\mathbf{q}$-independent scattering potential, $\mathbf{q}$-independent dispersion, $\mathbf{K}$-independent wavefunctions), depending on the particular physical system and conditions considered.

Fröhlich Scattering. In the case of Fröhlich scattering (3.42) becomes:

$$C^2_{pop} \int dz \int dz' \, \zeta^{(n')*}_{\mathbf{K}'}(z) \, \zeta^{(n')}_{\mathbf{K}'}(z') \, \zeta^{(n)}_{\mathbf{K}}(z) \, \zeta^{(n)*}_{\mathbf{K}}(z') \int \frac{dq_z}{2\pi} \frac{e^{iq_z(z-z')}}{q_z^2 + Q^2}, \tag{3.46}$$

where $Q = (|\mathbf{K} - \mathbf{K}'|^2 + q^2_{DH})^{1/2}$. The integral over q_z above is easily evaluated:

$$\mathscr{I}^{(2D)}_{pop} = \int \frac{dq_z}{2\pi} \frac{e^{iq_z(z-z')}}{q_z^2 + Q^2} = \pi \frac{e^{-|z-z'|Q}}{Q}, \tag{3.47}$$

so that the scattering rate can be written as:

$$\begin{aligned} \frac{1}{\tau_n^{(pop)}(\mathbf{K})} &\approx \frac{2\pi}{\hbar} \sum_{jn'}^{*} C^2_{pop} \frac{L(w_j)}{|\nabla_{2D} E_j|} \frac{1}{2Q_j} \\ &\times \int dz \int dz' \, \zeta^{(n')*}_{\mathbf{K}_j}(z) \, \zeta^{(n')}_{\mathbf{K}_j}(z') \, e^{-|z-z'|Q_j} \, \zeta^{(n)}_{\mathbf{K}}(z) \, \zeta^{(n)*}_{\mathbf{K}}(z'), \end{aligned} \tag{3.48}$$

where $Q_j = (|\mathbf{K}-\mathbf{K}_j|^2 + q_{DH}^2)^{1/2}$. Writing $g_{\mathbf{KK}'nn'}(z) = \zeta_{\mathbf{K}'}^{(n')*}(z)\zeta_{\mathbf{K}}^{(n)}(z)$, the overlap factor (the last factor in the equation above) can be written as:

$$\int dz \int dz' \; g_{\mathbf{KK}'nn'}(z') \, e^{-|z-z'|Q_j} \; g^*_{\mathbf{KK}'nn'}(z)$$

$$= 2\,\mathrm{Re}\left\{ \int_0^\infty dz \; g_{\mathbf{KK}'nn'}(z) \, e^{-Qz} \int_0^z dz' \; g^*_{\mathbf{KK}'nn'}(z') \, e^{Qz'} \right\}, \qquad (3.49)$$

expression which shows explicitly that the overlap factor is real and provides a form more amenable to numerical integration. Indeed we shall discuss below after (3.58) that the numerical advantage provided by (3.49) consists in the fact that the 'inner' integral over z' depends on z only via the upper integration limit as the integrand does not depend on z. Of course one can approximate $g_{\mathbf{KK}'nn'}(z)$ with $g_{\mathbf{K}_0\mathbf{K}_0nn'}(z)$, where $\mathbf{K}_0$ is the location of the band extremum, in order to reduce the number of integrals to be evaluated and tabulated.

Impurity Scattering. Similarly, in the case of impurity scattering the term (3.42) becomes:

$$C_{imp}^2 \int dz \int dz' \; \zeta_{\mathbf{K}'}^{(n')*}(z) \; \zeta_{\mathbf{K}'}^{(n')}(z') \; \zeta_{\mathbf{K}}^{(n)}(z) \; \zeta_{\mathbf{K}}^{(n)*}(z') \int \frac{dq_z}{2\pi} \frac{e^{iq_z(z-z')}}{(q_z^2+Q^2)^2}. \qquad (3.50)$$

The evaluation of the integral over q_z is a bit more involved. Let's write this integral as:

$$\mathscr{I}_{imp}^{(2D)} = \frac{1}{2\pi} \int dx \, \frac{e^{ipx}}{(x^2+Q^2)^2}, \qquad (3.51)$$

having set $p = z - z'$ and having renamed x the dummy integration variable q_z. Let's consider the case $p > 0$ and let's integrate by parts:

$$\mathscr{I}_{imp}^{(2D)} = \frac{1}{2\pi} \int dx \, \frac{1}{(x-iQ)^2} \frac{e^{ipx}}{(x-iQ)^2} = \frac{1}{2\pi} \frac{-1}{x-iQ} \frac{e^{ipx}}{(x+iQ)^2} \Bigg|_{-\infty}^{\infty}$$

$$+ \frac{1}{2\pi} \int dx \, \frac{1}{x-iQ} \frac{d}{dx} \left[\frac{e^{ipx}}{(x+iQ)^2} \right]. \qquad (3.52)$$

The first term on the right-hand side vanishes and we are left with an integrand which for $\mathrm{Im}(x) > 0$ now has as singularity only a pole at $x = iQ$:

$$\mathscr{I}_{imp}^{(2D)} = \frac{1}{2\pi} \int dx \, \frac{e^{ipx}}{x-iQ} \, \frac{ip(x+iQ)-2}{(x+iQ)^3}. \qquad (3.53)$$

Since we have assumed $p > 0$, we can integrate over the upper half of the complex plane enclosing the single pole $x = iQ$, and obtain:

$$\mathscr{I}_{imp}^{(2D)} = \frac{1}{4} \frac{e^{-pQ}}{Q^3} (pQ+1). \qquad (3.54)$$

The case $p < 0$ can be treated similarly by integrating over the lower half of the complex plane. Thus, we can express the impurity scattering rate as:

$$\frac{1}{\tau_n^{(imp)}(\mathbf{K})} \approx N_{imp} \frac{2\pi}{\hbar} \sum_{jn'}^{*} C_{imp}^2 \frac{L_{n'}(w_j)}{|\nabla_{2D} E_{n'j}|} \frac{1}{4Q_j^3} \times \int dz \int dz' \, \zeta_{\mathbf{K}_j}^{(n')*}(z)\, \zeta_{\mathbf{K}_j}^{(n')}(z') \times e^{-|z-z'|Q_j} \left[|z-z'|Q_j + 1\right] \zeta_{\mathbf{K}}^{(n)}(z)\, \zeta_{\mathbf{K}}^{(n)*}(z'), \quad (3.55)$$

where N_{imp} is the concentration of impurities. The overlap factor entering the last equation can be split into a term identical to the factor appearing in (3.48) – which can be recast in the form of (3.49) – and another term,

$$\int dz \int dz' \, \zeta_{\mathbf{K}_j}^{(n')*}(z)\, \zeta_{\mathbf{K}_j}^{(n')}(z')\, e^{-|z-z'|Q_j}\, |z-z'|Q_j\, \zeta_{\mathbf{K}}^{(n)}(z)\, \zeta_{\mathbf{K}}^{(n)*}(z'), \quad (3.56)$$

which can be rewritten as:

$$Q_j \int dz \int dz' \, g_{\mathbf{KK}'nn'}(z')\, e^{-|z-z'|Q_j}\, |z-z'|\, g_{\mathbf{KK}'nn'}^{*}(z)$$
$$= 2\, Q_j \, \mathrm{Re}\left\{ \int_0^{\infty} dz\, g_{\mathbf{KK}'nn'}(z)\, e^{-Qz} \int_0^{z} dz'\, g_{\mathbf{KK}'nn'}^{*}(z')\, e^{Qz'} (z-z') \right\}, \quad (3.57)$$

or:

$$= 2\, Q_j \, \mathrm{Re}\left\{ \int_0^{\infty} dz\, g_{\mathbf{KK}'nn'}(z)\, z\, e^{-Qz} \int_0^{z} dz'\, g_{\mathbf{KK}'nn'}^{*}(z')\, e^{Qz'} \right\}$$
$$- 2Q_j \, \mathrm{Re}\left\{ \int_0^{\infty} dz\, g_{\mathbf{KK}'nn'}(z)\, e^{-Qz} \int_0^{z} dz'\, g_{\mathbf{KK}'nn'}^{*}(z')\, z'\, e^{Qz'} \right\}. \quad (3.58)$$

The advantage of expressing the overlap factor in terms of (3.58) is that the inner integral over z' depends on z *only* via its upper integration limit. Thus, as we saw in (3.49), it can be evaluated as a discrete sum storing partial results and recalling these partial results when performing the 'outer' integration over z. Thus, evaluating the double integral (3.58) actually requires the same number of operations required to perform two 1D integrals. In other words, the computational effort scales with N_z (the number of z points) rather than N_z^2, as it may at first appear from a look at (3.55), or $N_z \times N_{q_z}$, as in (3.42).

3.1.3 Electron Transport in Thin Si Inversion Layers: Monte Carlo Results

As an example of the implementation of this general scheme to a concrete case of technological interest, we consider the 12-cell-thick Si layer mentioned above. This may be viewed as a prototypical thin body of SOI FETs, of a FinFET or other double-gate device. The major difference is that the layer is bounded by vacuum, rather than by an insulator, so we expect a slightly stronger confinement due to the larger Si-vacuum barrier (that is, workfunction, $\sim 4.5\,\text{eV}$).

Zunger's pseudopotential without spin-orbit interaction are used to calculate the band structure. An external potential inducing a sheet electron charge of 10^{13} electrons cm^{-2} is applied. Starting from a classical Poisson solution for the external potential and taking its Fourier transform as in (3.2), (3.4) is solved. From the wavefunctions, (3.35), averaged over a cell on the (x,y)-plane (i.e., on the plane of the surfaces), and their equilibrium Fermi occupation, the total charge can be obtained by summing over the (discretized) 2D BZ and Poisson solution may then be solved in real space. The process is iterated until convergence is reached. The resulting band structure is shown in Fig. 3.29, left. Note once more the rather complicated structure near the $\bar{X}$ symmetry point. We have already emphasized this feature recalling the previous observation by Esseni and Palestri [13]. Here we will see the role this structure plays in determining the high-field transport properties.

Nonpolar scattering rates with acoustic and optical phonons are computed at 300 K using either the full (3.44) – which accounts for the dependence of the overlap factor $\mathscr{F}^{(2D)}_{\mathbf{K}'\mathbf{K}nn'}$ on the initial and final wavevectors, $\mathbf{K}$ and $\mathbf{K}'$, respectively – or on the simplified expression (3.45) – where $\mathbf{K}_0$ is the wavevector at the valley minima, $\bar{\Gamma}$ for the 'unprimed' ladder or $\bar{X}$ for the primed ladder of states. Scattering with ionized impurities, fixed charges and surface roughness are ignored, wishing to obtain the intrinsic band-structure-dependent transport properties. The scattering rates calculated with these two methods and averaged over electron energy are shown in Fig. 3.29, right. Note the smaller rates obtained when using the full $(\mathbf{K},\mathbf{K}')$-dependent overlap factor $\mathscr{F}^{(2D)}_{\mathbf{K}'\mathbf{K}nn'}$, since this accounts also for the overlap between the initial and final Bloch states.

An ensemble Monte Carlo method, similar to the scheme described in [17] is employed to obtain the energy vs. field and drift-velocity vs. field characteristics shown in Fig. 3.30. Two features should be noted: First, when using either

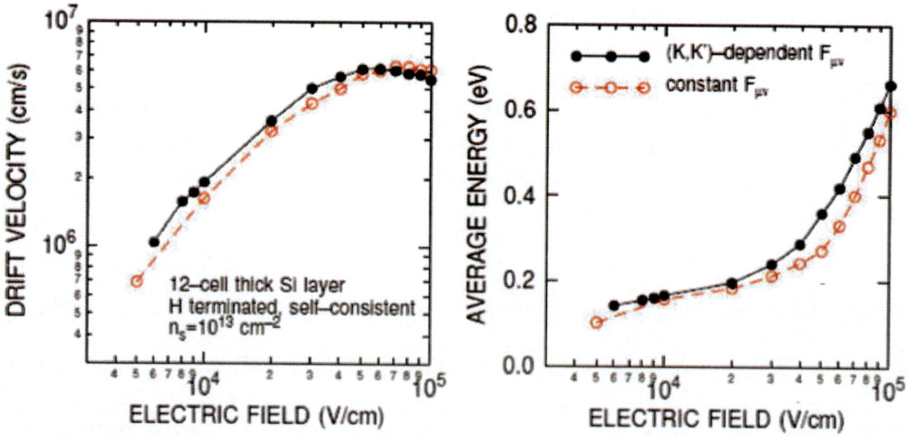

Fig. 3.30 Electron average energy (*right*) and drift velocity (*left*) as a function of field on the plane of the layer using the two models for the scattering rates as in Fig. 3.29. In either case the electron velocity saturates at a value much smaller then in bulk Si, consistently with experimental results so far left unexplained by models based on the effective-mass subband structure

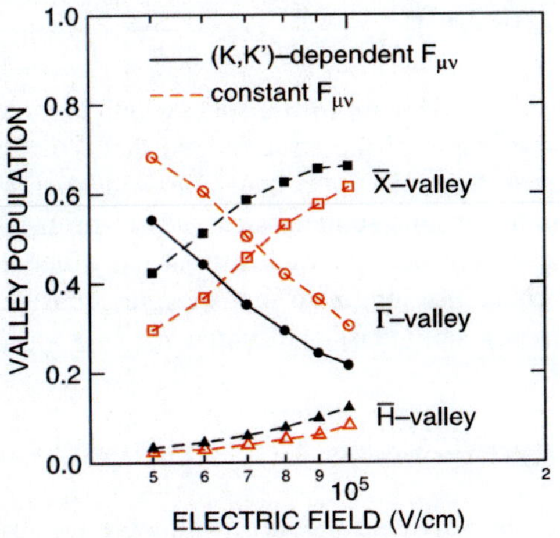

Fig. 3.31 Population of – somewhat arbitrarily defined – regions of the 2D BZ around the $\bar{\Gamma}$ and $\bar{X}$ symmetry points, corresponding roughly to the unprimed and primed subband ladders. The larger occupation of the primed states – which exhibit a significantly larger conductivity mass – yields the lower saturated velocity and higher average energy observed at high fields in Fig. 3.30

model to compute the scattering rates, the electron velocity saturates at a value (~ 5–$6 \times 10^6\,\mathrm{cm\,s^{-1}}$) much lower than the value of the saturated velocity of electrons in bulk Si at 300 K ($\approx 10^7\,\mathrm{cm\,s^{-1}}$). As discussed at length in [17], several experimental results have yielded such low values for the saturated velocity, but no study based on the effective-mass approximation, even with nonparabolic corrections, has been able to reproduce the experimental observations. Second, the use of the lower scattering rates obtained when using the full $(\mathbf{K},\mathbf{K}')$-dependent overlap factor, (3.44), results in a higher electron average energy (see right panel of Fig. 3.30), but in an even lower saturated electron velocity. As shown in Fig. 3.31, this is due to an enhanced population of the 'primed' states near the higher-mass $\bar{X}$ symmetry point. This constitutes a significant example of how the accurate evaluation of the band structure of small confined structures can produce results significantly different from those one may obtain using the effective-mass approximation.

We should note in passing that the Monte Carlo technique employed here can be easily extended to study 2D or 1D transport in other small structures. The main problems we must face consists in determining the correct physical models to be employed to handle electronic scattering, especially accounting for the presence of localized phonons in small structures, for scattering with interface or surface roughness, charges, and excitations and, of course, for long- and short-range carrier-carrier scattering.

3.2 Quantum Transport

So far we have considered the case of transport which is incoherent over distances much shorter than the length of the structure, so that we could consider different 'slices' of our device as decoupled, treat each one as independent of the neighboring slices, and reduce transport to a semiclassical, albeit lower-dimensionality, formulation. In this section we consider the opposite case in which the devices are shorter than the electronic coherence length so that we must consider the device globally and treat transport in non-semi-classical fashion.

3.2.1 A Mixed Supercell/Envelope Method for Open Systems

As discussed above, the supercell method is suitable to study confined systems. However, its periodic boundary conditions render it unsuitable to study a-periodic systems, such as those involved when dealing with (open boundary conditions) electronic transport problems. In this case one possible way to handle transport within a full-band-structure formalism is to employ a mixed scheme coupling the supercell method to an approximation which we shall call the 'envelope' approximation [34]. Its major advantage consists in its 'differential equation' nature which allows us to specify arbitrary boundary conditions at the 'contacts' (with open boundary conditions we shall discuss below). Its major drawback originates from the fact that, as its name implies, only the envelope of the full Bloch wavefunction can be calculated and, as an even more restrictive constrain, the envelope (and so also the external potential applied to the structure) must vary smoothly over the structure, so smoothly, actually, that they we are allowed to consider them constants over the transport direction of the cell (or supercell, as we shall see below).

Thus, the extension of the study to open systems described by an empirical pseudopotential band structure will be performed as follows (to fix the ideas, we consider two typical specific scenarios):

(1) A thin-body semiconductor channel (so, with confinement in one direction, say along the z axis) bounded by a gate and a substrate insulator, with source-to-drain flow along the x axis, and a device wide enough along the y direction as to allow us to consider the system homogeneous along the y axis (the typical situation of 2D device simulations). We start by 'slicing' the channel into 'vertical' (y,z) planes at discrete x locations. In each plane we consider a supercell extended along the x direction and consider the Fourier components $\Phi_{G_z}(x)$ of the confining potential (to be determined self-consistently). Writing the wavefunction as $e^{ik_y y}\sum_{\mathbf{G}} \phi_{\mathbf{G}}(x)e^{i\mathbf{G}\cdot\mathbf{r}}$, the 'slowly varying' envelope $\phi_{\mathbf{G}}(x)$ obeys the differential equation:

$$\sum_{\mathbf{G}'}\left\{\left[-\frac{\hbar^2}{2m}\frac{d^2}{dx^2}-i\frac{\hbar^2}{m}G_x\frac{d}{dx}\right]\delta_{\mathbf{G},\mathbf{G}'}+W^{(2D)}_{\mathbf{G}\mathbf{G}'}(x,k_y)\right\}\phi_{k_y\mathbf{G}'}(x) = E(k_y)\phi_{k_y\mathbf{G}}(x), \tag{3.59}$$

where the total 'potential'

$$W^{(2D)}_{\mathbf{GG}'}(x,k_y) \; = \; V^{(lat)}_{\mathbf{G}-\mathbf{G}'} + \Phi_{G_z-G'_z}(x)\;\delta_{\mathbf{G}_\parallel,\mathbf{G}'_\parallel} \; + \; \left[\frac{\hbar^2(G^2+k_y^2)}{2m} + \frac{\hbar^2 G_y k_y}{m}\right]\delta_{\mathbf{G},\mathbf{G}'} \tag{3.60}$$

results from mixing the supercell method along the 'closed boundary-conditions' z axis and the envelope method along the 'open boundary-conditions' x axis.

(2) A nanowire or nanotube with axis along the z direction. Once again, we slice the wire at discrete z locations and consider the Fourier components of the Hartree potential $\Phi_{\mathbf{G}_\parallel}(z)$ (which now depend on both the in-plane components of the in-plane reciprocal-space wavevector $\mathbf{G}_\parallel$). We then express the wavefunctions as $\sum_{\mathbf{G}} \phi_{\mathbf{G}}(z) e^{i\mathbf{G}\cdot\mathbf{r}}$ with a slowly-varying envelope $\phi_{\mathbf{G}}(z)$ obeying the differential equation:

$$\sum_{\mathbf{G}'}\left\{\left[-\frac{\hbar^2}{2m}\frac{d^2}{dz^2} - i\frac{\hbar^2}{m}G_z\frac{d}{dz}\right]\delta_{\mathbf{G},\mathbf{G}'} + W^{(1D)}_{\mathbf{G},\mathbf{G}'}(z)\right\}\phi_{\mathbf{G}'}(z) = E\phi_{\mathbf{G}}(z), \tag{3.61}$$

where

$$W^{(1D)}_{\mathbf{GG}'}(z) \; = \; V^{(lat)}_{\mathbf{G}-\mathbf{G}'} + \Phi_{\mathbf{G}_\parallel-\mathbf{G}'_\parallel}(z)\;\delta_{G_z,G'_z} \; + \; \frac{\hbar^2 G^2}{2m}\;\delta_{\mathbf{G},\mathbf{G}'} \tag{3.62}$$

in analogy with (3.59). In either case, (3.59) or (3.61) are differential equation valid as long as $\Phi_{G_z}(x)$ or $\Phi_{\mathbf{G}_\parallel}(z)$ do not vary too fast with x or z and will provide envelope wavefunctions $e^{ik_y y}\sum_{\mathbf{G}} \phi_{k_y\mathbf{G}}(x)e^{i\mathbf{G}\cdot\mathbf{r}}$ and $\sum_{\mathbf{G}} \phi_{\mathbf{G}}(z)e^{i\mathbf{G}\cdot\mathbf{r}}$ which can be used to obtain the charge density and solve Poisson equation (in real space) self-consistently. Of course, the boundary conditions supplementing (3.59) and (3.61) will have to be a generalization of the 'Quantum Transmitting Boundary Method' [39] by replacing plane waves with bulk Bloch functions. In the following we shall derive the proper open boundary conditions using the complex band structure of the contact. The review paper by Pecchia and Di Carlo [49] gives a general discussion of the problem within the tight-binding context, the paper by Brandbyge et al. [7] within the context of the 'TranSIESTA' code based on ab-initio pseudopotentials), while the remarkable work by Choi and Ihm [10] deals with the calculation of the ballistic conductance in 1D systems, also within a first-principles, nonlocal framework. Our treatment is similar to their approach (especially in the spirit by which supercell and z-dependence are treated, notwithstanding some major differences), but exploits the simplifying benefits afforded by the nature of local empirical pseudopotentials.

3.2.2 Boundary Conditions

In addition to the numerical complexity of the problem at hand (which we will briefly discuss below), the open nature of the boundary conditions presents a problem. To see this, consider the simple case of the effective-mass approximation and use Frensley's method [20] to determine the boundary conditions for an open system. Let's assume a one-dimensional system which occupies the region $(0,L)$ along the z axis with left and right 'contacts' (or 'reservoirs') occupying the regions $z < 0$ and $z > L$, respectively, defined as fully absorbing elements attempting to maintain thermal equilibrium and charge neutrality at the $z = 0$ and $z = L$ boundaries. Discretizing the z axis into N intervals $z_l = \Delta(l-1)$ with $\Delta = L/(N_1)$, the finite difference Schrödinger equation will provide the values ψ_l of the wavefunction $\psi(z)$ at z_l, but we will require knowledge of the wavefunction 'just outside the device', that is, the values ψ_0 and ψ_{N+1} of ψ at $z = -\Delta$ and $z = L+\Delta$, respectively. Considering the left reservoir, for an electron injected at energy E we have for $z < 0$ $\psi(z) = e^{ikz} + re^{-ikz}$, where r is the reflection coefficient, $k = (2m^*E)^{1/2}/\hbar$ is the wavevector, and m^* is the effective mass. Therefore, since $\psi_1 = 1 + r$, we have $\psi_0 = e^{-ik\Delta} + (\psi_1 - 1)e^{ik\Delta}$ and we have a simple expression connecting the magnitude and phase of ψ at $z = -\Delta$ to the magnitude and phase of ψ at $z = 0$. (A similar analysis applies to the $z = L$ boundary.) When moving to full band, we lose much of this simplicity: Even considering electrons injected from the left reservoir in a single band and with a given **k**-vector, reflections can occur into any of the (propagating or evanescent towards the left) states $\mathbf{k}'$ and bands n such that $E_n(\mathbf{k}') = E$. Therefore, relating the magnitude and phase of ψ at the device-reservoir boundary to its magnitude and phase 'just inside' the reservoir becomes a more complicated affair. In general we will also require electrons to be injected into a superposition of traveling states at a given energy, thus requiring an extension of the 'Quantum Transmitting Boundary Method' [39] to the full-band case. Here we will limit our discussion to the injection of a single carrier with a specified wavevector and band. The generalization to arbitrary injected states is algebraically cumbersome but straightforward.

Let's consider for now the case of 1D-transport case described by (3.61). The wavefunction inside the 'device' (assumed to span the interval $(0,L)$ along the z axis) for a (sub)band n and at energy E will have the form:

$$\psi_n(\mathbf{r}) = \psi_n(\mathbf{R},z) = \sum_{\mathbf{G}} \phi_{\mathbf{G}}^{(n)}(z)\, e^{i\mathbf{G}\cdot\mathbf{r}}. \tag{3.63}$$

Let's assume that the 'wire' (or nanoribbon or nanotube) is in contact with left $(z < 0)$ and right $(z > L)$ reservoirs consisting of semi-infinite extensions of the same structure held at chemical potentials 0 and eV, respectively. Let's assume that we know the complex band-structure of the reservoirs, so that we have knowledge of the Bloch eigenvectors $\phi_{\mathbf{G}}^{(L,p)}$ and $\phi_{\mathbf{G}}^{(R,p)}$ in the left and right reservoir, respectively, where p is the band index running over the same set of $N_{\mathbf{G}}$ bands we consider in the wire. Clearly, a complex band structure employing $N_{\mathbf{G}}$ plane waves will yield $2N_{\mathbf{G}}$ bands, but we consider only propagating waves from left to right and evanescent waves decaying to the left for the left contact and vice versa for the right contact.

Thus, assuming for now a 1D contact, we consider an injected wave $\psi_{nk_L}(\mathbf{R},z)$ in (sub)band n at energy $E_n(k_L) = E$ propagating from the left with wavenumber k_L partially reflected into a wave $\psi_r(\mathbf{R},z)$ into the left contact and partially transmitted into a wave $\psi_t(\mathbf{R},z)$ into the right contact. The reflected and transmitted waves must be linear superpositions of all propagating and evanescent waves at energy E and E+eV, respectively. Thus in the region $z < 0$ we have:

$$\begin{aligned}\psi_L(\mathbf{R},z) &= \psi_{nk_L}(\mathbf{R},z) + \psi_r(\mathbf{R},z) e^{ik_L z} \sum_{\mathbf{G}} \phi_{\mathbf{G},k_L}^{(L,n)} \, e^{i\mathbf{G}\cdot\mathbf{r}} \\ &= + \sum_p \alpha_p(E) \, e^{-ik_{Lp}z} \sum_{\mathbf{G}} \phi_{\mathbf{G},k_{Lp}}^{(L,p)} \, e^{i\mathbf{G}\cdot\mathbf{r}}, \end{aligned} \tag{3.64}$$

where, for each band p, k_{Lp} satisfies the condition $E_p(-k_{Lp}) = E$. As mentioned before, among the $2N_{\mathbf{G}}$ solutions of the equation $E = E_p(-k_{Lp})$ we select only the $N_{\mathbf{G}}$ solutions k_{Lp} which are real and such that $v_z^{(L)}(-k_{Lp}) = (1/\hbar) dE_p(k_{Lp})/dz < 0$ (reflected waves propagating towards the left inside the left reservoir), and those for which $\mathrm{Im}(k_{Lp}) < 0$ (reflected waves decaying into the left reservoir). Similarly, in the region $z > L$ we have:

$$\psi_R(\mathbf{R},z) = \psi_t(\mathbf{R},z) = \sum_p \beta_p(E) \, e^{ik_{Rp}z} \sum_{\mathbf{G}} \phi_{\mathbf{G},k_{Rp}}^{(R,p)} \, e^{i\mathbf{G}\cdot\mathbf{r}}, \tag{3.65}$$

where, for each band p, k_{Rp} satisfies the condition $E_p(k_{Rp}) = E$+eV and we select only those real k_{Rp} such that $v_z^{(R)}(k_{Rp}) = (1/\hbar) dE_p(k_{Rp})/dz > 0$ (transmitted waves propagating to the right inside the right contact), and those complex k_{Rp} for which $\mathrm{Im}(k_{Rp}) > 0$ (transmitted waves decaying into the right contact).

Discretizing (3.61) in the interval $(0,L)$ employing a uniform mesh of N points and interval $\Delta = L/(N-1)$, we have at each point $z_l = (l-1)\Delta$ for $l = 1, N$:

$$\begin{aligned} -\frac{\hbar^2}{2m\Delta^2} (\phi_{\mathbf{G},l+1} + \phi_{\mathbf{G},l-1}) - i\frac{\hbar^2}{2m\Delta} G_z (\phi_{\mathbf{G},l+1} - \phi_{\mathbf{G},l-1}) + \frac{\hbar^2}{m\Delta^2} \phi_{\mathbf{G},l} \\ + \sum_{\mathbf{G}'} W_{\mathbf{GG}'}^{(1D)}(z_l) \, \phi_{\mathbf{G}',l} = E \, \phi_{\mathbf{G},l}, \end{aligned} \tag{3.66}$$

where, of course, $\phi_{\mathbf{G},l} = \phi_{\mathbf{G}}(z_l)$. This gives rise to a Hamiltonian matrix similar to the 'usual' tri-diagonal form of effective-mass approximations, the major differences being (1) the rank of the matrix, as it is now a block tri-diagonal matrix with blocks of rank $N_{\mathbf{G}}$; (2) the complicated (off diagonal) external potential now entering both diagonal and off-diagonal terms of the matrices $\hat{\mathbf{D}}^{(l)}$ below, and (3) the boundary conditions to be imposed on (3.66) for $l = 1$ and $l = N$. (Note that the quantities α_p, β_p, $\phi_{\mathbf{G},l}$, etc. depend on the injection band-index n. We shall suppress this index in the following to simplify a notation already encumbered by too many superscripts and subscripts).

In order to translate the boundary conditions to a specific form of the matrix, we must consider the discretized form of (3.66) at the left and right contact/device boundaries and impose continuity of the wavefunctions and of their derivatives. At the device/left-reservoir boundary (for $l = 1$) we need an expression for $\phi_{\mathbf{G},0}$. From (3.64), comparing the coefficients of the expansion over $e^{i\mathbf{G}\cdot\mathbf{r}}$ term-by-term (since the equality must hold for every **R**), we have:

$$\phi_{\mathbf{G},0} = e^{-ik_L\Delta}\,\phi^{(L,n)}_{\mathbf{G},k_L} + \sum_p \alpha_p(E)\,e^{ik_{Lp}\Delta}\,\phi^{(L,p)}_{\mathbf{G},k_{Lp}}\,. \tag{3.67}$$

Let's define the matrix $\mathcal{M}^{(L)}$ with $(\mathbf{G},p)$-matrix-elements $\phi^{(L,p)}_{\mathbf{G},k_{Lp}}$ (thinking of each **G** as identified by an integer), and the 'partial reflection amplitudes' at $z = 0$:

$$r_{\mathbf{G}} = \sum_p \mathcal{M}^{(L)}_{\mathbf{G}p}\,\alpha_p. \tag{3.68}$$

and the coefficients (that is, the 'partial reflection amplitudes' at $z = -\Delta$):

$$\rho_{\mathbf{G}} = \sum_p \mathcal{M}^{(L)}_{\mathbf{G}p}\,\alpha_p\,e^{ik_{Lp}\Delta}\,. \tag{3.69}$$

Note that $\mathcal{M}^{(L)}$ is the matrix which maps the bands p of the left reservoir into the plane waves **G** of the device. Thus, (3.67) becomes:

$$\phi_{\mathbf{G},0} = e^{-ik_L\Delta}\,\phi^{(L,n)}_{\mathbf{G},k_L} + \rho_{\mathbf{G}}. \tag{3.70}$$

Assuming now the continuity of the wavefunction at $z = 0$ (i.e., for $l = 1$), we have the equation relating the coefficients $r_{\mathbf{G}}$ to the unknowns $\phi_{\mathbf{G},1}$ and to the inhomogeneous term $\phi^{(L,n)}_{\mathbf{G},k_L}$:

$$\phi_{\mathbf{G},1} = \phi^{(L,n)}_{\mathbf{G},k_L} + \sum_p \alpha_p(E)\,\phi^{(L,p)}_{\mathbf{G},k_{Lp}} = \phi^{(L,n)}_{\mathbf{G},k_L} + r_{\mathbf{G}}, \tag{3.71}$$

The coefficients $\rho_{\mathbf{G}}$ can be expressed in terms of the partial reflection amplitudes $r_{\mathbf{G}}$: From (3.68), inverting the matrix $\mathcal{M}^{(L)}$ we can express the coefficients α_p in terms of the $r_{\mathbf{G}}$'s:

$$\alpha_p = \sum_{\mathbf{G}} \mathcal{M}^{(L)-1}_{p\mathbf{G}}\,r_{\mathbf{G}}. \tag{3.72}$$

Inserting this expression into (3.69) we obtain the second required set of relations which link the coefficients $\rho_{\mathbf{G}}$ to the unknowns $\phi_{\mathbf{G},1}$ and to the inhomogeneous terms $\phi^{(L,n)}_{\mathbf{G},k_L}$:

$$\rho_{\mathbf{G}} = \sum_p \mathcal{M}^{(L)}_{\mathbf{G}p} \sum_{\mathbf{G}'} \mathcal{M}^{(L)-1}_{p\mathbf{G}'}\,r_{\mathbf{G}'}\,e^{ik_{Lp}\Delta}$$

$$= \sum_p \mathscr{M}^{(L)}_{\mathbf{G}p} \sum_{\mathbf{G}'} \mathscr{M}^{(L)-1}_{p\mathbf{G}'} \left[\phi_{\mathbf{G}',1} - \phi^{(L,n)}_{\mathbf{G}',k_L} \right] e^{ik_{Lp}\Delta}. \tag{3.73}$$

Note that the first form of this equation, expressing the (complex) phase difference between $r_{\mathbf{G}}$ and $\rho_{\mathbf{G}}$, is the crucial information about the nature of the left 'contact'.

Considering now (3.66) at $l = 1$, we have:

$$-\frac{\hbar^2}{2m\Delta^2} \left(\phi_{\mathbf{G},2} + \phi_{\mathbf{G},0}\right) - i\frac{\hbar^2}{2m\Delta} G_z \left(\phi_{\mathbf{G},2} - \phi_{\mathbf{G},0}\right) + \left(\frac{\hbar^2}{m\Delta^2} - E\right) \phi_{\mathbf{G},1}$$
$$+ \sum_{\mathbf{G}'} W^{(1D)}_{\mathbf{G}\mathbf{G}'}(z_1)\, \phi_{\mathbf{G}',1} = 0. \tag{3.74}$$

Substituting (3.70) for $\phi_{\mathbf{G},0}$ with the coefficients $\rho_{\mathbf{G}}$ expressed in terms of the unknowns $\phi_{\mathbf{G},1}$ and 'injected' wave $\phi^{(Ln)}_{\mathbf{G},k_L}$ using (3.73), we obtain the equation required to define the term 'contact self energy' and the inhomogeneous term (i.e., the right-hand-side) of the linear problem:

$$-\left(\frac{\hbar^2}{2m\Delta^2} + i\frac{\hbar^2}{2m\Delta} G_z\right) \phi_{\mathbf{G},2} + \left(\frac{\hbar^2}{m\Delta^2} - E\right) \phi_{\mathbf{G},1} + \sum_{\mathbf{G}'} \left(W^{(1D)}_{\mathbf{G}\mathbf{G}'}(z_1) + \Sigma^{(L)}_{\mathbf{G}\mathbf{G}'}\right) \phi_{\mathbf{G}',1}$$
$$= \left(\frac{\hbar^2}{2m\Delta^2} - i\frac{\hbar^2}{2m\Delta} G_z\right) e^{-ik_L\Delta} \phi^{(L,n)}_{\mathbf{G},k_L} + \sum_{\mathbf{G}'} \Sigma^{(L)}_{\mathbf{G}\mathbf{G}'}\, \phi^{(L,n)}_{\mathbf{G}',k_L}. \tag{3.75}$$

where the term

$$\Sigma^{(L)}_{\mathbf{G}\mathbf{G}'} = -\left(\frac{\hbar^2}{2m\Delta^2} - i\frac{\hbar^2}{2m\Delta} G_z\right) \sum_p \mathscr{M}^{(L)}_{\mathbf{G}p} \mathscr{M}^{(L)-1}_{p\mathbf{G}'}\, e^{ik_{Lp}\Delta} \tag{3.76}$$

can be viewed as the device/left-contact 'self-energy' matrix. It is the only term telling us how the structure of the left reservoirs affects the wavefunctions inside the device. Finally, using Feynman's theorem, the reflection probability R can be extracted from the coefficients $r_{\mathbf{G}} = \phi_{\mathbf{G},1} - \phi^{(L,n)}_{\mathbf{G},k_L}$ as follows:

$$R_n(E) = \sum_p{}' |\alpha_p(E)|^2\, v_{z,p}(k_{Lp}) = \sum_p{}' \left| \sum_{\mathbf{G}} \mathscr{M}^{(L)-1}_{p\mathbf{G}} \left(\phi_{\mathbf{G},1} - \phi^{(L,n)}_{\mathbf{G},k_L}\right) \right|^2 v_{z,p}(k_{Lp}), \tag{3.77}$$

where the 'prime' over the summation symbol indicates that the sum extends only over the propagating (i.e., non evanescent) waves.

The term which must be added to the Hamiltonian matrix in order to describe the device/right-contact interaction can be set up in a similar way: At the device/right-reservoir boundary (for $l = N$) we need to express $\phi_{\mathbf{G},N+1}$ using (3.65):

$$\phi_{\mathbf{G},N+1} = \sum_p \beta_p(E)\, e^{ik_{Rp}(L+\Delta)}\, \phi^{(R,p)}_{\mathbf{G},k_{Rp}}. \tag{3.78}$$

Let's consider the matrix $\mathcal{M}^{(R)}$ with matrix elements $\phi^{(R,p)}_{\mathbf{G},k_{Rp}}$ and define:

$$t_{\mathbf{G}} = \sum_p \mathcal{M}^{(R)}_{\mathbf{G}p} \beta_p, \tag{3.79}$$

and:

$$\tau_{\mathbf{G}} = \sum_p \mathcal{M}^{(R)}_{\mathbf{G}p} \beta_p \, e^{ik_{Rp}(L+\Delta)}. \tag{3.80}$$

Using this definition (3.78) becomes:

$$\phi_{\mathbf{G},N+1} = \tau_{\mathbf{G}}. \tag{3.81}$$

Imposing the continuity of the wavefunction at $z = L$, we obtain the equation relating the coefficients $t_{\mathbf{G}}$ to the unknowns $\phi_{\mathbf{G},N}$:

$$\phi_{\mathbf{G},N} = \sum_p \beta_p(E) \, \phi^{(R,p)}_{\mathbf{G},k_{Rp}} = t_{\mathbf{G}}. \tag{3.82}$$

The coefficients $\tau_{\mathbf{G}}$ can be expressed in terms of the 'partial transmission amplitudes' $t_{\mathbf{G}}$: From (3.79), inverting the matrix $\mathcal{M}^{(R)}$ we can express the coefficients β_p in terms of the $t_{\mathbf{G}}$'s:

$$\beta_p = \sum_{\mathbf{G}} \mathcal{M}^{(R)-1}_{p\mathbf{G}} \, t_{\mathbf{G}}. \tag{3.83}$$

Inserting this expression into (3.80) we obtain the second required set of relations which link the coefficients $\tau_{\mathbf{G}}$ to the unknowns $\phi_{\mathbf{G},N}$:

$$\begin{aligned}\tau_{\mathbf{G}} &= \sum_p \mathcal{M}^{(R)}_{\mathbf{G}p} \sum_{\mathbf{G}'} \mathcal{M}^{(R)-1}_{p\mathbf{G}'} \, t_{\mathbf{G}'} \; e^{ik_{Rp}(L+\Delta)} \\ &= \sum_p \mathcal{M}^{(R)}_{\mathbf{G}p} \sum_{\mathbf{G}'} \mathcal{M}^{(R)-1}_{p\mathbf{G}'} \, \phi_{\mathbf{G}',N} \; e^{ik_{Rp}(L+\Delta)}. \end{aligned} \tag{3.84}$$

Considering now (3.66) at $l = N$, we have:

$$\begin{aligned} -\frac{\hbar^2}{2m\Delta^2}\left(\phi_{\mathbf{G},N+1} + \phi_{\mathbf{G},N-1}\right) - i\frac{\hbar^2}{2m\Delta} G_z \left(\phi_{\mathbf{G},N+1} - \phi_{\mathbf{G},N-1}\right) + \left(\frac{\hbar^2}{m\Delta^2} - E\right) \phi_{\mathbf{G},N} \\ + \sum_{\mathbf{G}'} W^{(1D)}_{\mathbf{G}\mathbf{G}'}(z_N) \, \phi_{\mathbf{G}',N} = 0. \end{aligned} \tag{3.85}$$

Substituting (3.81) for $\phi_{\mathbf{G},N+1}$ with the coefficients $\tau_{\mathbf{G}}$ expressed in terms of the unknowns $\phi_{\mathbf{G},N}$, we obtain the following equation required to account for the device/right-contact interaction:

$$
-\left(\frac{\hbar^2}{2m\Delta^2} - i\frac{\hbar^2}{2m\Delta}G_z\right)\,\phi_{\mathbf{G},N-1} + \left(\frac{\hbar^2}{m\Delta^2} - E\right)\,\phi_{\mathbf{G},N}
$$

$$
+\sum_{\mathbf{G}'}\left(W^{(1D)}_{\mathbf{GG}'}(z_N) + \Sigma^{(R)}_{\mathbf{GG}'}\right)\,\phi_{\mathbf{G}',N} = 0, \tag{3.86}
$$

where:

$$
\Sigma^{(R)}_{\mathbf{GG}'} = -\left(\frac{\hbar^2}{2m\Delta^2} + i\frac{\hbar^2}{2m\Delta}G_z\right)\sum_p \mathscr{M}^{(R)}_{\mathbf{G}p}\,\mathscr{M}^{(R)-1}_{p\mathbf{G}'}\;e^{ik_{Rp}(L+\Delta)} \tag{3.87}
$$

is the device/right-contact self-energy matrix. Finally, the transmission probability T can be extracted from the coefficients $t_{\mathbf{G}} = \phi_{\mathbf{G},\mathbf{N}}$:

$$
T_n(E) = \sum_p{}' |\beta_p(E)|^2\,v_{z,p}(k_{Rp}) = \sum_p{}' \left|\sum_{\mathbf{G}} \mathscr{M}^{(R)-1}_{p\mathbf{G}}\;\phi_{\mathbf{G},N}\right|^2 v_{z,p}(k_{Rp}). \tag{3.88}
$$

To summarize our result, it is convenient to express the open-boundary-conditions linear system to be solved in full matrix form:

$$
\begin{bmatrix}
\widehat{\mathbf{D}}^{(1)}+\Sigma^{(L)} & \widehat{\mathbf{T}}^{(+)} & . & . & . & . & . & . & . \\
\widehat{\mathbf{T}}^{(-)} & . & . & . & . & . & . & . & . \\
. & . & \widehat{\mathbf{T}}^{(-)} & \widehat{\mathbf{D}}^{(l-1)} & \widehat{\mathbf{T}}^{(+)} & \mathbf{0} & \mathbf{0} & . & . \\
. & . & \mathbf{0} & \widehat{\mathbf{T}}^{(-)} & \widehat{\mathbf{D}}^{(l)} & \widehat{\mathbf{T}}^{(+)} & \mathbf{0} & . & . \\
. & . & \mathbf{0} & \mathbf{0} & \widehat{\mathbf{T}}^{(-)} & \widehat{\mathbf{D}}^{(l+1)} & \widehat{\mathbf{T}}^{(+)} & . & . \\
. & . & . & . & . & . & . & . & \widehat{\mathbf{T}}^{(+)} \\
. & . & . & . & . & . & . & \widehat{\mathbf{T}}^{(-)} & \widehat{\mathbf{D}}^{(N)}+\Sigma^{(R)}
\end{bmatrix}
\begin{bmatrix}
\phi^{(1)} \\ . \\ \phi^{(l-1)} \\ \phi^{(l)} \\ \phi^{(l+1)} \\ . \\ \phi^{(N)}
\end{bmatrix}
$$

$$
= \begin{bmatrix}
(A\mathbf{I}+\Sigma^{(L)})\,\phi^{(Ln)}_{k_L} \\ . \\ \mathbf{0} \\ \mathbf{0} \\ \mathbf{0} \\ . \\ \mathbf{0}
\end{bmatrix}, \tag{3.89}
$$

where each $\phi^{(l)} = \phi(z_l)$ is a column-vector with N_G components. The discretized differential operators ($N_G \times N_G$ difference operators) $\widehat{\mathbf{D}}^{(l)}$, $\widehat{\mathbf{T}}^{(+)}$ and $\widehat{\mathbf{T}}^{(-)}$ take the form:

$$\widehat{\mathbf{D}}^{(l)} = \begin{bmatrix} D^{(l)}_{\mathbf{G}_1} & W^{(1D)}_{\mathbf{G}_1,\mathbf{G}_2}(z_l) & W^{(1D)}_{\mathbf{G}_1,\mathbf{G}_3}(z_l) & \dots \\ W^{(1D)}_{\mathbf{G}_2,\mathbf{G}_1}(z_l) & D^{(l)}_{\mathbf{G}_2} & W^{(1D)}_{\mathbf{G}_2,\mathbf{G}_3}(z_l) & \dots \\ W^{(1D)}_{\mathbf{G}_3,\mathbf{G}_1}(z_l) & W^{(1D)}_{\mathbf{G}_3,\mathbf{G}_2}(z_l) & D^{(l)}_{\mathbf{G}_3} & \dots \\ . & . & . & \dots \end{bmatrix}, \tag{3.90}$$

$$\widehat{\mathbf{T}}^{(+)} = \begin{bmatrix} T_{\mathbf{G}_1} & 0 & 0 & \dots \\ 0 & T_{\mathbf{G}_2} & 0 & \dots \\ 0 & 0 & T_{\mathbf{G}_3} & \dots \\ . & . & . & \dots \end{bmatrix}, \tag{3.91}$$

and $\widehat{\mathbf{T}}^{(-)} = \widehat{\mathbf{T}}^{(+)\dagger}$, with

$$D^{(l)}_{\mathbf{G}} = \frac{\hbar^2}{m\Delta^2} - E + W^{(1D)}_{\mathbf{GG}}(z_l), \tag{3.92}$$

and

$$T_{\mathbf{G}} = -\frac{\hbar^2}{2m\Delta^2} - i\frac{\hbar^2}{2m\Delta}G_z. \tag{3.93}$$

Finally, the quantity A appearing in the right-hand-side vector is

$$A = \left(\frac{\hbar^2}{2m\Delta^2} - i\frac{\hbar^2}{2m\Delta}G_z\right) e^{-ik_L\Delta}, \tag{3.94}$$

and $\mathbf{I}$ is the $N_G \times N_G$ identity matrix.

3.3 Numerical Considerations

We should consider two main numerical difficulties with the formulation just presented: The complexity of the solution of the extremely large linear system given by (3.89) and the calculation of the self-energy terms expressing the open boundary conditions. Regarding the first problem, we simply note that the form of the Hamiltonian matrix in (3.89) is exactly of the form considered by Polizzi [52] and, as such, despite its huge rank, it may lead to an efficient numerical technique to find the solution of the linear system. Regarding the calculation of the self-energy terms, we should note that the discussion of the previous subsection is fully general and 'exact', in the sense that it guarantees unitarity within the framework of the local empirical pseudopotential description of the device and of the reservoirs. The major problem with this framework lies in its computational burden: In order to obtain the eigenvectors $\phi^{(L,p)}_{\mathbf{G},k_{Lp}}$ and $\phi^{(Rp)}_{\mathbf{G},k_{Rp}}$ and set up the matrices $\mathcal{M}^{(L)}$ and $\mathcal{M}^{(R)}$, we have to generate the complex band structure of the device. Moreover, having obtained

$\mathcal{M}$, we must invert this huge matrix. And huge matrix inversion is something to be avoided at all costs!

The calculation of the complex band structure can be performed following a well-known standard procedure: The original (supercell) eigenvalue problem whose solution yields the (real) band structure of the contacts is (see (3.4)):

$$\sum_{\mathbf{G}'}\left[H^{(L,R)}_{\mathbf{G}\mathbf{G}'}(k_{L,R}) - E_p(k_{L,R})\,\delta_{\mathbf{G},\mathbf{G}'}\right]\phi^{(L,R)p}_{\mathbf{G}',k_{L,R}} = 0, \tag{3.95}$$

where the index p labels the N_G bands, as usual. In our 1D-transport case (2D supercell), the Hamiltonian matrix $H^{(L,R)}_{\mathbf{G}\mathbf{G}'}$ is rewritten as a sum of a term proportional to k_z^2, one term proportional to k_z, and a term independent of k_z as follows:

$$\mathbf{H}^{(L,R)} = \frac{\hbar^2}{2m}\left[k_z^2\,\mathbf{I} + \mathbf{H}^{(L,R)(1)}k_z + \mathbf{H}^{(L,R)(0)}(E)\right]. \tag{3.96}$$

Thus, the Hermitian $N_G \times N_G$ eigenvalue problem in E, (3.95), is recast in the form of a non-Hermitian $2N_G \times 2N_G$ eigenvalue problem in k_z:

$$\begin{bmatrix} 0 & \mathbf{I} \\ -\mathbf{H}^{(L,R)(0)}(E) & -\mathbf{H}^{(L,R)(1)} \end{bmatrix}\begin{bmatrix} \phi^{(L,R)} \\ k_z\phi^{(L,R)} \end{bmatrix} = k_z\begin{bmatrix} \phi^{(L,R)} \\ k_z\phi^{(L,R)} \end{bmatrix}, \tag{3.97}$$

whose solutions, in general complex eigenvalues, provide the complex dispersion $k_{z,p}(E)$. The solution of this eigenvalue problem is numerically challenging: First, its rank is twice as large as the rank of the original Hamiltonian $\mathbf{H}^{(L,R)}$. For large supercell calculations of rank $\sim 10^4$, this is a nontrivial complication. Second, and more important, the eigenvalue problem (3.97) is non Hermitian, so that its solution requires significant additional computational efforts. A third consideration also shows that the complexity of the problem may actually be excessive for our scope: When dealing with electronic and transport-related properties of nanostructures, it is often sufficient to obtain only a few eigenpairs, E_p and $\phi^{(L,R)p}$, so that we can take full advantage of efficient numerical techniques aimed at extracting eigenpairs only in a small window of eigenvalues of interest. But (3.97) and its application to the open-bc problem discussed above require the calculation of *all* eigenpairs.

A possible solution to this numerical problem may rely on the following consideration: In many cases the evanescent waves (corresponding to eigenvalues k_z with an imaginary component) may contribute to the total reflected or transmitted wavefunctions by a small, perhaps negligible, amount. Therefore it is tempting to select only those M_p eigenpairs (out of the total number N_G of them) whose eigenvalues have small imaginary part or even only those which, having vanishing imaginary part, are already known from the solution of the original eigenvalue problem (3.95). The main idea is that, while in doing so we will not be able to solve the linear systems (3.68) and (3.79) exactly, on the other hand ignoring evanescent waves will not change the systems appreciably and will lead us to an approximate solution. We shall look for the 'best' approximate solution (in the 'least squares fit' sense) by minimizing the error in the linear systems. The net effect is that the matrices $\mathcal{M}^{(L,R)}$ will be replaced by *rectangular* $N_G \times M_p$ matrices.

One immediately see that this creates a physical and a mathematical problem, as expected: Physically, we give up unitarity but, as it will be discussed shortly, we hope that the deviations from the correct wavefunction is not too severe as long as evanescent waves belong to bands energetically strongly separated from the band to which the injected wave belongs. Mathematically, the linear systems (3.68) and (3.79) are over-determined (i.e., we have N_G equations in only $M_p < N_G$ unknowns) and the inverse matrices $\mathcal{M}^{(L,R)}$ required to obtain the solutions, (3.72) and (3.83), do not exist. This is a well-know mathematical problem which admits a 'pseudo-solution' consisting in finding sets of coefficients α_p and β_p which minimize the deviation from the ideal solution of (3.68) and (3.79). The matrix $\mathcal{M}^+$ which yields these sets and also minimizes this error is often called the Moore-Penrose pseudo-inverse of $\mathcal{M}$ and, when $\mathcal{M}$ is of full rank (that is, when the column-vectors constituting $\mathcal{M}$ are linearly independent) is given by [11,50]:

$$\mathcal{M}^+ = (\mathcal{M}^\dagger \mathcal{M})^{-1} \mathcal{M}^\dagger \tag{3.98}$$

and we may simply replace $\mathcal{M}^{(L,R)-1}$ with $\mathcal{M}^{(L,R)+}$ in the previous section.

The numerical advantage is obvious: The matrix $\mathcal{M}^\dagger \mathcal{M}$ which we must invert (...when it can actually be inverted, see Courrieu's paper [11] on how to proceed in general) in (3.98) to obtain the pseudo-inverse $\mathcal{M}^+$ is actually a smaller $M_p \times M_p$ matrix, so that obtaining the coefficients α_p and β_p and the self-energies $\Sigma^{(L,R)}$ involves only the inversion of this small matrix and a few matrix multiplications. Clearly, the real issue lies on establishing how badly unitarity is violated.

To summarize this approach, let consider the explicit form of the matrix $\mathcal{M}^{(L)}$, self-energy $\Sigma^{(L)}$, and reflection amplitudes α_p: The matrix $\mathcal{M}^{(L)}$ is the $N_G \times M_p$ matrix:

$$\mathcal{M}^{(L)} = \begin{bmatrix} \phi_{\mathbf{G}_1}^{(L,1)} & \phi_{\mathbf{G}_1}^{(L,2)} & \dots & \phi_{\mathbf{G}_1}^{(L,M_p)} \\ \phi_{\mathbf{G}_2}^{(L,1)} & \phi_{\mathbf{G}_2}^{(L,2)} & \dots & \phi_{\mathbf{G}_2}^{(L,M_p)} \\ \dots & \dots & .. & \dots \\ \phi_{\mathbf{G}_{N_G}}^{(L,1)} & \phi_{\mathbf{G}_{N_G}}^{(L,2)} & \dots & \phi_{\mathbf{G}_{N_G}}^{(L,M_p)} \end{bmatrix} . \tag{3.99}$$

Thus:

$$(\mathcal{M}^{(L)\dagger} \mathcal{M}^{(L)})_{pp'} = \sum_{\mathbf{G}} \phi_{\mathbf{G}}^{(L,p)*} \phi_{\mathbf{G}}^{(L,p')}, \tag{3.100}$$

is the small $M_p \times M_p$ matrix we must invert to obtain the pseudoinverse $\mathcal{M}^{(L)+}$ of $\mathcal{M}^{(L)}$ which is the $M_p \times N_G$ matrix:

$$\mathcal{M}_{p\mathbf{G}}^{(L)+} = \sum_{p'=1}^{M_p} (\mathcal{M}^{(L)\dagger} \mathcal{M}^{(L)})_{pp'}^{-1} \phi_{\mathbf{G}}^{(L,p')*}. \tag{3.101}$$

The left-contact self-energy is the $N_G \times N_G$ matrix:

$$\Sigma^{(L)}_{\mathbf{GG}'} = -\left(\frac{\hbar^2}{2m\Delta^2} - i\frac{\hbar^2}{2m\Delta}G_z\right)\sum_{pp'} \phi^{(L,p)}_{\mathbf{G}} \left(\mathscr{M}^{(L)\dagger}\mathscr{M}^{(L)}\right)^{-1}_{pp'} \phi^{(L,p')*}_{\mathbf{G}'} \, e^{ik_{L,p}\Delta}, \tag{3.102}$$

and, finally:

$$\alpha_p = \sum_{\mathbf{G}}\sum_{p'} \left(\mathscr{M}^{(L)\dagger}\mathscr{M}^{(L)}\right)^{-1}_{pp'} \phi^{(L,p')*}_{\mathbf{G}} \left[\phi_{\mathbf{G}}(z=0) - \phi^{(L,n)}_{\mathbf{G},k_L}\right]. \tag{3.103}$$

A similar expression holds for the right reservoir quantities leading to:

$$\beta_p = \sum_{\mathbf{G}}\sum_{p'} \left(\mathscr{M}^{(R)\dagger}\mathscr{M}^{(R)}\right)^{-1}_{pp'} \phi^{(R,p')*}_{\mathbf{G}} \, \phi_{\mathbf{G}}(z=L), \tag{3.104}$$

keeping in mind the sum over bands p of the right reservoir may be extend over a set of bands different (usually larger for a positive bias V) from the set of bands employed for left reservoir.

As an example of this procedure taken to its limit, suppose that we inject at a very low energy E from the left reservoir, such that $E < E_p(k_z)$ for any conduction band $p > 1$ in the contact. Then there is only one band ($p = 1$) such that the injected wave can be reflected into a propagating (as opposite to 'evanescent') wave. The matrix $\mathscr{M}^{(L)}$ now is simply the $N_G \times 1$ matrix (or 'vector', in this case):

$$\mathscr{M}^{(L)}_{\mathbf{G},p=1} = \phi^{(L,1)}_{\mathbf{G}}, \tag{3.105}$$

the self-energy matrix $\Sigma^{(L)}$ reduces to:

$$\Sigma^{(L)}_{\mathbf{GG}'} = -\left(\frac{\hbar^2}{2m\Delta^2} - i\frac{\hbar^2}{2m\Delta}G_z\right) \frac{\phi^{(L,1)}_{\mathbf{G}}\phi^{(L,1)*}_{\mathbf{G}'}}{\sum_{\mathbf{G}} \phi^{(L,1)*}_{\mathbf{G}}\phi^{(L,1)}_{\mathbf{G}}} \, e^{ik_{L,1}\Delta}, \tag{3.106}$$

(where $k_{L,1}$ is the only real wavenumber such that $E_1(-k_{L,1}) = E$, which can be determined from the known real band-structure), and the coefficient $\alpha_1(E)$ is expressed in terms of the coefficients $r_{\mathbf{G}} = \phi_{\mathbf{G}}(z=0) - \phi^{(L,1)}_{\mathbf{G}}$ as the error-minimizing solution of (3.72):

$$\alpha_1 = \left[\sum_{\mathbf{G}} \phi^{(L,1)*}_{\mathbf{G}}\phi^{(L,1)}_{\mathbf{G}}\right]^{-1} \sum_{\mathbf{G}} \phi^{(L,1)*}_{\mathbf{G}} \left[\phi_{\mathbf{G}}(z=0) - \phi^{(L,1)}_{\mathbf{G}}\right] \tag{3.107}$$

Clearly, this 'solution' violates unitarity, but only in the 'best possible way' when using only one reflected wave. As the number of reflected waves M_p increases, the violation becomes less and less severe until it vanishes for $M_p = N_G$. The 'acceptable' error has to be determined with numerical simulations. In practice, we may include exactly only propagating waves into the matrix $\mathscr{M}$. Evanescent waves belonging to bands with energy much larger or smaller then the injection energy

E may safely be ignored, but 'slowly' decaying evanescent waves belong to bands with energy sufficiently close to E may be required to approximate unitarity in a satisfactory way. These waves may be accounted for approximately (so, without the need to compute the complex band structure) assuming parabolic behavior in the gap (so, assuming the same 'real band' curvature effective mass) to compute the imaginary k_{Lp} or k_{Rp}, and using $\mathbf{k}\cdot\mathbf{p}$ perturbation theory (using only a few energetically 'adjacent' bands) to obtain the approximate eigenvectors $\phi_{\mathbf{G}}^{(L,p)}$ or $\phi_{\mathbf{G}}^{(R,p)}$ needed to build the matrix $\mathscr{M}$.

3.3.1 The Master Equation

The last and most challenging step in the program outlined here is the introduction of dissipation in the quantum transport formulation. Non-equilibrium Green's Function (NEGF) methods [30, 32] constitute a formidable challenge in this respect, although progress is being made [49]. However, abandoning the information provided by the 'two-times' Green's function – and so considering the density matrix – and neglecting the off-diagonal elements of the density matrix itself [62] – thus limiting ourselves to small devices – allows us to use a simpler formulation of the problem based on the Pauli Master equation [18, 19]. Coupling it to the full-band scheme we have discussed so far is still 'work in progress', so we present here selected results obtained embracing the much simpler effective-mass approximation, following closely [22].

To briefly review the method (discussed at length in [18, 19] for 1D simulation with scattering, and in [37] for the 2D ballistic case), the devices is assumed to be in contact with reservoirs which act as boundary condition for Poisson equation as well as particle reservoirs. Having obtained a first solution of the Poisson equation (such as a classical solution), the open-boundary-condition Schrödinger equation (either using an effective-mass approximation or the mixed supercell-envelope wave equation (3.59) or (3.61)) is solved obtaining a basis of states injected from the reservoirs. Representing the density matrix ρ on this basis of scattering states [18, 19, 37], the equation describing the dynamic of the density matrix becomes the 'simple' Master equation:

$$\frac{\partial \rho_i}{\partial t} = \sum_j (W_{ij}\rho_j - W_{ji}\rho_i) + \left[\frac{\partial (f_i - \rho_i)}{\partial t}\right]_{res}, \tag{3.108}$$

where i and j are the indices labeling the scattering states. The first and second terms on the right hand side can be considered respectively as Master and contact operators acting on the density matrix. The transition rate W_{ij} from j to i can be evaluated using Fermi golden rule as described above:

$$W_{ij} = \frac{2\pi}{\hbar} | < j|H_{int}|i > |^2 \delta(E_i - E_j \pm \hbar\omega_q), \tag{3.109}$$

where H_{int} is the interaction Hamiltonian, E_i and E_j are the total energies for a charge carrier in states i and j, respectively, while ω_q is either the frequency of the scattering excitation or zero for elastic scattering processes.

The Master equation, (3.108) can be solved either by direct inversion of the Master operator [19] or by a rather conventional Monte Carlo algorithm [18]. The latter method closely resembles the algorithm used to solve the semiclassical Boltzmann transport equation (BTE) since the structure of (3.108) does indeed resemble the structure of the BTE, the major difference consisting in the absence of the driving term due to the electrostatic potential, since this is assumed to have been diagonalized exactly when setting up the basis of the scattering states. The full problem is tackled by solving self-consistently the PME coupled with the Schrödinger equation (full-band or effective-mass), Poisson equation, and current continuity (required to maintain charge neutrality at the device/contact boundaries) [18].

3.4 Examples in the Effective Mass Approximation

As stated above, the full-band implementation of the method is work in progress. Therefore we have so far employed a simple effective-mass model which assumes a single parabolic valley with an effective mass of 0.98 m_0 similar to the Si longitudinal mass and a DOS mass (employed in the calculation of the scattering rates) comparable to the 6-valley DOS Si mass ($\sim$1.08 m_0, where m_0 is the free electron mass). Among the various scattering processes, we have considered nonpolar scattering with acoustic phonons in the elastic, equipartition approximation (see (3.44) for a similar expression), inelastic scattering with optical phonons (mimicking intervalley scattering), and, as discussed below, coherent multiple scattering with ionized impurities. The acoustic deformation potential has been taken as 10 eV, the optical deformation potential as 5×10^8 eV cm^{-1} and phonon energy of 60 meV, all other material parameters as appropriate for Si.

3.4.1 One-Dimensional Simulations

An n-i-n Diode. The first structure we have considered is a heavily doped (10^{20} cm^{-3}) one-dimensional *n-i-n* resistor, each region (*n* and *i*) 15 nm long. In Fig. 3.32 we compare the current obtained suppressing or accounting for phonon scattering shows that electron–phonon collisions depress the current by as much as 60% at high bias. Figure 3.33 illustrates the broadening of the spectrally-resolved electron density due to the loss of coherence caused by collisions.

A Resonant Tunnel Diode. The second one-dimensional structure we have studied is a double-barrier resonant tunnel diode (RTD) with a 10.0 nm-long *n* source and drain regions, two 0.5 nm-thick SiO_2 barriers 'sandwiching' a 2.0 nm-thick intrinsic

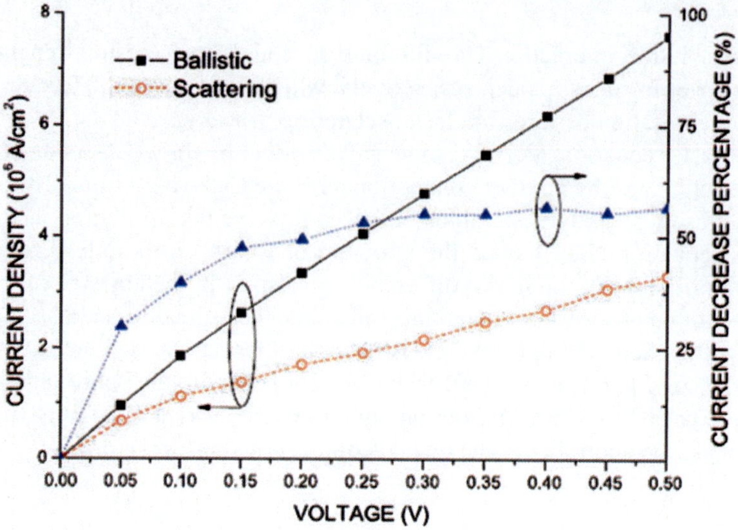

Fig. 3.32 Comparison of the calculated I–V characteristics of a 1D *n-i-n* resistor operating ballistically (*solid symbols*) or subjected to phonon scattering (*open symbols*). The *dotted* (*blue*) *line* shows the scattering-induced percentage reduction of the current

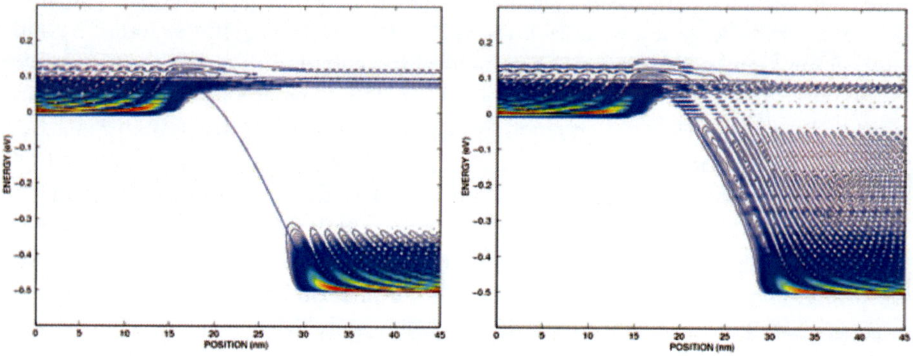

Fig. 3.33 Coherent and non-coherent transport energy spectrum of electron density of 1D *n-i-n* resistor at a bias of 0.5 V

well. Figure 3.34 shows the current–voltage characteristics of the diode emphasizing the effect of scattering. The effect of collisions on the current is dramatic, as well as on the reduction of the histeresis/bistabilty (not shown), mainly because collisions break the coherence necessary to sustain the quantum resonance. Comparing the energy spectra in the case of ballistic transport (Fig. 3.35, left) and accounting for scattering (right), we see the significant energy loss as electrons scatter from the second quasi-bound state in the well to the quasi-ground-state. Scattering states below the potential energy in the source (cathode) are due to the evanescent waves

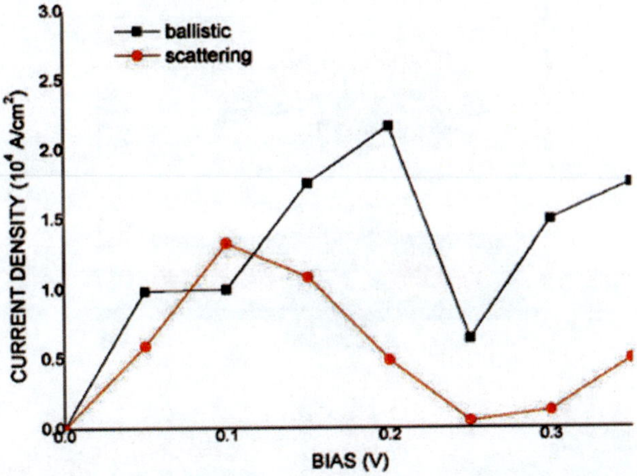

Fig. 3.34 Comparison of the calculated I–V characteristics in the ballistic (*black solid line*) and scattering limit (*red line*)

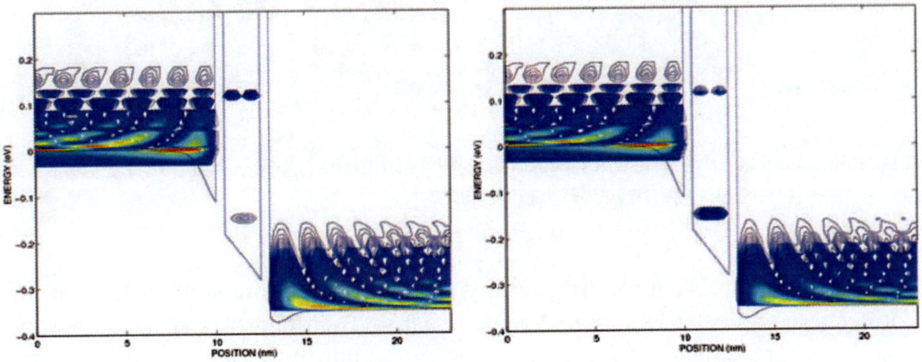

Fig. 3.35 Coherent and non-coherent transport energy spectrum of electron density of 1D RTD. At *right* note how inelastic scattering enhances the occupation of the ground-state in the well

introduced in the ballistic case, a numerical artifacts required to obtain charge neutrality near the device/contact boundary, as discussed by Frensley [20].

3.4.2 Two-Dimensional Simulations

Access Geometry in Double-Gate FETs. Extending our work to two-dimensional cases (that is, solving Poisson equation in 2D), we have considered thin-body double-gate FETs with different 'access' geometries, straight, tapered (referred to as 'taper' in the following) and dog-bone, exactly as those studied by Laux [37] in order to emphasize the geometry-induced quantum access resistance. In Fig. 3.36 we

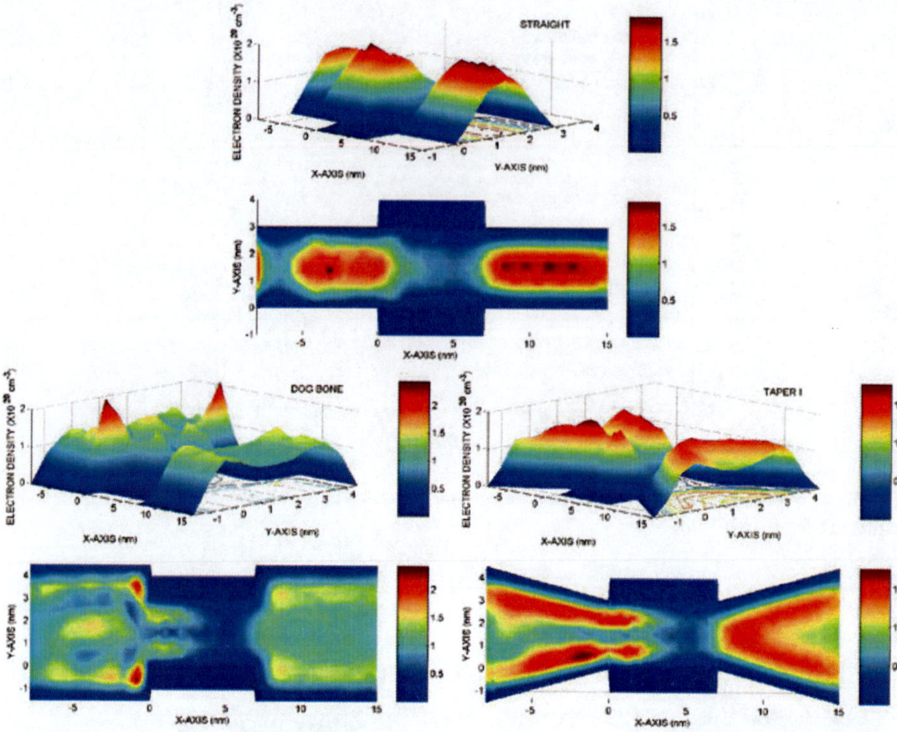

Fig. 3.36 Ballistic self-consistent electron density ($\times 10^{20}\,\mathrm{cm}^{-3}$) at $V_{GS} = 0.3\,\mathrm{V}$, $V_{DS} = 0.5\,\mathrm{V}$ for the three access geometries, as in the previous figure

show the self-consistent electron density under the conditions of ballistic transport for each geometry at $V_{GS} = 0.3\,\mathrm{V}$, $V_{DS} = 0.5\,\mathrm{V}$. The drain-current vs. drain-bias, I_{DS}-V_{DS}, characteristics, are shown in Fig. 3.37 both in the ballistic limit as well as when accounting for nonpolar electron–phonon scattering. The conclusion we draw, in agreement with [37], is that the straight access geometry yields the highest current, as it minimizes diffractions and reflections, while the dog-bone geometry presents the worst case, as it instead maximizes the reflections of the electron wavefunctions at the sharp edges. The tapered geometry appears to be an intermediate case.

The Effect of Impurity Scattering in the Source. These results seem to indicate that the quantum access resistance is dominated by geometric effects. However, since electron coherence within the source (and, to a much smaller extent, in the drain) is responsible for the magnitude of the effect, it is reasonable to ask to what extent phase randomization due to scattering with ionized impurities in the (usually very heavily doped) source and drain regions will destroy – or at least reduce – this 'waveguide' effect. In order to address this question, we have introduced point-like charges in the source and drain regions of the devices, following the work by Gilbert and Ferry [26]. In our 2D simulations these scattering centers consist of line-

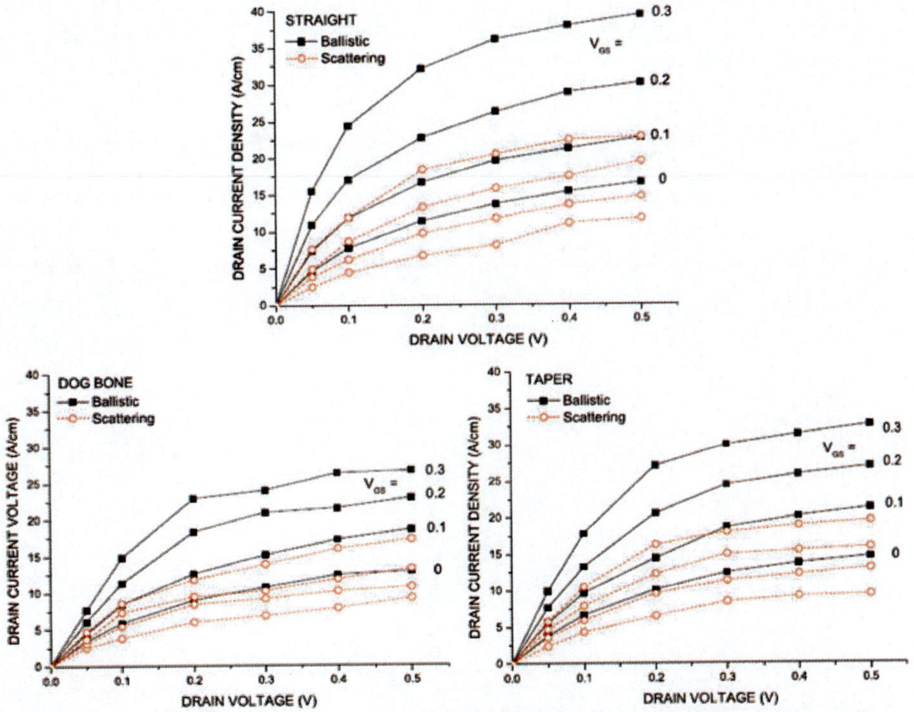

Fig. 3.37 Drain-current vs. drain-bias characteristics for DGFETs with three different access geometry (straight at *left*, taper at *center*, dog-bone at *right*), as in [37]. Both the ballistic current (*black solid symbols, solid lines*) as well as the current calculated accounting for phonon scattering (*red open symbols, dashed lines*) are shown. Note how the access resistance dominates in the ballistic regime, but its effect is reduced by scattering

charges with a linear ($\sim e/L_{TF}$, where L_{TF} is the Thomas-Fermi screening length) and areal density dictated by the requirement that we reproduce the actual volume density of the ionized impurities. In our simulated devices this requires that we place ten such scatterers in the devices, five each in the source and drain regions, while leaving the channel undoped. Note also that these scattering centers constitute non-phase-breaking scatterers, since decoherence and dissipation will emerge only after performing an average over their configurations [35]. The self-consistent algorithm we employ here also accounts automatically for the dielectric screening of their potential. This can be seen in Fig. 3.39 showing the self-consistent electron density for the straight, dog bone and tapered access geometries of the devices at equilibrium: The 'spikes' seen in the electron density occur at the location of the scattering centers and show directly the effect of dielectric screening as free electrons crowd the region around these attractive (donor) scattering centers.

We have repeated the calculations for four different randomly chosen configurations of the dopants. Having calculated the current–voltage characteristics for each configuration and averaging the result, we have obtained the current–voltage characteristics shown in Fig. 3.38. Despite the large fluctuations of the current for each

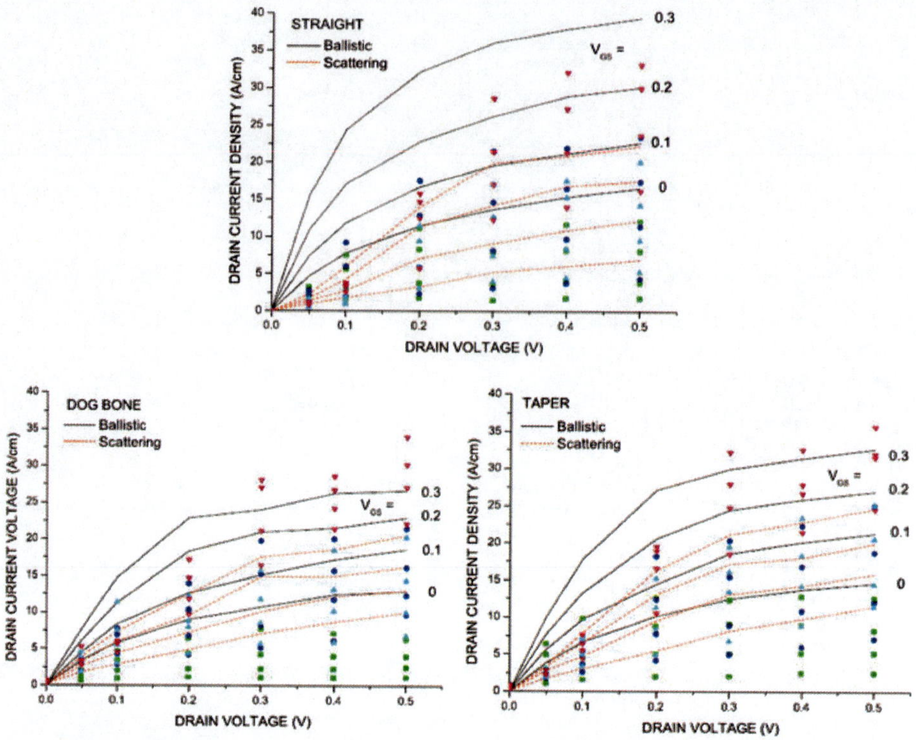

Fig. 3.38 Calculated I_{DS}-V_{DS} characteristics at various V_{GS} in the ballistic limit (*solid line*) and in the impurity scattering limit (*dashed line*) with ten random distribution of dopants averaged from four different spatial configurations of the dopants (different shape and color dots)

individual configuration (purple dots), one can clearly see that the average current of the three geometries exhibit differences which are significantly reduced with respect to the ballistic case whose results are shown in Fig. 3.37 in the absence of impurities, thus confirming the original suspicion that impurity scattering depresses the coherence of electron transport in the heavily doped source region. Of interest is also the observation that for some configurations the current in the presence of dopants can exceed the ballistic current. This is due to the existence of resonant states in the screening potential well. This has been discussed by Gilbert [26], who showed that discrete dopants modify the potential profile so drastically that resonant levels may induce 'spikes' in the current–voltage characteristics.

The Effect of Phonon Scattering. Finally we have applied the PME framework to the study of transport in the presence of optical and acoustic phonon scattering in 2D. The results, illustrated by the open-symbols/dashed-lines curves in Fig. 3.37, show that, similarly to what found in the ballistic case, the straight geometry yields the largest current, the dog-bone geometry the smallest. However, both the magnitude of the current as well as the difference caused by the various geometries are greatly reduced. This is due to the fact that scattering processes destroy the electron coherence and so reduce – but do not eliminate it altogether – the effects caused

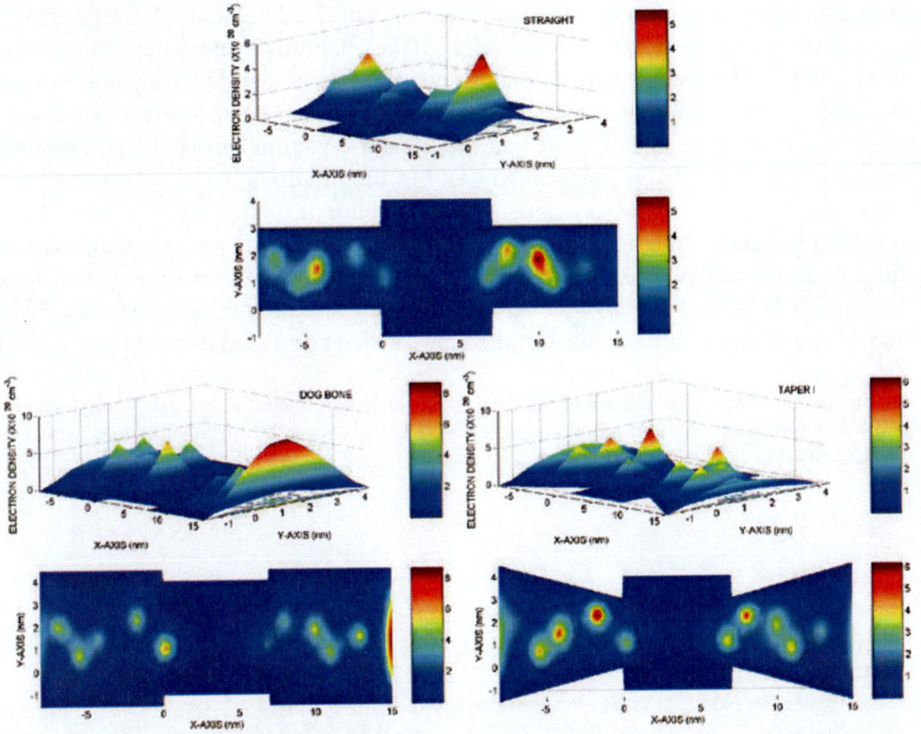

Fig. 3.39 Calculated electron density at equilibrium in DG FETs with ten 'dopants' introduced at random positions in the source and drain region of the devices

by the access geometry. In conclusion, the access geometry is still found to play a role in mesoscopic device design, although scattering (both phase-breaking and non-phase-breaking after configuration-averaging) reduces its importance.

4 Conclusions

We have presented a comprehensive discussion of a possible approach to handle transport, both semiclassical and quantum, within a full-band framework. We have argued that (local) empirical pseudopotentials represent the best compromise between accuracy and efficiency to obtain the band-structure of systems of current technological interest and have presented results regarding thin Si bodies, III–V hetero-layers, Si nanowires, graphene, graphene nanoribbons, and carbon nanotubes, discussing critically our results keeping ab-initio results from the literature – when available – as a benchmark. Moving to electronic transport, at the semiclassical level we have presented as a first example the case of high-field transport in thin Si inversion layers at high density, showing that the accurate band structure is required to obtain the low electron saturated velocity observed experimentally but never predicted theoretically. At the quantum level we have presented a mixed supercell/envelope approximated scheme to treat quantum transport in open systems, we have discussed in depth the problem of the open boundary conditions, and, finally, we have briefly reviewed the Master equation approach and reviewed results – unfortunately at present only within the effective mass approximation – regarding the effect of the accesses geometry, phonon scattering, and non-phase-breaking impurity scattering employing one- and two-dimensional simulations. Our final goal, obviously, remains the implementation of a Master equation scheme within the full-band framework, so that we may investigate the performance of sub-10 nm devices. The scheme outlined here promises to lead us towards this goal.

Acknowledgements One of the authors (MVF) would like to thank Steve Laux, Seonghoon Jin, and Eric Polizzi for help and stimulating discussions. This work has been supported in part by SRC, MARCO/MSD FCRP, and Samsung Electronics Corporation, Ltd.

References

1. H. Ajiki and T. Ando, Jap. J. Appl. Phys. **62**, 1255 (1993).
2. T. Ando, A. B. Fowler, and F. Stern, Rev. Mod. Phys. **54**, 437 (1982).
3. V. Barone, O. Hod, and G. Scuseria, Nano Letters **6**, 2748 (2006).
4. L. Bellaiche, S.-H. Wei, and A. Zunger, Phys. Rev. B **54**, 17568 (1996).
5. L. Bellaiche, L.-W. Wang, S.-H. Wei, and A. Zunger, Appl. Phys. Lett. **74**, 1842 (1999).
6. X. Blase, L. X. Benedict, E. L. Sherly, and S. G. Louie, Phys. Rev. Lett. **72**, 1878 (1994).
7. M. Brandbyge, J.-L. Mozos, P. Ordejón, J. Taylor, and K. Strokbo, Phys. Rev. B **65**, 165401 (2002).
8. L. Brey and H. A. Fertig, Phys. Rev. B **73**, 235411 (2006).
9. L. G. Bulusheva, A. V. Okotrub, D. A. Romanov, and D. Tomanek, J. Phys. Chem. A **102**, 975 (1998).
10. H. J. Choi and J. Ihm, Phys. Rev. B **59**, 2267 (1999).
11. P. Courrieu, Neural Inf. Processing - Letters and Reviews **8**, 25 (2005).
12. G. Dresselhaus, M. Dresselhaus, and J. G. Madrovies, Carbon **4**, 433 (1966).
13. D. Esseni and P. Palestri, Phys. Rev. B **72**, 165342 (2005).
14. M. Ezawa, Phys. Rev. B **73**, 045432 (2006).
15. M. Ezawa, Phys. Stat. Sol. (c) **4**, 489 (2007).
16. M. V. Fischetti and S. E. Laux, Phys. Rev. B **38**, 9721 (1988).
17. M. V. Fischetti and S. E. Laux, Phys. Rev. B **48**, 2244 (1993).
18. M. V. Fischetti, J. Appl. Phys. **83**, 270 (1998).
19. M. V. Fischetti, Phys. Rev. B **59**, 4901 (1999).
20. W. R. Frensley, Rev. Mod. Phys. **63**, 215 (1991).
21. J. T. Frey and D. J. Doren, "TubGen 3.3 Web Interface", http://turin.nss.udel.edu/research/tubegenonline.html
22. Bo Fu and M. V. Fischetti, *Dissipative Quantum Transport Using the Pauli Master Equation*, in Proc. International Workshop on Computational Electronics (2009), http://ieeexplore.ieee.org/stamp/stamp.jsp?tp=&arnumber=5091106&isnumber=5091070.
23. M. Fujita, K. Wakabayashi, K. Nakada, and K. Kusakabe, J. Phys. Soc. Jpn. **65**, 1920 (1996).
24. P. Giannozzi *et al.*, J. Phys.: Cond. Matter **21**, 395502 (2009).
25. G. Gilat and L. J. Raubenheimer, Phys. Rev. **144**, 390 (1966).
26. M. J. Gilbert and D. K. Ferry, IEEE Trans. Nanotechnol. **4** 355, (2005).

27. O. Gulseren, T. Yildirim, and S. Caraci, Phys. Rev. B **65**, 153405 (2002).
28. D. Gunlycke, P. A. Areshkin, J. Li, J. M. Mintmire, and C. T. White, Nano Letters **7**, 3608 (2007).
29. L. H. Hemstreet Jr., C. Y. Fong, and M. L. Cohen, Phys. Rev. B **2**, 2054 (1970).
30. L. P. Kadanoff and G. Baym, *Quantum Statistical Mechanics* (Benjamin, New York, 1962).
31. E. Kan, Z. Li, J. Yang, and J. G. Hou, J. Am. Chem. Soc. **130**, 4224 (2008).
32. L. V. Keldysh, Zh. Éksp. Teor. Fiz. **47**, 1515 (1964) [Sov. Phys. JEPT **20**, 1018 (1965)].
33. B. Khoshnevisan and Z. S. Tabatabaean, Appl. Phys. A **92**, 371 (2008).
34. M. Kohn and J. M. Luttinger, Phys. Rev. **98**, 915 (1955).
35. W. Kohn and J. M. Luttinger, Phys. Rev. **108**, 590 (1957).
36. Y. Kurokawa, S. Nomura, T. Takemori, and Y. Aoyagi, Phys. Rev. B **61**, 12616 (2000).
37. S. E. Laux, A. Kumar and M. V. Fischetti, J. Appl. Phys. **95**, 5545 (2004).
38. Y. Lee, T. Nagata, K. Kakushima, K. Shiraishi, and H. Iwai, "A Study on Electronic Structure of Silicon Nanowires with Diverse Diameters and Orientations for High Performance FET", Proc. *International Workshop on Density Functional Theory*, Tokyo, November, p. 83. 2008.
39. C. S. Lent and D. J. Kirkner, J. Appl. Phys. **67**, 6353 (1990).
40. K. Mäder and A. Zunger, Phys. Rev. B **50**, 17393 (1994).
41. A. Mayer, Carbon **42**, 2057 (2004).
42. T. Miyake and S. Saito, Phys. Rev. B **68**, 155424 (2003).
43. T. Miyake and S. Saito, Phys. Rev. B **72**, 073404 (2005).
44. K. Nehari, N. Cavassilas, J. L. Autran, M. Bescond, D. Munteanu, and M. Lannoo, Solid-State Electron. **50**, 716 (2006).
45. N. Neophytou, A. Paul, M. S. Lundstrom, and G. Klimeck, IEEE Trans. Electr. Dev. **55**, 1286 (2008).
46. T. W. Odom, J. Huang, P. Kim, and C. M. Lieber, Nature (London) **391**, 62 (1998).
47. T. W. Odom, J. Huang, P. Kim, and C. M. Lieber, J. Phys. Chem. B **104**, 2794 (2000).
48. M. Ouyang, J. Huang, C. L. Cheung, and C. M. Lieber, Science **292**, 702 (2001).
49. A. Pecchia and A. Di Carlo, Rep. Prog. Phys. **67**, 1497 (2004).
50. R. Penrose, Proc. Cambridge Phil. Soc. **51**, 406 (1955).
51. L. Pisani, J. A. Chan, B. Montanari, and N. M. Harrison, Phys. Rev. B **75**, 064418 (2007).
52. E. Polizzi, Phys. Rev. B **79**, 115112 (2009).
53. S. Reich and C. Thomsen, Phys. Rev. B **65**, 155411 (2002).
54. F. Sacconi, M. P. Persson, M. Povolotsky, L. Latessa, A. Pecchia, A. Gagliardi, A. Balint, T. Fraunheim, and A. Di Carlo, J. Comp. Electron. **6**, 329 (2007).
55. K.-I. Sasaki, S. Murakami, and R. Saito, J. Phys. Soc. Jpn. **75**, 074713 (2006).
56. W. Saslow, T. K. Bergstresser, and M. L. Cohen, Phys. Rev. Lett. **16**, 354 (1966).
57. H. Scheel, S. Reich, and C. Thomsen, Phys. Stat. Sol. (b) **242**, 2474 (2005).
58. H. Sevincli, M. Topsakal, and S. Ciraci, Phys. Rev. B **78**, 245402 (2008).
59. M. Sharma, A. Tiwari, and U. S. Sharma, "Ab-initio study of electronic band structure of zigzag single wall carbon nanotubes", in Proc. "International Workshop on New Trends in Science and Technology", Ankara, Turkey, Nov. 3-4, 2008, http://ntst08.cankaya.edu/proceedings/proceedings/Manoj/SharmaPaper.doc.
60. Y.-W. Son, M. L. Cohen, and S. G. Louie, Phys. Rev. Lett. **97**, 216803 (2006).
61. P. E. Trevisanutto, C. Giorgetti, L. Reining, M. Ladisa, and V. Olevano, Phys. Rev. Lett. **101**, 226405 (2008).
62. L. Van Hove, Physica **21**, 517 (1955).
63. L.-W. Wang and A. Zunger, J. Phys. Chem. **98**, 2158 (1994).
64. L. Yang, C.-H. Park, Y.-W. Son, M. L. Cohen, and S. G. Louie, Phys. Rev. Lett. **99**, 186801 (2007).
65. S. B. Zhang, C.-Y. Yeh, and A. Zunger, Phys. Rev. B **48**, 11204 (1993).
66. D. Zhang and E. Polizzi, J. Comp. Electr. **7**, 427 (2008).

Chapter 4
Quantum Master Equations in Electronic Transport

B. Novakovic and I. Knezevic

Abstract In this chapter we present several quantum master equations (QMEs) that describe the time evolution of the density matrix at various levels of approximations. We emphasize the similarity between the single-particle QME and the Boltzmann transport equation (BTE), starting from truncating the BBGKY chain of equations and ending with similar Monte-Carlo approaches to solve them stochastically and show what kind of boundary conditions are needed to solve the single-particle QME in the light of the *open nature* of modern electronic devices. The Pauli master equation (PME) and a QME in the perturbation expansion are described and compared both with one another and with the BTE. At the level of the reduced many-particle density matrix, we show several approaches to derive many-particle QMEs starting from the formal Nakajima–Zwanzig equation and ending with the partial-trace-free time-convolutionless equation of motion with memory dressing. Using those results we derive the correct distribution functions of the Landauer-type, for a small, ballistic open system attached to two large reservoirs with ideal black-body absorption characteristics.

Keywords Quantum transport · Master equation · Density matrix · Distribution function · Transient

1 Introduction

Electronic devices are many-particle objects. Therefore, they must be analyzed within the realm of statistical mechanics, with the goal to describe the time evolution of the full set of degrees of freedom belonging to a particular device.[1] Considering

[1] This gives an exact solution for the device's dynamical behavior (transient or steady state), but is not always necessary, because suitable approximations may suffice.

I. Knezevic (✉)
University of Wisconsin-Madison, 3442 Engineering Hall, 1415 Engineering Drive, Madison, WI 53706-1691
e-mail: knezevic@engr.wisc.edu

D. Vasileska and S.M. Goodnick (eds.), *Nano-Electronic Devices: Semiclassical and Quantum Transport Modeling*, DOI 10.1007/978-1-4419-8840-9_4,

that every particle, an electron, a phonon or another particle of interest, such as an exciton or a plasmon, can be described by several degrees of freedom (classical or quantum), the choice of which depends on the particular problem, and that there might be many particles in a single device, the problem clearly becomes intractable. In reality, one has to apply suitable approximations in order to reduce the problem to the one that is, at least, numerically feasible. This proceeds by choosing the relevant degrees of freedom and reducing the system of equations to describe their evolution, while the rest of the system is included by applying some assumptions about the irrelevant degrees of freedom. By degrees of freedom we mean, for example, the position and momentum of each particle (classically), or quantum numbers that span the Hamiltonian eigenstates (momentum, spin...).

Roughly speaking, each major approximation applied leads to a certain method or class of methods that are standardly used by device physicists and engineers to calculate the device transport properties. One possible classification of methods is done by approximating just how many particles/states in the many-particle problem are considered, so we can speak of a one-body problem (single particle states), two-body problem, and so on... This is commonly done by truncating the BBGKY hierarchy of equations [1, 2], that are able to describe the many-particle problem exactly, with all the mutual interactions between many-particle subsets. Along with the assumption of how the many degrees of freedom per particle are treated exactly we arrive at the kinetic and hydrodynamic models, most commonly in use. Kinetic models are at the level of distribution functions defined on a single-particle phase space, therefore treating one-body problems with interactions exactly, while hydrodynamic models incorporate additional assumptions about the momentum, therefore not treating the momentum exactly [3]. Most often [4–6], we account for interparticle interactions in the single-electron picture through the mean-field approximation (Hartree approximation), by self-consistently solving the Poisson equation along with any single-particle transport equation. Essentially, what we do is to solve the Poisson equation with the nonlinear charge density calculated by using the transport equation. When this system of equations converges, all other quantities of interest (e.g. current) can be calculated separately.

Another criterion we can use to distinguish between different models is whether they are quantum or semiclassical [7], classical being irrelevant in the context of small electronic devices. The simplest quantum model relies on particles populating the eigenstates of the single-particle Hamiltonian, obtained by solving the time-independent Schrödinger equation. This model can account for quantum tunneling, interference effects, sharp potentials and other quantum mechanical features, but is unable to handle the time dynamics of far from equilibrium states in the presence of scattering and coupling to the contacts [3]. More advanced quantum models define mixed states allowing for spatial localization of particles due to their coupling to the surroundings. Among these methods we can mention the single-particle density matrix method where the central equation is the Liouville–von Neumann equation [8], the Wigner function method with the Wigner equation [9] and the non-equilibrium Green's function method with the Dyson equation [10, 11]. Usually, these are all quantum kinetic equations, with the Liouville–von Neumann equation being known as the quantum master equation (QME), since it is an equation of motion for the density matrix, either a single-particle (quantum kinetic level), or a full/reduced many-particle density matrix. In some situations one can use the single-particle Pauli master equation (PME) [12], which, by its ability to model dissipation of eigenstates, can be situated between the pure Schrödinger equation (eigenstates without dissipation) and the single-particle density matrix method (mixed states with dissipation). The Boltzmann transport equation (BTE) is semiclassical. Its solution is a distribution function in the phase space that, therefore, does not respect the uncertainty relations and represents electrons as point-like particles for the purpose of drift and diffusion, making features like the tunneling, resonances, interference, etc. impossible. On the other hand, electrons are represented by plane waves during collisions, which makes the BTE unable to capture sharp potential changes (of the order of electron's wavelength). The BTE can be formally obtained by truncating the BBGKY chain [13]. Alternatively, it can be obtained from the NEGF method in the strong scattering limit [10].

Today, integrated circuits are made of many small electronic devices connected by leads to large reservoirs that supply them with charged particles (or other kind of matter/information). The natural framework in which modern electronic devices should be studied is the open system formalism, providing the necessary mathematical tools for handling a large number of variables and focusing on the most relevant ones [14, 15]. It requires the use of the reduced many-particle density matrix, that stores the information about the relevant variables after all the others have been traced out (a single-particle density matrix is generally insufficient). Most generally, we can refer to the electronic device in question as *the system*, which contains all the relevant variables, while everything else is *the environment* (e.g. reservoirs spatially separated from the system; other particles, like phonons, that share the same volume as the system). Therefore, the object of research is now a composite system, consisting of two, or more, physically coupled subsystems. The accuracy and the relevancy of our model will depend on what assumptions we apply to the environment.

In Sect. 2 we give an introduction to the exact many-particle density matrix and the corresponding equation for its time evolution, the Liouville–von Neumann equation. Then, we introduce the approximate single particle QME and describe some of its properties in closed and open systems. As examples of single-electron QMEs, two equations are mentioned: the PME, as applied to small electronic devices (open systems) [16, 17], in the Born–Markov limit and Hartree approximation, and the single-electron/many-phonon QME for bulk (closed system) [18–20], in the perturbation expansion and beyond the Born–Markov approximation. Monte Carlo solutions for both equations are described and compared to the conventional ensemble Monte Carlo technique. In Sect. 3 we introduce the reduced many-particle density matrix formalism, by starting from the formal derivation of the Nakajima–Zwanzig equation. In the following various techniques are introduced in order to make the Nakajima–Zwanzig equation more tractable: the Born–Markov approximation, the conventional time-convolutionless equation of motion, the partial-trace-free time-convolutionless equation of motion and the memory-dressing approach. In the final section, we build on the previous section and, by using the coarse-graining procedure and the short-time expansion of the generator of the time evolution, ultimately arrive at the correct steady-state distribution functions of the Landauer type, for the ballistic open quantum system.

2 The Single-Particle Quantum Master Equation

The QME is an equation of motion for the density matrix. In the single-particle picture, with off-diagonal elements included, it is a kinetic equation, where diagonal elements provide information about the population of single-particle states, while off-diagonal elements represent coherences between different single-particle states, describing localized particles. The single-particle QME is approximate and can be formally derived by truncating the BBGKY chain of equations, similar to the BTE. It describes the time-irreversible, dissipative time evolution for the single-particle states. In this section, we will discuss the general form of the single-particle QME, as well as two particular equations, starting from the full many-particle density matrix and its equation of motion, the Liouville–von Neumann equation.

2.1 The Density Matrix and the Liouville–von Neumann Equation

The density matrix formalism was pioneered by John von Neumann in 1927 [21,22] and is used to describe a mixed ensemble of states of a physical system, where by mixed we have in mind an ensemble that contain at least two, or more, different states of a physical system. Two extremes would be a pure ensemble, where all the states are the same, described by some state ket $|\alpha\rangle$, and a completely randomized ensemble, with each one of N states described by a different state ket $|\alpha_i\rangle$, where $i = 1,...,N$. Here, the state $|\alpha\rangle$, or $|\alpha_i\rangle$, is, in general, a linear combination of the eigenstates of the Hamiltonian. For a physical system with many particles the most exact density matrix is the one that describes a mixed ensemble of a full set of many-particle states, taking into account all the mutual interactions between the particles in the system. Such a many-particle density matrix at some initial time 0 is defined as

$$\rho_{12\cdots N}(0) = \sum_{i=0}^{M} W^{(i)}_{12\cdots N} \left|\Psi^{(i)}_{12\cdots N}(0)\right\rangle \left\langle \Psi^{(i)}_{12\cdots N}(0)\right|, \tag{4.1}$$

where M is the maximum number of many-particle states in the ensemble and $W^{(i)}_{12\cdots N}$'s are real positive numbers, representing the probability of occupation of the many-particle states $|\Psi^{(i)}_{12\cdots N}(0)\rangle$, which are symmetrized or anti-symmetrized linear combinations of products of a complete set of single-particle states [23]. The density matrix in (4.1) is normalized with the condition $\mathrm{Tr}(\rho_{12\cdots N}(0)) = 1$. From (4.1) follows that ρ is also hermitian, $\rho^{\dagger}_{12\cdots N}(0) = \rho_{12\cdots N}(0)$.

The time-evolution of the states $|\Psi^{(i)}_{12\cdots N}(0)\rangle$ is given by the many-particle time-dependent Schrödinger equation

$$i\hbar \frac{d}{dt} \left|\Psi^{(i)}_{12\cdots N}(t)\right\rangle = H_{12\cdots N} \left|\Psi^{(i)}_{12\cdots N}(t)\right\rangle . \tag{4.2}$$

These states are not necessarily orthogonal. Since states $|\Psi^{(i)}_{12\cdots N}(0)\rangle$ in (4.1) evolve according to (4.2), we have that the many-particle density matrix at some later time t will be given by

$$\rho_{12\cdots N}(t) = \sum_{i=0}^{M} W^{(i)}_{12\cdots N} \left|\Psi^{(i)}_{12\cdots N}(t)\right\rangle \left\langle \Psi^{(i)}_{12\cdots N}(t)\right| . \tag{4.3}$$

By differentiating (4.3) with respect to time and making use of (4.2) we arrive at the most general form of the Liouville–von Neumann equation, describing the time evolution of the full many-particle density matrix for a closed system

$$i\hbar \frac{d}{dt}\rho_{12\cdots N}(t) = [H_{12\cdots N}, \rho_{12\cdots N}(t)] \equiv \mathcal{L}_{12\cdots N}\rho_{12\cdots N}, \tag{4.4}$$

where $\mathcal{L}_{12\cdots N}$ is defined as a commutator superoperator generated by the many-particle Hamiltonian $H_{12\cdots N}$. Because this equation was generated by the Schrödinger equation, it preserves the previously stated properties of the density matrix, namely the normalization and hermiticity. If we use a shorthand notation $|\Psi^{(i)}_{12\cdots N}(t)\rangle \equiv |\alpha_i\rangle$, the expectation value of an observable A in a mixed ensemble described by the initial condition (4.1) and by (4.4), is given by

$$\begin{aligned}\langle A\rangle &= \sum_{i=1}^{M} w_i \langle\alpha_i| A |\alpha_i\rangle = \sum_{i=1}^{M} w_i \langle\alpha_i|\alpha_i\rangle \langle\alpha_i| A |\alpha_i\rangle \\ &= \sum_{i=1}^{M} \langle\alpha_i| \rho_{12\cdots N} A |\alpha_i\rangle = \mathrm{Tr}\left(\rho_{12\cdots N} A\right),\end{aligned} \tag{4.5}$$

where we use the fact the many-particle states, $|\alpha_i\rangle$ are properly normalized.

2.2 *The BBGKY Chain and the Single-Particle QME*

Instead of one exact many-particle Liouville–von Neumann equation (4.4), we can construct N coupled equations for the reduced density matrices, ρ_1, ρ_{12}, ..., $\rho_{12\cdots N}$, that form the BBGKY chain of equations [2]. Similar to the way the BTE, as a single particle equation for the distribution function over a single-particle phase space $(\mathbf{r}, \mathbf{p})$, is derived by applying approximations to the BBGKY chain of equations [13], we can derive the single-particle QME for the time evolution of the single-particle density matrix. If we assume that the dissipation processes are sufficiently weak (the weak-coupling or Born approximation) and memoryless or Markovian (one collision is completed before the next one starts, so that collisions do not depend on their history or initial conditions), then we can consider that the transport consists of periods of "free flights" (generalized "free flights" generated by the single particle Hamiltonian) and temporally and spatially very localized collisions described by a linear collision operator. In this way we can obtain a Boltzmann like QME for the time evolution of the single-particle density matrix $\rho(t)$ [3]

$$\frac{d\rho}{dt} = \frac{1}{i\hbar}\mathcal{L}\rho + C\rho, \tag{4.6}$$

where C is the collision superoperator, which is usually used to describe electron/phonon or electron/impurity interactions, and $\mathcal{L}$ is a commutator superoperator (4.4) generated by the single-particle Hamiltonian H. H, for noninteracting particles of the same kind (usually we are interested in electrons), is a sum of the kinetic energy operator and the potential energy due to any external potential $V_{ext}(\mathbf{r})$, but if we couple the transport equation (4.6) with the Poisson equation it will also include the Hartree potential $V_H(\mathbf{r})$ (mean-field approximation). So, we have in total

$$H = -\frac{\hbar^2}{2m}\nabla^2 + V_{ext}(\mathbf{r}) + V_H(\mathbf{r}). \tag{4.7}$$

Equation (4.6) is a limiting case of a density matrix completely reduced down to the single-particle states, with the additional assumptions about the nature of interactions in the system, stated above. The consequence of this derivation is the introduction of the time-irreversibility into the evolution of the single-particle density matrix ρ in (4.6), starting from the time-reversible (4.4).

So far we have considered a closed physical system for which $\mathcal{L}$ in (4.6) is hermitian, i.e. with real eigenvalues. Therefore it will contribute with complex oscillatory solutions for ρ in (4.6). The collision operator C will introduce negative real parts of eigenvalues which will cause an exponential decay of ρ. Therefore, this time-irreversible system is stable and behaves in an expected way. $\mathcal{L}$ is hermitian as a consequence of the hermiticity of the single-particle Hamiltonian for a closed system, where the hermiticity is defined through [3, 24]

$$\int_V [\psi^*(H\psi) - (H\psi)^*\psi]\, d^3r = 0$$
$$= \int_S \left(\psi^*\frac{d\psi}{dn} - \frac{d\psi^*}{dn}\psi\right) d^2r = \int_S \mathbf{J}d\mathbf{s}, \tag{4.8}$$

where Green's identity was used, ψ is the wavefunction, H the single-particle Hamiltonian and $\mathbf{J}$ the current density. We see that, when the number of particles is conserved in the volume V (closed system), the current density flux given by the last term in (4.8) is zero according to the current continuity equation and H, as well as $\mathcal{L}$, are hermitian.

If, on the other hand, the system is open, so that it exchanges particles with the environment, the number of particles is not conserved in general and both H and $\mathcal{L}$ are non-hermitian. Therefore, the eigenvalues of $\mathcal{L}$ will have imaginary parts and only non-positive imaginary parts are permissible in order to avoid having growing exponentials. To ensure this, it was shown by Frensley [3] that the boundary conditions have to be carefully chosen. In particular it is necessary to use time-irreversible boundary conditions, which can be easily defined only in phase space.

For example, if we have a 1D problem with two contacts and a region of interest (open system) in between we can choose different boundary conditions at (x_L, p_x) than at $(x_R, -p_x)$, where x_L and x_R are the left and right spatial boundaries of our open system. Now, under the time inversion those boundary conditions will apply to $(x_L, -p_x)$ and (x_R, p_x), respectively, and the problem will not be the same anymore. These BCs mean that the occupations of positive and negative propagating states are fixed by the left and right contacts, respectively. Even if we disregard the fact that the time-irreversible BCs are needed to achieve stability, they are a natural choice in the context of the following statement in [3] "if one's objective is to develop useful models of physical systems with many dynamical variables, rather than to construct a rigorously deductive mathematical system, it is clearly most profitable to adopt the view that irreversibility is a fundamental law of nature." The BCs of this form are naturally to be used with the Wigner function method. To include this kind of boundary conditions in (4.6) we can formally specify a contribution to the time evolution of the density matrix due to the injection/extraction through the contacts, a source term, the form of which can be determined phenomenologically

$$\frac{d\rho}{dt} = \frac{1}{i\hbar}\mathcal{L}\rho + \mathcal{C}\rho + \left(\frac{\partial \rho}{\partial t}\right)_{\text{inj/extr}} . \tag{4.9}$$

2.3 The Pauli Master Equation

As already mentioned in Sect. 1, the PME describes the time evolution of the probabilities of occupation of the single-particle Hamiltonian's eigenstates. With $p_n(t) \equiv \rho_{nn}(t)$ and for a closed system it is given by

$$\frac{d}{dt}p_n(t) = \sum_m [A_{nm}p_m(t) - A_{mn}p_n(t)] . \tag{4.10}$$

Equation (4.10) is easily justifiable at a phenomenological level, in situations when the exact Hamiltonian is not known, or when it is too complicated [15]. Then, we can always set up a master equation of the previous form, to describe the dissipative transport in the system. Coefficients A_{mn} represent transition rates between the levels and they can be found in a standard way, by using the quantum mechanical perturbation theory (Fermi's golden rule), or from experimental data. Alternatively, the PME follows from (4.6) by using Fermi's golden rule for the collision superoperator and a basis that diagonalizes the single-particle Hamiltonian that generates $\mathcal{L}$, since then the term $\mathcal{L}\rho$ vanishes and there is only the collision operator, which corresponds to the right-hand side of (4.10). So, the PME is a closed equation for the diagonal elements of the single-particle density matrix in the eigenbasis of the single-particle Hamiltonian, obtained from (4.6) by using Fermi's golden rule to describe scattering. It will be a complete description of the problem in the case the off-diagonal elements in (4.6) can be neglected. We will say more on the conditions to satisfy that requirement in the following.

The simplicity of the PME (4.10) makes it attractive for applications to real problems of quantum transport in electronic devices. However, the major disadvantage of the PME is that it violates the current continuity, as shown by Frensley [3]. The reason for this is that open systems are inhomogeneous, making the eigenstates have different spatial distributions. Mathematically, if we combine the PME and the current continuity equation, with $\rho(x,x;t)$ being the electron density, we can obtain for the rate of change of the electron density due to transitions between two eigenstates ψ_m and ψ_n [3]

$$\begin{aligned}\frac{\partial}{\partial t}\rho(x,x;t) &= \frac{\partial p_m}{\partial t}|\psi_m(x)|^2 + \frac{\partial p_n}{\partial t}|\psi_n(x)|^2 \\ &= [A_{nm}p_m(t) - A_{mn}p_n(t)] \times \left[|\psi_n(x)|^2 - |\psi_m(x)|^2\right]. \qquad (4.11)\end{aligned}$$

The left-hand side of (4.11) must be zero, because the divergence of an eigenstate's current density is zero. Since the second term on the right-hand side is non-zero, due to different spatial distributions of different eigenstates, we need the first term on the right-hand side to be zero, which is true only in equilibrium when detailed balance is satisfied. The conclusion is that the PME alone (i.e. without considering the off-diagonal terms) may be used at or very near equilibrium and in steady state, when $\partial p_{m,n}/\partial t = 0$ and therefore $\partial\rho(x,x;t)/\partial t = 0$, as it should be because $\nabla\cdot\mathbf{J}_m = \nabla\cdot\mathbf{J}_n = 0$.

A good example of using the PME in modeling small electronic devices is the work done by Fischetti [16, 17]. There, the PME application to small devices was justified and the results of steady state simulations with [16] and without [17] the full band structure were compared with those obtained by using the BTE. Set-up is such that contacts to the device as well as phonons and other particles important for scattering belong to the environment, while the device region with electrons is the open system. The justification and conditions for using the PME go as follows:

- As shown by Van Hove [25] and Kohn and Luttinger [26], if one starts from a quasidiagonal initial state and in the weak-scattering limit the off-diagonal terms remain negligible. Quasidiagonal states satisfy the condition that the off-diagonal terms are nonvanishing only when mixing states with energy difference $\delta E_{th} \ll \delta E_D$, where δE_{th} is the thermal broadening of the states and δE_D is the energy scale over which the matrix elements of perturbing interactions are constant.
- If the size of the device is comparable or smaller than the dephasing length of the incoming electrons from the contacts, $L \ll \lambda_\phi$ ($\lambda_\phi \approx 30$–50 nm for Si at 300 K), then they appear as plane waves, i.e. the density matrix is diagonal in the momentum representation. Assuming the weak-scattering limit in the open system (device), we can say, with respect to the previous statement, that neither are off-diagonal elements injected from the contacts nor do they form in the device region, so that the PME is applicable.
- The PME is unable to model the femtosecond time dynamics, because that is a genuinely off-diagonal problem on time-scales of the order of collision durations and strong-scattering effects beyond Fermi's golden rule. The PME's areas of

applicability are steady state with the weak-scattering and long-time limits and "adiabatic" transients, when the number of particles in the system changes very slowly with time.

The PME with Fermi's golden rule can only be used to find occupation probabilities governed by scattering in the system, but not due to the coupling to the contacts. Following the work of Fischetti [16, 17] this coupling can be introduced at a phenomenological level through a source term in the PME. The form of that source term for a general multiterminal configuration is given by [17]

$$\left(\frac{\partial \rho_{\mu}^{(s)}}{\partial t}\right)_{\text{res}} = |C_{\mu}^{(s)}|^2 \upsilon_{\perp}(\mathbf{k}_{\mu s})\left[f^{(s)}(\mathbf{k}_{\mu s}) - \rho_{\mu}^{(s)}\right], \tag{4.12}$$

where s indicates the contact/terminal, $\upsilon_{\perp}$ is the injecting velocity, $f^{(s)}$ the s-th contact distribution function, μ the full set of quantum numbers describing the eigenstates in the open system/device and $C_{\mu}^{(s)}$ takes care of the proper normalization of the states. Additional assumption is that the injecting distributions are given by the drifted Fermi–Dirac distribution $f^{(s)}\left(\mathbf{k}_{\mu}^{(s)} - \mathbf{k}_{d}^{s}\right)$, where $\mathbf{k}_{d}^{s}$ is calculated from the semiclassical current in the contact s. This takes into account the fast relaxation in the contacts and ensures the charge neutrality near the contacts/device boundaries as well as the current continuity. With this source term we can write the final steady state equation of motion for populations as

$$\sum_{\mu' r}\left[A_{\mu s;\mu' r}\rho_{\mu'}^{(r)} - A_{\mu' r;\mu s}\rho_{\mu}^{(s)}\right] + |C_{\mu}^{(s)}|^2 \upsilon_{\perp}(\mathbf{k}_{\mu s})\rho_{\mu}^{(s)}$$
$$= |C_{\mu}^{(s)}|^2 \upsilon_{\perp}(\mathbf{k}_{\mu s}) f^{(s)}\left(\mathbf{k}_{\mu}^{(s)} - \mathbf{k}_{d}^{s}\right). \tag{4.13}$$

This is a set of equations over μ that has to be solved self-consistently with $\mathbf{k}_d$ by applying the condition of current continuity at the contact/device boundaries.

Some of the results of the full-band calculations with (4.13) are given in Fig. 4.1 for an *nin* Si diode at 77 K, biased at 0.25 V [17]. For comparison purposes, alongside them are the results of the simulation with the Monte Carlo BTE.

2.4 A Single-Particle QME Beyond the Born–Markov Approximation

A somewhat different QME to study semiconductors in a uniform electric field can be constructed using the perturbation expansion of the single-electron/many-phonon Liouville–von Neumann equation [18–20]. The difference with the previous one is that it was applied to homogeneous bulk problems (not devices), but on the other hand it makes no assumption about the electron–phonon coupling (it is beyond the Born–Markov or weak-scattering/long-time limit of the PME) and is able to simulate energy-nonconserving transitions, multiple collisions and intracollisonal field effects [27, 28].

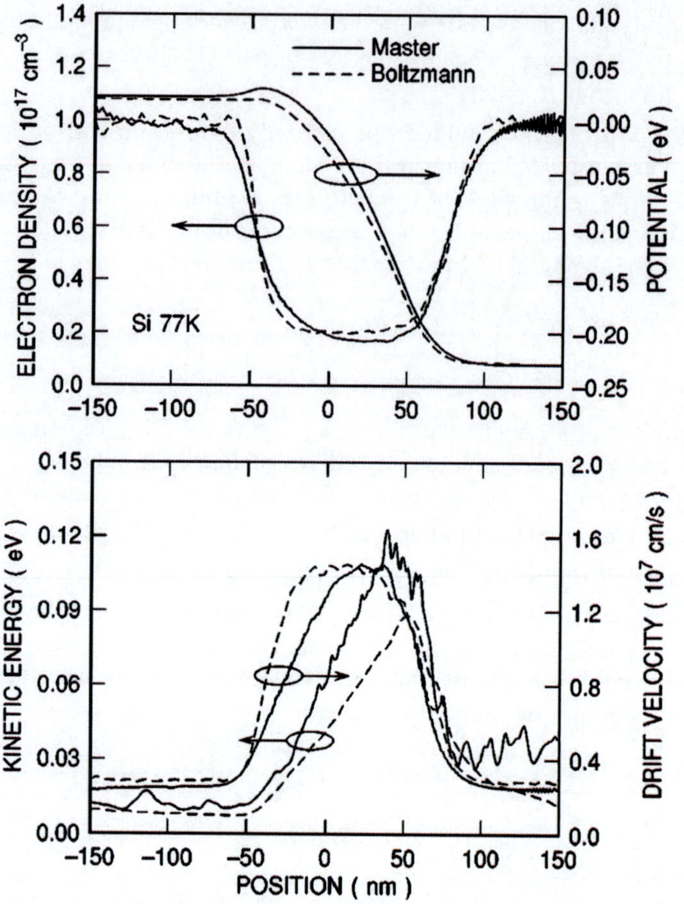

Fig. 4.1 *Top frame* – the electron charge density and potential energy for an *nin* Si diode at 77 K, biased at 0.25 V, where the *solid lines* are results of using the master equation (4.13), while the *dashed lines* are results of using the Monte Carlo BTE. *Bottom frame* – similar as the *top frame*, but with results for the average kinetic energy and drift velocity. Reprinted with permission from [17], M. V. Fischetti, Phys. Rev. B **59**, 4901 (1999). ©1999 The American Physical Society

The perturbation expansion to the Liouville–von Neumann equation for bulk semiconductors in a uniform electric field can be constructed as follows [18]. The Hamiltonian of this system in the effective mass approximation and with parabolic energy bands is a sum of several contributions

$$H = H_e + H_E + H_p + H_{e-ph} = H_0 + H_{e-ph}, \tag{4.14}$$

where

$$H_e = -\frac{\hbar^2}{2m^*}\nabla^2 \,, \quad H_E = e\mathbf{E}\mathbf{r} \,, \quad H_p = \sum_{\mathbf{q}} \hbar\omega_{\mathbf{q}} a^{\dagger}_{\mathbf{q}} a_{\mathbf{q}} \tag{4.15}$$

and H_{e-ph} is a standard Hamiltonian describing electron–phonon coupling and consisting of absorption and emission parts. H_0, describing the free and non-interacting electron gas, the equilibrium phonon distribution and the external homogeneous electric field is used to solve the time-dependent Schrödinger equation. Approximate solutions are the tensor products of the time-dependent accelerated plane waves (they would be accelerated Bloch waves beyond the effective mass approximation) normalized to 1 over the crystal volume V [29], and the many-body phonon states $|n_{\mathbf{q}},t\rangle$

$$|\mathbf{k}_0,n_{\mathbf{q}},t\rangle = \frac{1}{\sqrt{V}} e^{\mathrm{i}\mathbf{k}(t)\mathbf{r}} e^{-\mathrm{i}\int_0^t ds\omega(\mathbf{k}(s))} |n_{\mathbf{q}},t\rangle, \tag{4.16}$$

where $\mathbf{k}(t) = \mathbf{k}_0 - e\mathbf{E}t/\hbar$ and $\omega(\mathbf{k}(t)) = \hbar k^2/2m^*$.

If we use this basis set (whose time evolution is generated by H_0) for the density matrix, the Liouville–von Neumann equation contains only the interaction Hamiltonian

$$\mathrm{i}\hbar\frac{\partial}{\partial t}\rho(\mu,\mu',t) = \left[H_{e-ph}(t),\rho(t)\right]_{\mu,\mu'}, \tag{4.17}$$

where $\mu \equiv (\mathbf{k}_0,n_{\mathbf{q}})$. Upon the formal integration and perturbation expansion we obtain the following Dyson series for the diagonal elements of the density matrix $\rho(\mu,t) = \rho(\mu,\mu,t)$

$$\begin{aligned}\rho(\mu,t) &= \rho(\mu,0) + \int_0^t dt_1 \left[\widetilde{H}_{e-ph}(t_1),\rho(0)\right]_{\mu,\mu} \\ &\quad + \int_0^t dt_1 \int_0^{t_1} dt_2 \left[\widetilde{H}_{e-ph}(t_1),\left[\widetilde{H}_{e-ph}(t_2),\rho(0)\right]\right]_{\mu,\mu} + \cdots \\ &= \rho^{(0)}(\mu,t) + \rho^{(1)}(\mu,t) + \rho^{(2)}(\mu,t) + \cdots,\end{aligned} \tag{4.18}$$

where $\widetilde{H}_{e-ph} = (1/\mathrm{i}\hbar)H_{e-ph}$ and the initial condition is assumed to be diagonal and uncoupled, $\rho(\mu,\mu',0) = \rho(\mu,0) = \rho^{(0)}(\mu,t) = f_0(\mathbf{k}_0)P_{eq}(n_{\mathbf{q}})$, where f_0 and P_{eq} are the initial distribution functions of electrons and phonons, respectively.

We are only interested in the diagonal elements, whose time-evolution is given by (4.18), since, first, we want to evaluate expectation values of electronic quantities only and, second, they are diagonal in the electronic part of the wave function. Furthermore, (4.18) is a closed equation for the diagonal elements of $\rho(t)$, which is a consequence of a diagonal initial condition and the fact that there are only initial values of ρ at the right hand side of the perturbation expansion. Remember that we have mentioned a similar effect in a somewhat different context in Sect. 2.3, i.e. that the closed equation for the diagonal elements of the PME can be obtained from the general form of the single-particle QME (4.6) by working in the basis of the single-particle Hamiltonian and by approximating the collision superoperator with Fermi's golden rule. The fact that each term in the perturbation expansion starts from a diagonal state and have to end up in some other (or the same) diagonal state means that only even order terms in the expansion will survive. This can be explained by the fact that each interaction Hamiltonian (being linear in creation/destruction operators) will either create or destroy a phonon in that state (left or right) of the initial diagonal outer product of states (since in general $\rho = \sum |\alpha\rangle\langle\alpha|$) that is on the same side as that interaction Hamiltonian, after we expand the commutation relations. So to maintain the diagonalization we have to balance each absorption/emission at one of the sides by either the opposite process (emission/absorption) on the same side, or by the same process (absorption/emission) at the opposite side. This can only be achieved by having an even number of interaction Hamiltonians in a particular term in the perturbation expansion.

Equation (4.18) has several advantages over the steady state PME with Fermi's golden rule (of course within the limits of its applicability), beside the fact it can actually handle the transient regime. It is able to model quantum transitions of a finite duration and, because of the basis used, the acceleration of the plane waves during that time. The former ensures that the processes where the subsequent scattering effects begin before the previous ones have finished are accounted for (multiple collisions), while the latter ensures that the intracollisional field effect is not neglected. This approach also relaxes the constraint of the strict energy conservation during collisions, especially at short timescales. One of the disadvantages is that the trace over many-phonon degrees of freedom has to be taken in (4.18) [18].

2.5 *Monte Carlo Solution to the QME*

Using the Monte Carlo stochastic technique to solve the semiclassical BTE [30–33] is very common today, since it provides very accurate results (without using extensive approximations to make the problem numerically tractable), while the computational time is no more a bottleneck considering the availability of computing resources. The same idea of solving the semi-classical transport equation stochastically, instead of directly numerically, can be applied to the QME. In this section we will give a brief review of the ways this can be done in the case of a single-electron QME where we seek solutions (steady state and transient) to the diagonal elements of the density matrix. They will be algorithmically compared with the semiclassical Monte Carlo and shown to bear many similar characteristics, as far as the implementation is concerned.

2.5.1 The Steady-State PME for Inhomogeneous Devices

As has been shown in Sect. 2.3, the PME can be successfully applied to a certain class of problems which nowadays have high importance due to the down-scaling of electronic devices. The main equation of that section (4.13), which is a linear steady state equation for the occupations of levels with source terms modeling injection/extraction from the contacts, can be solved by using the Monte Carlo method [16]. For comparison purposes, let us write the standard BTE [33]

$$\frac{df(\mathbf{k},\mathbf{r},t)}{dt} + \frac{1}{\hbar}\nabla_{\mathbf{k}}E(\mathbf{k})\nabla_{\mathbf{r}}f(\mathbf{k},\mathbf{r},t) + \frac{\mathbf{F}}{\hbar}\cdot\nabla_{\mathbf{k}}f(\mathbf{k},\mathbf{r},t) = \left.\frac{\partial f(\mathbf{k},\mathbf{r},t)}{\partial t}\right|_{\text{Coll}}. \tag{4.19}$$

Diagonal elements of the density matrix from the PME (4.10), $p_n(t) = \rho_{n,n}(t)$ (n is a full set of basis quantum numbers), correspond to the distribution function $f(\mathbf{k},\mathbf{r},t)$ in (4.19), while the right hand side of (4.10) corresponds to the right hand side of (4.19). The main difference is in the drift and diffusion terms (due to the external field and spatial inhomogeneity) present in (4.19). Their absence from (4.10) is a consequence of a specific basis chosen for the density matrix, which diagonalizes the total potential consisting of the Hartree potential and the potential due to the external field. Although the BTE is most often used in the form given by (4.19), it can also be cast in the form without those two terms by a change in coordinates, from the phase space variables $(\mathbf{r},\mathbf{k})$ into the collision-free trajectories (path variables) [34]. So, to solve the PME we can use the conventional Monte Carlo procedure, used to solve the standard BTE (4.19), but without the free-flight part.

To better understand the relationship between (4.10) and (4.19) it can be shown that they are both limiting cases, but at the opposite ends of the domain [16]. As already mentioned in Sect. 2.3, the PME, being diagonal and therefore neglecting the off-diagonal elements, is justified for the quasidiagonal initial state. As shown by Van Hove [25], it is the state obtained by mixing the eigenstates of the unperturbed Hamiltonian, but only in a very narrow energy range (amplitudes are non-zero only for a very narrow range of energies of the states being mixed). Therefore, those states are highly delocalized. This physically corresponds to our assumption of devices much smaller than the dephasing length in the contacts, such that injecting electrons appear to them as spatially delocalized (but energetically very localized) wave packets, plane waves being the limiting case. There is one more group of states for which the diagonal form of the transport equation is justified and they are spatially very localized states, formed by linear combinations of eigenstates of the unperturbed Hamiltonian with amplitudes varying slowly with the energy. This opposite limit is satisfied by the BTE, which is therefore diagonal in the real space (the PME is diagonal in the wave vector space).

Finally, the implementation procedure would go as follows [16]:

- Electrons are initialized into the eigenstates $|\mu\rangle$, where μ is a full set of quantum numbers for the open system considered, according to the thermal equilibrium occupations as determined by the solution to the ballistic problem (no scattering).
- The time step is chosen and all transition probabilities are calculated. Scattering probability P_{scatter} is proportional to the transition rates determined by Fermi's golden rule, while injection/extraction probabilities (the processes that can change the number of particles in the open system) $P_{\text{in/out}}$ are proportional to the injection/extraction rates. Scattering or extraction events are selected according to the generated random number.
- If scattering is selected then the final state is chosen according to the final density of states and the matrix elements connecting the initial and final states, just like in the conventional Monte Carlo procedure. If extraction (exit through a contact)

is selected, the electron is simply removed. After all particles are processed, new particles are added to the states according to P_{in} and the drifted Fermi–Dirac distribution in the injecting contacts.

- After a few Monte Carlo steps the occupations of states, obtained from the Monte Carlo, are used to update the potential and wave functions with the Schrödinger/Poisson solver. The frequency of this update is determined by the plasma frequency of the whole device. The new potential is treated as a sudden perturbation which redistribute electrons from the old states $|\mu^{(old)}\rangle$ to the new states $|\mu^{(new)}\rangle$ according to the probability given by $|\langle\mu^{(new)}|\mu^{(old)}\rangle|^2$.

2.5.2 A Single-Electron QME in Homogeneous Bulk

The explanation of the similarity of (4.18) with the BTE can proceed by remembering what we said in Sect. 2.5.1, about the BTE written in the path variables, when it has the following form (after the drift-diffusion terms have disappeared)

$$f(t) = f(0) + P_i f - P_o f = f_0 + P_i f_0 - P_0 f_0 + P_i P_i f_0 - P_i P_0 f_0 - P_0 P_i f_0 + P_0 P_0 f_0 + \cdots, \tag{4.20}$$

where P_i and P_o are the integral operators for scattering "in" and "out". This equation is of the same general form as (4.18) and so similar Monte Carlo procedures can again be used to solve both equations, as will be outlined below.

The Monte Carlo algorithm to solve (4.18) has several novelties comparing to the one explained in Sect. 2.5.1 [20]. Beside the initialization and the standard random selections of the type of the scattering process (in/out scattering and the type of scattering) like in the conventional Monte Carlo, here we have several new random selections due to the perturbation expansion. First, there is a selection of the perturbative order (just the even ones, as shown previously), second, the selection of $n/2$ times where the first interaction Hamiltonians of each quantum process (a quantum process is defined as a pair of $\widetilde{H}_{e-ph}$'s for a distinct $\mathbf{q}$) are to be evaluated and, third, as already pointed out the average over the phonon variables $\mathbf{q}$ have to be performed (equivalent of taking the trace over the phonon degrees of freedom), for which a separate random number is reserved. So far, this is the same for both (4.18) and (4.20). The additional steps for the quantum case would be to select the side of $\rho(0)$ where each process starts and the time for the second $\widetilde{H}_{e-ph}$ in the process.

The restoration of this quantum Monte Carlo algorithm to the standard one, consisting of periods of free flights interrupted by scattering events, can be achieved by introducing a quantum analog of the self-scattering in the standard Monte Carlo algorithm, that makes scattering rates constant [35,36]. That can be achieved by the following transformation [19,20]

$$\rho(t) \to \exp\left(\int_{t_0}^{t} \gamma(t_1)dt_1\right)\rho(t), \tag{4.21}$$

where ρ is understood to represent diagonal elements ρ_μ as before. For constant $\gamma = 1/\tau$ and $t_0 = 0$ we have $\rho \rightarrow e^{(t/\tau)}\rho$, which gives the following equation instead of (4.18)

$$\rho(t) = e^{-(t/\tau)}\left[\rho_0 + \left(\widetilde{H}\widetilde{H} + \frac{1}{2\tau}\right)\rho_0 - \widetilde{H}\rho_0\widetilde{H} - \widetilde{H}\rho_0\widetilde{H} + \rho_0\left(\widetilde{H}\widetilde{H} + \frac{1}{2\tau}\right) + \cdots\right]. \tag{4.22}$$

In this concise notation the integral signs as well as argument lists and subscripts are dropped, and the commutation relations are expanded. This equation is actually equal to (4.18), since the damping factor $e^{-(t/\tau)}$ is going to cancel with all the factors $1/2\tau$ when all the integration and summations are performed. Nevertheless, this form makes the quantum Monte Carlo very similar to the standard ensemble Monte Carlo, consisting of periods of free flights interrupted by scattering events. The change to the previously explained algorithm is that the times selected for the first $\widetilde{H}$ in each process is separated by a constant time τ, the "free-flight" time, but only a few events will actually be quantum processes (scattering events) with a definite **q**. Although this procedure does not really contribute to the physical side of the problem, the fact that it is made similar to the semiclassical approach makes comparison with it much more transparent.

A representative result of the application of this algorithm and a comparison with the semiclassical Monte Carlo is shown in Fig. 4.2 [20]. We see a clear discrepancy

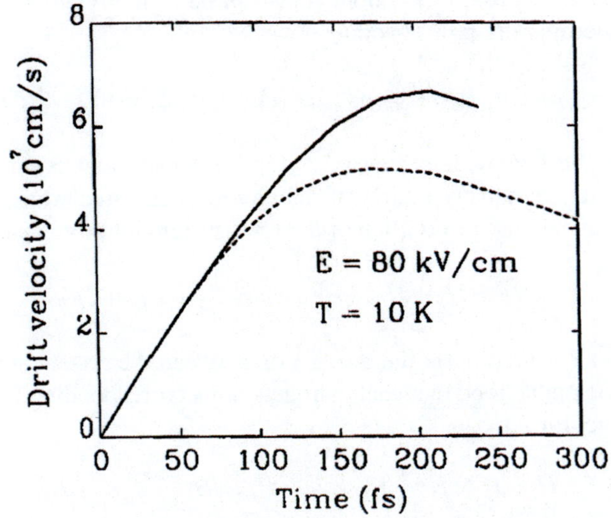

Fig. 4.2 Drift velocity overshoot in silicon. The result of the quantum Monte Carlo technique is shown with the *solid line*, while the semiclassical result is shown with the *dashed line*. Reprinted with permission from [20], C. Jacoboni, Semicond. Sci. Technol. **7**, B6 (1992). ©1992 IOP Publishing Ltd

in the drift velocity overshoot between the two techniques, which is attributed to the intracollisional field effect favoring transitions oriented along the field direction, comparing with the standard isotropic cross section.

3 Reduced Many-Particle QMEs

The reduced many-particle density matrix and the corresponding QME by its complexity fall between the single-particle and the full many-particle cases. This contributes to its flexibility, allowing us to find the optimal balance between the accuracy of modeling important physical processes in the open system and the computational complexity that results from including a large number of degrees of freedom. In this section we will first derive the formal, exact equation of motion for the reduced density matrix, the Nakajima–Zwanzig equation, and then introduce several approaches that make this equation more tractable for practical applications.

3.1 *The Nakajima–Zwanzig Equation*

Here, we will formally derive the Nakajima–Zwanzig equation for an exact reduced many-particle system. As already mentioned in Sect. 1 we are only interested in the time evolution of *the system*. Therefore, starting from (4.4) we need to trace out all *the environmental* degrees of freedom. This can be formally done by introducing a projection superoperator pair $\mathcal{P}$ and $\mathcal{Q}$

$$\mathcal{P}\rho(t) = \rho_E \otimes Tr_E(\rho(t)) = \rho_E \otimes \rho_S(t), \quad \mathcal{Q}\rho(t) = \rho(t) - \mathcal{P}\rho(t), \tag{4.23}$$

where $\rho(t)$ is the total density matrix, $\rho_S(t)$ the density matrix of the system and $\rho_E(t)$ represents the density matrix of the environment. Accordingly, we can split the Hamiltonian and the Liouvillian of the total system into three parts

$$H = H_S + H_E + H_I\,, \quad \mathcal{L} = \mathcal{L}_S + \mathcal{L}_E + \mathcal{L}_I, \tag{4.24}$$

where by index I we represent the interaction between the system and environment. Here, it is to be understood that each part acts in its corresponding Hilbert space (or Liouville space, for $\mathcal{L}$), e.g.

$$H = I_E \otimes H_S + H_E \otimes I_S + H_I\,, \quad H_I = \sum_i A_i \otimes B_i, \tag{4.25}$$

where I_α is the identity operator in the α-subspace, and A and B are operators that act on the environment and system Hilbert spaces, respectively. The form of interaction in (4.25) is the most general one. By acting with projection operators (4.23) on (4.4) we get a system of two equations, one for $P\rho$ and one for $Q\rho$. Upon formally solving it for the relevant part $P\rho$ we arrive at the formally exact equation of motion for the density matrix, the Nakajima–Zwanzig equation[2] [14, 37, 38]

$$\frac{d}{dt}\mathcal{P}\rho(t) = -\,\mathrm{i}\mathcal{P}\mathcal{L}(t)\mathcal{P}\rho(t) - \int_0^t ds\mathcal{K}(t,s)\mathcal{P}\rho(s) \\ -\,\mathrm{i}\mathcal{P}\mathcal{L}(t)\mathcal{G}(t,0)\mathcal{Q}\rho(0), \tag{4.26}$$

where the convolution or memory kernel $\mathcal{K}$ is

$$\mathcal{K}(t,s) = \mathcal{P}\mathcal{L}(t)\mathcal{G}(t,s)\mathcal{Q}\mathcal{L}(s)\mathcal{P}\,, \quad \mathcal{G}(t,s) = T_{\leftarrow}\exp\left[-\mathrm{i}\int_s^t ds'\mathcal{Q}\mathcal{L}(s')\right], \tag{4.27}$$

with $T_{\leftarrow}$ being the time ordering operator which sorts the operators to the right of it according to increasing time argument from right to left.

Equation (4.26) is not very useful for practical applications in this form because it is very complex. It contains all orders of interaction H_I and some memory terms, which makes it an exact non-Markovian QME. Memory terms are incorporated through the non-local memory kernel, the integral over past times $[0,t]$ and through the explicit dependence on the initial conditions in the second and third term. In the next section we will show some common approximations that are used to derive an approximate (to the second order in interaction) Markovian QME. Further modification to (4.26) that is commonly done is to choose the projection operator $\mathcal{P}$ such that the third term is canceled in the situations when the initial state of the total system is uncoupled $\rho(0) = \rho_E(0)\otimes\rho_S(0)$. This is achieved if $\mathcal{P}\rho$ is induced by $\rho_E(0)$ in (4.23) because

$$\mathcal{Q}\rho(0) = \rho(0) - \mathcal{P}\rho(0) = \rho(0) - \rho_E(0)\otimes\rho_S(0) = 0. \tag{4.28}$$

Now (4.26) is just

$$\frac{d}{dt}\mathcal{P}\rho(t) = -\mathrm{i}\mathcal{P}\mathcal{L}(t)\mathcal{P}\rho(t) - \int_0^t ds\mathcal{K}(t,s)\mathcal{P}\rho(s). \tag{4.29}$$

To finally obtain the reduced dynamics described by $\rho_S(t)$ we have to take the trace over environmental variables $Tr_E(\mathcal{P}\rho(t))$.

3.2 The Born–Markov Approximation

Now, we will briefly sketch how to derive an approximate Markovian QME that ultimately lead to a QME whose time-evolution generator (equivalent to $\mathcal{L}$ in (4.4))

[2] In the following we set $\hbar = 1$.

satisfies the quantum dynamical semigroup property, meaning that if we define a dynamical map $\mathcal{W}(t)$ as

$$\rho_S(t) = \mathcal{W}(t)\rho_S(0), \tag{4.30}$$

its property is

$$\mathcal{W}(t_1)\mathcal{W}(t_2) = \mathcal{W}(t_1 + t_2). \tag{4.31}$$

This defines a Markovian evolution and the necessary microscopic conditions for it will be stated in the following. The generator of this dynamical map can be defined as

$$\begin{aligned} \mathcal{W}(t) &= \exp(\mathcal{F}t), \\ \frac{d}{dt}\rho_S(t) &= \mathcal{F}\rho_S(t), \end{aligned} \tag{4.32}$$

from which it follows that the time evolution generator must be time-independent in order to have a Markovian QME.

The Born approximation is justified for weak coupling. This coupling is characterized by the interaction Hamiltonian H_I, which may refer to the coupling to reservoirs, phonons and everything else that can be encountered in real electronic devices. Since we assume that the coupling is weak we can keep only terms up to the second order in H_I in (4.29). Higher order interactions are contained in the memory term $\mathcal{K}$ in the integral in (4.29) and in order to keep just the second order term we need to have $\mathcal{L}_I$ in $\mathcal{K}$ appearing twice at most. To achieve that we can approximate the propagator $\mathcal{G}(t,s)$ with

$$\mathcal{G}(t,s) = T_{\leftarrow}exp\left[\int_s^t ds' \mathcal{Q}\left(\mathcal{L}_S(s') + \mathcal{L}_E(s')\right)\right], \tag{4.33}$$

which corresponds to leaving only zeroth order term in $\mathcal{L}_I(t)$. The Born approximation may be restated in several equivalent ways, depending on the way of derivation of final equations. The most obvious way, just mentioned, is to explicitly keep terms only up to the second order in interaction [15]. Equivalently, we can assume that, due to the weak-coupling, the density matrix of the system is always factorized during the evolution as [14]

$$\rho_S(t) = \rho_E \otimes \rho_S(t) \tag{4.34}$$

and that the density matrix of the reservoir is only negligibly affected by the interaction. The third way is somewhat less formal and is connected to the quantum mechanical scattering theory [22]. A variation of the Neumann series method, known as the Born series in this context, is used to approximate the form of the wave function after the scattering. This is also used in Fermi's golden rule, to calculate the transition rates which are valid in the weak-coupling and long-time limits.

The Markovian approximation would proceed by first replacing $\mathcal{P}\rho(s)$ by $\mathcal{P}\rho(t)$ in (4.29), thus removing any dependence at time t on the past states, for $s < t$,

$$\frac{d}{dt}\mathcal{P}\rho(t) = -\mathrm{i}\mathcal{P}\mathcal{L}(t)\mathcal{P}\rho(t) - \int_0^t ds\mathcal{K}(t,s)\mathcal{P}\rho(t). \tag{4.35}$$

This equation (in other forms and/or specific basis) is called the Redfield equation [14, 15, 39]. Second, there is an integral left which depends on the initial conditions, or in other words the interval between the present and initial states. To get rid of this we make a simple substitution $s \rightarrow t - s$ and let the upper limit of integration go to infinity, which gives us

$$\frac{d}{dt}\mathcal{P}\rho(t) = -\mathrm{i}\mathcal{P}\mathcal{L}(t)\mathcal{P}\rho(t) - \int_0^\infty ds\mathcal{K}(t,t-s)\mathcal{P}\rho(t). \tag{4.36}$$

These two approximations, that make up the Markovian approximation, are possible provided $\tau_E \ll \tau_S$, where τ_E is the environmental relaxation rate and τ_S the open system relaxation rate. This means that the time evolution can be coarse-grained such that $\rho_S(t)$ is almost constant during τ_E, while the integral in (4.36) vanishes fast with decreasing $t-s$ and, therefore, the Markovian approximation is justified.

Proceeding with some further less significant modifications to (4.36) we arrive at the most general form of the generator of the quantum dynamical semigroup [14, 15]. It constitutes the Lindblad form of the QME for an open system [40]

$$\frac{d}{dt}\rho_S(t) = -\mathrm{i}[H,\rho_S(t)] + \sum_k \gamma_k \left(A_k\rho_s A_k^\dagger - \frac{1}{2}A_k^\dagger A k\rho_S - \frac{1}{2}\rho_S A_k^\dagger A_k \right), \tag{4.37}$$

where H is the Hamiltonian that generates a unitary evolution, consisting of the system Hamiltonian and corrections due to the system–environment coupling, and A_k's are the Lindblad operators that describe the interaction with the environment in the Born–Markov limit.

3.3 The Conventional Time-Convolutionless Equation of Motion

The Nakajima–Zwanzig equation (4.26), that relies upon the use of the projection-operator technique, has several shortcomings that are the motivation for the following sections. Various variants of the projection-operators have been used in the past to study a range of physical systems. Argyres and Kelley [41] applied it to a theory of linear response in spin-systems, Barker and Ferry [42] to quantum transport in very small devices, Kassner [43] to relaxation in systems with initial system-bath coupling, Sparpaglione and Mukamel [44] to electron transfer in polar media, followed by a study of condensed phase electron transfer by Hu and Mukamel [45], while Romero-Rochin and Oppenheim [46] studied relaxation of two-level systems weakly coupled to a bath. However, this approach is limited by two computationally intensive operations needed to arrive at the final, reduced, density matrix of the open system: the time-convolution integral containing the memory kernel and the partial trace over environmental variables, $\mathrm{Tr}_E(\mathcal{P}\rho)$. Specifically, these limits would be lifted by applying the Markov and Born approximations of Sect. 3.2, respectively, because then the time-convolution disappears and the trace is a trivial operation since the equation for $\mathcal{P}\rho$ is already well factorized into the environmental and system parts.

Going beyond the Born–Markov approximation we have to think of different methods of leveraging the computational burden. In line with that, Tokuyama and Mory [47] proposed a time-convolutionless equation of motion in the Heisenberg picture. This was extended to the Schrödinger picture by Shibata et al. [48, 49] after which a stream of research appeared. Saeki analyzed the linear response of an externally driven systems coupled to a heat bath [50] and systems coupled to a stochastic reservoir [51, 52]. Ahn extended the latter to formulate the quantum kinetic equations for semiconductors [53], and a theory of optical gain in quantum-well lasers [54]. Later, he treated noisy quantum channels [55] and quantum information processing [56]. Chang and Skinner [57] applied the time-convolutionless approach to analyze relaxation of a two-level system strongly coupled to a harmonic bath, while Golosov and Reichmann [58] analyzed condensed-phase charge-transfer process. In the following, we will give a brief derivation of the time-convolutionless equation of motion and point out some of its shortcomings, resulting from the fact that it is still based on the projection-operator technique.

Let us choose some arbitrary, but proper and constant in time, environmental density matrix $\widetilde{\rho}_E$ as a generator for the time-independent projection operator (4.23). This means that $\mathrm{Tr}_E(\widetilde{\rho}_E) = 1$ and therefore

$$\mathrm{Tr}_E(\mathcal{P}\rho) = \mathrm{Tr}_E(\widetilde{\rho}_E) \cdot \mathrm{Tr}_E(\rho) = \mathrm{Tr}_E(\rho) = \rho_S. \tag{4.38}$$

The two equations for the projection operators $\mathcal{P}$ and $\mathcal{Q}$ are

$$\frac{d}{dt}(\mathcal{P}\rho(t)) = -\mathrm{i}\mathcal{P}\mathcal{L}(t)\rho(t) = -\mathrm{i}\mathcal{P}\mathcal{L}(t)\mathcal{P}\rho(t) - \mathrm{i}\mathcal{P}\mathcal{L}(t)\mathcal{Q}\rho(t), \tag{4.39}$$

$$\frac{d}{dt}(\mathcal{Q}\rho(t)) = -\mathrm{i}\mathcal{Q}\mathcal{L}(t)\rho(t) = -\mathrm{i}\mathcal{Q}\mathcal{L}(t)\mathcal{Q}\rho(t) - \mathrm{i}\mathcal{Q}\mathcal{L}(t)\mathcal{P}\rho(t). \tag{4.40}$$

A formal solution of (4.40) is

$$\mathcal{Q}\rho(t) = -\mathrm{i}\int_0^t dt' \mathcal{G}(t,t')\mathcal{Q}\mathcal{L}(t')\mathcal{P}\mathcal{U}(t',t)\rho(t) + \mathcal{G}(t,0)\mathcal{Q}\rho(0), \tag{4.41}$$

where for $t > t'$

$$\begin{aligned} \mathcal{G}(t,t') &= T_{\leftarrow} exp\left(-\mathrm{i}\int_{t'}^{t} ds \mathcal{Q}\mathcal{L}(s)\mathcal{Q}\right), \\ \mathcal{U}(t',t) &= T_{\rightarrow} exp\left(\mathrm{i}\int_{t'}^{t} ds \mathcal{L}(s)\right). \end{aligned} \tag{4.42}$$

The superoperator $\mathcal{U}(t,t')$ is defined by

$$\rho(t) = \mathcal{U}(t,t_0)\rho(t_0),$$

$$\mathcal{U}(t,t') = \Theta(t-t')T_{\leftarrow}\exp\left(-\mathrm{i}\int_{t'}^{t} ds\mathcal{L}(s)\right) + \Theta(t'-t)T_{\rightarrow}\exp\left(\mathrm{i}\int_{t}^{t'} ds\mathcal{L}(s)\right). \tag{4.43}$$

By using it we make (4.41) time-local, which is the essence of this approach. Equation (4.41) can be rearranged in the following way

$$\mathcal{D}(t;0)\mathcal{Q}\rho(t) = [1-\mathcal{D}(t;0)]\mathcal{P}\rho(t) + \mathcal{G}(t,0)\mathcal{Q}\rho(0), \tag{4.44}$$

where $\mathcal{D}(t;0)$ is defined as

$$\mathcal{D}(t;0) = 1 + \mathrm{i}\int_0^t dt'\mathcal{G}(t,t')\mathcal{Q}\mathcal{L}(t')\mathcal{P}\mathcal{U}(t',t). \tag{4.45}$$

Assuming that $\mathcal{D}(t;0)$ is invertible, (4.41) finally becomes

$$\mathcal{Q}\rho(t) = \left[\mathcal{D}(t;0)^{-1} - 1\right]\mathcal{P}\rho(t) + \mathcal{D}(t;0)^{-1}\mathcal{G}(t,0)\mathcal{Q}\rho(0). \tag{4.46}$$

Using the last equation in (4.39) we obtain

$$\frac{d}{dt}(\mathcal{P}\rho(t)) = -\mathrm{i}\mathcal{P}\mathcal{L}(t)\mathcal{D}(t;0)^{-1}\mathcal{P}\rho(t) - \mathrm{i}\mathcal{P}\mathcal{L}(t)\mathcal{D}(t,0)^{-1}\mathcal{G}(t,0)\mathcal{Q}\rho(0). \tag{4.47}$$

The last step that is left to obtain the conventional time-convolutionless equation of motion is to take the trace over environmental variables of (4.47), which gives us

$$\begin{aligned}
\frac{d}{dt}\rho_S(t) &= -\mathrm{i}\mathrm{Tr}_E\left[\mathcal{P}\mathcal{L}(t)\mathcal{D}(t;0)^{-1}\mathcal{P}\rho(t)\right] - \mathrm{i}\mathrm{Tr}_E\left[\mathcal{P}\mathcal{L}(t)\mathcal{D}(t;0)^{-1}\mathcal{G}(t,0)\mathcal{Q}\rho(0)\right] \\
&= -\mathrm{i}\mathrm{Tr}_E\left[\mathcal{L}(t)\mathcal{D}(t;0)^{-1}\widetilde{\rho}_E\otimes\rho_S(t)\right] - \mathrm{i}\mathrm{Tr}_E\left[\mathcal{L}(t)\mathcal{D}(t;0)^{-1}\mathcal{G}(t,0)\mathcal{Q}\rho(0)\right] \\
&= -\mathrm{i}\mathrm{Tr}_E\left[\mathcal{L}(t)\mathcal{D}(t;0)^{-1}\widetilde{\rho}_E\right]\rho_S(t) - \mathrm{i}\mathrm{Tr}_E\left[\mathcal{L}(t)\mathcal{D}(t;0)^{-1}\mathcal{G}(t,0)\mathcal{Q}\rho(0)\right].
\end{aligned} \tag{4.48}$$

This conventional form of the time-convolutionless equation of motion has three shortcomings. First, it explicitly depends on the choice of $\widetilde{\rho}_E$ that induces the projection operator, although the final result will not depend on it. Second, we have to evaluate complicated matrices $\mathcal{U}$, $\mathcal{G}$ and $\mathcal{D}$ involving all the degrees of freedom in the system+environment, but at the end we will extract only those degrees belonging to the system, by taking the trace. Third, this approach depends on invertibility of $\mathcal{D}$, which might be difficult to fulfill. These issues will be addressed in the following sections.

3.4 The Eigenproblem of the Projection Operator

The projection operator, as defined in (4.38), is idempotent ($\mathcal{P}^2 = \mathcal{P}$) because

$$\mathcal{P}^2\rho = \mathcal{P}(\mathcal{P}\rho) = \widetilde{\rho}_E \otimes \mathrm{Tr}_E\left[\widetilde{\rho}_E \otimes \mathrm{Tr}_E(\rho)\right]$$
$$= \widetilde{\rho}_E \otimes \mathrm{Tr}_E(\widetilde{\rho}_E)\,\mathrm{Tr}_E\left[\mathrm{Tr}_E(\rho)\right] = \widetilde{\rho}_E \otimes \mathrm{Tr}_E(\rho) = \mathcal{P}\rho. \quad (4.49)$$

Therefore, it has two eigenvalues, 0 and 1, and since they are both real we can conclude that $\mathcal{P}$ is also hermitian, $\mathcal{P} = \mathcal{P}^\dagger$. In analogy with the notion that system states are members of the respective Hilbert space, while operators (like ρ) act on it, we can introduce a Liouville space whose members are operators acting on the Hilbert space, while superoperators (like $\mathcal{L}$) act on it. To complete the definition we have to define the inner product which is conveniently done as $(A,B) = \mathrm{Tr}\left(A^\dagger B\right)$, where A and B are some operators belonging to the Liouville space. So, if the Hilbert spaces are $\mathcal{H}_S$, $\mathcal{H}_E$ and the composite space $\mathcal{H}_{S+E} = \mathcal{H}_E \otimes \mathcal{H}_S$, the respective Liouville spaces are $\mathcal{H}_S^2$, $\mathcal{H}_E^2$ and $\mathcal{H}_{S+E}^2$, where the dimensionality of Liouville spaces with respect to that of the corresponding Hilbert spaces is obvious. It follows that $\mathcal{P}$ is a superoperator acting on $\mathcal{H}_{S+E}^2$, which is $d_E^2 d_S^2$-dimensional. By construction (4.23) the image space of $\mathcal{P}$ corresponds to $\mathcal{H}_S^2$, so that the subspace of $\mathcal{P}$ spanned by the degenerate eigenvalue 1 is isomorphic to $\mathcal{H}_S^2$. We can write

$$\mathcal{H}_{S+E}^2 = \left(\mathcal{H}_{S+E}^2\right)_{\mathcal{P}=1} \oplus \left(\mathcal{H}_{S+E}^2\right)_{\mathcal{P}=0}, \quad (4.50)$$

where $\left(\mathcal{H}_{S+E}^2\right)_{\mathcal{P}=1}$ is the d_S^2-dimensional unit subspace and $\left(\mathcal{H}_{S+E}^2\right)_{\mathcal{P}=0}$ is the $d_S^2\left(d_E^2-1\right)$-dimensional zero subspace of the eigenspace of $\mathcal{H}_{S+E}^2$.

We can always arrange the eigenbasis of $\mathcal{P}$, $\left\{|n\rangle\,|n = 1,\ldots,d_E^2 d_S^2\right\}$, such that the first d_S^2 basis vectors span $\left(\mathcal{H}_{S+E}^2\right)_{\mathcal{P}=1}$ and therefore

$$\mathcal{P} = \sum_{n=1}^{d_S^2} |n\rangle\langle n|. \quad (4.51)$$

The eigenstates of the composite space $\mathcal{H}_{S+E}$ are constructed as $|i\alpha\rangle = |i\rangle \otimes |\alpha\rangle$, from which follows that the eigenstates $|n\rangle$ of $\mathcal{H}_{S+E}^2$ can be written by using four quantum numbers, i.e. as linear combinations of $|i\alpha, j\beta\rangle$. Here, states $|i\rangle$ belong to the environment, while states $|\alpha\rangle$ to the system. Furthermore, if we define $\mathcal{P}$ by using a uniform density matrix

$$\overline{\rho}_E = d_E^{-1} \cdot \mathbf{1}_{d_E \times d_E}, \quad (4.52)$$

we can avoid mixing states with different α and β to obtain a given $|n\rangle$ [59]. One finds that the states defined as

$$\left|\overline{\alpha\beta}\right\rangle = \frac{1}{\sqrt{d_E}} \sum_{i=1}^{d_E} |i\alpha, i\beta\rangle \tag{4.53}$$

constitute an orthonormal basis within the unit subspace of $\mathcal{P}$, i.e.

$$\mathcal{P}\left|\overline{\alpha\beta}\right\rangle = \left|\overline{\alpha\beta}\right\rangle, \quad \left\langle\overline{\alpha\beta}|\overline{\sigma\gamma}\right\rangle = \delta_{\alpha\sigma}\delta_{\beta\gamma}. \tag{4.54}$$

Finally, we can write

$$\mathcal{P} = \sum_{\alpha,\beta=1}^{d_S} \left|\overline{\alpha\beta}\right\rangle \left\langle\overline{\alpha\beta}\right| = \frac{1}{d_E} \sum_{\alpha,\beta=1}^{d_S} \left(\sum_{i=1}^{d_E} |i\alpha, i\beta\rangle\right) \left(\sum_{j=1}^{d_E} \langle j\alpha, j\beta|\right). \tag{4.55}$$

Since

$$\rho = \sum_{i,j=1}^{d_E} \sum_{\alpha,\beta=1}^{d_S} \rho_{j\beta}^{i\alpha} |i\alpha\rangle \langle j\beta| = \sum_{i,j=1}^{d_E} \sum_{\alpha,\beta=1}^{d_S} \rho^{i\alpha, j\beta} |i\alpha, j\beta\rangle, \tag{4.56}$$

we now have representations for both $\mathcal{P}$ and ρ, which allows us to explicitly calculate $\mathcal{P}\rho$ (with the help of $\langle i\alpha, j\beta|p\sigma, q\nu\rangle = \delta_{ip}\delta_{jq}\delta_{\alpha\sigma}\delta_{\beta\nu}$) as

$$\mathcal{P}\rho = \frac{1}{\sqrt{d_E}} \sum_{\alpha,\beta=1}^{d_S} (\mathrm{Tr}_E\rho)^{\alpha\beta} \left|\overline{\alpha\beta}\right\rangle = \sum_{\alpha,\beta=1}^{d_S} (\mathcal{P}\rho)^{\overline{\alpha\beta}} \left|\overline{\alpha\beta}\right\rangle, \tag{4.57}$$

where

$$(\mathcal{P}\rho)^{\overline{\alpha\beta}} = \frac{(\mathrm{Tr}_E\rho)^{\alpha\beta}}{\sqrt{d_E}}. \tag{4.58}$$

Equation (4.58) defines an isomorphism between $\left(\mathcal{H}_{S+E}^2\right)_{\mathcal{P}=1}$ and $\mathcal{H}_S^2$ that allows us to calculate the trace over environmental variables by effectively doing the basis transformation (4.53).

The conclusion of the previous paragraph is that by working in the eigenbasis of $\mathcal{P}$, as one of the possible eigenbasis of $\mathcal{H}_{S+E}^2$ (4.50), from the beginning we can avoid explicitly taking the trace over environmental variables at the end. In that eigenbasis, given by (4.53) and completed for $\left(\mathcal{H}_{S+E}^2\right)_{\mathcal{P}=0}$ (details in [59]), the total density operator can be written as a $d_S^2 d_E^2$-dimensional column vector

$$\rho = \begin{bmatrix} \rho_1 \\ \rho_2 \end{bmatrix}, \tag{4.59}$$

where ρ_1 is d_S^2-dimensional and ρ_2 is $d_S^2\left(d_E^2-1\right)$-dimensional, while the projection operators as $d_S^2 d_E^2 \times d_S^2 d_E^2$ matrices

$$\mathcal{P} = \begin{bmatrix} \mathbf{1}_{d_S^2\times d_S^2} & \mathbf{0}_{d_S^2\times d_S^2(d_E^2-1)} \\ \mathbf{0}_{d_S^2(d_E^2-1)\times d_S^2} & \mathbf{0}_{d_S^2(d_E^2-1)\times d_S^2(d_E^2-1)} \end{bmatrix},$$

$$\mathcal{Q} = \begin{bmatrix} \mathbf{0}_{d_S^2\times d_S^2} & \mathbf{0}_{d_S^2\times d_S^2(d_E^2-1)} \\ \mathbf{0}_{d_S^2(d_E^2-1)\times d_S^2} & \mathbf{1}_{d_S^2(d_E^2-1)\times d_S^2(d_E^2-1)} \end{bmatrix}. \tag{4.60}$$

We see that $\rho_S = \mathrm{Tr}_E(\rho)$ is given just by (using (4.58))

$$\rho_S = \sqrt{d_E}\cdot\rho_1. \tag{4.61}$$

Similarly, any superoperator $\mathcal{A}$ acting on $\mathcal{H}^2_{S+E}$ is represented by

$$\mathcal{A} = \begin{bmatrix} \mathcal{A}_{11} & \mathcal{A}_{12} \\ \mathcal{A}_{21} & \mathcal{A}_{22} \end{bmatrix}. \tag{4.62}$$

Additionally, if an operator is a system operator, i.e. $\mathcal{A}_{sys} = \mathbf{1}_E \otimes \mathcal{A}_S$, then it commutes with $\mathcal{P}$

$$\mathcal{P}\mathcal{A}_{sys}\rho = \overline{\rho}_E \otimes \mathrm{Tr}_E\left[(\mathbf{1}_E\otimes\mathcal{A}_S)\rho\right] = \overline{\rho}_E\otimes\mathcal{A}_S\mathrm{Tr}_E\rho \tag{4.63}$$

$$= (\mathbf{1}_E\otimes\mathcal{A}_S)(\overline{\rho}_E\otimes\mathrm{Tr}_E\rho) = \mathcal{A}_{sys}\mathcal{P}\rho, \tag{4.64}$$

which means that it is block-diagonal in the eigenbasis of $\mathcal{P}$. Furthermore, it is easily shown that the upper left block matrix is just $\mathcal{A}_S$ (see Appendix B of [61]), so that

$$\mathcal{A} = \begin{bmatrix} \mathcal{A}_S & \mathbf{0} \\ \mathbf{0} & \mathcal{A}_2 \end{bmatrix}. \tag{4.65}$$

The above mentioned isomorphism between $\left(\mathcal{H}^2_{S+E}\right)_{\mathcal{P}=1}$ and $\mathcal{H}^2_S$ and the decomposition of $\mathcal{H}^2_{S+E}$ according to (4.50) are graphically shown in Fig. 4.3. Because of the isomorphism (4.58, 4.61) density matrices of the form

$$\rho = \begin{bmatrix} \rho_1 \\ \mathbf{0} \end{bmatrix} \tag{4.66}$$

are called "purely system states", because they are completely determined by the state of the system S and depend on the environment only in an average sense (through the trace operation). On the other hand, density matrices for which $\rho_1 = 0$ and $\rho_2 \neq 0$ we call "entangled states" because they carry microscopic connections to the environmental states, beyond the point of easy separability like in the case of "purely system states". This can be seen by explicitly deriving the part of the basis for $\left(\mathcal{H}^2_{S+E}\right)_{\mathcal{P}=0}$, with the help of the Gram-Schmidt procedure (see Appendix of [59] and, for more compact and explicit form, Appendix A of [62]).

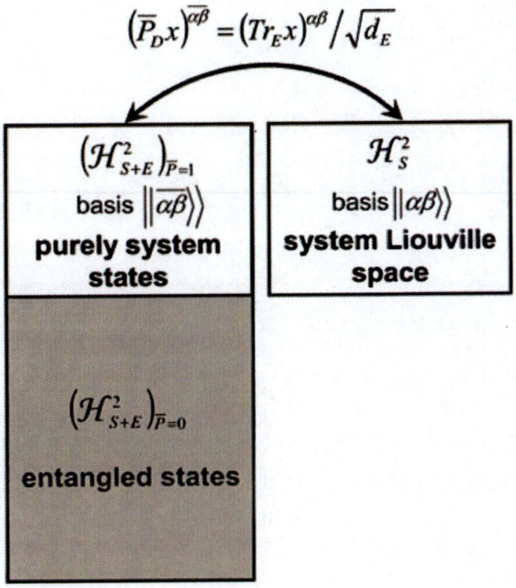

Fig. 4.3 Decomposition of the total Liouville space $\mathcal{H}^2_{S+E}$ into the subspaces of the projection operator $\mathcal{P}$ and the isomorphism between the unit subspace $(\mathcal{H}^2_{S+E})_{\mathcal{P}=1}$ and $\mathcal{H}^2_S$ for an operator x acting on $\mathcal{H}_{S+E}$. Reprinted with permission from [60], I. Knezevic and D. K. Ferry, Phys. Rev. A **69**, 012104 (2004). ©2004 The American Physical Society

3.5 A Partial-Trace-Free Equation of Motion

We proceed by writing the conventional time-convolutionless equation of motion from Sect. 3.3 in the basis of $\mathcal{P}$ derived in Sect. 3.4. The Liouville operator and the time-evolution operator are given by the following block forms

$$\mathcal{L}(t) = \begin{bmatrix} \mathcal{L}_{11}(t) & \mathcal{L}_{12}(t) \\ \mathcal{L}_{21}(t) & \mathcal{L}_{22}(t) \end{bmatrix}, \quad \mathcal{U}(t,t') = \begin{bmatrix} \mathcal{U}_{11}(t,t') & \mathcal{U}_{12}(t,t') \\ \mathcal{U}_{21}(t,t') & \mathcal{U}_{22}(t,t') \end{bmatrix}. \tag{4.67}$$

The Liouville–von Neumann and equation for the time-evolution now have the following forms

$$\begin{aligned} \frac{d\rho_1}{dt} &= -\mathrm{i}\mathcal{L}_{11}(t)\rho_1(t) - \mathrm{i}\mathcal{L}_{12}(t)\rho_2(t), \\ \frac{d\rho_2}{dt} &= -\mathrm{i}\mathcal{L}_{21}(t)\rho_1(t) - \mathrm{i}\mathcal{L}_{22}(t)\rho_2(t) \end{aligned} \tag{4.68}$$

and

$$\begin{aligned} \rho_1(t) &= \mathcal{U}_{11}(t,t')\rho_1(t') + \mathcal{U}_{12}(t,t')\rho_2(t'), \\ \rho_2(t) &= \mathcal{U}_{21}(t,t')\rho_1(t') + \mathcal{U}_{22}(t,t')\rho_2(t'). \end{aligned} \tag{4.69}$$

The block matrix forms of $\mathcal{G}$ and $\mathcal{D}$ from (4.42) and (4.45) are

$$\mathcal{G}(t,t') = T_{\leftarrow}\exp\left(-\mathrm{i}\int_{t'}^{t} ds\mathcal{Q}\mathcal{L}(s)\mathcal{Q}\right) = \begin{bmatrix} 1 & 0 \\ 0 & T_{\leftarrow}\exp\left(-\mathrm{i}\int_{t'}^{t} ds\mathcal{L}_{22}(s)\right) \end{bmatrix}, \tag{4.70}$$

$$\mathcal{D}(t;0) = 1 + \mathrm{i}\int_0^t dt' \begin{bmatrix} 1 & 0 \\ 0 & \mathcal{G}_{22}(t,t') \end{bmatrix} \begin{bmatrix} 0 & 0 \\ \mathcal{L}_{21}(t') & 0 \end{bmatrix} \begin{bmatrix} \mathcal{U}_{11}(t',t) & \mathcal{U}_{12}(t',t) \\ \mathcal{U}_{21}(t',t) & \mathcal{U}_{22}(t',t) \end{bmatrix}$$
$$= \begin{bmatrix} 1 & 0 \\ \mathrm{i}\int_0^t dt' \mathcal{G}_{22}(t,t')\mathcal{L}_{21}(t')\mathcal{U}_{11}(t',t) & 1 + \mathrm{i}\int_0^t dt' \mathcal{G}_{22}(t,t')\mathcal{L}_{21}(t')\mathcal{U}_{12}(t',t) \end{bmatrix}. \tag{4.71}$$

Since we need $\mathcal{D}^{-1}(t;0)$, from (4.71) we obtain

$$\mathcal{D}^{-1}(t;0) = \begin{bmatrix} 1 & 0 \\ -\mathcal{D}_{22}^{-1}(t;0)\mathcal{D}_{21}(t;0) & \mathcal{D}_{22}^{-1}(t;0) \end{bmatrix}. \tag{4.72}$$

As a final step we use all previously defined block forms of necessary operators and superoperators, along with the equation of motion for $\mathcal{P}\rho$ (4.47) and the isomorphism (4.58) to obtain

$$\frac{d\rho_S(t)}{dt} = -\mathrm{i}\left[\mathcal{L}_{11}(t) - \mathcal{L}_{12}(t)\mathcal{D}_{22}^{-1}(t;0)\mathcal{D}_{21}(t;0)\right]\rho_S(t)$$
$$- \mathrm{i}\sqrt{d_E}\mathcal{L}_{12}(t)\mathcal{D}_{22}^{-1}(t;0)\mathcal{G}_{22}(t,0)\rho_2(0). \tag{4.73}$$

Equation (4.73) is a partial-trace-free time-convolutionless equation of motion for the reduced density matrix $\rho_S(t)$. It describes the evolution of the representation basis of ρ_S. Working with representation matrices is a necessary condition of this method and might help in the case when one is interested in numerical implementation. The increased transparency of working with representation forms may also help when introducing various approximations in the exact equation of motion. Out of those three problems, mentioned at the end of Sect. 3.3, there is still one remaining. Namely, we still have the problem of evaluating the inverse of potentially large matrix $\mathcal{D}_{22}^{-1}(t;0)$ (if it exists at all). The solution to that problem will be discussed, among other things, in the next section.

3.6 *Memory Dressing*

Let us explicitly write the equations of motion for the density operator ρ in the eigenbasis of $\mathcal{P}$ from the previous section, i.e. within the partial-trace-free approach. By using (4.47) and (4.46), or directly (4.73) for ρ_1, we obtain

$$\frac{d\rho_1(t)}{dt} = -\mathrm{i}\left[\mathcal{L}_{11}(t) - \mathcal{L}_{12}(t)\mathcal{D}_{22}^{-1}(t;0)\mathcal{D}_{21}(t;0)\right]\rho_1(t)$$
$$-\mathrm{i}\mathcal{L}_{12}(t)\mathcal{D}_{22}^{-1}(t;0)\mathcal{G}_{22}(t,0)\rho_2(0),$$
$$\rho_2(t) = -\mathcal{D}_{22}^{-1}(t;0)\mathcal{D}_{21}(t;0)\rho_1(t) + \mathcal{D}_{22}^{-1}(t;0)\mathcal{G}_{22}(t,0)\rho_2(0), \tag{4.74}$$

where from (4.70) and (4.71) and by formally differentiating $\mathcal{D}(t;0)$'s submatrices with respect to time we have

$$
\begin{aligned}
\mathcal{G}_{22}(t,0) &= T_{\leftarrow}\exp\left(-\mathrm{i}\int_0^t ds\mathcal{L}_{22}(s)\right),\\
\frac{d\mathcal{D}_{21}(t;0)}{dt} &= -\mathrm{i}\mathcal{L}_{22}(t)\mathcal{D}_{21}(t;0)+\mathrm{i}\mathcal{D}_{21}(t;0)\mathcal{L}_{11}(t)+\mathrm{i}\mathcal{D}_{22}(t;0)\mathcal{L}_{21}(t),\\
\frac{d\mathcal{D}_{22}(t;0)}{dt} &= -\mathrm{i}\mathcal{L}_{22}(t)\mathcal{D}_{22}(t;0)+\mathrm{i}\mathcal{D}_{22}(t;0)\mathcal{L}_{22}(t)+\mathrm{i}\mathcal{D}_{21}(t;0)\mathcal{L}_{12}(t),\\
\mathcal{D}_{21}(0;0) &= 0\,,\quad \mathcal{D}_{22}(0;0)=1,
\end{aligned}
\tag{4.75}
$$

where in the last line the initial conditions are given. Taking the time derivative of the equation of motion for $\rho_1(t)$ in (4.69) and comparing those two equations with (4.74) we obtain the following relations for the representation of time evolution operator $\mathcal{U}(t,0)$

$$
\begin{aligned}
\frac{d\mathcal{U}_{11}(t,0)}{dt} &= -\mathrm{i}\left[\mathcal{L}_{11}(t)-\mathcal{L}_{12}(t)\mathcal{D}_{22}^{-1}(t;0)\mathcal{D}_{21}(t;0)\right]\mathcal{U}_{11}(t,0),\\
\frac{d\mathcal{U}_{12}(t,0)}{dt} &= -\mathrm{i}\left[\mathcal{L}_{11}(t)-\mathcal{L}_{12}(t)\mathcal{D}_{22}^{-1}(t;0)\mathcal{D}_{21}(t;0)\right]\mathcal{U}_{12}(t,0)\\
&\quad -\mathrm{i}\mathcal{L}_{12}(t)\mathcal{D}_{22}^{-1}(t;0)\mathcal{G}_{22}(t,0),\\
\mathcal{U}_{21}(t,0) &= -\mathcal{D}_{22}^{-1}(t;0)\mathcal{D}_{21}(t;0)\mathcal{U}_{11}(t,0),\\
\mathcal{U}_{22}(t,0) &= \mathcal{D}_{22}^{-1}(t;0)\left[\mathcal{G}_{22}(t,0)-\mathcal{D}_{21}(t;0)\mathcal{U}_{12}(t,0)\right].
\end{aligned}
\tag{4.76}
$$

These are generic time-convolutionless equations of motions, the form of which results from using the specific basis within the partial-trace-free-approach. They have the general feature of time-convolutionless equations that $\mathcal{U}_{21}$ and $\mathcal{U}_{22}$ are expressed in terms of $\mathcal{U}_{11}$ and $\mathcal{U}_{12}$. Formally, by solving (4.76) (for which we first have to solve (4.75)) we arrive at the final solution for the equation of motion of the reduced density operator ρ_S. However, this is a very difficult problem due to the sizes of the block matrices (the largest are at the position (2,2), being $d_S^2(d_E^2-1)\times d_S^2(d_E^2-1)$-dimensional) and because we need to evaluate the inverse of the matrix $\mathcal{D}_{22}$ which is in turn the solution of coupled equations for $\mathcal{D}_{21}$ and $\mathcal{D}_{22}$.

By inspection of (4.76) we see that we do not need all three large matrices $\mathcal{G}_{22}$, $\mathcal{D}_{21}$ and $\mathcal{D}_{22}$ separately, but only the following combinations of them (we designate each of them with a new letter)

$$
\begin{aligned}
\mathcal{R}(t) &= \mathcal{D}_{22}^{-1}(t;0)\mathcal{D}_{21}(t;0),\\
\mathcal{S}(t;0) &= \mathcal{D}_{22}^{-1}(t;0)\mathcal{G}_{22}(t,0),
\end{aligned}
\tag{4.77}
$$

where we left out the initial time in the argument list of $\mathcal{R}(t;0)$ for convenience. By using (4.75) we can derive the equations of motion for the matrices $\mathcal{R}$ and $\mathcal{S}$

$$\frac{d\mathcal{R}(t)}{dt} = -\mathrm{i}\mathcal{L}_{22}(t)\mathcal{R}(t) - \mathrm{i}\mathcal{R}(t)\mathcal{L}_{12}(t)\mathcal{R}(t) + \mathrm{i}\mathcal{R}(t)\mathcal{L}_{11}(t) + \mathrm{i}\mathcal{L}_{21}(t), \quad \mathcal{R}(0) = 0\,;$$
$$\frac{d\mathcal{S}(t;0)}{dt} = -\mathrm{i}\left[\mathcal{L}_{22}(t) + \mathrm{i}\mathcal{R}(t)\mathcal{L}_{12}(t)\right]\mathcal{S}(t;0)\,, \quad \mathcal{S}(0;0) = 1. \tag{4.78}$$

Since we are really interested in the evolution of ρ_1, due to its direct connection with ρ_S via (4.61), we only need the time evolution matrices $\mathcal{U}_{11}(t,0)$ and $\mathcal{U}_{12}(t,0)$. So, by starting from some initial state $\rho(0) = \left[\rho_1(0) \;\; \rho_2(0)\right]^T$, we have a new system of equations completely describing the time evolution of the reduced density operator ρ_S, consisting of (4.78) and

$$\frac{d\mathcal{U}_{11}(t,0)}{dt} = -\mathrm{i}\left[\mathcal{L}_{11}(t) - \mathcal{L}_{12}(t)\mathcal{R}(t)\right]\mathcal{U}_{11}(t,0)\,, \quad \mathcal{U}_{11}(0,0) = 1\,;$$
$$\frac{d\mathcal{U}_{12}(t,0)}{dt} = -\mathrm{i}\left[\mathcal{L}_{11}(t) - \mathcal{L}_{12}(t)\mathcal{R}(t)\right]\mathcal{U}_{12}(t,0) - \mathrm{i}\mathcal{L}_{12}(t)\mathcal{S}(t;0)\,, \; \mathcal{U}_{12}(0,0) = 0. \tag{4.79}$$

We see that by introducing $\mathcal{R}(t)$ and $\mathcal{S}(t;0)$ there is no more problem with the cumbersome inverse matrix $\mathcal{D}_{22}^{-1}(t;0)$. The equations for $\mathcal{U}_{21}(t,0)$ and $\mathcal{U}_{22}(t,0)$, which we do not need here, but are sometimes important, for example in calculating two-time correlation functions in electronic transport where $\mathcal{U}(t,t')$ for $t' \neq 0$ are required [63–65], are

$$\mathcal{U}_{21}(t,0) = -\mathcal{R}(t)\mathcal{U}_{11}(t,0)\,, \quad \mathcal{U}_{22}(t,0) = \mathcal{S}(t;0) - \mathcal{R}(t)\mathcal{U}_{12}(t,0). \tag{4.80}$$

The concept of memory dressing from the title of this section refers to $\mathcal{R}(t)$. This is because $\mathcal{R}(t)$ always goes along with $\mathcal{L}_{12}(t)$, which is the term representing physical interaction (as follows from the representation form (4.68)), in the "quasi-Liouvillian" $\mathcal{L}_{11}(t) - \mathcal{L}_{12}(t)\mathcal{R}(t)$. So, it is a memory dressing of the physical interaction. The self-contained non-linear equation of motion for the memory dressing $\mathcal{R}(t)$ (first of (4.78)) is a matrix Riccati equation, often encountered in control systems theory [66, 67]. It can be solved for $\mathcal{R}$ to an arbitrary order by using the perturbation expansion, which also allows for a convenient diagrammatic representation [60].

4 Coarse-Graining for the Steady State Distribution Function

The purpose of this section is to derive the steady state distribution function for the open system, by solving for $\rho_S(t)$ in a ballistic device (no scattering) that is attached to ideal contacts. We will show that, under these conditions, the distribution function is of Landauer-type. It says that the occupation of incoming states is fixed by the respective contact, while that of outgoing states by the open system alone. Furthermore, since there is no scattering in the open system, the occupation will remain the one determined by the contacts. We will use a coarse-graining procedure to approximate the exact non-Markovian time evolution towards the steady state. At the end, an interaction Hamiltonian, suitable for ideal contacts, will be constructed and used to solve the approximate Markovian equation of motion.

4.1 The Exact Dynamics and the Coarse-Graining Procedure

By using (4.61) and (4.77) in (4.74), we get the following form for the exact equation of motion for the reduced density matrix

$$\frac{d\rho_S(t)}{dt} = -\mathrm{i}\left[\mathcal{L}_{11} - \mathcal{L}_{12}\mathcal{R}(t)\right]\rho_S(t) - \mathrm{i}\sqrt{d_E}\mathcal{L}_{12}\mathcal{S}(t;0)\rho_2(0). \tag{4.81}$$

We will restrict our attention to the problems for which the initial density matrix is not correlated, i.e.

$$\rho(0) = \rho_E(0) \otimes \rho_S(0). \tag{4.82}$$

We see that when $\rho_E(0) = \overline{\rho}_E$ then $\rho_2(0) = 0$ and there exists a subdynamics (ρ_S does not depend on $\rho_2(0)$). This is because $\mathcal{P}$ is also generated by $\overline{\rho}_E$, so that $\rho(0)$ is an eigenstate of $\mathcal{P}$ and is of the form (4.66). Here, even though the environmental density matrix is not uniform, it can be proven that the following connecting relation holds

$$\rho_2(0) = \mathcal{M}\rho_1(0) = {d_E}^{-1/2}\mathcal{M}\rho_S(0), \tag{4.83}$$

where $\mathcal{M}$ in the eigenbasis of $\rho_E(0)$ is given by (see Appendix A of [62])

$$\mathcal{M}^i = \sqrt{\frac{d_E(d_E+1-i)}{d_E-1}}\left(\rho_E^i(0) - \frac{1}{d_E+1-i}\sum_{j=1}^{d_E}\rho_E^j(0)\right). \tag{4.84}$$

So, in this more general case (for arbitrary $\rho_E(0)$) there still exists the subdynamics in the following form

$$\rho_S(t) = [\mathcal{U}_{11}(t,0) + \mathcal{U}_{12}(t,0)\mathcal{M}]\rho_S(0) = \mathcal{W}(t,0)\rho_S(0), \tag{4.85}$$

which is in agreement with the statement made by Lindblad [68] that the subdynamics exists for an uncorrelated initial state. We can get a differential form of (4.85) by combining (4.74) and (4.83)

$$\frac{d\rho_S(t)}{dt} = -\mathrm{i}\left[\mathcal{L}_{11} - \mathcal{L}_{12}\mathcal{R}(t)\right]\rho_S(t) - \mathrm{i}\mathcal{L}_{12}\mathcal{S}(t)\mathcal{M}\rho_S(0). \tag{4.86}$$

In general we can write

$$\mathcal{W}(t,0) = T_{\leftarrow} \exp \left[\int_0^t \mathcal{F}(s) ds \right], \tag{4.87}$$

where $\mathcal{F}(t)$ is the generator of $\mathcal{W}(t,0)$.

It is very difficult to solve for the reduced system dynamics (4.85), because of the difficulties in obtaining $\mathcal{W}(t,0)$. We can either be content with a Markovian approximation in the weak-coupling and van Hove limits [69], or by an expansion up to the second or fourth orders in the interaction if we need a non-Markovian approximation [14]. Although the weak-coupling limit has been used before to study tunneling structures in the Markovian approximation [70, 71], it is not generally applicable to nanostructures [70]. Here, we will apply an approximation beyond the weak-coupling limit, by approximating the exact reduced system dynamics using coarse-graining over the environmental relaxation time τ [72, 73]. This limits the area of applicability to the open systems for which $\tau \ll \tau_S$, where τ_S is the open system relaxation time, which is still a pretty wide area. For example, in typical small semiconductor devices (quasi-ballistic), with highly doped contacts at room temperature, the major energy relaxation mechanism is electron–electron scattering in the contacts (relaxation time for electron–electron scattering is about 10 fs for GaAs at 10^{19} cm^{-3} and room temperature [74], while about 150 fs for polar optical phonon scattering [36]). Electron–electron relaxation will drive the environmental distribution function to a drifted Fermi–Dirac distribution in a time interval $\tau \approx 10-100$ fs, which is much shorter than the typical open system relaxation time for these devices $\tau_S \approx 1-10$ ps.

The coarse-graining procedure proceeds by splitting the total evolution time interval $[0,t]$ into segments of length τ, $[1,2,\ldots,n] \times \tau$, and defining the average of the generator of $\mathcal{W}(t,0)$ over each interval

$$\overline{\mathcal{F}}_j = \frac{1}{\tau} \int_{j\tau}^{(j+1)\tau} \mathcal{F}(s) ds. \tag{4.88}$$

This leads to the following connection between successive, discretized reduced density operators

$$\rho_{S,j+1} = \exp\left(\tau \overline{\mathcal{F}}_j\right) \rho_{S,j}, \tag{4.89}$$

which gives

$$\frac{\rho_{S,j+1} - \rho_{S,j}}{\tau} = \overline{\mathcal{F}}_j \rho_{S,j} \tag{4.90}$$

after expanding the exponent for small τ. This is just a discretized version of the exact equation of motion.

There are three approximations applied in deriving (4.90). First, we don't have the information about the time evolution inside each τ-interval, but only at its ends. Second, we cut the series after the first order in the expansion of $\exp\left(\tau\overline{\mathcal{F}}_j\right)$ in order to get (4.90). Third, the time ordering from the exact equation (4.87) is violated at the τ time scale, which can be shown explicitly by using the Dyson series to represent (4.87).

Finally, we will assume that the environmental state is nearly the same after every interval τ during the transient, in other words $\overline{\mathcal{F}}_0 = \overline{\mathcal{F}}_\tau \approx \overline{\mathcal{F}}_1 \approx \cdots \approx \overline{\mathcal{F}}_n$. This is also the most trivial way of ensuring that the coarse grained generators $\overline{\mathcal{F}}_i$'s commute (commute in an average sense). For this to be satisfied we have to ramp up the excitation (e.g. bias) to the system in small enough increments with sufficiently long time between two increments so that the open system is able to reach steady state, in the form of a drifted Fermi–Dirac distribution, after each small increment. This condition is more a thought experiment than a real constraint, because we are only interested in the steady state here. As the last step, we expand the discrete equation (4.90) to the continuum (since τ is a small parameter) to obtain the final equation

$$\frac{d\rho_S}{dt} = \overline{\mathcal{F}}_\tau \rho_S(t). \tag{4.91}$$

This equation is an approximate Markovian (because the generator $\overline{\mathcal{F}}_\tau$ is constant in time) QME for the reduced density matrix in the limit of small-increments/long-pauses kind of ramping up the bias and we will use it to obtain the steady state distribution function for arbitrarily large bias.

4.2 The Short-Time Expansion of $\overline{\mathcal{F}}_\tau$

The practical value of (4.91) is in the fact that $\overline{\mathcal{F}}_\tau$ can be calculated using the expansion of $\mathcal{F}(t)$ in the small parameter τ around zero. By introducing a definition $\mathcal{F}(t) = -\mathrm{i}\mathcal{L}_{\mathrm{eff}} - \mathcal{G}(t)$, where $\mathcal{L}_{\mathrm{eff}}$ is an effective system Liouvillian and $\mathcal{G}$ a correction due to the system-environment coupling, expanding (4.85) and (4.86) up to the second order in time and comparing coefficients it can be shown that (see Appendix B of [62])

$$\mathcal{F}(t) = -\mathrm{i}\mathcal{L}_{\mathrm{eff}} - 2\Lambda t + O(t^2), \tag{4.92}$$

where $\mathcal{L}_{\mathrm{eff}}$ is a commutator superoperator generated by $H_S + \langle H_{\mathrm{int}}\rangle$, while Λ in the basis $\alpha\beta$ of the system's Liouville space is given by

$$\Lambda^{\alpha\beta}_{\alpha'\beta'} = \frac{1}{2}\Bigg\{ \langle H^2_{\mathrm{int}}\rangle^{\alpha}_{\alpha'} \delta^{\beta'}_{\beta} + \langle H^2_{\mathrm{int}}\rangle^{\beta'}_{\beta} \delta^{\alpha}_{\alpha'} - 2\sum_{j,j'} (H_{\mathrm{int}})^{j'\alpha}_{j\alpha'} \rho^{j}_{E} (H_{\mathrm{int}})^{j\beta'}_{j'\beta}$$
$$- \left(\langle H_{\mathrm{int}}\rangle^2\right)^{\alpha}_{\alpha'} \delta^{\beta'}_{\beta} + 2\langle H_{\mathrm{int}}\rangle^{\alpha}_{\alpha'} \langle H_{\mathrm{int}}\rangle^{\beta'}_{\beta} - \left(\langle H_{\mathrm{int}}\rangle^2\right)^{\beta'}_{\beta} \delta^{\alpha}_{\alpha'} \Bigg\}, \tag{4.93}$$

where ρ_E^j are the eigenvalues of $\rho_E(0)$ and $\langle\cdots\rangle \equiv \mathrm{Tr}_E\left(\rho_E(0)\cdots\right)$. Λ contains important information on the directions of coherence loss. It has been implicitly defined previously [75], but only in the interaction (not Schrödinger) picture and for $\langle H_{\mathrm{int}}\rangle = 0$.

If the following condition holds

$$\|\Lambda\|\tau \ll \|\mathcal{L}_{\mathrm{eff}}\|, \tag{4.94}$$

then the short-time expansion of $\mathcal{F}$ is valid and

$$\overline{\mathcal{F}}_{\tau} = -\mathrm{i}\mathcal{L}_{\mathrm{eff}} - \Lambda\tau, \tag{4.95}$$

which gives the final coarse-grained Markovian QME for the reduced density matrix

$$\frac{d\rho_S(t)}{dt} = (-\mathrm{i}\mathcal{L}_{\mathrm{eff}} - \Lambda\tau)\,\rho_S(t). \tag{4.96}$$

We have already said that this coarse-grained Markovian approximation is valid if the environmental relaxation time τ is much smaller than the system relaxation time (corresponding to $1/\|\Lambda\|\tau$), or

$$\|\Lambda\|\tau^2 \ll 1, \tag{4.97}$$

which, along with (4.94), gives in total

$$\|\Lambda\|\tau^2 \ll \min\left\{1, \|\mathcal{L}_{\mathrm{eff}}\|\tau\right\}. \tag{4.98}$$

4.3 *Steady State in a Two-Terminal Ballistic Nanostructure*

In this section we will apply the main equation (4.96) to calculate the steady state distribution function for a two-terminal ballistic nanostructure attached to ideal contacts. By ideal contacts we mean contacts that behave like black bodies with respect to the emission/absorption of electrons. Therefore, they absorb all electrons coming from the open system. The consequence is that, as already mentioned, the occupation of states coming from the contacts is fixed by them, while the occupation of outgoing states is fixed by the open system. This gives a Landauer-type distribution function and specifically here, since the open system region is ballistic, the occupation of the incoming and outgoing states is the same and fixed by the injecting contact.

4.3.1 The Open System Model

Schematic of our two-terminal nanostructure is shown in Fig. 4.4. The device is biased negatively such that the negative polarity is at the left contact. All open

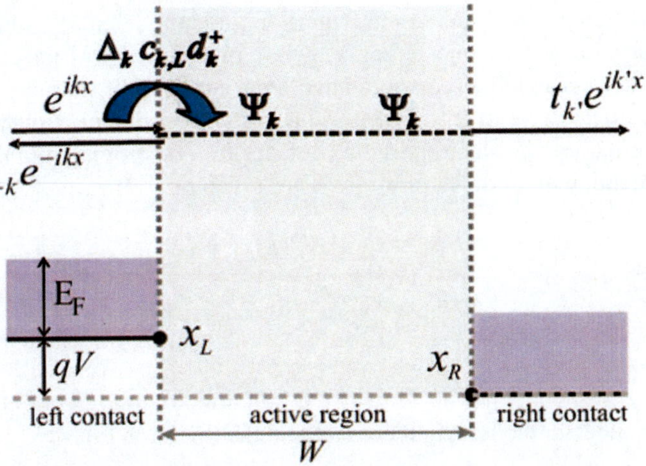

Fig. 4.4 Schematic of the two-terminal ballistic nanostructure, negatively biased at the *left* contact, with the boundaries between the open system and contacts shown at x_L and x_R, and with the graphical representation of the wave function injected from and the hoping type interaction with the *left* contact. It is similar for the wave function and interaction for the *right* contact

system eigenenergies ε_k above the bottom of the left contact have two eigenfunctions (double-degeneracy), one for the positive wave vector (ψ_k, injected from the left contact) and one for the negative wave vector (ψ_{-k}, injected from the right contact). The rest of the energy levels, made up of quasibound states that lay between the bottoms of the two contacts, have only one wave function for the states injected from the right contact and completely reflected. For doubly-degenerate scattering states we have the following asymptotic wave functions (assuming that the active region between x_L and x_R is wide enough)

$$\psi_k(x) = \begin{cases} e^{ikx} + r_{-k,L} e^{-ikx}, & x < x_L \\ t_{k',L} e^{ik'x}, & x > x_R \end{cases},$$

$$\psi_{-k}(x) = \begin{cases} e^{-ik'x} + r_{k',L} e^{ik'x}, & x > x_R \\ t_{-k,L} e^{-ikx}, & x < x_L \end{cases}, \qquad (4.99)$$

where t and r are the transmission and reflection coefficients, respectively, while k and k' are the wave vectors for the same energy level ε_k measured from the bottom of the left and right contacts, respectively.

In the formalism of second quantization the non-interacting many-body Hamiltonian of the open system is given by (considering only scattering states in the following)

$$H_S = \sum_{k>0} \omega_k \left(d_k^\dagger d_k + d_{-k}^\dagger d_{-k} \right), \qquad (4.100)$$

where $\omega_k = \varepsilon_k/\hbar$ and $d_{\pm k}$ and $d_{\pm k}^\dagger$ are the destruction and creation operators, respectively, for the open system states $\psi_{\pm k}$. The many-body effect that this Hamiltonian is able to model is the Pauli exclusion principle. Considering that the contacts are ideal, as explained above, the interaction Hamiltonian is modeled as a one-way coupling (for the particles that are injected only), while in the other way the electrons are free to propagate, because the contacts have ideal absorption characteristics. In other words, we do not have to enforce the Pauli exclusion principle (to "make room") by explicitly creating one electron in the contacts after destroying it in the open system region. So, the interaction Hamiltonians are given by

$$H_{\text{int}}^{L} = \sum_{k>0} \Delta_k d_k^\dagger c_{k,L} + \text{h.c.},$$
$$H_{\text{int}}^{R} = \sum_{k>0} \Delta_{-k} d_{-k}^\dagger c_{-k,R} + \text{h.c.}, \tag{4.101}$$

where $c_{\pm k,L/R}$ and $c^\dagger_{\pm k,L/R}$ are the destruction and creation operators, respectively, for the $\pm k$ states in the left/right contact and the injection rates are given by

$$\Delta_k = \frac{\hbar k}{m\,\|\psi_k\|^2}, \quad \Delta_{-k} = \frac{\hbar k'}{m\,\|\psi_{-k}\|^2}, \tag{4.102}$$

where $\|\psi_k\|^2 = \int_{x_L}^{x_R} |\psi_k(x)|^2 dx$.

In Fig. 4.4 there are only H_{int}^L and ψ_k graphically represented, but the situation is similar for the right contact.

4.3.2 Steady State Distribution Functions

Since the interaction Hamiltonians (4.101) are linear in the contact creation and destruction operators, we conclude that $\langle H_{\text{int}}^{L/R} \rangle = 0$, which gives us the following equations

$$\mathcal{L}_{\text{eff}} = \mathcal{L}_S,$$
$$\left(\Lambda^{L/R}\right)^{\alpha\beta}_{\alpha'\beta'} = \frac{1}{2}\left[\langle (H_{\text{int}}^{L/R})^2\rangle^{\alpha}_{\alpha'}\delta^{\beta'}_{\beta} + \langle (H_{\text{int}}^{L/R})^2\rangle^{\beta'}_{\beta}\delta^{\alpha}_{\alpha'}\right]$$
$$-\sum_{j,j'} (H_{\text{int}}^{L/R})^{j'\alpha}_{j\alpha'}\rho^{j}_{L/R}(H_{\text{int}}^{L/R})^{j\beta'}_{j'\beta}. \tag{4.103}$$

The quantities that we need to evaluate first are (for the left contact)

$$\langle (H_{\text{int}}^L)^2\rangle = \sum_{k>0} \Delta_k^2 \left[f_k^L d_k d_k^\dagger + (1-f_k^L)\, d_k^\dagger d_k\right], \tag{4.104}$$

which gives a contribution of the form $\Lambda^{\alpha\beta}_{\alpha\beta}$, and

$$\sum_{j,j'} (H_{\text{int}}^L)^{j'\alpha}_{j\alpha'}\rho^j_L (H_{\text{int}}^L)^{j\beta'}_{j'\beta} = \sum_{k>0} \Delta_k^2 \left[(1-f_k^L)\,(d_k^\dagger)^{\beta'}_{\beta}(d_k)^{\alpha}_{\alpha'} + f_k^L (d_k)^{\beta'}_{\beta}(d_k^\dagger)^{\alpha}_{\alpha'}\right], \tag{4.105}$$

which gives a contribution of the form $\Lambda^{\alpha\alpha}_{\beta\beta}$. It is similar for the right contact.

Quantities $f_{\pm k}^{L/R}$ define the drifted Fermi–Dirac distribution function in the contacts, as a consequence of the current flowing through the device. They take into account the feedback of the device under applied bias on the contacts and ensure that the charge neutrality and current continuity are satisfied near the device/contacts boundaries [16, 17]. The left contact distribution function is given by

$$f_{\pm k}^{L} = \frac{1}{\exp\left\{\frac{\hbar^2\left[(\pm k - k_d)^2 - k_F^2\right]}{2mk_BT}\right\} + 1}, \tag{4.106}$$

where k_d is the drift wave vector. Here, there is a common k_d for both contacts (since they carry the same current), but in a more general multi-terminal case there will be one drift wave vector for each contact, defined by the current density J_l through the l-th contact by $k_d^l = mJ_l/n_l q\hbar$, where n_l is the charge density of the l-th contact. This is an additional parameter that has to be determined self-consistently, by an additional equation $J_l^{dev} = J_l^{contact}$ that ensures the current continuity across the device/contacts boundaries (but not on a state-by-state basis). Here, J_l^{dev} is the current density due to the injection from the l-th contact only. There is a similar equation to (4.106) for the right contact with the following changes: $k \to k'$ in the denominator and $L \to R$. Detailed Monte Carlo–molecular dynamics simulations in bulk semiconductors show that when electron–electron scattering is the dominant relaxation mechanism the distribution function is very close to the one given by (4.106) [74, 76, 77].

Since $\Lambda = \sum_k \Lambda_k$, according to (4.104) and (4.105), it is just a sum of independent modes. Each mode can be represented with a two-state basis: one state for a particle being in the state ψ_k ("+" state) and another state for a particle being absent from it ("−" state). The reduced density matrix in this basis is a column vector with four elements, $\rho_{S,k} = (\rho_{S,k}^{++}, \rho_{S,k}^{+-}, \rho_{S,k}^{-+}, \rho_{S,k}^{--})^T$, and the equation of motion is

$$\frac{d\rho_{S,k}}{dt} = \left[-\mathrm{i}\mathcal{L}_{S,k} - \Lambda_k \tau\right]\rho_{S,k}, \tag{4.107}$$

where

$$\mathcal{L}_{S,k} = \begin{bmatrix} 0 & 0 & 0 & 0 \\ 0 & 2\omega_k & 0 & 0 \\ 0 & 0 & -2\omega_k & 0 \\ 0 & 0 & 0 & 0 \end{bmatrix}, \tag{4.108}$$

$$\Lambda_k = \begin{bmatrix} A_k & 0 & 0 & -B_k \\ 0 & C_k & 0 & 0 \\ 0 & 0 & C_k & 0 \\ -A_k & 0 & 0 & B_k \end{bmatrix}. \tag{4.109}$$

Quantities $A_k = \Delta_k^2\left(1 - f_k^L\right)$, $B_k = \Delta_k^2 f_k^L$ and $C_k = (A_k + B_k)/2 = \Delta_k^2/2$ are calculated using (4.104) and (4.105).

The elements $\rho_{S,k}^{+-}$ and $\rho_{S,k}^{-+}$ are zero in the steady state, because they decay as $\exp(\mp i2\omega_k - \tau C_k)t$. The two remaining elements, $\rho_{S,k}^{++} = f_k(t)$ and $\rho_{S,k}^{--} = 1 - f_k(t)$, give the following equation

$$\frac{f_k(t)}{dt} = -\tau(A_k + B_k) f_k(t) + \tau B_k = -\tau\Delta_k^2 f_k(t) + \tau\Delta_k^2 f_k^L, \tag{4.110}$$

where $f_k(t)$ is the distribution function of $+k$ states in the open system region. In the steady state, this gives just

$$\begin{aligned} f_k^{\infty} &= f_k^L, \\ f_{-k}^{\infty} &= f_{-k'}^R, \end{aligned} \tag{4.111}$$

where f_{-k}^{∞} is the steady state distribution function for $-k$ states in the open system region, which can be derived in a similar way, starting by evaluating $\langle (H_{\text{int}}^R)^2 \rangle$ and $\sum_{j,j'} (H_{\text{int}}^R)_{j\alpha'}^{j'\alpha} \rho_R^j (H_{\text{int}}^R)_{j'\beta}^{j\beta'}$. We see that the result is the distribution function for the scattering states of the ballistic open system determined by the injecting contact only, which is what it should be considering the problem that we were solving.

5 Conclusion

In this chapter we gave a review of several types of single-particle and reduced many-particle QMEs used in electronic transport. The density matrix is a quantum statistical concept introduced by John von Neumann in 1927 [21,22] and used to describe a mixed ensemble of states of some physical system. Since physical systems under consideration (electronic devices) are many-particle objects, it is extremely important to arrive at the form of the QME which is, on the one hand, sufficiently accurate in capturing important physical phenomena and, on the other hand, not too computationally complex for practical applications.

The single-particle QME of Sect. 2 is a special case of the reduced many-particle density matrix, where the reduction of the number of exactly described degrees of freedom is performed down to the single particle variables. It can be derived, similarly to the Boltzmann transport equation (BTE), by truncating the BBGKY chain of equations [2, 13]. In the case of electrons, this means that the transport is divided into periods of "free flight", whose evolutions are determined by the single-particle Hamiltonian (usually including the kinetic energy and energies due to the external and Hartree potentials), and collisions with phonons and impurities in the Born–Markov approximation, represented by a linear collision superoperator (4.6). Within the context of the open system formalism the time-irreversible boundary conditions are required to maintain the stability of solutions (i.e. no growing exponentials) [3]. They can be incorporated through an explicit source term, that describes additional dynamics due to the coupling to the contacts/reservoirs, whose form can only be determined phenomenologically (4.9). The PME (4.10), a closed single-particle QME for the diagonal elements of the density matrix in the basis of the single-particle Hamiltonian (with off-diagonal elements neglected), is applied to steady state transport in small devices [16, 17]. A single-electron/many-phonon QME within the perturbation expansion using the Dyson series (4.18), is applied to transients in bulk semiconductors beyond the Born–Markov approximation for scattering [18–20]. Because of the similarities between the single-particle QME and the BTE, the natural choice to solve the single-particle QME would be to use the Monte Carlo method, which is shown to be similar to the conventional ensemble Monte Carlo.

The reduced many-particle QME, as an equation of motion for the reduced many-particle density matrix within the open system formalism, provides a very good way to achieve the balance between the mathematical and physical rigor and practical applicability (computational complexity). The projection operator technique, applied to obtain the rigorous Nakajima–Zwanzig equation (4.26), is the starting point in this approach of Sect. 3. Several techniques are introduced that further modify the Nakajima–Zwanzig equation, making it more tractable. It is shown that in the Born–Markov approximation it yields the most general form of the generator of the quantum dynamical semigroup, the Lindblad form (4.37). The two most notable problems with the Nakajima–Zwanzig equation, the time-convoluted memory kernel and the need to carry all the degrees of freedom in the system through the calculation only to trace them out at the end, lead to the derivation of the conventional time-convolutionless equation of motion (4.48) and its further improvement, the partial-trace-free time-convolutionless equation of motion (4.73). The partial-trace-free approach is achieved at the expense of working in the specific basis, that diagonalizes the unity subspace of the projection operator, from the beginning. This is not a drawback since the numerical computation is our final goal. Using the partial-trace-free approach, at the end it was shown that by introducing the memory dressing $\mathcal{R}(t)$ (4.77), which can be solved using the perturbation expansion [60], the final system of equations for the time evolution (4.78) and (4.79) are much more tractable.

Using the results of Sect. 3, it is shown in Sect. 4 how the Landauer-type steady state distribution functions can be obtained within the reduced many-particle density matrix formalism. Working within the limits of initially separable states $[\rho(0) = \rho_E(0) \otimes \rho_S(0)]$ and by using the coarse-graining procedure, the approximate Markovian QME (4.91) is derived. Since the steady state distribution functions are required, so that the exact transient behavior is not important, (4.91) provides an opportunity of deriving the generator of the time evolution in the limit of the short-time expansion, by assuming that the bias is ramped up in small increments separated by sufficiently long time intervals. The many-particle model Hamiltonian for coupling between the small ballistic open system and two large, ideal ("black-body") reservoirs is developed and shown to yield the correct Landauer-type distribution functions for the open system, where the occupation of levels is set by the contacts only.

References

1. L.E. Reichl, *A Modern Course in Statistical Physics* (WILEY-VCH, Weinheim, 1980)
2. M. Bonitz, *Quantum Kinetic Theory* (Teubner, Stuttgart; Leipzig, 1998)
3. W.R. Frensley, Rev. Mod. Phys. **62**, 745 (1990)
4. N.C. Kluksdahl, A.M. Kriman, D.K. Ferry, C. Ringhofer, Phys. Rev. B **39**, 7720 (1989)
5. S.E. Laux, A. Kumar, M.V. Fischetti, IEEE Trans. Nanotechnol. **1**, 255 (2002)
6. W. Pötz, J. Appl. Phys. **66**, 2458 (1989)
7. A. Gehring, S. Selberherr, J. Comput. Electron. **3**, 149 (2004)
8. J. von Neumann, *Mathematical Foundations of Quantum Mechanics* (Princeton University Press, Princeton, 1955)
9. E. Wigner, Phys. Rev. **40**, 749 (1932)
10. L.P. Kadanoff, G. Baym, *Quantum Statistical Mechanics* (Benjamin, New York, 1962)
11. L.V. Keldysh, Sov. Phys. JETP **20**, 1018 (1965)
12. W. Pauli, *Festschrift zum 60. Geburtstage A. Sommerfeld* (Hirzel, Leipzig, 1928), p. 30
13. K. Huang, *Statistical Mechanics* (John Wiley & Sons, New York, 1987)
14. H.P. Breuer, F. Petruccione, *The Theory of Open Quantum Systems* (Oxford University Press, Oxford, 2002)
15. U. Weiss, *Quantum Dissipative Systems* (World Scientific, Singapore, 1999)
16. M.V. Fischetti, J. Appl. Phys. **83**, 270 (1998)
17. M.V. Fischetti, Phys. Rev. B **59**, 4901 (1999)
18. R. Brunetti, C. Jacoboni, F. Rossi, Phys. Rev. B **39**, 10781 (1989)
19. F. Rossi, C. Jacoboni, Europhys. Lett. **18**, 169 (1992)
20. C. Jacoboni, Semicond. Sci. Technol. **7**, B6 (1992)
21. J. von Neumann, Nachr. Ges. Wiss. p. 245 (1927)
22. J.J. Sakurai, *Modern Quantum Mechanics* (Addison–Wesley, USA, 1994)
23. A.L. Fetter, J.D. Walecka (eds.), *Quantum theory of many-particle systems* (McGraw-Hill, Inc., 1971)
24. A. Messiah, *Quantum Mechanics* (Dover, Mineola, 1999)
25. L. van Hove, Rev. Mod. Phys. **21**, 517 (1955)
26. W. Kohn, J.M. Luttinger, Phys. Rev. **108**, 590 (1957)
27. J.R. Barker, D.K. Ferry, Phys. Rev. Lett. **42**, 1779 (1979)
28. I.B. Levinson, Y. Yasevichyute, Sov. Phys. JETP **35**, 991 (1972)
29. W.V. Houston, Phys. Rev. **57**, 184 (1940)
30. P.J. PriceF, Semicond. Semimet. **14**, 249 (1979)
31. C. Jacoboni, L. Reggiani, Rev. Mod. Phys. **55**, 645 (1983)
32. M.V. Fischetti, S.E. Laux, P.M. Solomon, A. Kumar, J. Comput. Electron. **3**, 287 (2004)
33. D. Vasileska, S.M. Goodnick, Mater. Sci. Eng. **R 38**, 181 (2002)
34. H. Budd, Phys. Rev. **158**, 798 (1967)
35. R. Chambers, Proc. R. Soc. London **65**, 458 (1952)
36. M. Lundstrom, *Fundamentals of Carrier Transport* (Cambridge University Press, Cambridge, 2000)
37. S. Nakajima, Prog. Theor. Phys. **20**, 948 (1958)
38. R. Zwanzig, J. Chem. Phys. **33**, 1338 (1960)
39. A.G. Redfield, IBM J. Res. Dev. **1**, 19 (1957)
40. G. Lindblad, Commun. Math. Phys. **48**, 199 (1976)
41. P. Argyres, P. Kelley, Phys. Rev. **134**, A98 (1964)
42. J.R. Barker, D.K. Ferry, Solid-State Electron. **23**, 531 (1980)
43. K. Kassner, Phys. Rev. A **36**, 5381 (1987)
44. M. Sparpaglione, S. Mukamel, J. Chem. Phys. **88**, 3263 (1988)
45. Y. Hu, S. Mukamel, J. Chem. Phys. **91**, 6973 (1989)
46. V. Romero-Rochin, I. Oppenheim, Physica A **155**, 52 (1989)
47. M. Tokuyama, H. Mori, Prog. Theor. Phys. **55**, 411 (1975)
48. F. Shibata, Y. Takahashi, N. Hashitsume, J. Stat. Phys. **17**, 171 (1977)

49. N. Hashitsume, F. Shibata, M. Shingu, J. Stat. Phys. **17**, 155 (1977)
50. M. Saeki, Prog. Theor. Phys. **67**, 1313 (1982)
51. M. Saeki, Prog. Theor. Phys. **79**, 396 (1988)
52. M. Saeki, Prog. Theor. Phys. **89**, 607 (1993)
53. D. Ahn, Phys. Rev. B **51**, 2159 (1995)
54. D. Ahn, Prog. Quantum Electron. **21**, 249 (1997)
55. D. Ahn, J.H. Oh, K. Kimm, S. Hwang, Phys. Rev. A **61**, 052310 (2000)
56. D. Ahn, J. Lee, M.S. Kim, S.W. Hwang, Phys. Rev. A **66**, 012302 (2002)
57. T.M. Chang, J.L. Skinner, Physica A **193**, 483 (1993)
58. A.A. Golosov, D.R. Reichmann, J. Chem. Phys. **115**, 9849 (2001)
59. I. Knezevic, D.K. Ferry, Phys. Rev. E **66**, 016131 (2002)
60. I. Knezevic, D.K. Ferry, Phys. Rev. A **69**, 012104 (2004)
61. I. Knezevic, D.K. Ferry, Phys. Rev. E **67**, 066122 (2003)
62. I. Knezevic, Phys. Rev. B **77**, 125301 (2008)
63. D. Semkat, D. Kremp, M. Bonitz, Phys. Rev. E **59**, 1557 (1999)
64. D. Semkat, D. Kremp, M. Bonitz, J. Math. Phys. **41**, 7458 (2000)
65. K.M. et al., Phys. Rev. E **63**, 020102(R) (2001)
66. W.T. Reid, *Riccati Differential Equations* (Academic Press, New York, 1972)
67. S. Bittanti, A.J. Laub, J.C.Willems (eds.), *The Riccati Equation* (Springer-Verlag, Berlin, 1991)
68. G. Lindblad, J. Phys. A **29**, 4197 (1996)
69. E.B. Davies, Commun. Math. Phys. **39**, 91 (1974)
70. X.Q. Li, J.Y. Luo, Y.G. Yang, P. Cui, Y.J. Yan, Phys. Rev. B **71**, 205304 (2005)
71. J.N. Pedersen, A. Wacker, Phys. Rev. B **72**, 195330 (2005)
72. D. Bacon, D.A. Lidar, K.B. Whaley, Phys. Rev. A **60**, 1944 (1999)
73. D.A. Lidar, Z. Bihary, K.B. Whaley, Chem. Phys. **268**, 35 (2001)
74. A.M. Kriman, M.J. Kann, D.K. Ferry, R. Joshi, Phys. Rev. Lett. **65**, 1619 (1990)
75. R. Alicki, Phys. Rev. A **40**, 4077 (1989)
76. P. Lugli, D.K. Ferry, IEEE Trans. Electron Devices **32**, 2431 (1985)
77. P. Lugli, D.K. Ferry, Phys. Rev. Lett. **56**, 1295 (1986)

Chapter 5
Wigner Function Approach

M. Nedjalkov, D. Querlioz, P. Dollfus, and H. Kosina

Abstract The Wigner function formalism has been introduced with an emphasis on basic theoretical aspects, and recently developed numerical approaches and applications for modeling and simulation of the transport of current carriers in electronic structures. Two alternative ways: the historical introduction of the function on top of the operator mechanics, and an independent formulation of the Wigner theory in phase space which then recovers the operator mechanics, demonstrate that the formalism provides an autonomous description of the quantum world.

The conditions of carrier transport in nano-electronic devices impose to extend this coherent physical picture by processes of interaction with the environment. Relevant becomes the Wigner–Boltzmann equation, derived for the case of interaction with phonons and impurities. The numerical aspects focus on two particle models developed to solve this equation. These models make the analogy between classical and Wigner transport pictures even closer: particles are merely classical, the only characteristics which carries the quantum information is a dimensionless quantity – affinity or sign.

The recent ground-breaking applications of the affinity method for simulation of typical nano-devices as the resonant tunneling diode and the ultra-short DG-MOSFET firmly establish the Wigner–Boltzmann equation as a bridge between coherent and semi-classical transport pictures. It became a basic route to understand the nano-device operation as an interplay between coherent and de-coherence phenomena. The latter, due to the environment: phonon field, contacts or defects, attempts to recover the classical transport picture.

Keywords Wigner function · Wigner-Boltzmann equation · Monte Carlo · Quantum particles · De-coherence

M. Nedjalkov (✉)
Institute of Microelectronics, TU Vienna, Vienna, Austria
e-mail: mixi@iue.tuwien.ac.at

D. Vasileska and S.M. Goodnick (eds.), *Nano-Electronic Devices: Semiclassical and Quantum Transport Modeling*, DOI 10.1007/978-1-4419-8840-9_5,

1 Introduction

The Wigner picture of quantum mechanics constitutes a phase space formulation of the quantum theory. Both states and observables are represented by functions of the phase space coordinates. The Weyl transform attributes to any given operator of the wave mechanics a phase space counterpart which is a c-number. Furthermore, the Wigner function is both the phase space counterpart of the density matrix and the quantum counterpart of the classical distribution function. Basic notions of the classical statistical mechanics are retained in this picture. In particular the usual quantities of interest in operator quantum mechanics, i.e. mean values and probabilities, are evaluated in the phase space by rules resembling the formulae of the classical statistics. It is for these reasons that the Wigner function is often considered as a quasi-distribution. The phase space formulation of quantum mechanics has been established historically on top of the operator mechanics [1–3]. In this respect, it is natural to raise the question of whether the Wigner theory can be considered as an equivalent autonomous alternative of the operator mechanics. What outlines classical from quantum behavior in the phase space? In particular how to determine if a given function of the phase space coordinates is a possible quantum or classical state? These questions have been addressed by the inverse approach, which has been explored later [4, 5]. It provides an independent formulation of the Wigner theory and then recovers the operator mechanics, which completes the proof of the logical equivalence between the two theories.

Device modeling needs a conjunction of Wigner quantum mechanics of carrier – potential interactions with other interactions due to the environment. Physical models of the carrier kinetics taking into account the engineering characteristics of the device structure are developed, which are further approached by corresponding numerical methods. Models, algorithms and applications are mutually developed within the Wigner transport picture. This work is an effort to give a self-contained overview of the basic notions, and to point at some recent results in the field. Further details and a presentation of the recent advances can be found in [66].

We feel that here is the place to acknowledge the work of W. Frensley, D. K. Ferry and co-authors, C. Jacoboni and the Modena group and other important contributions, which are frequently cited in the sequel.

In the next section we will introduce the Wigner quantum mechanics by following the historical approach. Some concepts of statistical mechanics and Hermitian operators are recalled in a way to outline the mutual relationship between the classical and quantum counterparts. The operator ordering is discussed: actually there are alternative phase space formulations of the quantum mechanics which are associated with alternative ordering prescriptions. A particular ordering given by the Weyl transform introduces the Wigner function. The corresponding evolution equation is a central entity in this approach. The presented detailed derivation of the Wigner equation is based on the von Neumann equation for the density matrix. Fundamental concepts of the picture are discussed along with the characteristics of pure and mixed state Wigner functions.

We believe that it is important to introduce in parallel some basic notions of the inverse approach. Conditions determining whether a given phase space function is a possible quantum state are presented. Explicit expressions exist which associate to given phase space pure or mixed quantum state the corresponding wave function or density matrix. Results which establish the equivalence between the operator and Wigner quantum mechanics are summarized. Behind the abstract mathematical aspects, these results allow to understand and solve practical for semiconductor device community problems, encountered when bound states exist in the physical system.

A strong advantage of the Wigner formalism of quantum transport is its ability to include all relevant scattering mechanisms. Though the full quantum treatment of scattering is difficult to apply to practical situations, namely for the description of transport in realistic devices, it is shown that under some reasonable approximations, such as the fast and weak scattering limits, the Wigner collision operator simplifies into the well-known Boltzmann collision operator. This is demonstrated in Sect. 3 for the case of electron–phonon and electron–ionized impurity interactions. The Wigner transport equation thus reduces to the so-called Wigner–Boltzmann equation. In the latter form, the transport equation becomes very convenient for device simulation. It can benefit from all the knowledge acquired for many years in semiclassical device physics and especially in the physics of scattering.

Furthermore we show that the analogy between classical and Wigner transport pictures become even closer. Particle models are associated with the Wigner-quantum transport in Sect. 4. The Wigner potential is interpreted as a source which, in addition to the common classical parameters, associates to each particle a new dimensionless quantity which, depending on the model, could be affinity or sign. This quantity is the only characteristic carrying the quantum information for the system. It is taken into account in the computation of the physical averages.

Two numerical techniques of Monte Carlo device simulation are described in Sect. 4. They may be seen as a generalization of the well-known Monte Carlo method for semi-classical device simulation.

Finally, in Sect. 5, the device simulation is applied to some typical nano-devices, namely the resonant tunneling diode (RTD) and the ultra-short double-gate (DG) metal-oxide-semiconductor field-effect transistor (MOSFET). Quantum and decoherence effects taking place in these are emphasized.

The occurrence of quantum de-coherence in devices of a size smaller than the electron wave length and mean free path is becoming an important subject of experimental and theoretical research [6–9]. The theory of de-coherence has shown that the semi-classical behavior of a quantum system may emerge from the interaction with its environment. For electrons in a nano-device, the environment likely to induce de-coherence may be the phonon field, the contacts or defects.

In this final section the theory of de-coherence is briefly introduced through an academic example of the free evolution of a Gaussian wave packet and the phonon scattering-induced de-coherence is investigated in a typical nano-device, the RTD. The Wigner–Boltzmann formalism is proved to be an appropriate framework for such analysis [10]. One of its major advantage lies in the fact that it offers a straightforward access to the off-diagonal elements of the density matrix which provides a clear visualization of de-coherence phenomena.

The Wigner–Boltzmann equation may also become – in establishing a link between semi-classical and quantum transport – a ground-breaking route to understanding nano-device behavior. We focus in particular on the case of the ultra-small DG-MOSFET with gate length of 6 nm through comparison between quantum (Wigner–Boltzmann) and semi-classical simulations. Beyond the analysis of direct source-drain tunneling and quantum reflections on the steep potential drop at the drain-end of the channel, the results emphasize the role of scattering which remains surprisingly important in such a small device in spite of significant quantum coherence effects.

2 Wigner Quantum Mechanics

2.1 Classical Distribution Function

A single particle of mass m is considered to move with a potential energy $V(x)$. The phase space is defined by the Cartesian product of the particle position x and momentum p. Physical quantities are dynamical functions $A(x,p)$ of the phase space coordinates, such as the kinetic and potential energies and their sum giving the Hamiltonian $H(x,p)$. The state of the single particle at given time is presented by a point in the phase space. Provided that the initial particle coordinates are known, the novel coordinates $x(t),p(t)$ at time t are obtained from the Hamilton equations

$$\dot{x} = \frac{\partial H(x,p)}{\partial p} = \frac{p}{m}; \qquad \dot{p} = -\frac{\partial H(x,p)}{\partial x} = -\frac{\partial V(x)}{\partial x} \tag{5.1}$$

The function $A(t)$ describes how physical quantities change in time. Two ways are possible: (a) $A(t) = A(x(t),p(t))$ is the old function in the novel coordinates; (b) $A(t) = A(t,x,p)$ is a new function of the old coordinates. In the first case we postulate that the laws of mechanics do not change with time: A remains the same function for the old and the new coordinates. Then, with the help of (5.1) we obtain the equation of evolution for A:

$$\dot{A} = \frac{\partial A(x,p)}{\partial x}\frac{\partial H(x,p)}{\partial p} - \frac{\partial A(x,p)}{\partial p}\frac{\partial H(x,p)}{\partial x} = [A,H]_P; \qquad [x,p]_P = 1 \tag{5.2}$$

A basic notion between the dynamical functions is endowed with the Poisson bracket $[\cdot,\cdot]_P$. It gives rise to an automorphic (conserving the algebraic structure) mapping of the set of such functions.

Alternatively, in the second case we have to postulate a law for the evolution of $A(t,x,p)$. If it is imposed according to (5.2), the automorphism consistently leads to the conservation of the mechanical laws: the new function in the old coordinates is the old function in the novel coordinates: $A(t,x,p) = A(x(t),p(t))$!

A statistical description is introduced if the coordinates of the point cannot be stated exactly, but with some probability. According to the basic postulate of classical statistical mechanics, the state of the particle system is completely specified by a function $f(x,p)$, with the following properties:

$$f(x,p) \geq 0 \qquad \int dxdp f(x,p) = 1 \tag{5.3}$$

Physical quantities A are then described by the corresponding mean values:

$$\langle A \rangle(t) = \int dxdp A(t,x,p) f(x,p) \tag{5.4}$$

This equation is not convenient since it requires calculation of the evolution of any particular quantity A. However, due to the automorphism of the Poisson bracket, it is possible to change the variables so that time is transferred to the distribution function f [11]. Equation (5.4) modifies to:

$$\langle A \rangle(t) = \int dxdp A(x,p) f(x,p,t) \tag{5.5}$$

The evolution equation for f can be derived with the help of (5.1) and (5.2):

$$\left(\frac{\partial}{\partial t} + \frac{p}{m} . \frac{\partial}{\partial x} + F(x) \frac{\partial}{\partial p} \right) f(x,p,t) = \left(\frac{\partial f}{\partial t} \right)_c \tag{5.6}$$

Here the force $F = -\nabla_x V$ is given by the derivative of the potential energy V. The characteristics of the differential operator in the brackets, called Liouville operator, are classical Newton's trajectories, obtained from (5.1). Over such trajectories the left hand side of (5.6) becomes a total time derivative. In the case of no interaction with the environment, $\left(\frac{\partial f}{\partial t} \right)_c = 0$, i.e. trajectories carry a constant value of f. Otherwise the particles are redistributed between the trajectories and the right hand side of (5.6) is equal to the net change of the particle density due to collisions. In the rest of this section we derive a quantum analog of (5.3), (5.5) and the Boltzmann equation (5.6).

2.2 *Quantum Operators*

We recall the principles of the operator quantum mechanics, which will be used to reformulate the formalism in the phase space. Physical quantities in quantum mechanics are presented by Hermitian operators $\hat{A}$:

$$\hat{A}|\phi_n\rangle = a_n|\phi_n\rangle; \quad \langle \phi_n|\phi_m\rangle = \delta_{mn} \qquad \sum_n |\phi_n\rangle\langle\phi_n| = \hat{1} \tag{5.7}$$

Such operators have real eigenvalues and a complete system of orthonormal eigenvectors which form an abstract Hilbert space. The states of the system are specified by the elements $|\Psi_t\rangle$ of the Hilbert space $\mathscr{H}$ which are square integrable and normalized with respect to the L_2 norm in $\mathscr{H}$. In wave mechanics it is postulated that the evolution of $|\Psi_t\rangle$ is provided by the Schrödinger equation

$$\hat{H}|\Psi_t\rangle = i\hbar \frac{\partial |\Psi_t\rangle}{\partial t} \qquad \langle \Psi_t|\Psi_t\rangle = 1 \qquad |\Psi_t\rangle = \sum_n c_n(t)|\phi_n\rangle \tag{5.8}$$

The state can be decomposed in the complete basis of an observable A. Also, it can be shown that during the evolution the state remains normalized. This property is often called conservation of probability.

According to the correspondence principle, to classical position and momentum variables correspond the Hermitian operators $\hat{x}$ and $\hat{p}$, satisfying a quantum counterpart of the Poisson bracket:

$$x \to \hat{x} \qquad p \to \hat{p} \qquad \hat{x}\hat{p} - \hat{p}\hat{x} = [\hat{x}, \hat{p}]_- = i\hbar\hat{1} \tag{5.9}$$

Wave mechanics uses only half of the phase space – coordinate or momentum representation – for the description of the physical system. We assume a coordinate representation; according to (5.7) and (5.9) it holds that

$$\hat{x}|x\rangle = x|x\rangle \qquad \int dx|x\rangle\langle x| = \hat{1} \qquad \hat{p} = -i\hbar\frac{\partial}{\partial x} \tag{5.10}$$

Finally, we recall the equation for the averaged value of a physical quantity:

$$\langle A\rangle(t) = \langle \Psi_t|\hat{A}|\Psi_t\rangle = \int dx \langle \Psi_t|x\rangle\langle x|\hat{A}|\Psi_t\rangle \tag{5.11}$$

The operator formulation of the quantum mechanics looks too abstract when compared to the familiar classical concepts. Nevertheless it is possible to reformulate the ideas of the quantum mechanics in the phase space. The first step is to evaluate the actual number of variables involved in (5.11). With the help of (5.7) and (5.10) it holds:

$$\langle x|\hat{A}|\Psi_t\rangle = \int dx' \sum_n a_n \langle x|\phi_n\rangle\langle\phi_n|x'\rangle\langle x'|\Psi_t\rangle = \int dx' \alpha(x,x')\Psi_t(x')$$

where $\Psi_t(x) = \langle x|\Psi_t\rangle$. A substitution in (5.11) shows that the physical average is actually evaluated in a "double half" of the phase space:

$$\langle A\rangle(t) = \int dx' \int dx \alpha(x,x')\rho_t(x',x) = Tr(\hat{\rho}_t\hat{A}) \tag{5.12}$$

with ρ_t and $\hat{\rho}_t$ the density matrix and density operator:

$$\rho_t(x,x') = \Psi_t^*(x')\Psi_t(x) = \langle x|\Psi_t\rangle\langle\Psi_t|x'\rangle = \langle x|\hat{\rho}_t|x'\rangle \qquad \rho_t = \sum_{m,n} c_m^*(t)c_n(t)|\phi_n\rangle\langle\phi_m| \tag{5.13}$$

2.3 Weyl Transform

Equation (5.12) resembles (5.5) provided that one of the spatial variables is replaced by a momentum variable. A proper transform for such a replacement is needed. The important consequence is that the transformed density matrix can be interpreted as the quantum counterpart of the classical distribution function. Pursuing a proper rule, we consider how an operator $\hat{A}$ can be associated to a given physical quantity. $\hat{A}$ can be obtained explicitly with the help of (5.9) and the knowledge of $A(x,p)$: the Taylor expansion, for example, can be used to establish the rule:

$$A(x,p) = \sum_{i,j} b_{i,j} x^i p^j \quad \rightarrow \quad A(\hat{x},\hat{p}) = \sum_{i,j} b_{i,j} \hat{x}^i \hat{p}^j$$

For the Hamiltonian of a particle in a potential field, $H(x,p) = \frac{p^2}{2m} + V(x)$, this rule leads to a consistent result. However, for general functions A the procedure is not well defined, since the operators $\hat{p}$ and $\hat{x}$ do not commute. First, non-Hermitian operators can appear. Second, even for Hermitian operators there is ambiguity in the correspondence: let us consider two equivalent expressions for the function $A(x,p)$:

$$A_1 = px^2p = A_2 = \frac{1}{2}(p^2x^2 + x^2p^2)$$

The substitution of x and p by $\hat{x}$ and $\hat{p}$ gives rise to the following operators:

$$A_1 \rightarrow \hat{A}_1 = \hat{p}\hat{x}^2\hat{p} \qquad A_2 \rightarrow \hat{A}_2 = \frac{1}{2}(\hat{p}^2\hat{x}^2 + \hat{x}^2\hat{p}^2)$$

Now, while $A_1 = A_2$, the obtained operators differ by $\hbar^2$: $\hat{A}_1 = \hat{A}_2 + \hbar^2$. The example shows how different operator functions are mapped into the same function of the phase space coordinates: the relation (5.9) is not sufficient to establish a unique correspondence between A and $\hat{A}$. A certain rule must be applied in order to remove this ambiguity. We will make use of the fact that an arbitrary function $f(x,p)$ can be obtained from the generating function $F(s,q) = e^{i(sx+qp)}$ as follows:

$$f(x,p) = f\left(\frac{1}{i}\nabla_s, \frac{1}{i}\nabla_q\right) F(s,q)_{s=0,q=0} = \frac{1}{(2\pi)^2}\int dsdqdldm f(l,m) e^{-i(ls+mq)} F(s,q)$$

It remains to consider possible operator generalizations of F, e.g.

$$\hat{F}_1 = e^{i(s\hat{x})}e^{i(q\hat{p})}; \qquad \hat{F}_2 = e^{i(q\hat{p})}e^{i(s\hat{x})}; \qquad e^{i(s\hat{x}+q\hat{p})}$$

which represent the standard order where the positions precede the momenta, the anti-standard order, where the momenta come before the positions, and the Weyl order. The fully symmetric Weyl order bears some of the basic properties of a characteristic function of a probability distribution [4] and will be used henceforth to establish the rule of correspondence. The choice of alternative orders leads to alternative quasi-distributions. It should be noted that once postulated, the correspondence rule must be consistently applied to all notions of the operator mechanics in order to ensure conservation of the values of the physical averages (5.11). The Weyl transform reads:

$$A(x,p) = W(\hat{A}(\hat{x},\hat{p})) = \frac{\hbar}{(2\pi)}\int dsdq Tr\left(\hat{A}(\hat{x},\hat{p}) e^{i(s\hat{x}+q\hat{p})}\right) e^{-i(sx+qp)} \tag{5.14}$$

Equivalently, as discussed in the appendix, it holds:

$$\hat{A} = \hat{A}(\hat{x}, \hat{p}) = \int dsdq\beta(s,q)e^{i(s\hat{x}+q\hat{p})} \tag{5.15}$$

Here β is adjoint to A via the Fourier transform:

$$A(x,p) = \int dsdq\beta(s,q)e^{i(sx+qp)} \qquad \beta(s,q) = \frac{1}{(2\pi)^2}\int dxdpA(x,p)e^{-i(sx+qp)} \tag{5.16}$$

The Wigner function is defined as the transform of the density operator, multiplied by the normalization factor $(2\pi\hbar)^{-1}$. The Weyl map W provides the algebra of phase space functions with a non-commutative *-product defined as:

$$W(\hat{A}) * W(\hat{B}) = A(x,p) * B(x,p) = W(\hat{A}\,\hat{B}) \tag{5.17}$$

Basic notions of the operator quantum mechanics are formulated in the phase space with the help of the *-product.

2.4 *Wigner Function for Pure State*

Equation (5.8) and its adjoint equation give rise to the von Neumann equation of motion for the pure state density matrix ρ_t (5.13).

$$\begin{aligned} i\hbar\frac{\partial\rho(x,x',t)}{\partial t} &= \langle x|[\hat{H},\hat{\rho}_t]_-|x'\rangle \\ &= \left\{-\frac{\hbar^2}{2m}\left(\frac{\partial^2}{\partial x^2} - \frac{\partial^2}{\partial x'^2}\right) + \left(V(x) - V(x')\right)\right\}\rho(x,x',t) \end{aligned} \tag{5.18}$$

The variables are changed with the help of a center of mass transform:

$$x_1 = (x+x')/2, \quad x_2 = x - x'$$

$$\begin{aligned} &\frac{\partial\rho(x_1+x_2/2,x_1-x_2/2,t)}{\partial t} \\ &= \frac{1}{i\hbar}\left\{-\frac{\hbar^2}{m}\frac{\partial^2}{\partial x_1\partial x_2} + (V(x_1+x_2/2) - V(x_1-x_2/2))\right\}\rho(x_1+x_2/2,x_1-x_2/2,t) \end{aligned} \tag{5.19}$$

As shown in the appendix, the Wigner function is obtained by Fourier transform with respect to x_2:

$$f_w(x_1,p,t) = \frac{1}{(2\pi\hbar)}\int dx_2\rho(x_1+x_2/2,x_1-x_2/2,t)e^{-ix_2\cdot p/\hbar} \tag{5.20}$$

We note that, due to the Wigner transform, x_1 and p are independent variables. It is easy to show that the corresponding operators commute. Thus x_1 and p define a phase space – the Wigner phase space.

The Fourier transform of the right hand side of (5.19) gives rise to two terms which are evaluated as follows. It is convenient to introduce the abbreviation $\rho(+,-,t)$ for $\rho(x_1+x_2/2, x_1-x_2/2, t)$:

$$I = -\frac{1}{i\hbar}\frac{\hbar^2}{m(2\pi\hbar)}\int dx_2 e^{-ix_2.p/\hbar}\frac{\partial^2\rho(+,-,t)}{\partial x_1\partial x_2}$$
$$= -\frac{1}{m(2\pi\hbar)}p.\frac{\partial}{\partial x_1}\int dx_2 e^{-ix_2.p/\hbar}\rho(+,-,t) = -\frac{1}{m}p.\frac{\partial f_w(x_1,p,t)}{\partial x_1}$$

where we have integrated by parts and used the fact that the density matrix tends to zero at infinity: $\rho \to 0$ if $x_2 \to \pm\infty$.

$$II = \frac{1}{i\hbar(2\pi\hbar)}\int dx_2 e^{-ix_2.p/\hbar}(V(x_1+x_2/2)-V(x_1-x_2/2))\rho(+,-,t)$$
$$= \frac{1}{i\hbar(2\pi\hbar)}\int dx_2\int dx' e^{-ix_2.p/\hbar}(V(x_1+x_2/2)-V(x_1-x_2/2))$$
$$\times\,\delta(x_2-x')\rho(x_1+x'/2, x_1-x'/2, t)$$

After a substitution of the delta function with the integral

$$\delta(x_2-x') = \frac{1}{(2\pi\hbar)}\int dp' e^{i(x_2-x')p'/\hbar}. \tag{5.21}$$

the following is obtained:

$$II = \frac{1}{i\hbar(2\pi\hbar)}\int dp'\int dx_2 e^{-ix_2.(p-p')/\hbar}(V(x_1+x_2/2)-V(x_1-x_2/2))$$
$$\times\frac{1}{(2\pi\hbar)}\int dx' e^{-ix'p'/\hbar}\rho(x_1+x'/2, x_1-x'/2, t)$$
$$= \int dp' V_w(x_1, p-p') f_w(x_1, p', t)$$

We summarize the results of these transformations. Equation (5.19) gives rise to the Wigner equation:

$$\frac{\partial f_w(x,p,t)}{\partial t} + \frac{p}{m}.\frac{\partial f_w(x,p,t)}{\partial x} = \int dp' V_w(x, p-p') f_w(x, p', t) \tag{5.22}$$

where V_w is the Wigner potential.

$$V_w(x,p) = \frac{1}{i\hbar(2\pi\hbar)} \int dx' e^{-ix'p/\hbar}(V(x+x'/2) - V(x-x'/2)) \tag{5.23}$$

A change of the sign of x' reveals the antisymmetry of the Wigner potential.

2.5 Properties of the Wigner Function

We first outline the equivalence between the Schrödinger equation and the Wigner equation in the case of a pure state. From Ψ_t we can obtain ρ and thus f_w. The opposite is also true: it can be shown that, if we know f_w we can obtain Ψ_t up to a phase factor.

Comparing this with the Boltzmann equation (5.6), we can recognize on the left hand side of the Wigner equation the field-less Liouville operator. Furthermore, it is easy to see that the Wigner potential is a real quantity, $V_w = V_w^*$. It follows that, being a solution of an equation with real coefficients, f_w is real. The Wigner function conserves the probability in time:

$$\int dx \int dp f_w(x,p,t) = \int dx \int dx_2 \rho(x+x_2/2, x_1 - x_2/2, t)\delta(x_2) = \int dx \langle x|\hat{\rho}_t|x\rangle = 1 \tag{5.24}$$

In a similar way it can be demonstrated that the position or momentum probability distributions are obtained after integration over momentum p or position x respectively:

$$\int dp f_w(x,p,t) = |\Psi_t(x)|^2 \qquad \int dx f_w(x,p,t) = |\Psi_t(p)|^2 \tag{5.25}$$

The most important property of the Wigner picture is that the mean value $\langle A\rangle(t)$ of any physical quantity is given by

$$\langle A\rangle(t) = \int dx \int dp f_w(x,p,t) A(x,p) \tag{5.26}$$

where $A(x,p)$ is the classical function (5.16). This is proven in the appendix.

Our goal to derive a quantum analog of (5.3), (5.5) and (5.6) has been attained to a large extent. Equation (5.24) corresponds to the second equation in (5.3) and the Wigner function is real. Equation (5.26) is equivalent to (5.5). The left hand sides of the Wigner equation (5.22) and the Boltzmann equation are given by the Liouville operator. Classical and quantum pictures become very close.

Nevertheless, there are basic differences. The Wigner function allows negative values and thus is not a probability function. It cannot be interpreted as a joint distribution of particle position and momentum. Actually, the Wigner function can have nonzero values in domains where the particle density is zero. As follows from (5.25), a physical interpretation is possible only after an integration.

The quantum character of the Wigner function is underlined by the following remarkable result. If the spectrum of $\hat{A}$, (5.7), is non-degenerate, then the corresponding to an eigenvector Wigner function $f_{w(n)} = (2\pi\hbar)^{-1} W(|\phi_n\rangle\langle\phi_n|)$ satisfies the following equation:

$$f_{w(n)}(x,p) * A(x,p) = a_n f_{w(n)}(x,p); \quad A(x,p) * f_{w(n)}(x,p) = a_n f_{w(n)}(x,p) \tag{5.27}$$

The probability P that a measurement of the observable corresponding to a given generic operator $\hat{A}$ yields the value a_n in a state $f_w(x,p,t)$ is:

$$P(a_n) = (2\pi\hbar) \int dx dp f_w(x,p,t) f_{w(n)}(x,p) \tag{5.28}$$

2.6 Classical Limit of the Wigner Equation

We discuss the classical limit of (5.22) by considering the case when the potential V is a linear or a quadratic function of the position:

$$V\left(x \pm \frac{x'}{2}\right) = V(x) \pm \frac{\partial V(x)}{\partial x}\frac{x'}{2} + \cdots = V(x) \mp F(x)\frac{x'}{2} + \cdots$$

where the dots stand for the quadratic term. The force F can be at most a linear function of the position. As the even terms of the Taylor expansion of V cancel in (5.23), the Wigner potential becomes:

$$V_w(x,p) = \frac{i}{\hbar(2\pi\hbar)} \int dx' e^{-ix'p/\hbar} F(x) x'$$

The right hand side of (5.22) becomes

$$\begin{aligned}
\int dp' V_w(x,p-p') f_w(x,p',t) &= \frac{i}{\hbar(2\pi\hbar)} \int dp' \int dx' e^{-ix'(p-p')/\hbar} F(x) x' f_w(x,p',t) \\
&= \frac{-F(x)}{(2\pi\hbar)} \frac{\partial}{\partial p} \int dp' \int dx' e^{-ix'(p-p')/\hbar} f_w(x,p',t) \\
&= -F(x) \frac{f_w(x,p,t)}{\partial p}
\end{aligned} \tag{5.29}$$

where we have used the equality $ix' e^{-ix'(p-p')/\hbar} = -\hbar \frac{\partial}{\partial p} e^{-ix'(p-p')/\hbar}$. Then the Wigner equation reduces to the collisionless Boltzmann equation:

$$\frac{\partial f_w(x,p,t)}{\partial t} + \frac{p}{m} \cdot \frac{\partial f_w(x,p,t)}{\partial x} + F(x) \frac{\partial f_w(x,p,t)}{\partial p} = 0 \tag{5.30}$$

Now consider as an initial condition a minimum uncertainty wave-packet. The Wigner function of such a packet is a Gaussian of both position and momentum [12]. The latter can equally well be interpreted as an initial distribution of classical electrons. Provided that the force is a constant or linear function of the position, the packet evolves according to (5.30). The evolution resembles that of the classical distribution. Despite the spread in the phase space, the Gaussian components determine the general shape of the packet. f_w remains positive during the evolution.

However, stronger variations of the field with position introduce interference effects. Near band offsets the packet rapidly looses its shape and negative values appear.

2.7 Wigner Potential and Fourier Transform

In this section we discuss some properties of the Wigner potential in terms of the Fourier transform. For this purpose we express the momentum p through the wave number k as $p = \hbar k$. We introduce $\hat{V}(q)$, the Fourier transform of the potential. The Fourier transform and its inverse read

$$\hat{V}(q) = \int \mathrm{d}x V(x) \mathrm{e}^{-iqx}, \qquad V(x) = \frac{1}{2\pi} \int \mathrm{d}q \hat{V}(q) \mathrm{e}^{iqx}. \tag{5.31}$$

The result of the Fourier transform is in general a complex function, which can be expressed in polar form by its modulus and phase.

$$\hat{V}(q) = A(q) \mathrm{e}^{i\varphi(q)} \tag{5.32}$$

With the variable substitutions $s = x \pm x'/2$ the integrals in the definition (5.23) of the Wigner potential can be evaluated as

$$\int \mathrm{d}x' V\left(x + \frac{x'}{2}\right) \mathrm{e}^{-ikx'} = 2\mathrm{e}^{2ikx} \int \mathrm{d}s V(s) \mathrm{e}^{-2iks} = 2\mathrm{e}^{2ikx} \hat{V}(2k),$$
$$\int \mathrm{d}x' V\left(x - \frac{x'}{2}\right) \mathrm{e}^{-ikx'} = \left[2\mathrm{e}^{2ikx} \hat{V}(2k)\right]^*,$$

and the following relation between the Wigner potential (5.23) and the Fourier transform of the potential can be established.

$$V_\mathrm{w}(x, \hbar k) = \frac{1}{i\hbar(2\pi\hbar)} \left\{ 2\mathrm{e}^{2ikx} \hat{V}(2k) - \left[2\mathrm{e}^{2ikx} \hat{V}(2k)\right]^* \right\}$$

This expression can be simplified using the polar form (5.32).

$$V_\mathrm{w}(x, \hbar k) = \frac{2}{\pi\hbar^2} A(2k) \sin[\varphi(2k) + 2kx] \tag{5.33}$$

The x-dependence of the Wigner potential is given analytically by an undamped sine function, independent of the actual shape of the potential. This result also shows, that even for a well localized potential barrier the Wigner potential is fully delocalized in the coordinate space. In any numerical procedure, therefore, the Wigner potential needs to be truncated at some finite x-coordinate.

Another property of the Wigner potential can be derived by considering the function

$$\Delta(x,x') = V\left(x+\frac{x'}{2}\right) - V\left(x-\frac{x'}{2}\right). \tag{5.34}$$

The Wigner potential is defined as the Fourier transform of this function with respect to the argument x'. We note that

$$\Delta(x,-x') = -\Delta(x,x'). \tag{5.35}$$

Due to this antisymmetry, the substitution $\exp(-ikx') = \cos(kx') - i\sin(kx')$ in (5.23) readily yields the Fourier sine transform.

$$\begin{aligned} V_{\mathrm{w}}(x,\hbar k) &= \frac{1}{i\hbar(2\pi\hbar)} \int \mathrm{d}x'\, \Delta(x,x')\mathrm{e}^{-ikx'} \\ &= -\frac{1}{\hbar(2\pi\hbar)} \int \mathrm{d}x'\, \Delta(x,x')\sin(kx') \end{aligned} \tag{5.36}$$

In general the potential $V(x)$ is given within a finite simulation domain, representing, for instance, the active region of an electronic device. Outside of this domain the potential is continued by two constants, say V_L and V_R. This situation represents an active device region connected to semi-infinite leads on both sides, where the leads are assumed to be ideal conductors. Therefore, in practical cases Δ will have the asymptotic behavior,

$$\lim_{x'\to\pm\infty} \Delta(x,x') = \mp(V_L - V_R) \tag{5.37}$$

where $(V_L - V_R)$ is the potential difference between the left and the right lead. Since the integrand in (5.36) does not vanish at infinity, the Fourier integral will diverge at $q = 0$. From the asymptotic behavior of $\Delta(x,x')$ for $x' \to \infty$ we find the asymptotic behavior of $V_{\mathrm{w}}(x,\hbar k)$ for $k \to 0$.

$$\Delta(x,x') \simeq (V_R - V_L)\mathrm{sgn}(x'), \qquad x' \to \infty \tag{5.38}$$

$$V_{\mathrm{w}}(x,\hbar k) \simeq \frac{2(V_L - V_R)}{\hbar(2\pi\hbar)} \mathscr{P}\frac{1}{k}, \qquad k \to 0 \tag{5.39}$$

Here, sgn denotes the signum function and $\mathscr{P}$ the principal value. This consideration shows that if the potential difference is nonzero, there will be a pole in the Wigner potential at $k = 0$. Numerical methods for the Wigner equation generally use a k-space discretization, where the discrete k-points are located symmetrically around the origin and the point $k = 0$ is not included. In this way, no particular treatment of the singularity is needed.

2.8 Classical Force

The potential operator in (5.22) takes the form

$$Qf_{\mathrm{w}}(x,p) = \hbar \int \mathrm{d}q V_{\mathrm{w}}(x,\hbar q) f_{\mathrm{w}}(x,p-\hbar q), \tag{5.40}$$

if variables are changed according to $p - p' = \hbar q$. To deal with the singularity of V_{w}, one can define a small neighborhood around $q = 0$ and split the domain of integration as follows [13].

$$Qf_{\mathrm{w}}(x,p) \quad = \int\limits_{|q|\leq q_{\mathrm{c}}/2} + \int\limits_{|q|>q_{\mathrm{c}}/2} = Q_{cl} f_{\mathrm{w}} + Q_{qm} f_{\mathrm{w}} \tag{5.41}$$

Here q_{c} is some small wave number. In this way, we have split the potential operator Q in two parts, which we refer to as Q_{cl} and Q_{qm}. A linearization can be introduced in the integral over the small wave numbers.

$$Q_{cl} f_{\mathrm{w}}(x,p,t) = \hbar \int\limits_{|q|\leq q_{\mathrm{c}}/2} \mathrm{d}q V_{\mathrm{w}}(x,\hbar q) f_{\mathrm{w}}(x,p-\hbar q) \tag{5.42}$$

$$\simeq \hbar \int\limits_{|q|\leq q_{\mathrm{c}}/2} \mathrm{d}q V_{\mathrm{w}}(x,\hbar q) \left[f_{\mathrm{w}}(x,p) - \hbar q \frac{\partial f_{\mathrm{w}}(x,p)}{\partial p} \right] \tag{5.43}$$

$$= -\frac{\partial f_{\mathrm{w}}(x,p)}{\partial p} \hbar^2 \int\limits_{|q|\leq q_{\mathrm{c}}/2} \mathrm{d}q\, q V_{\mathrm{w}}(x,\hbar q) \tag{5.44}$$

In the second line the integral over f_{w} vanishes since V_{w} is an odd function in q. Substituting (5.33) into (5.44) gives

$$\begin{aligned}
-\hbar^2 \int\limits_{|q|\leq q_{\mathrm{c}}/2} \mathrm{d}q\, q V_{\mathrm{w}}(x,\hbar q) &= -\frac{2}{\pi} \int_{-q_{\mathrm{c}}/2}^{q_{\mathrm{c}}/2} \mathrm{d}q\, q A(2q) \sin[\varphi(2q) + 2qx] \\
&= -\frac{1}{2\pi} \int_{-q_{\mathrm{c}}}^{q_{\mathrm{c}}} \mathrm{d}q\, q A(q) \sin[\varphi(q) + qx] \\
&= \frac{\partial}{\partial x} \frac{1}{2\pi} \int_{-q_{\mathrm{c}}}^{q_{\mathrm{c}}} \mathrm{d}q A(q) \cos[\varphi(q) + qx] \\
&= \frac{\partial}{\partial x} \Re \left\{ \frac{1}{2\pi} \int_{-q_{\mathrm{c}}}^{q_{\mathrm{c}}} \mathrm{d}q A(q) \mathrm{e}^{i\varphi(q)} \mathrm{e}^{iqx} \right\} \\
&= \frac{\partial}{\partial x} \Re \left\{ \frac{1}{2\pi} \int_{-q_{\mathrm{c}}}^{q_{\mathrm{c}}} \mathrm{d}q \hat{V}(q) \mathrm{e}^{iqx} \right\} \\
&= \frac{\partial}{\partial x} V_{\mathrm{cl}}(x)
\end{aligned}$$

Here we introduced the classical potential component as

$$V_{\mathrm{cl}}(x) = \frac{1}{2\pi}\int_{-q_{\mathrm{c}}}^{q_{\mathrm{c}}} \mathrm{d}q\,\hat{V}(q)\mathrm{e}^{iqx}. \tag{5.45}$$

This function is real, as can be easily shown by substituting $\hat{V}(q)$.

$$V_{\mathrm{cl}}(x) = \frac{1}{2\pi}\int_{-q_{\mathrm{c}}}^{q_{\mathrm{c}}} \mathrm{d}q \int \mathrm{d}y V(y)\mathrm{e}^{iq(x-y)} = \int dy V(y)\frac{\sin[q_{\mathrm{c}}(x-y)]}{\pi(x-y)} \tag{5.46}$$

So we have a convolution of two real functions, the potential $V(x)$ and the $\sin(x)/x$ function.

According to its definition (5.45), the classical potential component shows a smooth spatial variation, as it is composed of long-wavelength Fourier components only. Equation (5.45) motivates the following spectral decomposition of the potential profile into a slowly varying, classical component (5.45) and a rapidly varying, quantum mechanical component.

$$V(x) = V_{\mathrm{cl}}(x) + V_{\mathrm{qm}}(x) \tag{5.47}$$

When the linearization described above is introduced in the classical component, this decomposition yields a Wigner equation including both a local classical force term and a nonlocal potential operator.

$$\left(\frac{\partial}{\partial t} + \frac{p}{m}\frac{\partial}{\partial x} - \frac{\partial V_{\mathrm{cl}}(x)}{\partial x}\frac{\partial}{\partial p}\right) f_{\mathrm{w}}(x,p,t) = \int \mathrm{d}p' V_{\mathrm{w}}^{\mathrm{qm}}(x,p') f_{\mathrm{w}}(x,p-p',t) \tag{5.48}$$

The Wigner potential is calculated from the quantum mechanical potential component, $V_{\mathrm{qm}} = V - V_{\mathrm{cl}}$. The two potential components have the following properties. The classical component accommodates the applied voltage. As it is treated through a classical force term, it does not induce any quantum reflections. The quantum mechanical component vanishes at infinity and has a smooth Fourier transform.

2.9 Quantum Statistics

The density operator $\hat{\rho}_t = |\Psi_t\rangle\langle\Psi_t|$, used to obtain the Wigner function, corresponds to a system in a pure state. The state of the system is often not known exactly. Assuming that a set of possible states $\hat{\rho}_t^i$ can be occupied with probabilities γ_i, the definition (5.13) of density operator can be generalized for a mixed state:

$$\hat{\rho}_t = \sum_i \gamma_i \hat{\rho}_t^i \qquad \sum_i \gamma_i = 1, \qquad \gamma_i > 0 \tag{5.49}$$

The mean value of a given physical quantity becomes a statistical average of "averages in states i". It is easy to see that the von Neumann equation (5.18) and the expression (5.12) hold also in this case. Accordingly, the mixed state Wigner function and equation are derived from $\hat{\rho}$ and its equation of motion as in the case of a pure state. Since the derivation is reversible, one can equivalently postulate $f_w(x,p,t)$ as a definition of the state of the system. Note that if the set γ_i is known, the density matrix can be obtained from (5.49). This is for example possible in models where γ_i are defined by the boundary conditions [12]. Then the problem is reduced to a set of pure state problems. However, for more complex physical systems, containing electrons which interact with other types of quasi-particles, γ_i are not known a priori. In this case $\hat{\rho}_i$ and γ_i are obtained with the help of the basic notions (5.18) and (5.12). We note that in the latter the Hamiltonian already contains the term accounting for the interaction with the quasi-particles, so that (5.18) must be augmented accordingly. Indeed the corresponding representation of the system is given by the basis vectors $|X_i\rangle|x\rangle$ where the additional degrees of freedom X describing the quasi-particles are assumed enumerable. Of particular interest are the electron averages, so that the operator $\hat{A}$ does not affect X_i. Equation (5.12) becomes:

$$\langle\hat{A}\rangle(t) = Tr(\hat{\rho}_t\hat{A}) = \sum_i \int dx\langle x|\langle X_i|\hat{\rho}_t\hat{A}|X_i\rangle|x\rangle = \int dx\langle x|\hat{\rho}_t^e\hat{A}|x\rangle = Tr_e(\hat{\rho}_t^e\hat{A}) \quad (5.50)$$

where $\hat{\rho}_t^e = \sum_i\langle X_i|\hat{\rho}_t|X_i\rangle$ is the electron, or reduced density operator. The set of probabilities γ_i and the set of electron density operators $\hat{\rho}_t^{e,i}$ are now introduced according to:

$$\gamma_i = Tr_e(\langle X_i|\hat{\rho}_t|X_i\rangle) \geq 0, \quad \sum_i \gamma_i = 1; \qquad \hat{\rho}_t^{e,i} = \frac{\langle X_i|\hat{\rho}_t|X_i\rangle}{Tr_e(\langle X_i|\hat{\rho}_t|X_i\rangle)}, \quad Tr_e(\hat{\rho}_t^{e,i}) = 1$$

These estimates follow from the fact that $\hat{\rho}_t$ is a positively defined operator and from the conservation of the probability. Hence, in a formal consistence with (5.49), it holds

$$\hat{\rho}_t^e = \sum_i \gamma_i\hat{\rho}_t^{e,i}$$

However, in order to obtain γ_i and $\hat{\rho}_t^{e,i}$ one needs $\hat{\rho}_t$ which entails solving the evolution equation for the whole system. Usually this is not possible, moreover we are not interested in the detailed information about the state of the quasi-particles. This implies to approximate the evolution equation to a closed equation for the electron subsystem. Alternatively this can be done in terms of the Wigner functions obtained after a Wigner transform of the corresponding density operators.

With the help of (5.13) and (5.49) it is obtained:

$$f_w(x,p,t) = \sum_{m,n}\left(\sum_i \gamma_i c_n(t)c_m^*(t)\right) f_{w(m,n)}(x,p)$$

This equation introduces the off-diagonal Wigner function

$$f_{w(m,n)} = (2\pi\hbar)^{-1} W(|\phi_n\rangle\langle\phi_m|) \tag{5.51}$$

If $|\psi_{t,1}\rangle$ and $|\psi_{t,2}\rangle$ are two states, solutions of (5.8), the off-diagonal Wigner function $f_{w(1,2)} = (2\pi\hbar)^{-1} W(|\psi_{t,1}\rangle\langle\psi_{t,2}|)$ is a solution of (5.22). Furthermore if $|\psi_1\rangle$ and $|\psi_2\rangle$ are two stationary energy eigenstates, corresponding to energy eigenvalues E_1 and E_2, it holds:

$$H(x,p) * f_{w(1,2)} = E_1 f_{w(1,2)} \qquad f_{w(1,2)} * H(x,p) = E_2 f_{w(1,2)} \tag{5.52}$$

where $H(x,p) = W(\hat{H}) = p^2/2m + V(x)$.

2.10 *Quantum Phase Space States*

It has been shown that the laws and relations of the operator quantum mechanics can be reformulated into the language of the phase space functionals. A systematic presentation of the inverse approach is not possible within this chapter, however we provide some selected ideas which help the reader to build up an initial impression.

A basic question which must be addressed is about the identification of the admissible quantum phase space functionals. Conditions have been derived, which specify the functionals in terms of pure or mixed quantum states and the rest of non-quantum states. A phase space function is an off-diagonal pure state if it can be presented in the form (5.51) for two complex valued, normalized functions $\langle x|\phi_{m,n}\rangle$. In particular, if $m = n$ the function is just a pure state. The first necessary and sufficient condition for a pure state has been introduced by Tatarskii [4], and will be formulated later. The condition has been generalized for off-diagonal pure states [5] as follows:

If $f_w(x,p,t)$ is square-integrable, and if Z, defined as

$$Z(x,x',t) = \int dp e^{ix'p/\hbar} f_w(x,p,t) \tag{5.53}$$

satisfies the following equation

$$\frac{\partial^2 lnZ(x,x',t)}{\partial x'^2} = \left(\frac{1}{2}\right)^2 \frac{\partial^2 lnZ(x,x',t)}{\partial x^2} \tag{5.54}$$

then f_w is a phase space function of the form (5.51):

$$f_{w(1,2)}(x,p,t) = \frac{1}{2\pi\hbar} \int dy e^{-iyp/\hbar} \psi_2^* \left(x - \frac{y}{2}, t\right) \psi_1 \left(x + \frac{y}{2}, t\right) \tag{5.55}$$

where $\psi_{1,2}$ are some complex square integrable functions. If, moreover, f_w is a real function, then it is a pure state Wigner function. On the other hand, if f_w is a pure state, or an off-diagonal pure state Wigner function, then it satisfies the above differential equation.

The proof presented in [5] is short and elegant: Equation (5.54) can be viewed as a wave equation with 'time' variable x', spatial variable x, and velocity $1/2$. The general solution, known as the one-dimensional case of d'Alembert's solution, is given by two arbitrary functions which are shifted in time to the left and right with the velocity used to define the equation. Thus:

$$lnZ(x,x',t) = ln\psi_2^*\left(x - \frac{\hbar x'}{2},t\right) + ln\psi_1\left(x + \frac{\hbar x'}{2},t\right)$$

where $ln\psi_{1,2}$ are two arbitrary functions. Then the evaluation of (5.55) is straightforward. Moreover $\psi_{1,2}$ are square integrable as f_w is square integrable. Besides, if f_w is real, then ψ_1 is proportional to ψ_2^*. The normalization of ψ follows from the normalization of f_w which is a pure state. The converse result is shown by direct calculations.

Equation (5.54) provides the pure state quantum condition. Physical states are presented by its real and normalized solutions, namely the pure state Wigner functions. The non-real off-diagonal solutions are relevant for the treatment of the mixed states. An important result follows [4]: Let us assume that f_w is a solution of (5.22) and satisfies the quantum condition at the initial time. Then f_w is a solution of (5.54) for all times. Namely, the Wigner evolution preserves the pure (possible off-diagonal) quantum condition. In contrast, it can be shown that this is not true if the evolution is provided by the classical limit (5.29). Moreover, as originally shown by Tatarskii, the quantum character of the evolution is not ensured solely by the Wigner equation: the initial condition must also be an admissible quantum state. In this way the pure state condition implicitly implies the Heisenberg uncertainty relation.

The wave functions can be explicitly constructed from the knowledge of $f_{w(1,2)}$. Namely, if f_w satisfies the conditions around (5.54), it takes the form (5.51). Then with the help of (5.53) it holds:

$$\psi_1(x) = N_1 Z\left(\frac{x}{2},x\right) \quad \psi_2(x) = N_2 Z^*\left(\frac{x}{2},-x\right) \quad N_1 = \psi_2^*(0)^{-1} \quad N_2 = \psi_1(0)^{-1}$$

A shift of the arguments of Z is assumed if one of the wave functions becomes zero at zero. These expressions are valid for stationary wave functions: in the time-dependent case they introduce an arbitrary time-dependent phase. For this case an alternative formula is suggested in [5].

The following result is important: Let us assume β to be such that $\hat{A}$, defined by (5.15), is a generic linear operator. Hence $A(x,p)$, defined in (5.16) satisfies the following equations:

$$A(x,p) * f_{w(m,n)}(x,p) = a_n f_{w(m,n)}(x,p) \qquad f_{w(m,n)} * A(x,p) = a_m f_{w(m,n)}(x,p) \tag{5.56}$$

Then $f_{w(m,n)}$ is a (off-diagonal) pure state, where the associated functions ϕ_n and ϕ_m satisfy the eigenvalue equations:

$$\hat{A}\phi_n(x) = a_n\phi_n(x) \qquad \hat{A}^*\phi_m(x) = a_m^*\phi_m(x)$$

The result holds in particular for the energy eigenvalue problem.

The above considerations make it possible to establish a one to one correspondence between the space of all real pure state functions $f_w(x,p)$ defined in the phase space – the functions satisfying the conditions around (5.54) and the Hilbert space of the physical states $\psi(x)$:

$$\psi \to f_w: \quad f_w(x,p) = \frac{1}{2\pi\hbar}\int dy e^{-iyp/\hbar}\psi^*\left(x-\frac{y}{2}\right)\psi\left(x+\frac{y}{2}\right)$$

$$f_w \to \psi: \quad \psi(x) = N\int dp e^{ipx/\hbar} f_w\left(\frac{x}{2},p\right)$$

where N is defined as a normalization phase factor constant.

Similar necessary and sufficient conditions are formulated for mixed phase space quantum states [5].

These considerations illustrate how the Wigner quantum mechanics can be introduced in an independent way, and used as a formalism to re-derive the standard operator quantum mechanics.

2.11 Summary

We summarize the basic notions used in the Wigner representation of quantum mechanics by taking into account the three dimensional nature of the space. The momentum variable will be replaced by the wave vector $\mathbf{k}$, as the latter is usually preferred for modeling of carrier transport in semiconductors and devices. This allows to skip $\hbar$ in the definitions (5.20):

$$f_w(\mathbf{r},\mathbf{k},t) = \frac{1}{(2\pi)^3}\int d\mathbf{r}'\rho(\mathbf{r}+\mathbf{r}'/2,\mathbf{r}-\mathbf{r}'/2,t)e^{-i\mathbf{r}'.\mathbf{k}}, \tag{5.57}$$

and to restate the Wigner equation and the Wigner potential as follows:

$$\frac{\partial f_w(\mathbf{r},\mathbf{k},t)}{\partial t} + \frac{\hbar\mathbf{k}}{m}\cdot\nabla_{\mathbf{r}} f_w(\mathbf{r},\mathbf{k},t) = \int dk' V_w(\mathbf{r},\mathbf{k}-\mathbf{k}')f_w(\mathbf{r},\mathbf{k}',t) \tag{5.58}$$

$$V_w(\mathbf{r},\mathbf{k}) = \frac{1}{i\hbar(2\pi)^3}\int d\mathbf{r}' e^{-i\mathbf{r}'\mathbf{k}}(V(\mathbf{r}+\mathbf{r}'/2) - V(\mathbf{r}-\mathbf{r}'/2)) \tag{5.59}$$

If one is interested in the properties of the system along a desired direction, in the general case the relevant Wigner function becomes (5.57), integrated over the obsolete variables. It is a special case when the task is separable into transversal and longitudinal modes: $\rho = \rho_x \rho_\perp$. Then (5.57) can be reduced to the single-dimensional definition after an integration over the transversal variables. It is also possible to consider a Wigner function of the type $f_x(x,k_x,\mathbf{k}_\perp)$ where the longitudinal variables come from the single-dimensional definition, imposed e.g. by the fact that the potential depends only on x, while the transversal variables are introduced by other parts of the Hamiltonian accounting e.g. for phonons.

2.12 The Bound-States Problem

If the state $\psi_n(\mathbf{r},t) = \psi_n(\mathbf{r},0)\exp(-E_n t/\hbar)$ of the physical system is a given energy eigenstate, the density matrix is time-independent, $\rho_{nn}(\mathbf{r}_1,\mathbf{r}_2,t) = \psi_n^*(\mathbf{r}_1,0)$ $\psi_n(\mathbf{r}_2,0)$. In this case the system Hamiltonian and the density operator commute, and the von Neumann equation (5.18) reduces to

$$i\hbar\frac{\partial\hat{\rho}}{\partial t} = [\hat{H},\hat{\rho}]_- = 0. \tag{5.60}$$

This equation does not contain the system Hamiltonian any longer, and cannot determine the bound-state density matrix, since any given bound-state density matrix, being time-independent, will satisfy this equation. Similar arguments hold for the Wigner equation, linked to (5.60) by the Weyl transform. As it has been shown in [14], bound states cannot be obtained from the ballistic Wigner equation (5.58).

The harmonic oscillator is an example clearly demonstrating this problem. If the potential is a quadratic function of position, $V(\mathbf{r}) = m^*\omega^2|\mathbf{r}|^2/2$, the Wigner equation (5.58) reduces to the collisionless Boltzmann equation, the three dimensional version of (5.30), with $F(\mathbf{r}) = -m^*\omega^2\mathbf{r}$ being the classical force. The equation propagates an initial distribution classically. This demonstrates that, in the spirit of Sect. 2.10, the single equation (5.58) is not completely equivalent to the Schrödinger equation. Two alternative solutions of this problem can be pursued.

The solutions of the Wigner equation have to be subjected to a necessary and sufficient condition which selects an allowed class of Wigner distributions describing quantum-mechanical pure states. The condition preceding (5.54) originally formulated [4] in terms of the density matrix is:

$$\nabla_{\mathbf{r}_1}\nabla_{\mathbf{r}_2}\ln\rho(\mathbf{r}_1,\mathbf{r}_2) = 0 \tag{5.61}$$

$$\rho(\mathbf{r}_1,\mathbf{r}_2) = \int f_w\left(\mathbf{k},\frac{\mathbf{r}_1+\mathbf{r}_2}{2}\right)e^{i\mathbf{k}\cdot(\mathbf{r}_1-\mathbf{r}_2)}\frac{d\mathbf{k}}{(2\pi)^3} \tag{5.62}$$

This restriction holds also for the initial condition, responsible for the correct physical foundation of the computational task. Thus bound states enter externally, via the initial establishment of the task. The system Hamiltonian does not provide further information via the Wigner equation: the only property of the latter is that a bound state remains unaffected during the evolution. For example, in the case of the harmonic oscillator, the quantization condition for the energy does not follow from the Wigner equation, but from a supplementary condition.

The alternative way is to incorporate bound states as a part of the computational task. Carruthers and Zachariasen [14] start from the Schrödinger equation and derive an adjoint Wigner equation. If this adjoint equation is considered in addition, the usual Schrödinger eigenvalue problem can be reconstructed from the two Wigner equations. The adjoint equation is obtained with the help of the anticommutator [14],

$$[\hat{H},\hat{\rho}]_{+} = \hat{H}\hat{\rho} + \hat{\rho}\hat{H} = 2E\hat{\rho},$$

and takes a form, consistent with (5.52) and (5.56):

$$\frac{\hbar^2}{2m^*}\left(|\mathbf{k}|^2 - \frac{1}{4}\nabla_{\mathbf{r}}^2\right) f_{w(m,n)}(\mathbf{k},\mathbf{r}) - \int \widetilde{V}_{\mathrm{w}}(\mathbf{k}-\mathbf{k}',\mathbf{r})\, f_{w(m,n)}(\mathbf{k}',\mathbf{r})\, d\mathbf{k}'$$
$$= \frac{E_m + E_n}{2} f_{w(m,n)}(\mathbf{k},\mathbf{r}) \tag{5.63}$$
$$\widetilde{V}_{\mathrm{w}}(\mathbf{q},\mathbf{r}) = \frac{1}{2i\hbar}\int \left\{V\left(\mathbf{r}+\frac{\mathbf{s}}{2}\right) + V\left(\mathbf{r}-\frac{\mathbf{s}}{2}\right)\right\} \mathrm{e}^{-i\mathbf{q}\cdot\mathbf{s}} \frac{d^3s}{(2\pi)^3}$$

For $m = n$ one obtains the bound-state Wigner functions, which are real valued. The case $m \neq n$ gives the off-diagonal functions (5.51). The entire set of $f_{w(m,n)}(\mathbf{k},\mathbf{r})$ form a complete orthonormal basis.

The fact that the Wigner equation alone cannot provide the bound-states of a closed system has some implications for the numerical solution methods. Consider a system in which quasi-bound states of long life time exist. In this case the energy levels have very little broadening, which indicates that the system is almost closed. Such a system would be a double barrier structure realized by a semiconductor heterostructure. The spacing between resonance energies is typically in the 10^{-2} eV range. For thick barriers the broadening of the resonances can be in the 10^{-9} eV range. To resolve such a resonance a highly non-uniform energy grid with extremely small spacing around the resonance peaks would be needed. The discrete Fourier transform utilized by a numerical Wigner equation solver, on the other hand, permits only equi-distant grids in momentum space. With such a grid the extremely narrow resonances cannot be resolved in practice, and the discrete Wigner equation would become ill-conditioned. From this discussion one can conclude that a numerical Wigner function approach is applicable only to sufficiently open systems, i.e., to systems with not too narrow resonances.

The bound state problem is inherent to the coherent picture imposed by the ballistic Wigner transport. Bound states can be equally well treated in the more realistic picture which accounts for de-coherence processes of interaction with the environment, introduced in the next section.

3 Wigner–Boltzmann Equation

3.1 Introduction

The Wigner function approach allows to handle open-boundary systems, (carrier exchange with the environment is actually the basic characteristic of an operating electronic device), under stationary, small signal, or transient conditions, in a natural way [15]. Early works investigate the theoretical and numerical properties of the coherent Wigner equation, appropriate for ballistic transport [4, 12, 16]. At that time it has been recognized that dissipative processes are not only a part of the world of device physics, but that neglecting the interplay between coherent and de-coherence phenomena may lead to unphysical behavior of the modeled system [17]. The reason for such behavior are quasi-bound, or 'notch' states which may be charged properly by the boundary conditions only via a dissipation mechanism.

Dissipative interactions have been approached by means of phenomenological models based on the relaxation time approximation, [15, 18, 19] and also by introducing an actual Boltzmann-like collision operator [17, 20]. The phonon collision operator acting upon the Wigner distribution has been initially suggested as an a priory assumption that 'is an adequate approximation at some level' [17]. Can the classical Boltzmann scattering operator and the quantum Wigner-potential operator reside in a common equation? The answer is not trivial: derivations from first principles and analysis of the assumptions and approximations have been provided only recently for interactions with ionized impurities [21] and with phonons [22]. Moreover the two approaches are very different.

Consider for instance the short-range Coulomb potential created by an ionized impurity $e^2 \exp(-\beta\,|\mathbf{r}-\mathbf{r}_i|)/4\pi\varepsilon\,|\mathbf{r}-\mathbf{r}_i|$, where ε is the semiconductor permittivity and β is the screening factor in the static screening approximation. The demonstration starts with the derivation of the Wigner potential associated with this Coulomb potential, from which a quantum evolution term is derived. After some tedious but straightforward calculations, considering a large number of dopants and within the fast collision approximation, the electron–impurity collision term finally takes exactly the same form as commonly derived for the Boltzmann collision operator with continuous doping density [21].

The semiclassical phonon collision is derived from the equation for the generalized Wigner function [23, 24]. Along with the electron coordinates, the function depends on the occupation number of the phonon states in the system. Of interest is the electron, or reduced, Wigner function obtained from the generalized Wigner function by a trace over the phonon coordinates. A closed equation for the reduced Wigner function can be derived after a hierarchy of approximations, which includes the weak scattering limit and assumes that the phonon system is in equilibrium [22]. They concern the interaction with the phonons, while the potential operator remains exact. The phonon interaction in the resulting equation, being nonlocal in both space and time is yet quantum. The Wigner–Boltzmann equation is obtained after a classical limit in the phonon term, leading to the instantaneous, local in position Boltzmann collision operator.

The effects neglected by this limit can be studied from the homogeneous form of the equation for the reduced Wigner function. In this case the latter reduces to the Levinson equation [25], or equivalently to the Barker–Ferry equation, [26] with infinite electron lifetime. It should be noted that the inclusion of a finite lifetime requires a refined set of approximations in the generalized Wigner equation [27].

Effects of time dependent collisional broadening (CB) and retardation of phonon replicas have been investigated theoretically and experimentally in homogeneous semiconductors [28–32]. These effects are related to the lack of energy conservation and the memory character of the electron–phonon dynamics, and are due to the finite duration of the interaction process. The effect of the action of the electric field during the process of collision – the intra-collisional field effect (ICFE) – has attracted the scientific attention for quite some time [33–35]. Numerical studies demonstrate the CB, CR and ICFE effects in the case of ultrafast and/or high field transport in semiconductors and insulators [24,36–40] and in the case of photo-excited semiconductors [31, 32]. The solutions of the Levinson equation show the establishment of the classical, energy conserving delta function for long times. Semiclassically forbidden states are occupied at early evolution times [22, 32]. The first experimental evidence of memory effects and energy non-conserving transitions in the relaxation of hot carrier distributions have been reported a decade ago [29]. At higher times, which are above few hundred femtoseconds for GaAs, the Boltzmann limit dominates in the carrier evolution. A theoretical analysis [41] supports this result: the classical limit and the first order correction of the equation have been derived by using a small parameter. The latter requires that the product of the time scale and the phonon frequency scale to become much larger than unity, which gives rise to coarse graining in time. Thus, for long evolution times, the quantum effects in the electron–phonon interaction can be neglected. Consequently, the intra-collisional field effect is not important in stationary high field transport in semiconductors [38]. Rather, the effect must be sought in the time domain of the early time evolution, which precedes the formation of the classical energy conserving δ-function [39, 42]. We note that the above considerations hold in the weak collision limit, where the next interaction begins well after the completion of the current one.

The above considerations show that the inclusion of the Boltzmann collision operator in the Wigner equation requires that the dwell time of the carriers inside the device, and hence the device itself, must be sufficiently large. On the contrary, the application of the Wigner potential operator is reasonable for small device domains, where the potential changes over a region comparable with the coherence length of the electron. These requirements are not contradictory, since common devices are composed by an active quantum domain attached to large contact regions.

3.2 Electron–Phonon Interaction

We consider the dynamics of a single electron, subject to the action of the electric potential and interacting with the lattice vibrations. The description of the system is provided by both electron and phonon coordinates. The Wigner function and the Wigner equation for such a coupled electron–phonon system are defined as follows. The Hamiltonian of the system is given by

$$\begin{aligned} H &= H_0 + V + H_p + H_{e-p} \\ &= -\frac{\hbar^2}{2m}\nabla_{\mathbf{r}} + V(\mathbf{r}) + \sum_{\mathbf{q}} b^\dagger_{\mathbf{q}} b_{\mathbf{q}} \hbar\omega_{\mathbf{q}} + i\hbar \sum_{\mathbf{q}} C(\mathbf{q})(b_{\mathbf{q}} e^{i\mathbf{q}\mathbf{r}} - b^\dagger_{\mathbf{q}} e^{-i\mathbf{q}\mathbf{r}}) \end{aligned} \tag{5.64}$$

where the free electron part is H_0, the structure potential is $V(\mathbf{r})$, the free-phonon Hamiltonian is given by H_p and the electron–phonon interaction is H_{e-p}. In the above expressions $b^\dagger_{\mathbf{q}}$ and $b_{\mathbf{q}}$ are the creation and annihilation operators for the phonon mode $\mathbf{q}$, $\omega_{\mathbf{q}}$ is the energy of that mode and $\mathscr{C} = i\hbar C(\mathbf{q})$ is the electron–phonon coupling element, which depends on the type of phonon scattering analyzed. The state of the phonon subsystem is presented by the set $\{n_{\mathbf{q}}\}$ where $n_{\mathbf{q}}$ is the occupation number of the phonons in mode $\mathbf{q}$. Then the representation is given by the vectors $|\{n_{\mathbf{q}}\},\mathbf{r}\rangle = |\{n_{\mathbf{q}}\}\rangle|\mathbf{r}\rangle$. The generalized Wigner function [23] is defined by:

$$f_w(\mathbf{r},\mathbf{k},\{n_{\mathbf{q}}\},\{n_{\mathbf{q}}\}',t) = \frac{1}{(2\pi)^3}\int d\mathbf{r}' e^{-i\mathbf{k}\mathbf{r}'} \langle \mathbf{r}+\mathbf{r}'/2,\{n_{\mathbf{q}}\}|\hat{\rho}_t|\{n_{\mathbf{q}}\}',\mathbf{r}-\mathbf{r}'/2\rangle$$

The equation of motion of f_w is derived [43] with the help of (5.18):

$$\frac{\partial f_w(\mathbf{r},\mathbf{k},\{n_{\mathbf{q}}\},\{n_{\mathbf{q}}\}',t)}{\partial t} = \frac{1}{i\hbar}\int d\mathbf{r}' e^{-i\mathbf{k}\mathbf{r}'} \langle \mathbf{r}+\mathbf{r}'/2,\{n_{\mathbf{q}}\}|\,[H,\hat{\rho}_t]_-\,|\{n_{\mathbf{q}}\}',\mathbf{r}-\mathbf{r}'/2\rangle$$

The right hand side of this equation is shortly denoted by $WT(H)$. In the following we evaluate $WT(H)$ for each term of the Hamiltonian (5.64). $WT(H_0+V(\mathbf{r}))$ can be readily evaluated by using the steps applied after (5.18). The free phonon term is evaluated as:

$$WT(H_p) = \frac{1}{i\hbar}\left(\varepsilon(\{n_{\mathbf{q}}\}) - \varepsilon(\{n'_{\mathbf{q}}\}\right) f_w(\mathbf{r},\mathbf{k},\{n_{\mathbf{q}}\},\{n_{\mathbf{q}}\}',t)$$

where $\varepsilon(\{n_{\mathbf{q}}\}) = \sum_{\mathbf{q}} n_{\mathbf{q}}\hbar\omega_{\mathbf{q}}$. The transform $WT(H_{e-p})$ gives rise to four terms. By inserting $\int d\mathbf{r}''|\mathbf{r}''><\mathbf{r}''|$ in the first one it is obtained:

$$\begin{aligned} &\int d\mathbf{r}' \int d\mathbf{r}'' e^{-i\mathbf{k}\mathbf{r}'} \left\langle \mathbf{r}+\frac{\mathbf{r}'}{2},\{n_{\mathbf{q}}\}|b_{\mathbf{q}'}e^{i\mathbf{q}'\mathbf{r}''}|\mathbf{r}''\right\rangle \left\langle \mathbf{r}''|\hat{\rho}_t|\{n'_{\mathbf{q}}\},\mathbf{r}-\mathbf{r}'/2\right\rangle \\ &= \sqrt{n_{\mathbf{q}'}+1}\int d\mathbf{r}' e^{-i\mathbf{k}\mathbf{r}'} e^{i\mathbf{q}'(\mathbf{r}+\mathbf{r}'/2)} \left\langle \mathbf{r}+\frac{\mathbf{r}'}{2},\{n_1,\ldots,n_{\mathbf{q}'}+1,\ldots\}|\hat{\rho}_t|\{n'_{\mathbf{q}}\},\mathbf{r}-\frac{\mathbf{r}'}{2}\right\rangle \\ &= \sqrt{n_{\mathbf{q}'}+1}e^{i\mathbf{q}'\mathbf{r}} f_w\left(\mathbf{r},\mathbf{k}-\frac{\mathbf{q}'}{2},\{n_1,\ldots,n_{\mathbf{q}'}+1,\ldots\},\{n'_{\mathbf{q}}\},t\right) \end{aligned}$$

where the ortho-normality relation $\langle\mathbf{r}|\mathbf{r}'\rangle = \delta(\mathbf{r}-\mathbf{r}')$ has been used along with the fact that $b_{\mathbf{q}}$ becomes a creation operator when operating to the left. The remaining terms are evaluated in a similar way.

We are now ready to formulate the generalized Wigner equation:

$$
\begin{aligned}
&\left(\frac{\partial}{\partial t}+\frac{\hbar\mathbf{k}}{m}\cdot\nabla_{\mathbf{r}}\right) f_w(\mathbf{r},\mathbf{k},\{n_{\mathbf{q}}\},\{n'_{\mathbf{q}}\},t)\\
&=\frac{1}{i\hbar}\left(\varepsilon(\{n_{\mathbf{q}}\})-\varepsilon(\{n'_{\mathbf{q}}\})\right) f_w(\mathbf{r},\mathbf{k},\{n_{\mathbf{q}}\},\{n'_{\mathbf{q}}\},t)\\
&+\int d\mathbf{k}' V_w(\mathbf{r},\mathbf{k}-\mathbf{k}')f_w(\mathbf{r},\mathbf{k}',\{n_{\mathbf{q}}\},\{n'_{\mathbf{q}}\},t)+\sum_{\mathbf{q}'}C(\mathbf{q}')\\
&\times\left\{ e^{i\mathbf{q}'\mathbf{r}}\sqrt{n_{\mathbf{q}'}+1}f_w\left(\mathbf{r},\mathbf{k}-\frac{\mathbf{q}'}{2},\{n_{\mathbf{q}}\}^+_{\mathbf{q}'},\{n'_{\mathbf{q}}\},t\right)\right.\\
&-e^{-i\mathbf{q}'\mathbf{r}}\sqrt{n_{\mathbf{q}'}}f_w\left(\mathbf{r},\mathbf{k}+\frac{\mathbf{q}'}{2},\{n_{\mathbf{q}}\}^-_{\mathbf{q}'},\{n'_{\mathbf{q}}\},t\right)\\
&-e^{i\mathbf{q}'\mathbf{r}}\sqrt{n'_{\mathbf{q}'}}f_w\left(\mathbf{r},\mathbf{k}+\frac{\mathbf{q}'}{2},\{n_{\mathbf{q}}\},\{n'_{\mathbf{q}}\}^-_{\mathbf{q}'},t\right)\\
&\left.+e^{-i\mathbf{q}'\mathbf{r}}\sqrt{n'_{\mathbf{q}'}+1}f_w\left(\mathbf{r},\mathbf{k}-\frac{\mathbf{q}'}{2},\{n_{\mathbf{q}'}\},\{n'_{\mathbf{q}}\}^+_{\mathbf{q}'},t\right)\right\}
\end{aligned}
\tag{5.65}
$$

where we denoted by $\{n_{\mathbf{q}}\}^+_{\mathbf{q}'}$ ($\{n_{\mathbf{q}}\}^-_{\mathbf{q}'}$) the states of the phonon subsystem, obtained from $\{n_{\mathbf{q}}\}$ by increasing (decreasing) the number of phonons in the mode $\mathbf{q}'$ by unity. Furthermore we observe that the last two terms in the curly brackets can be obtained from the first ones by the following rule: (a): the argument of the exponent changes its sign; (b): the phonon number in the mode determined by the summation index ($\mathbf{q}'$) is changed in the right state instead in the left state; (c): in the square roots $n_{\mathbf{q}'}$ is replaced by $n'_{\mathbf{q}'}$. In what follows we denote the last two terms by *i.c.*.

The generalized Wigner equation couples an element $f_w(\dots,\{n\},\{m\},t)$ to four neighborhood elements for any phonon mode $\mathbf{q}$. For any such mode $n_{\mathbf{q}}$ can be any integer between 0 and infinity and the sum over $\mathbf{q}$ couples all modes.

In accordance with Sect. 2.9 and (5.50) of interest is the reduced Wigner function, which is obtained from the generalized Wigner function by taking the trace over the phonon states. An exact equation for the reduced Wigner function can not be obtained since the trace operation does not commute with the electron–phonon interaction Hamiltonian. In what follows we derive a model, which approximates the generalized Wigner equation, but is closed with respect to the reduced Wigner function. The model is general enough to account for the quantum character of the interaction with the phonons. The electron-device potential part of the transport is treated on a rigorous quantum level. A classical limit in the electron–phonon operators gives rise to the Wigner–Boltzmann equation. The derivation introduces a consistent hierarchy of assumptions and simplifications.

3.2.1 Weak Coupling

We begin with the assumptions which simplify (5.65) towards a model equation set for the electron Wigner function. Of interest are the diagonal elements of the generalized WF. The evolution of an initial state of the system defined at time $t = 0$ is considered. The state is assumed diagonal with respect to the phonon coordinates, which corresponds to the evolution process of an initially decoupled electron–phonon system.

$$\left(\frac{\partial}{\partial t}+\frac{\hbar\mathbf{k}}{m}\cdot\nabla_{\mathbf{r}}\right) f_w(\mathbf{r},\mathbf{k},\{n_{\mathbf{q}}\},\{n_{\mathbf{q}}\},t)=\int d\mathbf{k}' V_w(\mathbf{r},\mathbf{k}-\mathbf{k}') f_w(\mathbf{r},\mathbf{k}',\{n_{\mathbf{q}}\},\{n_{\mathbf{q}}\},t)$$
$$+\sum_{\mathbf{q}'} C(\mathbf{q}')\left\{ e^{i\mathbf{q}'\mathbf{r}}\sqrt{n_{\mathbf{q}'}+1} f_w\left(\mathbf{r},\mathbf{k}-\frac{\mathbf{q}'}{2},\{n_{\mathbf{q}}\}^{+}_{\mathbf{q}'},\{n_{\mathbf{q}}\},t\right)\right.$$
$$\left. -e^{-i\mathbf{q}'\mathbf{r}}\sqrt{n_{\mathbf{q}'}} f_w\left(\mathbf{r},\mathbf{k}+\frac{\mathbf{q}'}{2},\{n_{\mathbf{q}}\}^{-}_{\mathbf{q}'},\{n_{\mathbf{q}}\},t\right)+i.c.\right\} \tag{5.66}$$

A diagonal element is linked to so called first-off-diagonal elements, which are diagonal in all modes but the current mode $\mathbf{q}'$ of the summation. In this mode the four neighbors of $n_{\mathbf{q}'}, n_{\mathbf{q}'}$ namely $n_{\mathbf{q}'} \pm 1, n_{\mathbf{q}'}$ and $n_{\mathbf{q}'}, n_{\mathbf{q}'} \pm 1$ are concerned. This is schematically presented on Fig. 5.1.

The auxiliary equation for the first-off-diagonal element in (5.66) is obtained by the help of (5.65):

$$\left(\frac{\partial}{\partial t}+\frac{\hbar(\mathbf{k}-\frac{\mathbf{q}'}{2})}{m}\cdot\nabla_{\mathbf{r}}\right) f_w\left(\mathbf{r},\mathbf{k}-\frac{\mathbf{q}'}{2},\{n_{\mathbf{q}}\}^{+}_{\mathbf{q}'},\{n_{\mathbf{q}}\},t\right)$$
$$= -i\omega_{\mathbf{q}'} f_w\left(\mathbf{r},\mathbf{k}-\frac{\mathbf{q}'}{2},\{n_{\mathbf{q}}\}^{+}_{\mathbf{q}'},\{n_{\mathbf{q}}\},t\right)$$

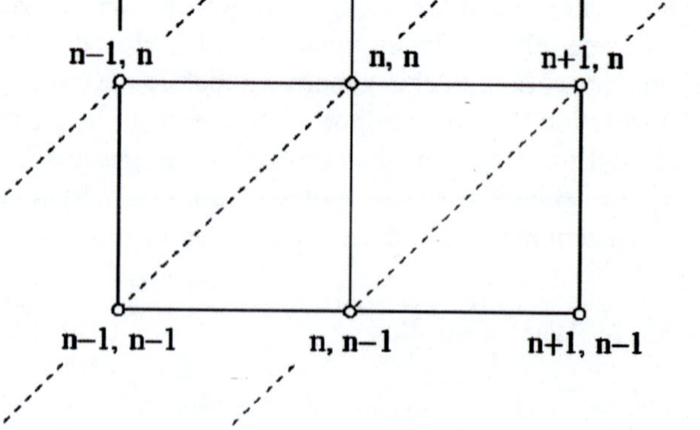

Fig. 5.1 Diagonal and first-off-diagonal elements

$$+\int d\mathbf{k}' V_w(\mathbf{r},\mathbf{k}-\mathbf{k}') f_w\left(\mathbf{r},\mathbf{k}-\frac{\mathbf{q}'}{2},\{n_\mathbf{q}\}^+_{\mathbf{q}'},\{n_\mathbf{q}\},t\right) + \sum_{\mathbf{q}''} C(\mathbf{q}'')$$

$$\times\left\{ e^{i\mathbf{q}''\mathbf{r}}\sqrt{n_{\mathbf{q}''}+1} f_w\left(\mathbf{r},\mathbf{k}-\frac{\mathbf{q}'}{2}-\frac{\mathbf{q}''}{2},\{\{n_\mathbf{q}\}^+_{\mathbf{q}'}\}^+_{\mathbf{q}''},\{n_\mathbf{q}\},t\right)\right.$$

$$-e^{-i\mathbf{q}''\mathbf{r}}\sqrt{n_{\mathbf{q}''}} f_w\left(\mathbf{r},\mathbf{k}-\frac{\mathbf{q}'}{2}+\frac{\mathbf{q}''}{2},\{\{n_\mathbf{q}\}^+_{\mathbf{q}'}\}^-_{\mathbf{q}''},\{n_\mathbf{q}\},t\right)$$

$$-e^{i\mathbf{q}''\mathbf{r}}\sqrt{n_{\mathbf{q}''}} f_w\left(\mathbf{r},\mathbf{k}-\frac{\mathbf{q}'}{2}+\frac{\mathbf{q}''}{2},\{n_\mathbf{q}\}^+_{\mathbf{q}'},\{n_\mathbf{q}\}^-_{\mathbf{q}''},t\right)$$

$$\left.+e^{-i\mathbf{q}''\mathbf{r}}\sqrt{n_{\mathbf{q}''}+1} f_w\left(\mathbf{r},\mathbf{k}-\frac{\mathbf{q}'}{2}-\frac{\mathbf{q}''}{2},\{n_\mathbf{q}\}^+_{\mathbf{q}'},\{n_\mathbf{q}\}^+_{\mathbf{q}''},t\right)\right\} \quad (5.67)$$

Accordingly, the first-off-diagonal elements are linked to elements which in general are placed further away from the diagonal ones by increasing or decreasing the phonon number in a second mode, $\mathbf{q}''$, by unity. These are the second-off-diagonal elements. The only exception is provided by two contributions which recover diagonal elements. They are obtained when the running index $\mathbf{q}''$ coincides with $\mathbf{q}'$ due to: (a): $(\{\{n_\mathbf{q}\}^+_{\mathbf{q}'}\}^-_{\mathbf{q}''},\{n_\mathbf{q}\})$ in the term in the fifth row of (5.67). We note that in this case $n''_\mathbf{q} = n'_\mathbf{q} + 1$ in the square root in front of f_w. (b): $(\{n_\mathbf{q}\}^+_{\mathbf{q}'},\{n_\mathbf{q}\}^+_{\mathbf{q}''})$ in the last row of (5.67).

Next we observe that each link of two elements corresponds to a multiplication by the factor C. Thus the next assumption is that C is a small quantity. While the first-off-diagonal elements give contributions to (5.66) by order of C^2, the second-off-diagonal elements give rise to higher order contributions and are neglected. The physical meaning of the assumption is that the interaction with a phonon in mode $\mathbf{q}'$ which begins from a diagonal element completes at a diagonal element by another interaction with the phonon in the same mode, without any interference with phonons of other modes. The assumption allows to truncate the considered elements to those between the two lines parallel to the main diagonal on Fig. 5.1. As a next step we need to solve the truncated equation, which can be done explicitly after further approximations related to the Wigner potential. For this it is sufficient to consider the classical force according (5.30). Such a model is able to account for correlations between electric field and scattering – the ICFE. As we aim at derivation of a Boltzmann type of collisions, we entirely neglect the Wigner potential term:

$$\left(\frac{\partial}{\partial t} + \frac{\hbar\left(\mathbf{k}-\frac{\mathbf{q}'}{2}\right)}{m}\cdot\nabla_\mathbf{r} + i\omega_{\mathbf{q}'}\right) f_w\left(\mathbf{r},\mathbf{k}-\frac{\mathbf{q}'}{2},\{n_\mathbf{q}\}^+_{\mathbf{q}'},\{n_\mathbf{q}\},t\right)$$

$$= C(\mathbf{q}')e^{-i\mathbf{q}'\mathbf{r}}\sqrt{n_{\mathbf{q}'}+1}\left(-f_w(\mathbf{r},\mathbf{k},\{n_\mathbf{q}\},\{n_\mathbf{q}\},t) + f_w\left(\mathbf{r},\mathbf{k}-\mathbf{q}',\{n_\mathbf{q}\}^+_{\mathbf{q}'},\{n_\mathbf{q}\}^+_{\mathbf{q}'},t\right)\right) \quad (5.68)$$

We consider the trajectory

$$\mathbf{k}(t') = \mathbf{k} - \frac{\mathbf{q}'}{2}; \qquad \mathbf{R}(t',\mathbf{q}') = \mathbf{r} - \int_{t'}^{t} d\tau \frac{\hbar\mathbf{k}(\tau)}{m} = \mathbf{r} - \frac{\hbar(\mathbf{k}-\mathbf{q}'/2)}{m}(t-t'); \quad (5.69)$$

initialized at time t by $\mathbf{k} - \frac{\mathbf{q}'}{2}$, $\mathbf{r}$ and the function

$$f_w(\mathbf{R}(t',\mathbf{q}'),\mathbf{k}(t'),\{n_\mathbf{q}\}^+_{\mathbf{q}'},\{n_\mathbf{q}\},t)e^{i\omega'_\mathbf{q}t'} \quad (5.70)$$

The total time derivative of this function, taken at time $t' = t$ gives the left hand side of (5.68). Then we consider a form of this equation, obtained by a parameterization by t' with the help of (5.69), and a multiplication by the exponent. A final integration in the time interval $0,t$ gives rise to:

$$f_w\left(\mathbf{r},\mathbf{k}-\frac{\mathbf{q}'}{2},\{n_\mathbf{q}\}^+_{\mathbf{q}'},\{n_\mathbf{q}\},t\right) = C(\mathbf{q}')\int_0^t dt' \mathrm{e}^{-i\omega_{\mathbf{q}'}(t-t')}\mathrm{e}^{-i\mathbf{q}'\mathbf{R}(t',\mathbf{q}')}\sqrt{n_{\mathbf{q}'}+1}$$

$$\left(f_w(\mathbf{R}(t',\mathbf{q}'),\mathbf{k}-\mathbf{q}',\{n_\mathbf{q}\}^+_{\mathbf{q}'},\{n_\mathbf{q}\}^+_{\mathbf{q}'},t') - f_w(\mathbf{R}(t',\mathbf{q}'),\mathbf{k},\{n_\mathbf{q}\},\{n_\mathbf{q}\},t')\right) \quad (5.71)$$

where we used the fact that the initial condition is zero due to the assumption for an initially decoupled system.

The corresponding equation for the second first-off-diagonal element is obtained in the same fashion:

$$f_w\left(\mathbf{r},\mathbf{k}+\frac{\mathbf{q}'}{2},\{n_\mathbf{q}\}^-_{\mathbf{q}'},\{n_\mathbf{q}\},t\right) = C(\mathbf{q}')\int_0^t dt' \mathrm{e}^{i\omega_{\mathbf{q}'}(t-t')}\mathrm{e}^{i\mathbf{q}'\mathbf{R}(t',-\mathbf{q}')}\sqrt{n_{\mathbf{q}'}}$$

$$\left(f_w(\mathbf{R}(t',-\mathbf{q}'),\mathbf{k},\{n_\mathbf{q}\},\{n_\mathbf{q}\},t') - f_w(\mathbf{R}(t',-\mathbf{q}'),\mathbf{k}+\mathbf{q}',\{n_\mathbf{q}\}^-_{\mathbf{q}'},\{n_\mathbf{q}\}^-_{\mathbf{q}'},t')\right)$$
$$(5.72)$$

The remaining two elements, which compose the *i.c.* term in (5.66) give rise to two integral equations which are complex conjugate to the first two. In this way the relevant information is provided by (5.66), (5.71) and (5.72), which can be unified as follows:

$$\left(\frac{\partial}{\partial t} + \frac{\hbar\mathbf{k}}{m}\cdot\nabla_\mathbf{r}\right)f_w(\mathbf{r},\mathbf{k},\{n_\mathbf{q}\},\{n_\mathbf{q}\},t) = \int \mathrm{d}\mathbf{k}' V_w(\mathbf{r},\mathbf{k}-\mathbf{k}')f_w(\mathbf{r},\mathbf{k}',\{n_\mathbf{q}\},\{n_\mathbf{q}\},t)$$

$$+2Re\sum_{\mathbf{q}'} C^2(\mathbf{q}')\int_0^t dt' \left\{(n_{\mathbf{q}'}+1)\mathrm{e}^{i\frac{\varepsilon(\mathbf{k})-\varepsilon(\mathbf{k}-\mathbf{q}')-\hbar\omega_{\mathbf{q}'}}{\hbar}(t-t')}\right.$$

$$\left(f_w(\mathbf{R}(t',\mathbf{q}'),\mathbf{k}-\mathbf{q}',\{n_\mathbf{q}\}^+_{\mathbf{q}'},\{n_\mathbf{q}\}^+_{\mathbf{q}'},t') - f_w(\mathbf{R}(t',\mathbf{q}'),\mathbf{k},\{n_\mathbf{q}\},\{n_\mathbf{q}\},t')\right)$$

$$-n_{\mathbf{q}'}\mathrm{e}^{i\frac{\varepsilon(\mathbf{k})-\varepsilon(\mathbf{k}+\mathbf{q}')+\hbar\omega_{\mathbf{q}'}}{\hbar}(t-t')}\left(f_w(\mathbf{R}(t',-\mathbf{q}'),\mathbf{k},\{n_\mathbf{q}\},\{n_\mathbf{q}\},t')\right.$$

$$\left.\left.-f_w(\mathbf{R}\ (t',-\mathbf{q}'),\mathbf{k}+\mathbf{q}',\{n_\mathbf{q}\}^-_{\mathbf{q}'},\{n_\mathbf{q}\}^-_{\mathbf{q}'},t')\right)\right\} \quad (5.73)$$

where we have used the equalities:

$$\pm\mathbf{q}'\mathbf{r} \mp \omega_{\mathbf{q}'}(t-t') \mp \mathbf{q}'\mathbf{R}(t',\pm\mathbf{q}') = \frac{\varepsilon(\mathbf{k}) - \varepsilon(\mathbf{k}\mp\mathbf{q}') \mp \hbar\omega_{\mathbf{q}'}}{\hbar}(t-t')$$

The model involves only diagonal elements, so that the double counting of the phonon coordinates becomes obsolete and one set of phonon numbers may be omitted.

3.2.2 Equilibrium Phonons

The obtained equation set (5.73) is still infinite with respect to the phonon coordinates, which are to be eliminated by the trace operation. The next assumption is that the phonon system is a thermostat for the electrons, i.e. the phonon distribution remains in equilibrium during the evolution:

$$P(n_{\mathbf{q}},t') = \int d\mathbf{r} \int d\mathbf{k} \sum_{\{n_{\mathbf{q}'}\}}{}' f_w(\mathbf{r},\mathbf{k},\{n_{\mathbf{q}'}\},\{n_{\mathbf{q}'}\},t') = P_{eq}(n_{\mathbf{q}}) = \frac{e^{-\hbar\omega_{\mathbf{q}}n_{\mathbf{q}}/kT}}{n(\mathbf{q})+1} \tag{5.74}$$

Here $P(n_{\mathbf{q}},t')$ is the probability for finding $n_{\mathbf{q}}$ phonons in mode $\mathbf{q}$ at time t', the $\sum'$ denotes summation over all phonon coordinates but the one in mode $\mathbf{q}$, and n($\mathbf{q}$) is the mean equilibrium phonon number (Bose distribution):

$$n(\mathbf{q}) = \sum_{n_{\mathbf{q}}=0}^{\infty} n_{\mathbf{q}} P_{eq}(n_{\mathbf{q}}) = \frac{1}{e^{\hbar\omega_{\mathbf{q}}/kT}-1}; \qquad \sum_{n_{\mathbf{q}}=0}^{\infty} P_{eq}(n_{\mathbf{q}}) = 1 \tag{5.75}$$

The condition (5.74) is equivalent to the assumption that at any time $0 \le t' \le t$ it holds

$$f_w(\mathbf{r},\mathbf{k},\{n_{\mathbf{q}}\},\{n_{\mathbf{q}}\},t') = f(\mathbf{r},\mathbf{k},t') \prod_{\mathbf{q}} P_{eq}(n_{\mathbf{q}}) \tag{5.76}$$

where $f(\mathbf{r},\mathbf{k},t')$ reduced or electron Wigner function. Accordingly, the four terms in the curly brackets of (5.73) become dependent on the phonon coordinates by the following factors:

$$(n_{\mathbf{q}'}+1)P_{eq}(n_{\mathbf{q}'}+1)\prod_{\mathbf{q}}{}' P_{eq}(n_{\mathbf{q}}); \qquad (n_{\mathbf{q}'}+1)P_{eq}(n_{\mathbf{q}'})\prod_{\mathbf{q}}{}' P_{eq}(n_{\mathbf{q}})$$

$$n_{\mathbf{q}'}P_{eq}(n_{\mathbf{q}'})\prod_{\mathbf{q}}{}' P_{eq}(n_{\mathbf{q}}); \qquad n_{\mathbf{q}'}P_{eq}(n_{\mathbf{q}'}-1)\prod_{\mathbf{q}}{}' P_{eq}(n_{\mathbf{q}}) \tag{5.77}$$

Now the trace operation, namely the sum over $n_{\mathbf{q}}$ for all modes $\mathbf{q}$, can be readily done with the help of (5.75) and the following equalities [44]:

$$n(\mathbf{q}) = \sum_{n_{\mathbf{q}}} (n_{\mathbf{q}}+1)P_{eq}(n_{\mathbf{q}}+1); \quad n(\mathbf{q})+1 = \sum_{n_{\mathbf{q}}} n_{\mathbf{q}} P_{eq}(n_{\mathbf{q}}-1);$$

The factors depending on the phonon coordinates are replaced by the following numbers:

$$n(\mathbf{q}'); \qquad (n(\mathbf{q}')+1); \tag{5.78}$$

$$n(\mathbf{q}'); \qquad (n(\mathbf{q}')+1) \tag{5.79}$$

This is an important step which allows to close the equation set for the electron Wigner function, transforming it into a single equation.

However, it is important to clarify what the physical side of the formal assumption (5.76) is. The peculiarities of the model (5.73) in conjunction with (5.76) can be conveniently analyzed from the integral form of the equation set, written for a homogeneous system, where the space dependence appears due to the initial condition only, which is of a decoupled electron–phonon system. The integral form is obtained within the following steps: $\mathbf{R}(t', \pm\mathbf{q}')$ is replaced from (5.69), introduced is another trajectory, initialized by $\mathbf{r}, \mathbf{k}, T$,

$$\mathbf{k}_T(t) = \mathbf{k}; \qquad \mathbf{R}_T(t) = \mathbf{r} - \int_t^T d\tau \frac{\hbar \mathbf{k}_T(\tau)}{m} = \mathbf{r} - \frac{\hbar \mathbf{k}}{m}(T-t); \tag{5.80}$$

where T now becomes the evolution time. $\mathbf{k}, \mathbf{r}$ are replaced on both sides of (5.73) by $\mathbf{k}_T(t), Rv_T(t)$ and the equation is integrated on t in the limits $0, T$. The initial condition in the form (5.76) appears explicitly:

$$\begin{aligned}
&f_w(\mathbf{r}, \mathbf{k}, \{n_\mathbf{q}\}, T) = f(\mathbf{r}, \mathbf{k}, 0) \prod_\mathbf{q} P_{eq}(n_\mathbf{q}) \\
&+2Re \sum_{\mathbf{q}'} C^2(\mathbf{q}') \int_0^T dt \int_0^t dt' \Bigg\{ (n_{\mathbf{q}'}+1) e^{i\frac{\varepsilon(\mathbf{k})-\varepsilon(\mathbf{k}-\mathbf{q}')-\hbar\omega_{\mathbf{q}'}}{\hbar}(t-t')} \\
&\Bigg(f_w(\mathbf{R}_T(t) - \frac{\hbar(\mathbf{k}-\mathbf{q}'/2)}{m}(t-t'), \mathbf{k}-\mathbf{q}', \{n_\mathbf{q}\}^+_{\mathbf{q}'}, t') \\
&-f_w(\mathbf{R}_T(t) - \frac{\hbar(\mathbf{k}-\mathbf{q}'/2)}{m}(t-t'), \mathbf{k}, \{n_\mathbf{q}\}, t') \Bigg) - n_{\mathbf{q}'} e^{i\frac{\varepsilon(\mathbf{k})-\varepsilon(\mathbf{k}+\mathbf{q}')+\hbar\omega_{\mathbf{q}'}}{\hbar}(t-t')} \\
&\Bigg(f_w(\mathbf{R}_T(t) - \frac{\hbar(\mathbf{k}+\mathbf{q}'/2)}{m}(t-t'), \mathbf{k}, \{n_\mathbf{q}\}, t') \\
&-f_w\ (\mathbf{R}_T(t) - \frac{\hbar(\mathbf{k}+\mathbf{q}'/2)}{m}(t-t'), \mathbf{k}+\mathbf{q}', \{n_\mathbf{q}\}^-_{\mathbf{q}'}, t') \Bigg) \Bigg\}
\end{aligned} \tag{5.81}$$

The phonon system can be at any state with certain set of numbers $\{n_\mathbf{q}\}$, however now the initial condition assigns a probability to this set. A replacement of the equation into itself presents the solution as consecutive iterations of the initial condition. We fix a set of numbers $\{n_\mathbf{q}\}$, corresponding to the phonon state of interest, and consider the first iteration of the term in the third row of (5.81). Until time t' the arguments of the initial condition are $\mathbf{r}_{t'} = \mathbf{R}_T(t) - \frac{\hbar(\mathbf{k}-\mathbf{q}'/2)}{m}(t-t'), \mathbf{k}_{t'} = \mathbf{k} - \mathbf{q}'$ which we may think as coordinates of a given particle, and the phonon system is in another state with an extra phonon in mode $\mathbf{q}'$ which contributes to the state of interest. At time t' the interaction begins by absorption of the half of the wave vector of a phonon in mode $\mathbf{q}'$, so that the particle appears with a wave vector $\mathbf{k}-\mathbf{q}'/2$, and moves along a trajectory determined by $\mathbf{r}_{t'} + \frac{\hbar(\mathbf{k}-\mathbf{q}'/2)}{m}(t-t')$ At time t the second half of the phonon is absorbed. The particle coordinates become $\mathbf{R}_T(t)$, $\mathbf{k}$ – exactly the right ones, which will bring it to $\mathbf{r},\mathbf{k}$ at time T. It contributes to the function on the left by the real part of the initial condition value at the starting point, multiplied by the pre factor in front of the considered term. The process corresponds to real absorption of a phonon: the phonons at the initial state are reduced by one. The next term describes a virtual process: the particle at $\mathbf{r}_{t'},\mathbf{k}$ first emits half of the wave vector of a phonon in mode $\mathbf{q}'$, but then, at time t it is absorbed back. Thus the initial phonon state does not change at the end of the interaction. The rest of the terms can be explained in the same way. We also note that the origin of the ICFE is the acceleration of the model particle along the trajectories.

The important message from this picture is the finite duration of the interaction process. We also expect the usual for a physical point of view existence of a mean interval with vanishing probabilities for large deviations from the mean. By recalling the fact that given interaction completes before another initiates, it follows that for a given evolution interval there is only a finite number of involved phonons. In accordance, the assumption for a thermostat means that the number of phonons is so huge that a given phonon mode can be involved only once in the interaction. A quantitative analysis can be found in [27].

3.2.3 Closed Model

The assumption for equilibrium allows to eliminate the phonon degrees of freedom from (5.73), which are now replaced by the numbers (5.79). The equation can be conveniently rewritten by relying on the symmetry of C, $n_{\mathbf{q}}$ and $\omega_{\mathbf{q}}$ with respect to the change the sign of the wave vector. We change the sign of $\mathbf{q}$ in the last two rows and introduce the variable $\mathbf{k}' = \mathbf{k} - \mathbf{q}'$.

$$\left(\frac{\partial}{\partial t}+\frac{\mathbf{k}}{m}\cdot\nabla_{\mathbf{r}}\right) f_w(\mathbf{r},\mathbf{k},t) = \int d\mathbf{k}' \,\{V_w(\mathbf{r},\mathbf{k}-\mathbf{k}') f_w(\mathbf{r},\mathbf{k}',t)$$
$$+\int_0^t dt' \left(S(\mathbf{k}',\mathbf{k},t,t') f_w(\mathbf{R}(t',\mathbf{q}'),\mathbf{k}',t') - S(\mathbf{k},\mathbf{k}',t,t') f_w(\mathbf{R}(t',\mathbf{q}'),\mathbf{k},t')\right)\} \quad (5.82)$$

$$S(\mathbf{k}',\mathbf{k},t,t') = \frac{2VC_{\mathbf{q}}^2}{(2\pi)^3}\left(n(\mathbf{q})cos(\Omega(\mathbf{k}',\mathbf{k},t,t')) + (n(\mathbf{q})+1)cos(\Omega(\mathbf{k},\mathbf{k}',t,t'))\right)$$

$$\Omega(\mathbf{k},\mathbf{k}',t,t') = \frac{\varepsilon(\mathbf{k}) - \varepsilon(\mathbf{k}') + \hbar\omega_{\mathbf{q}}}{\hbar}(t-t'); \quad \mathbf{q} = \mathbf{k} - \mathbf{k}'$$

The phonon interaction in this equation bears quantum character despite all simplifying assumptions. No approximations are introduced for the coherent part of the transport process: if the phonon interaction is neglected, the common Wigner equation for an electron in a potential field is recovered. The analysis of the physical processes involved is the same as for (5.81). The main peculiarity is the non-locality in the real space. The Boltzmann distribution function in point $\mathbf{r},\mathbf{k}$ at time t collects contributions only from the past of the real space part of the trajectory passing through this point. Since the finite duration of the phonon interaction, the solution of (5.83) can collect contributions from all points in the phase space and thus gives rise to a spatial non-locality. There is a lack of energy conservation even in the most simple homogeneous case, where the electric field is zero. The energy conserving delta function in the Boltzmann type of interaction is obtained after a limit which neglects the duration of the collision process.

3.2.4 Classical Limit: General Form of the Equation

We consider the classical limit of the electron–phonon interaction. The time integral in (5.83) is of the form:

$$\int_0^t \mathrm{d}\tau e^{\frac{i}{\hbar}\varepsilon\tau}\phi(\tau) \tag{5.83}$$

The following formal limit holds in terms of generalized functions:

$$\lim_{\hbar\to 0}\frac{1}{\hbar}\int_0^\infty \mathrm{d}\tau e^{\frac{i}{\hbar}\varepsilon\tau}\phi(\tau) = \phi(0)\left\{\pi\delta(\varepsilon)+i\mathscr{P}\frac{1}{\varepsilon}\right\} \tag{5.84}$$

The actual meaning of the process of encouraging a constant to approach zero is that the product of the energy and time scales becomes much larger than $\hbar$. The mathematical aspects of the derivation are considered in [41]. As applied to the right hand side of (5.83) the limit (5.84) leads to cancellation of all principal values $\mathscr{P}$. This is in accordance with the fact that (5.83) contains only real quantities. The energy and momentum conservation laws are incorporated in the obtained equation. We note that the time argument of the integrant is zero, which implies $t = t'$ and thus $\mathbf{R}(t',\mathbf{q}') = \mathbf{r}$.

The general form of the obtained Wigner–Boltzmann equation is

$$\left(\frac{\partial}{\partial t}+\frac{\mathbf{k}}{m}\cdot\nabla_{\mathbf{r}}\right)f_w(\mathbf{r},\mathbf{k},t) = \int \mathrm{d}\mathbf{k}' V_w(\mathbf{r},\mathbf{k}-\mathbf{k}')f_w(\mathbf{r},\mathbf{k}',t')$$
$$+\int \mathrm{d}\mathbf{k}'\left(f_w(\mathbf{r},\mathbf{k}',t)S(\mathbf{k}',\mathbf{k}) - f_w(\mathbf{r},\mathbf{k},t)S(\mathbf{k},\mathbf{k}')\right) \tag{5.85}$$

with the particular for the electron–phonon interaction scattering rate S:

$$S(\mathbf{k}',\mathbf{k}) = \frac{2\pi}{\hbar}\frac{V}{(2\pi)^3}\left\{|\mathscr{C}(\mathbf{q})|^2\delta(\varepsilon(\mathbf{k})-\varepsilon(\mathbf{k}')-\hbar\omega_{\mathbf{q}})n(\mathbf{q})\right.$$
$$\left.+|\mathscr{C}(\mathbf{q})|^2\delta(\varepsilon(\mathbf{k})-\varepsilon(\mathbf{k}')+\hbar\omega_{\mathbf{q}})(n(\mathbf{q})+1)\right\}$$

where $\mathbf{q}=\mathbf{k}-\mathbf{k}'$, and C has been replaced by the electron–phonon matrix element $\mathscr{C}$: $C^2 = |\mathscr{C}|^2/(\hbar)^2$.

The interaction with phonons is now treated classically while the interaction with the Wigner potential is considered on a rigorous quantum level. We conclude by noting that a classical limit in the potential term recovers the Boltzmann equation.

3.3 Electron–Impurity Interaction

Let us now see in more detail, as was already mentioned above, how the short-range scattering by ionized impurities may be included into the Wigner transport equation. For an assembly of dopant atoms j of position $\mathbf{r}_j$ the short-range interaction potential with electrons may be written in the form of a screened Coulomb potential

$$V_{e-ii} = \sum_j \frac{e^2 \exp\left(-\beta \left|\mathbf{r}-\mathbf{r}_j\right|\right)}{4\pi\varepsilon \left|\mathbf{r}-\mathbf{r}_j\right|} \tag{5.86}$$

where ε and β are the dielectric constant and the screening factor, respectively. The corresponding Wigner potential simply writes

$$\begin{aligned} V_w(\mathbf{r},\mathbf{k}) &= \frac{i}{\hbar(2\pi)^3}\frac{e^2}{4\pi\varepsilon}\sum_j \int d\mathbf{r}' e^{-i\mathbf{k}\mathbf{r}'}\left(\frac{e^{-\beta\left|\mathbf{r}-\frac{\mathbf{r}'}{2}-\mathbf{r}_j\right|}}{\left|\mathbf{r}-\frac{\mathbf{r}'}{2}-\mathbf{r}_j\right|} - \frac{e^{-\beta\left|\mathbf{r}+\frac{\mathbf{r}'}{2}-\mathbf{r}_j\right|}}{\left|\mathbf{r}+\frac{\mathbf{r}'}{2}-\mathbf{r}_j\right|}\right) \\ &= \frac{i}{\hbar(2\pi)^3}\frac{e^2}{4\pi\varepsilon}\sum_j \left(2^3\left(e^{-2i\mathbf{k}(\mathbf{r}-\mathbf{r}_j)} - e^{2i\mathbf{k}(\mathbf{r}-\mathbf{r}_j)}\right)\int d\mathbf{r}'' \frac{e^{-2i\mathbf{k}\mathbf{r}''}e^{-\beta|\mathbf{r}''|}}{|\mathbf{r}''|}\right) \\ &= \frac{i}{\hbar(\pi)^3}\frac{e^2}{\varepsilon}\sum_j \left(\left(e^{-2i\mathbf{k}(\mathbf{r}-\mathbf{r}_j)} - e^{2i\mathbf{k}(\mathbf{r}-\mathbf{r}_j)}\right)\frac{1}{4\mathbf{k}^2+\beta^2}\right) \end{aligned} \tag{5.87}$$

which leads to the quantum evolution term

$$\begin{aligned} Q f_w(\mathbf{r},\mathbf{k}) &= \frac{i}{\hbar(\pi)^3}\int d\mathbf{k}' f_w(\mathbf{r},\mathbf{k}')\frac{e^2}{\varepsilon} \\ &\times \sum_j \left(\left(e^{-2i(\mathbf{k}-\mathbf{k}')(\mathbf{r}-\mathbf{r}_j)} - e^{2i(\mathbf{k}-\mathbf{k}')(\mathbf{r}-\mathbf{r}_j)}\right)\frac{1}{4(\mathbf{k}-\mathbf{k}')^2+\beta^2}\right) \end{aligned} \tag{5.88}$$

In this section, we assume as a simplification the external field to be zero. It is in accordance with the similar assumption that was used in the previous section regarding electron–phonon scattering. Over a trajectory initialized by $\mathbf{r}(t) = \mathbf{r},\mathbf{k},t$, where the notation implies the meaning of a usual change of variables,

$$\mathbf{r}(t) = \mathbf{R}(t') + \frac{\hbar\mathbf{k}}{m}(t-t')$$

in the Wigner transport equation (5.58), the left hand side term $\frac{\partial f_w(\mathbf{r},\mathbf{k},t')}{\partial t'} + \frac{\hbar \mathbf{k}}{m}\frac{\partial f_w(\mathbf{r},\mathbf{k},t')}{\partial \mathbf{r}}$ simplifies into $(\frac{\partial f_w(\mathbf{R}(t'),\mathbf{k},t')}{\partial t'})_{\mathbf{R}(t')}$. By taking (5.88) into account, the Wigner transport equation thus becomes

$$\left(\frac{\partial f_w\left(\mathbf{R}\left(t'\right),\mathbf{k},t'\right)}{\partial t'}\right)_{\mathbf{R}(t')} = \frac{i}{\hbar\left(\pi\right)^3}\int d\mathbf{k}' f_w\left(\mathbf{R}\left(t'\right),\mathbf{k}'\right)\frac{e^2}{\varepsilon}$$
$$\times\sum_j\left(\left(e^{-2i(\mathbf{k}-\mathbf{k}')(\mathbf{R}(t')-\mathbf{r}_j)} - e^{2i(\mathbf{k}-\mathbf{k}')(\mathbf{R}(t')-\mathbf{r}_j)}\right)\frac{1}{4(\mathbf{k}-\mathbf{k}')^2+\beta^2}\right),$$

which may be integrated into

$$f_w\left(\mathbf{r},\mathbf{k}',t\right) = ic + \frac{e^2}{\varepsilon}\frac{i}{\hbar\left(\pi\right)^3}\int_0^t dt'\int d\mathbf{k}'' f_w\left(\mathbf{R}'\left(t'\right),\mathbf{k}'',t'\right)$$
$$\times\sum_j\left[\left(e^{-2i(\mathbf{k}'-\mathbf{k}'')(\mathbf{R}'(t')-\mathbf{r}_j)} - e^{2i(\mathbf{k}'-\mathbf{k}'')(\mathbf{R}'(t')-\mathbf{r}_j)}\right)\frac{1}{4(\mathbf{k}-\mathbf{k}')^2+\beta^2}\right] \tag{5.89}$$

where the prime of R prompts that the trajectory is now initialized by the arguments $\mathbf{r},\mathbf{k}',t$ of the left hand side of the equation. By choosing a time origin far enough from time t, the initial condition term vanishes. Substituting (5.89) into (5.88) leads to

$$Q f_w\left(\mathbf{r},\mathbf{k},t\right) = -\frac{1}{\hbar^2\left(\pi\right)^6}\frac{e^4}{\varepsilon^2}\int_0^t dt'\int d\mathbf{k}'\int d\mathbf{k}'' f_w\left(\mathbf{R}'\left(t'\right),\mathbf{k}'',t'\right)$$
$$\times\sum_j\left[\left(e^{-2i(\mathbf{k}-\mathbf{k}')(\mathbf{r}-\mathbf{r}_j)} - e^{2i(\mathbf{k}-\mathbf{k}')(\mathbf{r}-\mathbf{r}_j)}\right)\right.$$
$$\times\left(e^{-2i(\mathbf{k}'-\mathbf{k}'')(\mathbf{R}'(t')-\mathbf{r}_j)} - e^{2i(\mathbf{k}'-\mathbf{k}'')(\mathbf{R}'(t')-\mathbf{r}_j)}\right)$$
$$\left.\times\left(4\left(\mathbf{k}-\mathbf{k}'\right)^2+\beta^2\right)^{-1}\left(4\left(\mathbf{k}'-\mathbf{k}''\right)^2+\beta^2\right)^{-1}\right] \tag{5.90}$$

By developing the product of exponential functions, the non-cross terms give

$$S_1 = \left(4\left(\mathbf{k}-\mathbf{k}'\right)^2+\beta^2\right)^{-1}\left(4\left(\mathbf{k}'-\mathbf{k}''\right)^2+\beta^2\right)^{-1}$$
$$\times\sum_j e^{-2i(\mathbf{k}-\mathbf{k}')(\mathbf{r}-\mathbf{r}_j)}e^{-2i(\mathbf{k}'-\mathbf{k}'')(\mathbf{R}'(t')-\mathbf{r}_j)} + cc \tag{5.91}$$

$$S_1 = \left(4\left(\mathbf{k}-\mathbf{k}'\right)^2+\beta^2\right)^{-1}\left(4\left(\mathbf{k}'-\mathbf{k}''\right)^2+\beta^2\right)^{-1}$$
$$\times e^{-2i(\mathbf{k}-\mathbf{k}')(\mathbf{r})}\, e^{-2i(\mathbf{k}'-\mathbf{k}'')\left(\mathbf{r}-\frac{\hbar\mathbf{k}'}{m}(t-t')\right)}\sum_j e^{2i(\mathbf{k}-\mathbf{k}'')(\mathbf{r}_j)} \tag{5.92}$$

If the number of doping atoms in density N_D is assumed to be large enough the discrete sum in (5.90) can be replaced by an integral that takes the form

$$\sum_j e^{2i(\mathbf{k}-\mathbf{k}'')\mathbf{r}_j} \approx N_D \int d\mathbf{r}_j e^{2i(\mathbf{k}-\mathbf{k}'')\mathbf{r}_j} = N_D (2\pi)^3 \delta\left(2\left(\mathbf{k}-\mathbf{k}''\right)\right) \tag{5.93}$$

and then,

$$\begin{aligned} S_1 \approx & \left(4(\mathbf{k}-\mathbf{k}')^2+\beta^2\right)^{-1}\left(4(\mathbf{k}'-\mathbf{k}'')^2+\beta^2\right)^{-1} e^{-2i(\mathbf{k}-\mathbf{k}')(\mathbf{r})} \\ & \times e^{-2i(\mathbf{k}'-\mathbf{k}'')\left(\mathbf{r}-\frac{\hbar\mathbf{k}'}{m}(t-t')\right)} \pi^3 N_D\, \delta\left(\mathbf{k}-\mathbf{k}''\right) \end{aligned} \tag{5.94}$$

$$S_1 \approx \pi^3 \left[\frac{4}{4(\mathbf{k}-\mathbf{k}')^2+\beta^2}\right]^2 e^{-2i(\mathbf{k}'-\mathbf{k})\left(-\frac{\hbar\mathbf{k}'}{m}(t-t')\right)} N_D\, \delta\left(\mathbf{k}-\mathbf{k}''\right) \tag{5.95}$$

Similarly, the cross terms of the product of exponential functions in (5.90) may be written as

$$S_2 \approx \pi^3 \left[\frac{1}{(\mathbf{k}-\mathbf{k}'')^2+\beta^2}\right]^2 e^{\frac{2i}{4}(\mathbf{k}-\mathbf{k}'')\left(-\frac{\hbar(\mathbf{k}+\mathbf{k}'')}{m}(t-t')\right)} N_D\, \delta\left(\mathbf{k}-2\mathbf{k}'+\mathbf{k}''\right) + cc \tag{5.96}$$

Substituting (5.95) and (5.96) into (5.90) yields

$$\begin{aligned} Qf_w(\mathbf{r},\mathbf{k},t) = & -\frac{e^4 N_D}{\hbar^2 \pi^3 \varepsilon^2} \int_0^t dt' \Bigg\{ \int d\mathbf{k}' f_w\left(\mathbf{R}'(t'),\mathbf{k},t'\right) e^{-2i(\mathbf{k}'-\mathbf{k})\left(-\frac{\hbar\mathbf{k}'}{m}(t-t')\right)} \\ & \times \left(4(\mathbf{k}-\mathbf{k}')^2+\beta^2\right)^{-2} + cc \\ & - \int d\mathbf{k}'' f_w\left(\mathbf{R}'(t'),\mathbf{k}'',t'\right) e^{2i(\mathbf{k}^2-\mathbf{k}''^2)\left(\frac{\hbar}{m}(t-t')\right)} \\ & \left((\mathbf{k}-\mathbf{k}'')^2+\beta^2\right)^{-2} + cc \Bigg\} \end{aligned} \tag{5.97}$$

The change of variable $2\mathbf{k}' = \mathbf{k}+\mathbf{k}''$ in the first integral of (5.97) leads to

$$\begin{aligned} Qf_w(\mathbf{r},\mathbf{k},t) = & -\frac{e^4 N_D}{\hbar^2 (2\pi)^3 \varepsilon^2} \int_0^t dt' \Bigg\{ \int d\mathbf{k}'' \left((\mathbf{k}-\mathbf{k}'')^2+\beta^2\right)^{-2} \\ & \Big[f_w\left(\mathbf{R}'(t'),\mathbf{k},t'\right) e^{2i\frac{1}{4}(\mathbf{k}^2-\mathbf{k}''^2)\left(\frac{\hbar}{m}(t-t')\right)} \\ & - f_w\left(\mathbf{R}'(t'),\mathbf{k}'',t'\right) e^{2i\frac{1}{4}(\mathbf{k}^2-\mathbf{k}''^2)\left(\frac{\hbar}{m}(t-t')\right)} \Big] + cc \Bigg\} \end{aligned} \tag{5.98}$$

In the limit of fast collisions, as seen for electron–phonon scattering in (5.84), we finally find:

$$\left[\frac{\partial}{\partial t}+\frac{\hbar \mathbf{k}}{m}\frac{\partial}{\partial \mathbf{r}}\right] f_w(\mathbf{r},\mathbf{k},t) = \frac{e^4 N_D}{\hbar\,(2\pi)^2\,\varepsilon^2} \times \int d\mathbf{k}'' \left\{ \left((\mathbf{k}-\mathbf{k}'')^2+\beta^2 \right)^{-2} \right.$$
$$\left. \times\, \delta\left(E(\mathbf{k})-E(\mathbf{k}'')\right) \left[f_w(\mathbf{r},\mathbf{k}'',t)-f_w(\mathbf{r},\mathbf{k},t)\right] \right\} \tag{5.99}$$

This equation is of the form of (5.85), and is exactly the same, as the one commonly used to model the electron/ionized impurity scattering in the Boltzmann equation (see e.g. [45] (4.24)). Once again the Wigner function allows modeling of this scattering in an intuitive and familiar way that is ideal for electron device simulation.

4 Numerical Approaches: Particle Algorithms

The first applications of the Wigner function in computational electronics are already more than two decades old. Coherent transport in one-dimensional (1D) structures have been successfully approached within deterministic methods [16]. Addressed have been issues related to the correct impose of the boundary conditions which ensure the convergency of the method as well as the discretization scheme. Latter deterministic approaches [12, 15]. have been refined towards self-consistent schemes which take into account the Poisson equation, and dissipation processes have been included by using the relaxation time approximation. The importance of the dissipative processes for the correct distribution of the charge across the device has very soon turned the attention towards the Boltzmann collision term [17]. The three dimensional space of the before- and after- scattering wave vectors has been reduced with the help of an assumption for overall transversal equilibrium to wave vector components along the transport direction.

At that time it has been recognized that an extension of the deterministic approaches to more dimensions is prohibited by the enormous increase of the memory requirements, a fact which remains true even for today's computers. Indeed, despite the progress of the deterministic Boltzmann simulators which nowadays can consider even 3D problems, the situation with Wigner model remains unchanged. The reason is that, in contrast to the Boltzmann scattering matrix, which is relatively sparse due to the δ-functions introduced by the conservation laws, the counterpart provided by the Wigner potential operator is dense.

One of the main difficulty in the implementation of the deterministic solution comes from the discretization of the diffusion term $\nabla_{\mathbf{r}} f_w$ because of the typically rapid variations of the Wigner function in the phase-space. Though a second order discretization scheme is widely used, it has been shown that first, second, third and fourth order schemes lead to very different $I-V$ characteristics of RTDs [46]. In the case of nano-transistors, the third order is required to provide good results in subthreshold regime [47,48].

A basic property of the stochastic methods is that they turn the memory requirements of the deterministic counterparts into computation time requirements. The efforts towards development of stochastic methods for Wigner transport begun almost two decades ago [23,49–51]. As based on the formal analogy between the Wigner and Boltzmann equations, they have been inspired by the success of the classical device Monte Carlo methods, and thus brought the idea of numerical quantum particles.

Particle models are developed for computation of physical quantities in the framework of different kinetic theories. Actually, numerical particles emerged in the field due to the probabilistic transparency of the Boltzmann equation: the numerical concepts of the device Monte Carlo simulators are developed in accordance with the underlying physics of the transport of classical carriers. The most simple version of these simulators is built up on the free electron quasi-particle concepts of effective mass and energy dispersion. Expansions of the physical concepts with respect to the band structure, scattering mechanisms, Pauli exclusion principle etc, retain the picture of developing particles.

Further particle models are already introduced by numerical approaches. Sometimes these introduced for numerical purposes models can be used to interpret and explain the underlying physics even of pure quantum phenomena such as tunneling and interference.

Below we summarize some particle models starting with the direct application of the classical picture.

The smoothed effective potential approach, [52] utilizes classical particles to account for quantum mechanical size quantization effects. The effective potential is a smoothing of the real classical potential due to the finite size of the electron wave packet. It has been shown that the classical trajectories resulting from the effective potential have important details in common with the corresponding Bohm trajectories [53]. A further generalization of the approach replaces the action of the Hamiltonian on the wave function by the action of a classical Hamiltonian on particles with an appropriately modified potential. A set of coupled equations is obtained for the inhomogeneous equilibrium distribution function in the device and its first order correction. The effective potential, defined in terms of a pseudo-differential operator acting on the device potential, becomes also a function of the momenta of the classical particles [54,55].

Ultrafast phenomena in photo-excited semiconductors are described by a set of coupled equations where the distributions of the electrons and holes and the inter-band polarization are treated as independent dynamical variables. If interaction processes are treated on a semiclassical level, so that all transition functions become positive, the set of equations has the structure of rate equations which can be solved by a Monte Carlo method [56]. The remarkable fact that a particle model is associated with the evolution of the inter-band polarization, a complex quantity responsible for the coherence in the photo-generation processes, shows how the method has evolved beyond the understanding of a computer experiment which emulates natural processes.

Furthermore the positiveness of the transition functions is not a necessary condition for a Monte Carlo approach. It has been shown that the action of the Wigner potential, which is an antisymmetric quantity, gives rise to a Markov process which can be regarded as a scattering of a particle between consecutive points in the phase space [57].

Wigner trajectories have been defined by modified Hamilton equations, formulated with the help of a quantum force [50]. The latter is manifestly nonlocal in space and is expressed through the Wigner potential and function, and its derivative with respect to the momentum coordinate. The quantum force has singularities at the points where the momentum derivative of the Wigner function becomes zero. At these points trajectories can be created or destroyed [50]. Due to this Wigner trajectories can merely provide a pictorial explanation of the evolution of the quantum system and in particular nicely illustrate tunneling processes [58, 59].

In general, Wigner trajectories remain an auxiliary tool for modeling of quantum transport, unless the Wigner function in the quantum force term is assumed to be known. An appropriate approximation for a nearly equilibrium system is a displaced Maxwell–Boltzmann distribution function. It can be shown that such an assumption corresponds to the zeroth order correction in the effective potential approach. In this case the quantum force is defined everywhere except at the phase-space origin, and gives rise to an effective lowering of the peaks of the potential barriers [49]. The increase of the particle flow observed through the barriers is associated with tunneling processes.

Another particle model is introduced by Wigner paths [24, 37, 60]. It has been shown that a ballistic evolution of a δ-like contribution to the Wigner function carries its value following a classical trajectory [36]. The action of the Wigner potential operator is interpreted as scattering, which, along with the scattering by the phonons, links pieces of classical trajectories to Wigner paths. We note that, in this model, the phonon interaction is treated fully quantum mechanically according to the first-principle equation (5.65). That is, the scattering with phonons begins with exchange of half of the phonon momentum and completes after a finite time. During this time, an arbitrary number of interactions with other phonons can be initiated and/or completed. In comparison, Levinson's equation considers a single interaction with finite duration while Boltzmann scattering is instantaneous, so that the trajectory changes with the full phonon momentum. During the evolution particles accumulate a numerical quantity called weight, which carries the quantum information for the system. The weight is taken into account in the computation of the physical averages.

Next we introduce two particle models for solving the Wigner–Boltzmann equation. They unify classical and quantum regions within a single transport picture where the scattering occurs in the full wave vector space, and two dimensional devices can be considered [61–63].

Since it can take negative values in some regions of the phase space, it may look nonsensical to represent the Wigner function with particles which cannot have a "negative presence". Classically, electrons either do or do not exist. To solve this apparent inconsistency with a view to developing a statistical particle approach to the solution of the Wigner–Boltzmann equation it is necessary to give simulated particles the strange property to carry negative contributions. With this in mind, it has been suggested to describe the Wigner function as a sum of Dirac excitations still localized in the phase-space but weighted by an amplitude, called affinity in [64,65]. The particle affinities contain all the information on the quantum state of the electron system. They evolve continuously according to the local quantum evolution term of the Wigner–Boltzmann equation generated by the potential and can take negative values which are taken into account as weights in the reconstruction of the Wigner function and in the computation of all physical averages.

An alternative particle approach interprets the Wigner equation, with a Boltzmann scattering term as a Boltzmann equation with a generation term. The interaction with the Wigner potential gives rise to generation of particle pairs with opposite sign. The sign is the basic property which outlines the introduced numerical particles from classical quasi-particles. It is an important property, since positive and negative particles annihilate one another. The negative values of the Wigner function in certain phase space regions can be explained in a natural way by the accumulation of negative particles in these regions. The Wigner–Boltzmann transport process corresponds to drift, scattering, generation and annihilation of these particles.

These models present the state of the art in the field and will be described in detail in the rest of this section.

4.1 The Affinity Method

4.1.1 Principles

In this approach, the Wigner function is represented as a sum of Dirac excitations of the form

$$f_w(\mathbf{r},\mathbf{k},t) = \sum_j \delta(\mathbf{r}-\mathbf{r}_j(t))\;\delta(\mathbf{k}-\mathbf{k}_j(t))\;A_j(t) \tag{5.100}$$

In contrast to classical particles, these excitations are weighted by an amplitude A_j, called affinity, which evolves continuously under the action of the quantum evolution term of the Wigner–Boltzmann equation (5.58) which describes the non-local effect of the potential. Since the Wigner function can take negative values in the presence of quantum transport effects, the affinity may be negative too. Consistently with the Heisenberg inequalities, such excitations of negative weight cannot represent physical particles and will be called pseudo-particles. They should be considered as mathematical objects useful to solve the Wigner transport equation.

Let's remember the quantum evolution term of this equation, which writes by introducing the associated operator Q as

$$Q f_w(\mathbf{r},\mathbf{k},t) = \int d\mathbf{k}'\, V_w(\mathbf{r},\mathbf{k}-\mathbf{k}',t)\, f_w(\mathbf{r},\mathbf{k}',t) \tag{5.101}$$

with the Wigner potential defined by (5.59). Compared to the semi-classical Monte Carlo algorithm, one of the main changes consists of adding, at each time step, the update of the Wigner function and of the particle affinities. In a mesh of the phase-space $M(\mathbf{r},\mathbf{k})$ the quantum evolution term $Q f_w$ induces the change of the affinity of particles in the mesh according to

$$\sum_{i\in M(x,k)} \frac{dA_i}{dt} = Q f_w(\mathbf{r},\mathbf{k}) \tag{5.102}$$

which means that at each time step the affinity of all pseudo-particles in a mesh of the phase-space is updated according to the value of $Q f_w$ in this mesh. The non-local effect of the potential is thus fully applied to the affinity evolution, in contrast to the semi-classical case where the local effect of the potential gradient induces the change of wave vector. The simple idea on which is based this quantum simulation method now appears clearly. Along its trajectory a pseudo-particle scatter as a classical particle, and during a free flight the coordinates of the j-th particle obey, in the effective mass approximation,

$$\frac{d}{dt}\mathbf{r}_j = \frac{\hbar}{m}\mathbf{k}_j \tag{5.103}$$

$$\frac{d}{dt}\mathbf{k}_j = 0 \tag{5.104}$$

The wave vector of each pseudo-particle is thus constant during a free flight and can take a new value only after scattering. However, if the potential may be separated into slowly and rapidly varying parts, the slowly varying part may be treated semi-classically through the evolution of the particle wave vector under the influence of electric field while only the rapidly varying part is taken into account in the computation of the Wigner potential and in the affinity (5.102).

In the semi-classical limit, i.e. if the full potential is treated as a slowly varying quantity, the quantum evolution term $Q f_w$ is zero and the particle affinity is constant. The method turns out to be equivalent to the semi-classical Monte Carlo algorithm. It should be noted that the strong similarity and even compatibility of this technique with the conventional Monte Carlo solution of the Boltzmann equation is one of its highest advantage, which will be illustrated later.

We now detail some important specific features of the numerical implementation of the affinity method. Additional discussion may be found in [66]. Though so far the algorithm has been implemented for 1D transport problems only, i.e. with phase-space coordinates of the Wigner function reduced to (x,k), the discussion below is made in the general case of the full phase space (r, k).

4.1.2 Conservation of Affinity and Pseudo-Particle Injection

First of all, it should be reminded that in semi-classical device simulation with Ohmic contacts, the only condition of particle injection is the neutrality of real-space meshes adjacent to the Ohmic contacts. After each time step, if particles are missing in some "Ohmic" meshes with respect to the charge neutrality, the appropriate number of carriers (of affinity equal to 1) is injected in these meshes to recover the neutrality. Assuming these "Ohmic" regions to be in thermal equilibrium, an equilibrium distribution is used to select randomly their wave vector components. In this way the consistence between the distribution of potential and the average number and the distribution of particles in the device is reached. Obviously, this condition of particle injection should still be used in Wigner simulation of Ohmic contacts if the transport in the contact region is assumed to be essentially semi-classical. However, it is not enough to ensure the conservation of total affinity and charge within an algorithm in which the particle affinity evolves continuously.

Indeed, one of the most important difficulties in this MC method lies in the fact that even a particle with zero affinity may gain finite affinity through the quantum evolution term $Q f_w$ according to (5.102). It means that if there is no particle in a particular region of the phase space where the affinity should evolve, a significant error may occur with possible non-conservation of charge since the contribution of each particle to the total charge in the device is weighted by its affinity. This problem is very important for device simulation and should be fixed by implementing an appropriate algorithm to inject particles of convenient affinity.

The correct approach consists in filling the phase-space with pseudo-particles of zero affinity as follows. After each time step the quantum evolution term $Q f_w(\mathbf{r},\mathbf{k})$ is calculated in the full phase-space. If in a mesh $M(\mathbf{r},\mathbf{k})$ of the phase-space, even inside the device, the quantity $|Q f_w(\mathbf{r},\mathbf{k})|$ is finite, a pseudo-particle of zero affinity is injected in the mesh [65]. In summary, it is necessary to combine the "semi-classical injection" of particles of affinity equal to 1 at Ohmic contacts to guarantee the electrical neutrality near the contacts and the "quantum injection" of pseudo-particles of 0 affinity in all regions of the phase-space where particles are missing and where $Q f_w$ takes significant values.

4.1.3 Computation of the Wigner Potential and of the Affinity Evolution

A fundamental problem lies in the choice of the limits of integration for the calculation of the Wigner potential (5.59). There are two possible approaches depending on whether the contacts are assumed to be coherent or non-coherent. In the former case the integration is cut at a maximum size from the contact corresponding to the "coherence length" beyond which no quantum effect may occur [67], which raises the question of the relevant choice of the coherence length in the contact. In the latter case, the integration should be limited to positions $\mathbf{r}'$ such that both $\mathbf{r}-\mathbf{r}'/2$ and $\mathbf{r}+\mathbf{r}'/2$ belong to the device [68]. This approach is used in the model we have developed. However, we have checked that in the devices considered in the application section (RTD, MOSFET), all limits of integration larger than that corresponding to the hypothesis of decoherent contacts yield the same results. This insensitivity is certainly due to the fact that in these cases contact regions, or access regions, have a semi-classical behavior dominated by scattering.

To describe the time evolution of pseudo-particle affinities, a very stable discretization scheme is required. Indeed, we observed that due to the noise inherent to the technique, the MC simulation acts as a stiff problem, which tends to make the solution of (5.102) unstable. In our model, an implicit Backward Euler scheme was finally implemented:

$$A_i(t+dt) - A_i(t) = \frac{1}{N} dt \times Q f_w(\mathbf{r}, \mathbf{k}, t+dt) \tag{5.105}$$

where N is here the number of pseudo-particles in the mesh $M(\mathbf{r}, \mathbf{k})$ of the phase space. This backward Euler scheme is implicit. It may be implemented by matrix inversion of the quantum evolution operator Q, or by using a predictor/corrector technique of high order, at least fourth order. The two techniques give the same results but the predictor/corrector one is faster. All higher precision schemes were proved to be detrimental to the simulation stability and required longer simulation time to obtain good average quantities. In particular Cayley's scheme, known to be the best technique for the evaluation of the time derivative in the deterministic solution of the Wigner–Boltzmann equation, leads to unstable results in MC simulation.

4.2 The Particle Generation Method

Monte Carlo algorithms can be devised based on the notion that the terms on right hand side of the Wigner–Boltzmann equation represent gain and loss terms for the phase space density. To introduce the ideas we consider the semiclassical Boltzmann equation.

$$\left(\frac{\partial}{\partial t} + \mathbf{v}(\mathbf{k}) \cdot \nabla_{\mathbf{r}} + \frac{1}{\hbar} \mathbf{F}(\mathbf{r}) \cdot \nabla_{\mathbf{k}}\right) f(\mathbf{r}, \mathbf{k}, t) = \int d\mathbf{k}' f(\mathbf{r}, \mathbf{k}', t) S(\mathbf{k}', \mathbf{k}) - \lambda(\mathbf{k}) f(\mathbf{r}, \mathbf{k}, t) \tag{5.106}$$

$S(\mathbf{k}, \mathbf{k}')$ denotes the transition rate from initial state $\mathbf{k}'$ to final state $\mathbf{k}$, induced by the physical scattering processes, and λ is the total scattering rate.

$$\lambda(\mathbf{k}) = \int d\mathbf{k}' S(\mathbf{k}, \mathbf{k}') \tag{5.107}$$

We note that the positive term on the RHS of (5.106) is an integral operator representing a particle gain term. In a Monte Carlo algorithm transitions from state $\mathbf{k}'$ to $\mathbf{k}$ are selected randomly from the normalized transition probability $S(\mathbf{k}, \mathbf{k}')/\lambda(\mathbf{k}')$.

The negative term on the RHS of (5.106) is local in **k**-space. In a Monte Carlo algorithm the term $-\lambda f$ gives rise to the exponential distribution for the carrier free flight time.

The Wigner–Boltzmann equation has the same structure as (5.106). We use (5.48) and augment it by a Boltzmann scattering operator.

$$\left(\frac{\partial}{\partial t} + \mathbf{v}(\mathbf{k}) \cdot \nabla_{\mathbf{r}} + \frac{1}{\hbar} \mathbf{F}_{\mathrm{cl}}(\mathbf{r}) \cdot \nabla_{\mathbf{k}}\right) f_{\mathrm{w}}(\mathbf{r},\mathbf{k},t)$$
$$= \int d\mathbf{k}' \Gamma(\mathbf{k},\mathbf{k}')\mu(\mathbf{k}') f_{\mathrm{w}}(\mathbf{r},\mathbf{k}',t) - \mu(\mathbf{k}) f_{\mathrm{w}}(\mathbf{r},\mathbf{k},t) \tag{5.108}$$

The integral kernel Γ in this equation has the form

$$\Gamma(\mathbf{r},\mathbf{k},\mathbf{k}') = \frac{1}{\mu(\mathbf{r},\mathbf{k}')}\left[S(\mathbf{k}',\mathbf{k}) + V_{\mathrm{w}}(\mathbf{r},\mathbf{k}-\mathbf{k}') + \alpha(\mathbf{k},\mathbf{r})\,\delta(\mathbf{k}-\mathbf{k}')\right], \tag{5.109}$$
$$\mu(\mathbf{r},\mathbf{k}') = \lambda(\mathbf{r},\mathbf{k}') + \alpha(\mathbf{r},\mathbf{k}'), \tag{5.110}$$

where μ is the normalization factor. It holds

$$\int d\mathbf{k}'\, \Gamma(\mathbf{k},\mathbf{k}',\mathbf{r}) = 1. \tag{5.111}$$

In (5.109) a fictitious scattering mechanism

$$S_{\mathrm{self}}(\mathbf{k}',\mathbf{k}) = \alpha(\mathbf{r},\mathbf{k})\delta(\mathbf{k}-\mathbf{k}') \tag{5.112}$$

is introduced, referred to as self-scattering [69]. Mathematically, the related contributions in the gain and loss terms simply cancel and have no effect. Physically, because of the δ-function, this mechanism does not change the state of the electron and hence does not alter the free-flight trajectory. The choice of α offers a degree of freedom in the construction of a Monte Carlo algorithm, as shown below.

4.2.1 Integral Form of the Wigner–Boltzmann Equation

Equation (5.108) can be transformed into a path integral equation [70]. The adjoint integral equation, which will give rise to forward Monte Carlo algorithms, has the following integral kernel.

$$P(\mathbf{k}_f,t_f|\mathbf{k}_i,t_i) = \Gamma[\mathbf{k}_f,\mathbf{K}(t_f)]\;\mu[\mathbf{K}(t_f)]\exp\left\{-\int_{t_i}^{t_f} \mu[\mathbf{K}(\tau)]\,\mathrm{d}\tau\right\} \tag{5.113}$$

The kernel represents a transition consisting of a free flight starting at time t_i with initial state $\mathbf{k}_i$, followed by a scattering process to the final state $\mathbf{k}_f$ at time t_f. For the sake of brevity the **r**-dependences of Γ and μ are omitted in the following. In a

Monte Carlo simulation, the time of the next scattering event, t_f, is generated from the exponential distribution appearing in (5.113):

$$p_t(t_f, t_i, \mathbf{k}_i) = \mu[\mathbf{K}(t_f)]\exp\left\{-\int_{t_i}^{t_f} \mu[\mathbf{K}(\tau)]\,\mathrm{d}\tau\right\} \tag{5.114}$$

We denote by $\mathbf{k}'$ the state at the end of the free flight, $\mathbf{k}' = \mathbf{K}(t_f)$. A transition from the trajectory end point $\mathbf{k}'$ to the final state $\mathbf{k}_f$ is realized using the kernel Γ. In contrast to the classical case, where P would represent a transition probability, such an interpretation is not possible in the case of the Wigner equation because P is not positive semidefinite. The problem originates from the Wigner potential, which assumes positive and negative values.

Because of its antisymmetry with respect to $\mathbf{q}$, the Wigner potential can be reformulated in terms of one positive function V_{W}^{+}

$$V_{\mathrm{W}}^{+}(\mathbf{r},\mathbf{q}) = \max(0, V_{\mathrm{W}}(\mathbf{r},\mathbf{q})) \tag{5.115}$$

$$V_{\mathrm{W}}(\mathbf{r},\mathbf{q}) = V_{\mathrm{W}}^{+}(\mathbf{r},\mathbf{q}) - V_{\mathrm{W}}^{+}(\mathbf{r},-\mathbf{q}) \tag{5.116}$$

Then, the kernel Γ is rewritten as a sum of the following conditional probability distributions.

$$\Gamma(\mathbf{k},\mathbf{k}') = \frac{\lambda}{\mu} s(\mathbf{k},\mathbf{k}') + \frac{\alpha}{\mu}\delta(\mathbf{k}'-\mathbf{k}) + \frac{\gamma}{\mu}\left[w(\mathbf{k},\mathbf{k}') - w^{*}(\mathbf{k},\mathbf{k}')\right], \tag{5.117}$$

$$s(\mathbf{k}',\mathbf{k}) = \frac{S(\mathbf{k}',\mathbf{k})}{\lambda(\mathbf{k}')}, \qquad w(\mathbf{k},\mathbf{k}') = \frac{V_{\mathrm{W}}^{+}(\mathbf{k}-\mathbf{k}')}{\gamma}, \qquad w^{*}(\mathbf{k},\mathbf{k}') = w(\mathbf{k}',\mathbf{k}) \tag{5.118}$$

The normalization factor for the Wigner potential is

$$\gamma(\mathbf{r}) = \int d\mathbf{q} V^{+}(\mathbf{r},\mathbf{q}). \tag{5.119}$$

In the following, different variants of generating the final state $\mathbf{k}_f$ from the kernel Γ will be discussed.

4.2.2 The Markov Chain Method

We have now to decompose the kernel P into a transition probability p and the remaining function P/p. More details on the Markov chain method can be found in [71, 72]. With respect to (5.113), one could use the absolute value of Γ as a transition probability. Practically, it is more convenient to use the absolute values of the components of Γ, giving the following transition probability.

$$p(\mathbf{k}_f,\mathbf{k}') = \frac{\lambda}{\nu} s(\mathbf{k}_f,\mathbf{k}') + \frac{\alpha}{\nu}\delta(\mathbf{k}_f-\mathbf{k}') + \frac{\gamma}{\nu} w(\mathbf{k}_f,\mathbf{k}') + \frac{\gamma}{\nu} w^{*}(\mathbf{k}_f,\mathbf{k}') \tag{5.120}$$

The normalization factor is

$$\nu = \int d\mathbf{k}f\, p(\mathbf{k}_f,\mathbf{k}') = \lambda + \alpha + 2\gamma. \tag{5.121}$$

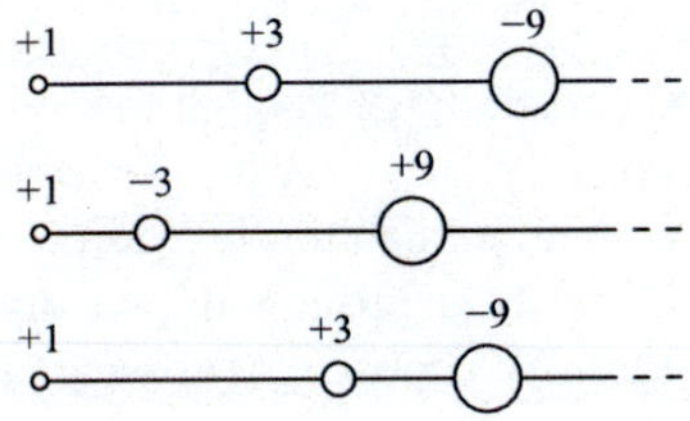

Fig. 5.2 With the Markov chain method, the number of numerical particles is conserved. The magnitude of the particle weight increases with each event, and the sign of the weight changes randomly according to a given probability distribution

In the first method considered here, the free-light time is generated from the exponential distribution (5.114). To generate the final state $\mathbf{k}_f$ for the given trajectory endpoint $\mathbf{k}'$, one of the four terms in (5.120) is selected with the associated probabilities λ/ν, α/ν, γ/ν, and γ/ν, respectively. Apparently, these probabilities sum up to one. If classical scattering is selected, $\mathbf{k}_f$ is generated from s. If self-scattering is selected, the state does not change and $\mathbf{k}_f = \mathbf{k}'$ holds. If the third or fourth term are selected, the particle state is changed by scattering from the Wigner potential and $\mathbf{k}_f$ is selected from w or w^*, respectively. The particle weight has to be multiplied by the ratio

$$\frac{\Gamma}{p} = \pm\frac{\nu}{\mu} = \pm\left(1+\frac{2\gamma}{\lambda+\alpha}\right), \tag{5.122}$$

where the minus sign applies if $\mathbf{k}_f$ has been generated from w^*. For instance, for a quantum mechanical system, where the classical scattering rate λ is less than the Wigner scattering rate γ, the self-scattering rate α can be chosen such that $\lambda + \alpha = \gamma$. Then, the multiplier (5.122) evaluates to ± 3. An ensemble of particles would evolve as shown schematically in Fig. 5.2. As the multiplier (5.122) is always greater than one, the absolute value of the particle weight will inevitably grow with the number of transitions on the trajectory.

4.2.3 Pair Generation Method

To solve the problem of growing particle weights, one can split particles. In this way, an increase in particle weight is transformed to an increase in particle number. The basic idea of splitting is refined so as to avoid fractional weights. Different interpretations of the kernel are presented, that conserve the magnitude of the particle weight [73]. Choosing the initial weight to be $+1$, all generated particles will have weight $+1$ or -1. This is achieved by interpreting the potential operator in the Wigner–Boltzmann equation as a generation term of positive and negative particles. We consider the kernel (5.117).

$$\Gamma(\mathbf{k}_f,\mathbf{k}') = \frac{\lambda}{\mu}s(\mathbf{k}_f,\mathbf{k}') + \frac{\alpha}{\mu}\delta(\mathbf{k}_f-\mathbf{k}') + \frac{\gamma}{\mu}\left[w(\mathbf{k}_f,\mathbf{k}') - w^*(\mathbf{k}_f,\mathbf{k}')\right] \tag{5.123}$$

If the Wigner scattering rate γ is larger than the classical scattering rate λ, the self-scattering rate α has to be chosen large enough to satisfy the inequality $\gamma/\mu \leq 1$. Typical choices are $\mu = \mathrm{Max}(\lambda,\gamma)$ or $\mu = \lambda + \gamma$. These expressions also hold for the less interesting case $\gamma < \lambda$, where quantum interference effects are less important than classical scattering effects. In the following, we discuss the case $\gamma > \lambda$, where quantum effects are dominant. We choose the self-scattering rate equal to $\alpha = \gamma$ and regroup the kernel as

$$\Gamma(\mathbf{k}_f,\mathbf{k}') = \frac{\lambda}{\mu} s(\mathbf{k}_f,\mathbf{k}') + \left(1 - \frac{\lambda}{\mu}\right) \left[\delta(\mathbf{k}_f - \mathbf{k}') + w(\mathbf{k}_f,\mathbf{k}') - w^*(\mathbf{k}_f,\mathbf{k}')\right]. \quad (5.124)$$

As in the classical Monte Carlo method, the distribution of the free-flight duration is given by the exponential distribution (5.114). At the end of a free flight, classical scattering is selected with probability $p_s = \lambda/\mu$. With the complementary probability, $1 - p_s$, a self-scattering event and a pair generation event occur. The weight of the state generated from w^* is multiplied by -1. The weights of the states generated from w and from self-scattering do not change. Therefore, the magnitude of the initial particle weight is conserved, as shown in Fig. 5.3. In this algorithm, classical scattering and pair generation are complementary events and thus cannot occur at the same time, as shown in Fig. 5.4. Different choices of the self-scattering rate α result in different variants of the Monte Carlo algorithm. A more detailed discussion can be found in [73].

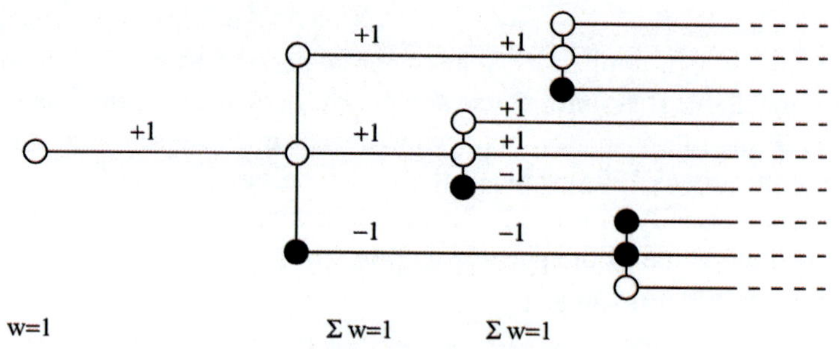

Fig. 5.3 With the pair generation method the magnitude of the particle weight is conserved, but one initial particle generates a cascade of numerical particles. At all times mass is exactly conserved

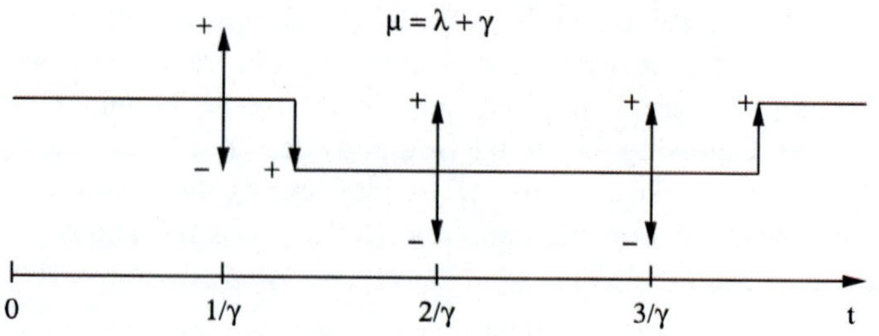

Fig. 5.4 Trajectory in k-space of a sample particle resulting from the pair-generation method. Discontinuities in the main trajectory indicate semi-classical scattering events, whereas *arrows* indicate instances when particle pairs are generated

In the pair-generation method described, the weights of the generated particles are ± 1, because the generation rate used equals 2γ (generation of one pair at a rate of γ). If a generation rate larger than 2γ or a fixed time-step less than $(2\gamma)^{-1}$ were used, the magnitude of the generated weight would be less than one. The resulting fractional weights are referred to as affinities. On the other hand, a generation rate less than 2γ would result in an under-sampling of the physical process. Then, the magnitude of the generated weights would be generally greater than one.

Instead of using $V_w^+(\mathbf{r},\mathbf{q})$ to generate the momentum transfer $\hbar\mathbf{q}$, one can construct a Monte Carlo algorithm which uses the amplitude of the Fourier transform, $A(\mathbf{q})$ in (5.33). The advantage is that the numerical representation of $A(\mathbf{q})$ only requires a discretization of the momentum coordinate, whereas for the Wigner potential $V_\mathrm{w}^+(\mathbf{q},\mathbf{r})$ both momentum and spatial coordinates need to be discretized.

We start with the potential operator (5.58) defined in the three-dimensional $\mathbf{k}$-space, change variables $\mathbf{q}=\mathbf{k}'-\mathbf{k}$ and $\mathbf{q}=\mathbf{k}-\mathbf{k}'$, and build a symmetrized expression.

$$\mathcal{Q}f_\mathrm{w}(\mathbf{r},\mathbf{k}) = \frac{1}{2}\int d\mathbf{q} V_\mathrm{w}(\mathbf{r},\mathbf{q})\left[f_\mathrm{w}(\mathbf{r},\mathbf{k}-\mathbf{q}) - f_\mathrm{w}(\mathbf{r},\mathbf{k}+\mathbf{q})\right]. \tag{5.125}$$

Expressing the Wigner potential through the three-dimensional Fourier transform of the potential,

$$V_\mathrm{w}(\mathbf{r},\mathbf{q}) = \frac{2}{\hbar\pi^3} A(2\mathbf{q})\sin\left[\varphi(2\mathbf{q}) + 2\mathbf{q}\cdot\mathbf{r}\right], \tag{5.126}$$

the potential operator (5.125) can be rewritten as

$$\mathcal{Q}f_\mathrm{w}(\mathbf{r},\mathbf{k}) = \frac{1}{\hbar}\int \frac{d\mathbf{q}}{(2\pi)^3} A(\mathbf{q})\,\sin\left[\varphi(\mathbf{q}) + \mathbf{q}\cdot\mathbf{r}\right]\left[f_\mathrm{w}\left(\mathbf{r},\mathbf{k}-\frac{\mathbf{q}}{2}\right) - f_\mathrm{w}\left(\mathbf{r},\mathbf{k}+\frac{\mathbf{q}}{2}\right)\right] \tag{5.127}$$

An advantage of this formulation is that no discretization of the spatial variable $\mathbf{r}$ is needed. The expression can be evaluated at the actual position $\mathbf{r}$ of a particle. The structure of (5.127) suggests the usage of a rejection technique. The normalization factor γ now is larger than the actual pair generation rate.

$$\gamma = \frac{1}{\hbar}\int \frac{d\mathbf{q}}{(2\pi)^3} A(\mathbf{q}) \tag{5.128}$$

The rate of γ is used as in the algorithms described above to randomly generate the times between two particle pair-generation events. From the distribution $A(\mathbf{q})$ one generates randomly the momentum transfer $\mathbf{q}$. Then the sine function is evaluated at the actual particle position $\mathbf{r}$.

$$s = \sin\left[\varphi(\mathbf{q}) + \mathbf{q}\cdot\mathbf{r}\right] \tag{5.129}$$

With probability $|s|$ the pair-generation event is accepted, otherwise a self-scattering event is performed. In the former case, two particle states are generated with momenta $\mathbf{k}_1 = \mathbf{k}-\mathbf{q}/2$ and $\mathbf{k}_2 = \mathbf{k}+\mathbf{q}/2$ and statistical weights $w_1 = w_0\mathrm{sign}(s)$ and $w_2 = -w_1$, respectively, where w_0 is the statistical weight of the initial particle.

4.2.4 The Negative Sign Problem

In the following, we analyze the growth rates of particle weights and particle numbers associated with the different Monte Carlo algorithms. In the Markov chain method discussed in Sect. 4.2.2, the weight increases at each scattering event by the multiplier (5.122). The growth rate of the weight can be estimated for the case of constant coefficients γ and μ. Because free-flight times are generated with rate μ, the mean free-flight time will be $1/\mu$. During a given time interval t, on average $n = \mu t$ scattering events will occur. The total weight is then estimated asymptotically for $t \gg 1/\mu$.

$$|W(t)| = \left(1 + \frac{2\gamma}{\mu}\right)^n = \left(1 + \frac{2\gamma t}{n}\right)^n \simeq \exp(2\gamma t) \tag{5.130}$$

This expression shows that the growth rate is determined by the Wigner scattering rate γ independently of the classical and the self-scattering rates. The growth rate 2γ is equal to the L_1 norm of the Wigner potential.

In the pair generation method, the potential operator

$$Qf_{\mathrm{w}}(\mathbf{k}) = \int d\mathbf{q} V^+(\mathbf{q}) \left[f_{\mathrm{w}}(\mathbf{k}-\mathbf{q}) - f_{\mathrm{w}}(\mathbf{k}+\mathbf{q})\right] \tag{5.131}$$

is interpreted as a generation term. It describes the creation of two new states, $\mathbf{k}-\mathbf{q}$ and $\mathbf{k}+\mathbf{q}$. The pair generation rate is equal to γ. When generating the second state, the sign of the statistical weight is changed. It should be noted that the Wigner–Boltzmann equation strictly conserves mass, as can be seen by taking the zeroth order moment of (5.108):

$$\frac{\partial n}{\partial t} + \operatorname{div} \mathbf{J} = 0 \tag{5.132}$$

Looking at the number of particles regardless of their statistical weights, that is, counting each particle as positive, would correspond to using the following potential operator:

$$Q^* f_{\mathrm{w}}(\mathbf{k}) = \int d\mathbf{q} V_{\mathrm{w}}^+(\mathbf{q}) \left[f_{\mathrm{w}}(\mathbf{k}-\mathbf{q}) + f_{\mathrm{w}}(\mathbf{k}+\mathbf{q})\right] \tag{5.133}$$

Using (5.133), a continuity equation for numerical particles is obtained as

$$\frac{\partial n^*}{\partial t} + \operatorname{div} \mathbf{J}^* = 2\gamma(\mathbf{r}) n^* \tag{5.134}$$

Assuming a constant γ, the generation rate in this equation will give rise to an exponential increase in the number of numerical particles N^*.

$$N^*(t) = N^*(0) \exp(2\gamma t) \tag{5.135}$$

This discussion shows that the appearance of an exponential growth rate is independent of the details of the particular Monte Carlo algorithm. It is a fundamental consequence of the non-positive kernel.

4.2.5 Particle Annihilation

The discussed particle models are instable, because either the particle weight or the particle number grows exponentially in time. Using the Markov chain method, it has been demonstrated that tunneling can be treated numerically by means of a particle model [74]. However, because of the exponentially increasing particle weight at the very short timescale $(2\gamma)^{-1}$, application of this algorithm turned out to be restricted to single-barrier tunneling and small barrier heights only. This method can be useful for devices where quantum effects are weak, and the potential operator is a small correction to the otherwise classical transport equation.

A stable Monte Carlo algorithm can be obtained by combining one of the particle generation methods with a method to control the particle number. One can assume that two particles of opposite weight and a sufficiently small distance in phase space annihilate each other. The reason is that the motions of both particles are governed by the same equation. Therefore, when they come close to each other at some time instant, the two particles have approximately the same initial condition. They can be considered a super particle of total weight zero, which indeed needs not be considered further in the simulation. In an ensemble Monte Carlo method, a particle removal step should be performed at given time steps. During the time step, the ensemble is allowed to grow to a certain limit, then particles are removed and the initial size of the ensemble is restored.

For a stationary transport problem a one-particle Monte Carlo method can be devised which annihilates numerical particles at the same rate as they are generated. For this purpose a phase space mesh can be utilized [75]. In the following we describe an algorithm which traces only one branch of the trajectory tree originating from a single particle injected at the contact.

After each generation event one has to deal with three particle states, namely the initial state $\mathbf{k}$ and the two generated states, $\mathbf{k}_1$ and $\mathbf{k}_2$. In a first step the weights of all three particles are stored on the annihilation mesh, that is, the statistical weight of each particle is added to a counter associated with the mesh element. Then one has to decide which of the three states is used to continue the trajectory. One may choose the weight of the continuing particle to have the same sign as the incoming one (Fig. 5.5). In this way the statistical weight along one trajectory does not change,

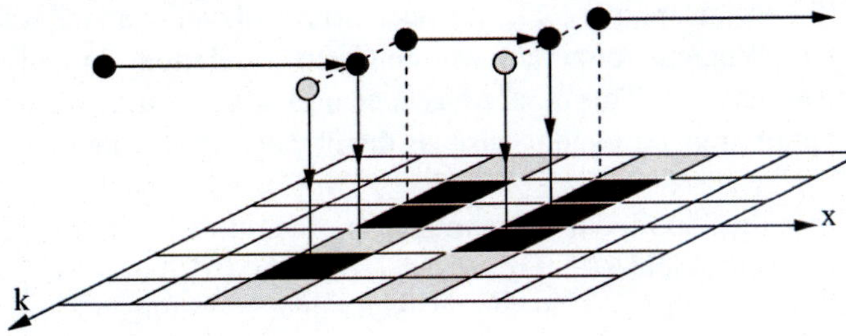

Fig. 5.5 The particle annihilation strategy attempts to minimize the weights stored in the mesh elements. The weights of the initial and continuing particle have the same sign to ensure current continuity. Particles and mesh elements carrying a positive weight are in *black*, the ones carrying a negative weight are in *grey*

which results in exact current conservation [76]. Note that because of the pair-wise generation of particles with weights ± 1 the algorithm also ensures exact mass conservation. If the initial state has a positive statistical weight, out of the three mesh elements the one with the largest stored weight is selected. Continuing from that element will reduce the weight of the element. Conversely, a negative trajectory is to be continued from the element with the smallest stored weight. A certain fraction of negative trajectories needs to be constructed in order to resolve the negative parts of the Wigner function. This rule for selecting the continuing particle is an attempt to minimize the weights stored in the three elements after each pair-generation event. The repeated execution of this rule in the Monte Carlo main loop results in a minimization of the stored weight on the whole annihilation mesh. Particle annihilation takes place when positive and negative particles are alternately stored in the same mesh element. Note that because of the mass conservation property of the transport equation and of the associated particle model, no net-charge can build up on the annihilation mesh. The weights stored on the mesh sum up to zero. The local weights on the mesh have to be kept small, as they are a measure for the numerical error of the method. This can be controlled by the fraction of negative trajectories, which has to be specified by the user.

5 Applications

For many years, the RTD was certainly the device operating at room temperature in which the wave-like behavior of electrons played the most prominent role. Thanks to the control of tunneling through the resonant state of a quantum well coupled to two electrodes via tunnel barriers, the RTD provides a negative differential resistance (NDR) in the $I - V$ characteristics. Since the pioneering works of Tsu and Esaki [77] and the first experimental evidence for NDR effect in an RTD at low temperature [78] and at room temperature [79], an intense research effort has been devoted to this fascinating device. Beyond its high potential of applications [80], the RTD is also an incomparable "toy" for fundamental physics and quantum device physics, in particular to understand the quantum features of shot noise, as in [81–87]. It is also a useful test device for new materials in which quantum transport is likely to occur, as [88–90]. Here, the RTD has been used to develop and validate the affinity technique of Wigner–Boltzmann Monte Carlo simulation. Some typical results are presented in Sect. 5.1. This device has been used also to study the impact of scattering on quantum transport and to discuss the physics of de-coherence, as reported in Sect. 5.2.

The model is then applied to the simulation of an ultra-short double-gate Metal-oxide-semiconductor Field-effect transistor (DG-MOSFET) in Sect. 5.3.

Among the new silicon-on-insulator (SOI) device architectures based on thin undoped channel controlled by multiple gates which are currently developed and envisioned to be the future of CMOS technology [91, 92], the double-gate planar configuration is one of the most promising [93] to overcome the limitations of conventional bulk-device towards further scaling, in particular the limitations linked with the multiple sources of leakage and variability [94–97]. Compared to the single-gate SOI transistor, a second back-gate is "introduced" underneath the channel [98–101] thanks to the molecular bonding of two substrates. The electrostatics of this architecture is excellent [102]. Its main issue is the self-alignment of both gates which is required for optimized performance [98, 103]. This challenge has been recently taken up by including metal gates, high-κ dielectrics, metallic source/drain with gate length down to 6 nm [104].

5.1 Application to Resonant Tunneling Diodes

As shown schematically in Fig. 5.6 the simulated GaAs/GaAlAs RTD consists of a 5 nm-thick quantum well sandwiched between two AlGaAs barriers 0.3 eV high and 3 nm wide. The quantum well, the barriers, and 9.5 nm-thick buffer regions surrounding the barriers are slightly doped to $10^{16}\,\mathrm{cm}^{-3}$. The 50 nm-long access regions are doped to $10^{18}\,\mathrm{cm}^{-3}$. The temperature is 300 K. The scattering mechanisms considered are those due to polar optical phonons, acoustic phonons and ionized impurities, in a single Γ band with effective mass of 0.06 m_0. The transport algorithm is self-consistently coupled with the 1D Poisson equation.

Current–voltage characteristics are plotted in Fig. 5.7. The result obtained from the Wigner–Boltzmann model including scattering (circles, solid line) is compared with that given by the ballistic simulation for which scattering mechanisms have been artificially deactivated (squares, solid line) and with that obtained using a well-established ballistic Green's function technique self-consistently coupled to Poisson's equation [87]. An excellent agreement was found between both ballistic results, which suggests that the Wigner–Boltzmann Monte Carlo approach correctly handles the quantum transport effects including the resonance on a quasi-bound state. It is also clearly seen here that scattering effects dramatically reduce the peak-to-valley ratio. It is thus essential to consider them properly for room-temperature simulation of RTDs.

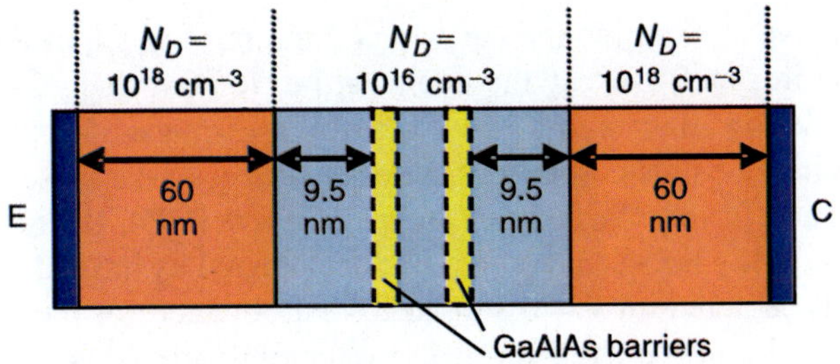

Fig. 5.6 Schematic cross-section of the simulated RTD. The GaAlAs barriers and the GaAs quantum well are 3 nm- and 5 nm-thick, respectively

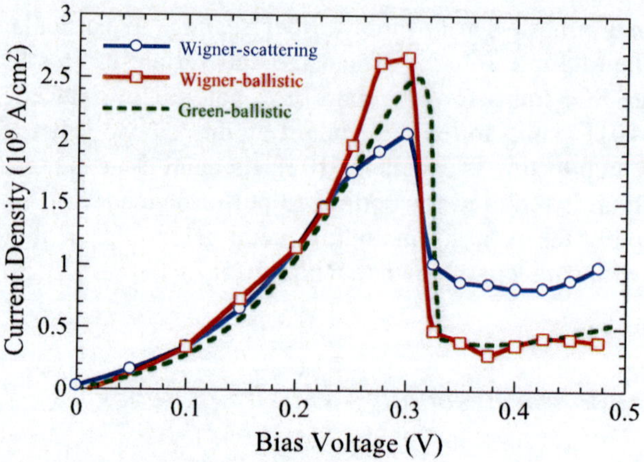

Fig. 5.7 $I-V$ characteristics of the RTD schematized in Fig. 5.6 obtained using Wigner MC simulation with scattering mechanisms activated (*circles, solid line*) or artificially deactivated (*squares, solid line*) and using ballistic Green's function simulation (*dashed line*)

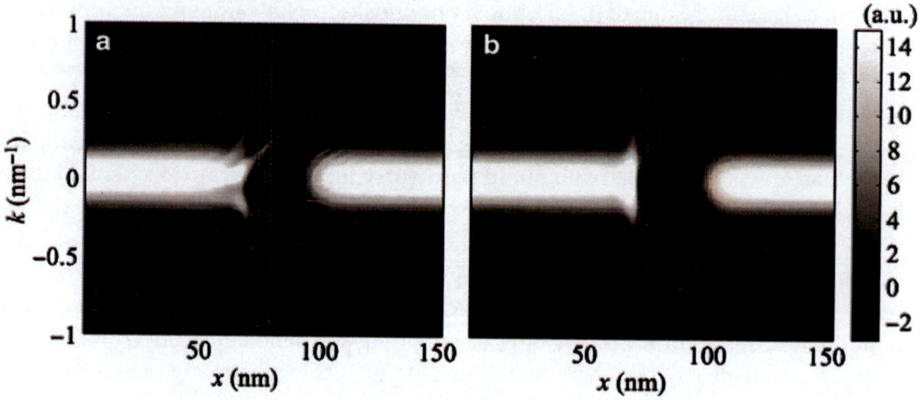

Fig. 5.8 Cartography in phase-space of the Wigner function computed (**a**) for a resonant state ($V = 0.3\,\text{V}$) and (**b**) a non-resonant state ($V = 0.475\,\text{V}$)

It is instructive to examine the cartography in phase-space of the Wigner function displayed in Fig. 5.8. Near the contacts, i.e. for $x < 40\,\text{nm}$ and $x > 120\,\text{nm}$, the Wigner function appears to be very close to a displaced Maxwellian function. The transport may be thus considered to be semi-classical in these regions. In contrast, the situation is very different in the quantum well. For the resonant state, i.e. $V = 0.3\,\text{V}$ (Fig. 5.8a), between the barriers schematized by dashed lines one can see a peak (a spot) centered on $k = 0$ similar to that obtained for the Wigner function associated with the first energy level of a quantum well. This peak is due to the contribution of electrons crossing the double-barrier through the resonant state in the well. For a non-resonant state (Fig. 5.8b) this peak vanishes and becomes almost invisible. It should also be noted that, in both cases, the oscillations of the Wigner function give rise to some negative values in small part of the phase-space (darkest shaded areas), which is the signature of quantum coherence.

The conduction band profiles plotted in Fig. 5.9 highlight the importance of the self-consistence for RTD simulation. In particular, when scattering is included a potential drop appears in the emitter region while the conduction band is flat in the ballistic case. This potential drop may induce an energy spreading of electrons, which modifies the resonant condition at $V = 0.3$ V for electrons reaching the double barrier and contributes to the suppression of current peak at the resonance.

As shown in Fig. 5.10, a peak of electron density appears in the quantum well under resonant bias ($V = 0.3$ V), which is in accordance with the spot observed on

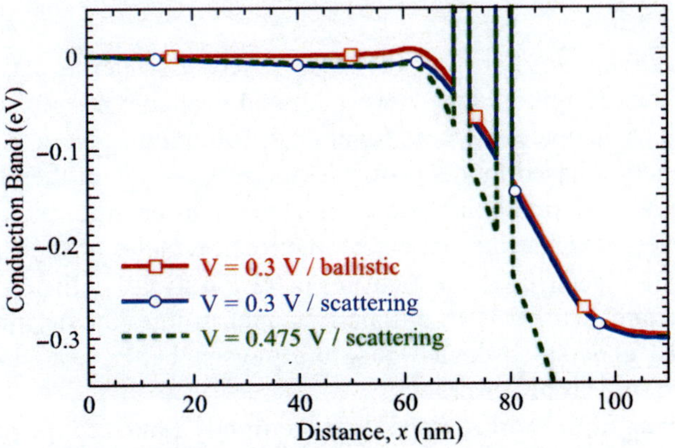

Fig. 5.9 Conduction band profile obtained by Wigner simulation, at peak ($V = 0.3$ V, *solid line, circles*) and valley ($V = 0.475$ V, *dashed line*) biases from simulation with scattering, and at peak bias (*solid line, squares*) from ballistic simulation

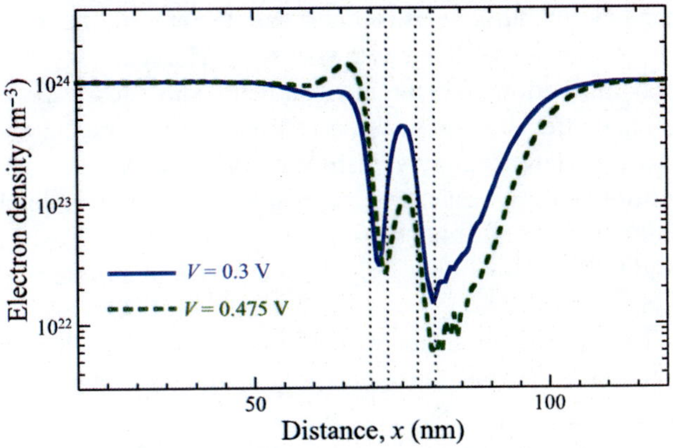

Fig. 5.10 Electron density in the RTD, obtained simulation at peak ($V = 0.3$ V, *solid line*) and valley ($V = 0.475$ V, *dashed line*) voltages from Wigner simulation with scattering

the Wigner function map (Fig. 5.8a). In off-resonance bias $V = 0.475\,\text{V}$, this peak suppresses and an electron accumulation is formed in front of the double-barrier as a consequence of its weak transparency.

Finally, it should be noted that the peak-to-valley ratio obtained for typical GaAlAs/GaAs RTDs at room temperature and 77 K have been found in good agreement with experimental data [65], which suggests this MC technique is actually able to provide realistic simulation results for nano-devices exhibiting quantum transport effects with significant rates of scattering.

5.2 *Interpretation of Device Behavior Through De-Coherence Theory*

Understanding quantum transport in the presence of scattering has always been a difficult problem. Originally, there were limited available models to approach this question in electron devices, where scattering is ubiquitous. Now, with the progress of Wigner function based models – as we have seen – and of the Green's function formalism, powerful simulation tools are starting to emerge, including relatively detailed physics of scattering. However, the interpretation of their results remains difficult. This is because we are tied in our vision to the collision-less picture of quantum mechanisms that is traditionally taught in introductory quantum mechanism class. To understand device quantum physics better, a novel point of view would be therefore highly desirable.

It is thus insightful to look in fields more tightly linked to quantum mechanics than electron devices for inspiration. Particularly, in atomic physics and quantum optics, de-coherence theory has been widely successful to understand the effect of an environment (source of scattering) on a quantum system. De-coherence theory studies how the intrication between a quantum system and its environment may emerge from their interaction. This tends to lead to a separation of the system states: two different system states can intricate differently with the environment. If the system was initially in a superposition of these two states, interference between them becomes impossible after intrication with the environment. This thus leads to a suppression of some coherence effects – and to the occurrence of a more classical behavior for the system since interference may vanish. With this point of view we can even see a sort of competition between quantum coherence, and scattering leading to classical behaviors. Many more details may be found in recent excellent textbooks like [105]. It is a good lead to see if this theory highly successful in atomic physics may apply to electron devices.

The Wigner function and the density matrix are used very often in atomic physics to study de-coherence. Besides, it is encouraging to realize that our derivation of the impact of phonon scattering in Sect. 3.2 is analogous to the models commonly used for de-coherence problems. Indeed, we considered a full system consisting of the system of interest (an electron) and its environment (a phonon mode). We performed advanced derivation on the full system and then went to a reduced Wigner function for the electron system only through a trace on the environment states (phonon numbers). It is thus very natural to look for phonon-induced de-coherence effects using our model.

In practical de-coherence studies, the density matrix is usually complementary to the Wigner function. Although our model computes a Wigner function, it is easy to switch from one formalism to another by appropriate Fourier transform. The emergence of semi-classical behaviors is very clear on the Wigner function due to the continuity between this formalism and Boltzmann's formalism. Quantum coherence is however more clearly identified in the density matrix elements.

5.2.1 Study of the Free Evolution of a Wave Packet in GaAs: Scattering-Induced De-Coherence

To understand how de-coherence occurs in electron devices we may start with a simple case: the propagation of a free wave packet. In collision-less quantum mechanics, wave packets tend to spread infinitely when propagating, becoming always more de-localized spatially, as seen in many textbooks. Is it the case in an electron device?

To answer this question we consider a simple Gaussian wave-packet

$$\psi(x) = N \exp\left[-\frac{(x-x_0)^2}{2\sigma^2}\right] \exp\left[ik_0 x\right] \tag{5.136}$$

the Wigner function of which is written

$$f_w(x,k) = N' \exp\left[-\frac{(x-x_0)^2}{\sigma^2}\right] \exp\left[-(k-k_0)^2\sigma^2\right] \tag{5.137}$$

where N and N' are normalization constants. Figures 5.11a and b show the cartography of the Wigner function and the density matrix (DM), respectively, associated with the initial state defined by $k_0 = 4 \times 10^8\,\mathrm{m}^{-1}$, $\sigma = 10\,\mathrm{nm}$. Figures 5.11c and e display the Wigner function of the wave packet after 130 fs of ballistic (no coupling with phonons) and diffusive (with phonon coupling) propagation, respectively. Phonon scattering tends to widespread the Wigner function over smaller wave vector and displacement values (Figs. 5.11e) than in the purely coherent case (Figs. 5.11c).

The density matrix allows us to analyze the situation in a smarter way. The DM associated with Wigner functions of Figs. 5.11c and e are plotted in Figs. 5.11d and f, respectively. In the ballistic case (Fig. 5.11d) all diagonal and off-diagonal elements grow from the initial state according to the natural coherent extension of the wave packet, as described in many textbooks of quantum mechanics. When including interactions with phonons (Fig. 5.11f), the result is very different. The diagonal elements still grow similarly but they extend over a larger range, as indicated by the

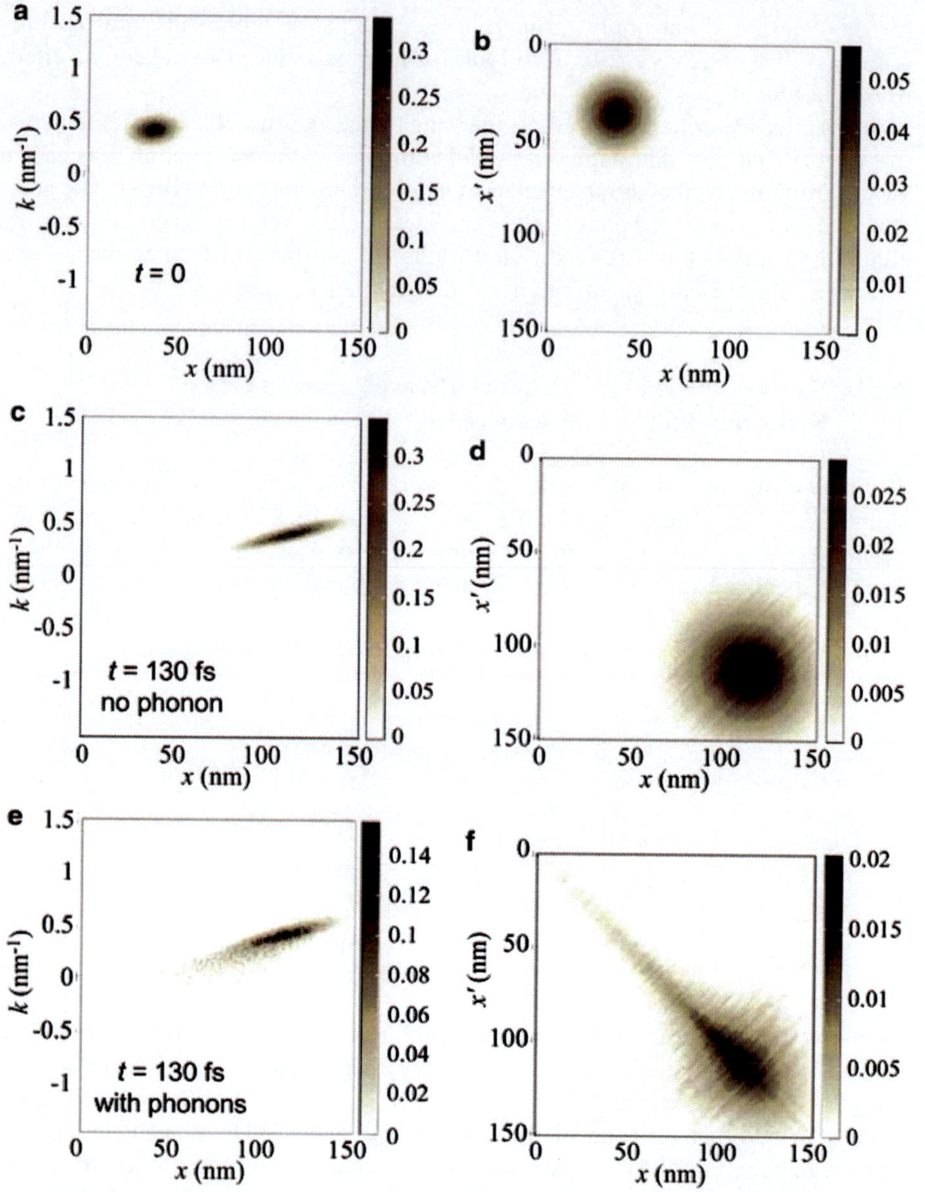

Fig. 5.11 Evolution of a free Gaussian wave packet coupled or uncoupled with a phonon bath at room temperature in GaAs. (**a**) Wigner function (WF) and (**b**) modulus of density matrix (DM) elements of the initial pure state. Simulated WF and DM after 130 fs without (**c**, **d**) or with (**e**, **f**) coupling to the phonon bath. DM elements are expressed in nm^{-1}

distribution tail at small x values. However, the off-diagonal elements do not extend as in the coherent case. They actually reduce as a function of time. It seems that actually the wave packet does not extend but splits into different wave packets which are not more de-localized than in the initial state. The quantum extension of the wave packet is inhibited by interactions with phonons. In other words, phonon scattering prevent the wave packet from de-localizing as in the case of free propagation. Many more details may be found in [10].

5.2.2 Reinterpretation of the RTD Behavior: De-Coherence and Quantum/Semi-Classical Transition

After this academic study of wave packets, we may turn to the simulation of the RTD presented in the previous section, including the same phonon and impurity scattering mechanisms.

Figure 5.8 shows the Wigner function of the RTD operating at peak voltage ($V = 0.3$ V). In a large part of access regions ($x < 30$ nm and $x > 120$ nm) the transport is essentially semi-classical and the Wigner function matches very well a semi-classical distribution function represented by a displaced Maxwellian function. Inside the quantum well the Wigner function around $k = 0$ is similar to that of the Wigner function of the first bound state in a square potential [106]. In the overall active region of the device, oscillations of the Wigner function reveal the presence of spatial coherence. Hence, there is apparently a transition between coherent quantum and semi-classical regions within the device. To understand better this behavior and the de-coherence effect, it is insightful to analyze the density matrix associated with the Wigner function for different strengths of electron–phonon scattering.

Accordingly, the density matrix is displayed in Figs. 5.12a–c for three different scattering situations. In Fig. 5.12a the transport is fully ballistic in the active region, which means that phonon scattering has been artificially switched off. In Fig. 5.12b standard scattering rates were used as for the Wigner function plotted in Fig. 5.8. In Fig. 5.12c phonon scattering rates have been artificially multiplied by five.

In the ballistic case a strong coherence is observed between electrons in the quantum well and in the emitter region. The amplitude of off-diagonal elements is even significant between electrons in collector and emitter regions, which is a clear indication of a coherent transport regime. When including standard scattering rates the off-diagonal elements are strongly reduced. When phonon scattering rates are artificially multiplied by five, the off-diagonal elements of the density matrix vanish, i.e. the coherence between electrons on left and right sides almost disappears. The process of double barrier tunneling is thus no longer fully resonant. Electrons can be seen as entering and leaving the quasi-bound state in distinct processes, with the possibility of energy exchange with the phonons. This illustrates the well-known coherent versus sequential tunneling situation.

This phonon-induced transition between coherent and sequential tunneling regimes manifests itself in the current–voltage characteristics of the RTD plotted in Fig. 5.13 for the three scattering situations. Phonon scattering tends to suppress the resonant tunneling peak while the valley current increases to such a point that the negative differential conductance effect almost disappears. The device tends to behave as two incoherent tunneling resistances connected in series for which a semi-classical-type description could be accurate enough.

Fig. 5.12 Density matrix of a RTD operating at peak voltage for three: (**a**) no scattering, (**b**) standard phonon scattering rates, (**c**) standard rates multiplied by 5

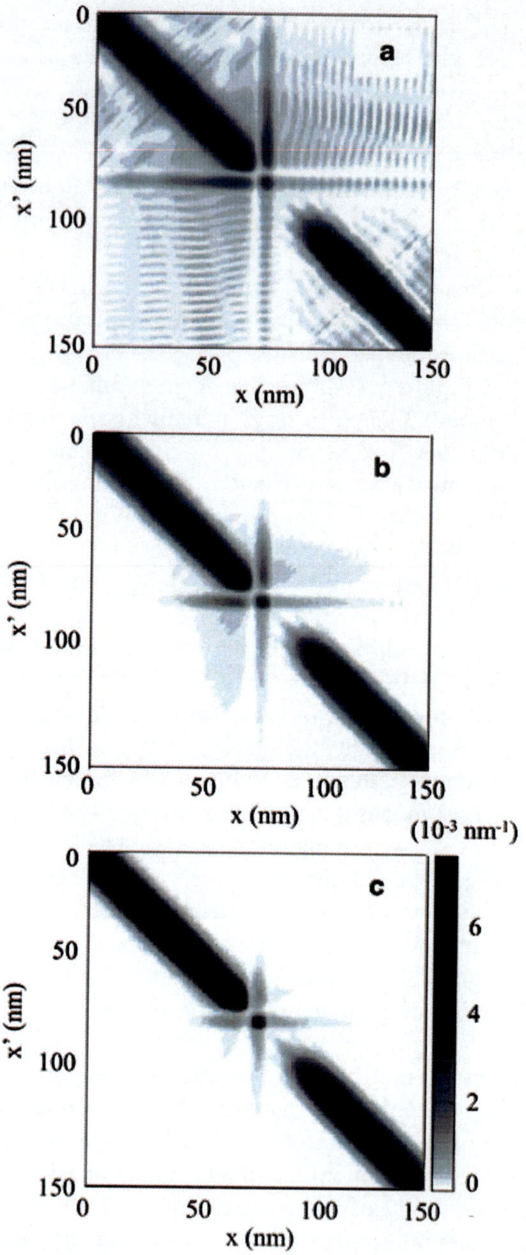

All these considerations give a clear view of how electrons are de-localized in the active part of the device and become more localized in the access region. As already observed from the Wigner function displayed in Fig. 5.8, this suggests a transition from "quantum" to "semiclassical" transport from the active region to the access ones. More advanced considerations about this transition may be found in [10].

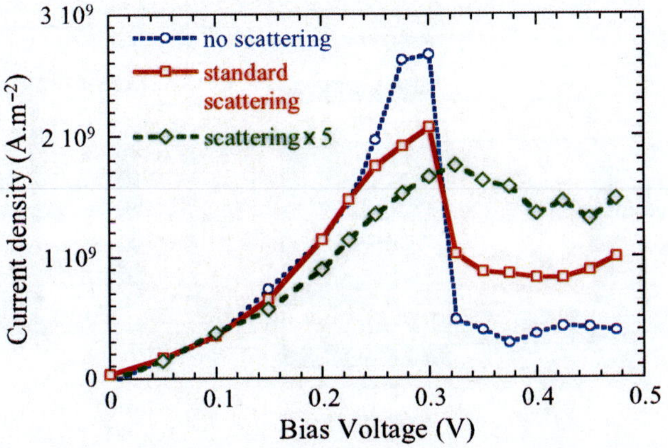

Fig. 5.13 $I-V$ characteristics for the RTD obtained from Wigner simulation, with scattering artificially deactivated (*empty circles*), with standard scattering (*squares*), and scattering rates artificially multiplied by 5 (*diamonds*)

5.3 Application to Nano-Scale Transistors

As a last illustration of application of the Wigner–Boltzmann MC method, we present here some results obtained for the ultra-small MOSFET with self-aligned double-gate, schematized in Fig. 5.14. It is inspired by the recommendations of the 2005 and 2007 ITRS Edition for the High-Performance 16 nm technology node [93] scheduled to be available in 2019. This DG-MOSFET structure is typical of a possible design for implementation in standard CMOS technology in the future.

The gate length is $L_G = 6$ nm, the silicon film thickness is $T_{Si} = 3$ nm and the equivalent gate oxide thickness is aggressively scaled to $EOT = 0.5$ nm. The source and drain access are 15 nm long and doped to $5 \times 10^{19}\,\mathrm{cm}^{-3}$. The gate metal work function is 4.36 eV and the supply voltage is $V_{DD} = 0.7$ V. The tunneling through gate oxide layers is not considered here. It is assumed indeed that silicon oxide may be replaced by high-κ material of same EOT and higher physical thickness to control this effect without degrading the interface quality. All simulations were performed at room temperature.

The DG-MOSFET is simulated here in the multi-sub-band mode-space approximation which decouples the gate-to-gate z direction and the xy plane parallel to interfaces. Assuming the potential V to be y-independent, the formation of uncoupled sub-bands may be simply deduced from the effective 1D Schrödinger's equation to be solved at each position x_i in the channel self-consistently with 2D Poisson's equation. Each resulting sub-band profile $E_n(x)$ is used as potential energy for the particle transport along the source-to-drain axis in the sub-band. The transport can be treated either semi-classically using the Boltzmann algorithm or in a quantum way using the Wigner–Boltzmann method. In this approach, the sub-bands

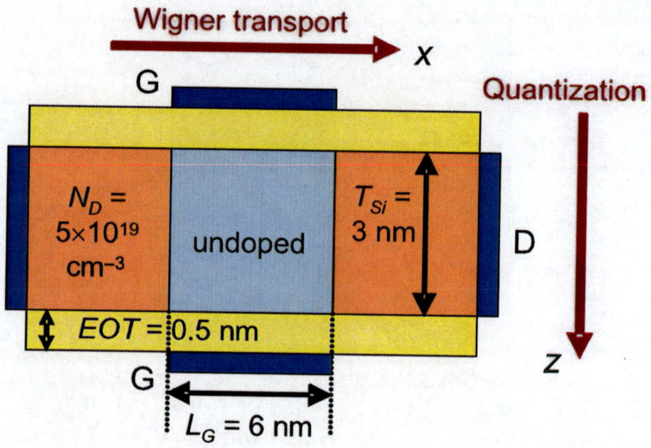

Fig. 5.14 Schematic cross-section of the simulated DG-MOSFET structure

are assumed to be independent and coupled only by scattering mechanisms. In its semi-classical form this technique has been developed in several groups [107–110].

To treat the 2D electron gas, the MC procedure makes use here of scattering rates calculated according to the envelope functions whose dependence on time and position generates an additional difficulty. In contrast to the case of standard Monte Carlo simulation, it is no longer possible to store the scattering rates in a look-up table prior to the simulation. They have to be regularly updated throughout the simulation. Phonon and ionized impurity scattering rates are derived as in [111] where 2D electron mobility in Si/SiGe heterostructures was calculated in good agreement with experimental data. The oxide interface roughness scattering rate is calculated by considering both the classical effect of electrostatic potential fluctuations [112] and the quantum effect on eigen-energies [113] which becomes significant for Si film thickness smaller than 5 nm [114]. Standard parameters, i.e. root-mean-square $\Delta_m = 0.5$ nm and correlation length $L_C = 1.5$ nm, are used to characterize the surface roughness.

5.3.1 Quantum Transport Effects

First of all, we look at the current–voltage characteristics of the transistor. The transfer characteristics $I_D - V_{GS}$ obtained at room temperature are plotted in Figs. 5.15 and 5.16 for low and high drain bias, respectively. In these figures the Wigner simulation results are systematically compared with that of two other mode-space approaches: (i) the comparison with the semi-classical Boltzmann MC model (triangles, solid lines) which includes scattering will show the impact of quantum transport and (ii) the comparison with a quantum ballistic model based on the non-equilibrium Green's function formalism (NEGF) (circles, dashed lines) [115] will show the impact of scattering.

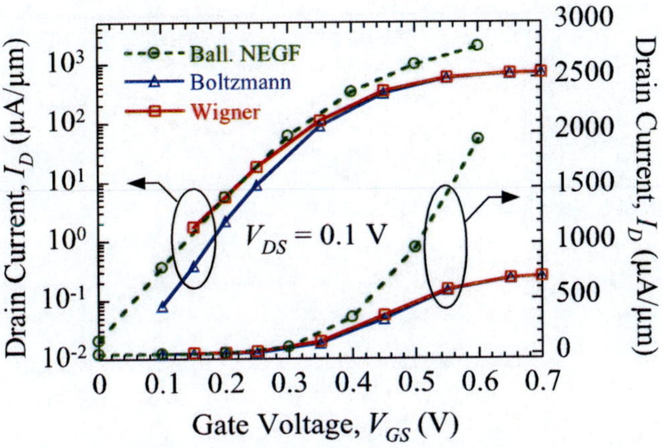

Fig. 5.15 Transfer characteristics obtained at $V_{DS} = 0.1\,\text{V}$ using three types of mode-space simulation, i.e. Wigner MC (*squares, solid lines*) Boltzmann MC (*triangles, solid lines*) and ballistic Green's function (*circles, dashed lines*). Both MC simulations include scattering. Results are displayed in both log and linear scale. $T = 300\,\text{K}$

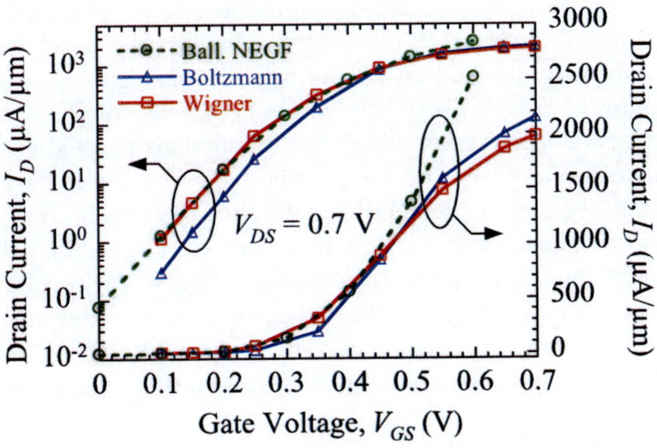

Fig. 5.16 Transfer characteristics obtained at $V_{DS} = 0.7\,\text{V}$ using three types of mode-space simulation, i.e. Wigner MC (*squares, solid lines*) Boltzmann MC (*triangles, solid lines*) and ballistic Green's function (*circles, dashed lines*). $T = 300\,\text{K}$

Let us first consider the results obtained at low V_{GS} (subthreshold regime) and low V_{DS} (see Fig. 5.15). Wigner and Boltzmann curves are very different in this regime. The semi-classical simulation gives a better subthreshold slope than the quantum approach (70 mV dec^{-1} vs 80 mV dec^{-1}) and an off-state current I_{OFF} (extrapolated at $V_{GS} = 0\,\text{V}$) five times smaller. The subthreshold current is thus strongly influenced by quantum transport at this ultra-small gate length, which may be easily understood. The additional current is nothing but a tunneling current of electrons flowing from the source to the drain through the gate-induced potential barrier.

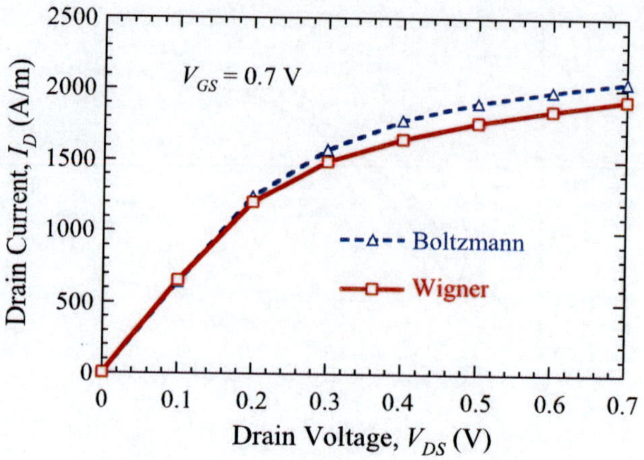

Fig. 5.17 Drain current as a function of drain voltage obtained at $V_{GS} = 0.7\,\text{V}$ using Wigner MC (*squares, solid lines*) and Boltzmann MC (*triangles, dashed lines*) simulation

Since the source-drain tunneling current is especially strong in the subthreshold regime, it is interesting to compare quantum models, i.e. Wigner MC and ballistic NEGF results. It is remarkable that they coincide closely, which confirms that scattering mechanisms have a very small impact in this regime.

The situation is dramatically different at high gate voltage. One observes in Figs. 5.15 and 5.16 that Wigner and Green simulations provide very different results, which means that scattering has an important influence on the current, both at low V_{DS} (Ohmic regime) and at high V_{DS} (saturation regime). In contrast, the Wigner current becomes quite close to the Boltzmann one and even similar at low V_{DS}. Surprisingly enough, by looking at the currents obtained at high V_{DS} (Fig. 5.16), one can observe that beyond a given gate voltage the Wigner current becomes smaller than the Boltzmann current [62]. To understand this behavior the $I_D - V_{DS}$ characteristics obtained at $V_{GS} = V_{DD} = 0.7\,\text{V}$ from both Wigner and Boltzmann models are plotted in Fig. 5.17.

As already remarked just above, both currents are very similar at low V_{DS}, which suggests that quantum transport effects are negligible in Ohmic regime. At higher drain voltage two quantum effects compete. In one hand the tunneling source-drain current tends to enhance the total drain current, but on the other hand quantum reflections may occur at high drain bias due to the sharp potential drop at the drain-end of the channel, which contributes to reducing the drain current. Actually, the height of the gate-induced barrier being small in this regime the contribution of the tunneling current becomes quite weak, which makes the reflection effect significant. More details on this effect may be found in [62].

To illustrate these quantum effects the phase-space cartography of the Wigner function in the first sub-band is compared to that of the Boltzmann function in Fig. 5.18 at given bias $V_{GS} = 0.45\,\text{V}$ and $V_{DS} = 0.7\,\text{V}$. Both functions are very similar in the source region. The main feature of the Boltzmann function in the channel is the stream of hot electrons which forms the ballistic peak (Fig. 5.18a). In contrast,

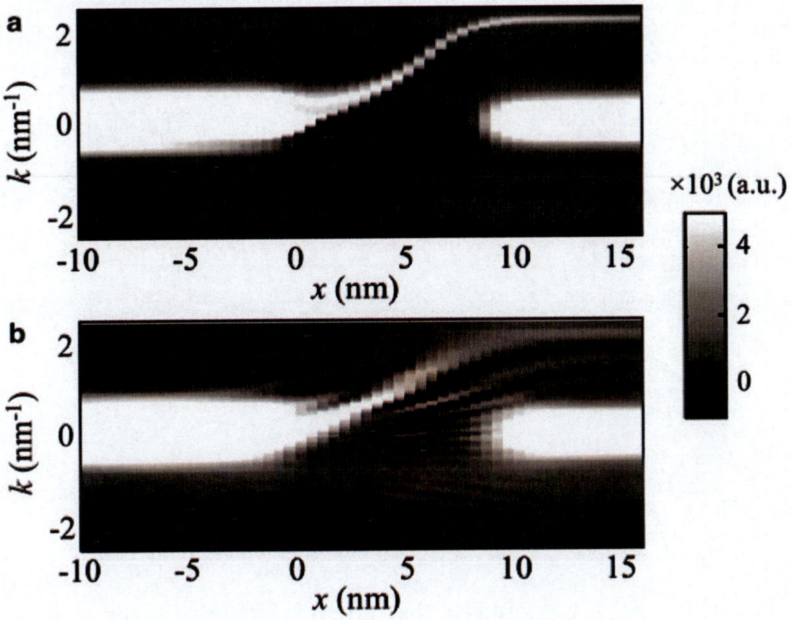

Fig. 5.18 Cartography of (**a**) Boltzmann and (**b**) Wigner functions of the first sub-band for VGS = 0.45 V and VDS = 0.7 V. The gated part of the channel extends from x = 0 to x = 6 nm

though this peak is still visible on the Wigner function (Fig. 5.18b), strong positive/negative oscillations of the Wigner function are observed where the quantum reflections occur, i.e. in the part of the channel falls abruptly, between the top of the barrier and the drain-end.

5.3.2 Impact of Scattering

We now examine the impact of scattering on device performance and operation above threshold voltage since it has been shown to be important at high gate voltage V_{GS}. In conventional MOSFET with long gate, the current is proportional to the carrier mobility in the channel. It is thus strongly dependent on scattering in the channel. In nano-transistors the channel resistance is reduced and may become comparable to that in the access regions. Hence, scattering in the access might have a significant influence on the device characteristics.

To understand the overall impact of scattering in the different parts of the device, transfer characteristics are compared in Fig. 5.19. Results of three types of simulation are plotted: (a) Ballistic Green's function method ("Ball. NEGF"), with ballistic transport in both access regions and in channel, (b) Wigner MC with scattering everywhere ("Wigner") and (c) Wigner MC with scattering activated in the access regions but deactivated in the channel ("Wigner–Ball. Channel").

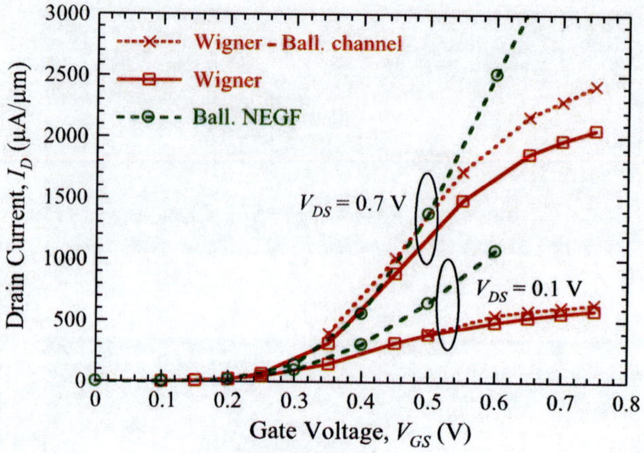

Fig. 5.19 Transfer characteristics for $V_{DS} = 0.1$ V and $V_{DS} = 0.7$ V. The results are shown for three types of simulation: ballistic NEGF, Wigner MC and Wigner MC with all scattering mechanisms deactivated in the gated part of the channel

In Ohmic regime, i.e. at low V_{DS}, the current is strongly limited by access resistances for $V_{GS} > 0.5$ V. For $V_{GS} = 0.6$ V the Wigner drain current is two times smaller than in the NEGF simulation. The source access resistance reaches $140\,\Omega\,\mu$m while the target value of ITRS 2005 was $60\,\Omega\,\mu$m only. This problem is critical in ultra-thin structures where T_{Si} is reduced to control short-channel effects. However, it should be noted that in the 2007 edition the ITRS target for HP16 node is raised to $145\,\Omega\,\mu$m, i.e. close to the simulated value.

At high V_{DS} the impact of scattering is less pronounced but still important. The transconductance $g_m = \partial I_D / \partial V_{GS}$ is frequently used as factor of merit to assess the transistor performance. Ballistic NEGF simulation strongly overestimates g_m which appears to be limited by scattering occurring both in the access and in the channel. With ballistic channel and scattering only in access regions, the transconductance is improved by 18% with respect to standard Wigner simulation ($7090\,\mu$A$\,\mu$m^{-1} instead of $5970\,\mu$A$\,\mu$m^{-1}) and the ON-current I_{ON} is enhanced by 16% ($2290\,\mu$A$\,\mu$m^{-1} instead of $1970\,\mu$A$\,\mu$m^{-1}). Thus, in spite of the strong part of ballistic transport in ultra-short MOSFET [116], scattering still has a significant influence, both in the channel and the highly-doped source access region. When artificially enhancing the scattering rates in DG-MOSFET, the detailed analysis of de-coherence has shown that scattering plays an important role in the emergence of the semi-classical behavior at longer gate length, i.e. of the localization of electrons [21].

Acknowledgments This work has been partially supported by the: **European Community** through Projects PULLNANO (FP6-IST-026828), SINANO (FP6-IST-506844) and NANOSIL (FP7-ICT-216171); **French Agence Nationale de la Recherche** through Project MODERN (ANR-05-NANO-02); **Austrian Science Fund** through Projects FWF-P21685, P17285-N02 and F2509 (SFB 025 IR-ON).

References

1. H. Weyl, "Quantenmechanik und Gruppentheorie," *Zeitschrift fr Physik*, vol. 46, pp. 1–46, 1927.
2. E. Wigner, "On the quantum corrections for thermodynamic equilibrium," *Physical Review*, vol. 40, pp. 749–759, 1932.
3. J. E. Moyal, "Quantum mechanics as a statistical theory," *Proceedings of the Cambridge Philosophical Society*, vol. 45, pp. 99–124, 1949.
4. V. I. Tatarskii, "The Wigner Representation of Quantum Mechanics," *Sov. Phys. Usp.*, vol. 26, pp. 311–327, 1983.
5. N. C. Dias and J. N. Prata, "Admissible states in quantum phase space," *Annals of Physics*, vol. 313, pp. 110–146, 2004.
6. D. K. Ferry, R. Akis, and J. P. Bird, "Einselection in action: decoherence and pointer states in open quantum dots," *Physical Review Letters*, vol. 93, p. 026803, 2004.
7. I. Knezevic, "Decoherence due to contacts in ballistic nanostructures," *Physical Review B*, vol. 77, p. 125301, 2008.
8. F. Buscemi, P. Bordone, and A. Bertoni, "Simulation of decoherence in 1D systems, a comparison between distinguishable- and indistinguishable-particle collisions," *Physica Status Solidi (c)*, vol. 5, pp. 139–142, 2008.
9. F. Buscemi, E. Cancellieri, P. Bordone, A. Bertoni, and C.Jacoboni, "Electron decoherence in a semiconductor due to electron-phonon scattering," *Physica Status Solidi (c)*, vol. 5, pp. 52–55, 2008.
10. D. Querlioz, J. Saint-Martin, A. Bournel, and P. Dollfus, "Wigner Monte Carlo simulation of phonon induced electron decoherence in semiconductor nanodevices," *Physical Review B*, vol. 78, p. 165306, 2008.
11. R. Balescu, *Equilibrium and Nonequilibrium Statistical Mechanics*. Wiley and Sons, 1975.
12. N. C. Kluksdahl, A. M. Kriman, D. K. Ferry, and C. Ringhofer, "Self-consistent study of resonant-tunneling diode," *Physical Review B*, vol. 39, pp. 7720–7734, 1989.
13. A. Gehring and H. Kosina, "Wigner-Function Based Simulation of Classic and Ballistic Transport in Scaled DG-MOSFETs Using the Monte Carlo Method," *Journal of Compuational Electronics*, vol. 4, pp. 67–70, 2005.
14. P. Carruthers and F. Zachariasen, "Quantum Collision Theory with Phase-Space Distributions," *Rev.Mod.Phys.*, vol. 55, no. 1, pp. 245–285, 1983.
15. B. Biegel and J. Plummer, "Comparison of self-consistency iteration options for the Wigner function method of quantum device simulation," *Physical Review B*, vol. 54, pp. 8070–8082, 1996.
16. W. Frensley, "Wigner-Function Model of Resonant-Tunneling Semiconductor Device," *Physical Review B*, vol. 36, no. 3, pp. 1570–1580, 1987.
17. W. Frensley, "Boundary conditions for open quantum systems driven far from equilibrium," *Reviews of Modern Physics*, vol. 62, no. 3, pp. 745–789, 1990.
18. K. Gullapalli, D. Miller, and D. Neikirk, "Simulation of quantum transport in memory-switching double-barrier quantum-well diodes," *Physical Review B*, vol. 49, pp. 2622–2628, 1994.
19. F. A. Buot and K. L. Jensen, "Lattice Weil-Wigner Formulation of Exact-Many Body Quantum-Transport Theory and Applications to Novel Solid-State Quantum-Based Devices," *Physical Review B*, vol. 42, no. 15, pp. 9429–9457, 1990.
20. R. K. Mains and G. I. Haddad, "Wigner function modeling of resonant tunneling diodes with high peak-to-valley ratios," *Journal of Applied Physics*, vol. 64, pp. 5041–5044, 1988.
21. D. Querlioz, H. N. Nguyen, J. Saint-Martin, A. Bournel, S. Galdin-Retailleau, and P. Dollfus, "Wigner-Boltzmann Monte Carlo approach to nanodevice simulation: from quantum to semiclassical transport," *Journal of Computational Electronics*, vol. 8, pp. 324–335, 2009.
22. M. Nedjalkov, "Wigner transport in presence of phonons: Particle models of the electron kinetics," in *From Nanostructures to Nanosensing Applications, Proceedings of the International School of Physics 'Enrico Fermi'* (A. P. A. D'Amico, G. Balestrino, ed.), vol. 160, (Amsterdam), pp. 55–103, IOS Press, 2005.

23. F. Rossi, C.Jacoboni, and M.Nedjalkov, "A Monte Carlo Solution of the Wigner Transport Equation," *Semiconductor Sci. Technology*, vol. 9, pp. 934–936, 1994.
24. P. Bordone, M. Pascoli, R. Brunetti, A. Bertoni, and C. Jacoboni, "Quantum transport of electrons in open nanostructures with the Wigner-function formalism," *Physical Review B*, vol. 59, no. 4, pp. 3060–3069, 1999.
25. I. Levinson, "Translational invariance in uniform fields and the equation for the density matrix in the Wigner representation," *Soviet Phys. JETP*, vol. 30, no. 2, pp. 362–367, 1970.
26. J. R. Barker and D. K. Ferry, "Self-Scattering Path-Variable Formulation of High Field, Time-Dependent Quantum Kinetic Equations for Semiconductor Transport in the Finite-Collision-Duration Regime," *Physical Review Letters*, vol. 42, no. 26, pp. 1779–1781, 1979.
27. M. Nedjalkov, D. Vasileska, D. Ferry, C. Jacoboni, C. Ringhofer, I. Dimov, and V. Palankovski, "Wigner transport models of the electron-phonon kinetics in quantum wires," *Physical Review B*, vol. 74, pp. 035311–1–035311–18, July 2006.
28. J. Schilp, T. Kuhn, and G. Mahler, "Electron-phonon quantum kinetics in pulse-excited semiconductors: Memory and renormalization effects," *Physical Review B*, vol. 50, no. 8, pp. 5435–5447, 1994.
29. C. Fuerst, A. Leitenstorfer, A. Laubereau, and R. Zimmermann, "Quantum Kinetic Electron-Phonon Interaction in GaAs: Energy Nonconserving Scattering Events and Memory Effects," *Physical Review Letters*, vol. 78, pp. 3733–3736, 1997.
30. P. Bordone, D. Vasileska, and D. Ferry, "Collision-Duration Time for Optical-Phonon Emission in Semiconductors," *Physical Review B*, vol. 53, no. 7, pp. 3846–3855, 1996.
31. T. Kuhn and F. Rossi, "Monte Carlo Simulation of Ultrafast Processes in Photoexcited Semiconductors: Coherent and Incoherent Dynamics," *Physical Review B*, vol. 46, pp. 7496–7514, 1992.
32. F. Rossi and T. Kuhn, "Theory of Ultrafast Phenomena in Photoexcited Semiconductors," *Reviews of Modern Physics*, vol. 74, pp. 895–950, July 2002.
33. K. Thornber, "High-field electronic conduction in insulators," *Solid-State Electron.*, vol. 21, pp. 259–266, 1978.
34. J. Barker and D. Ferry, "On the Physics and Modeling of Small Semiconductor Devices–I," *Solid-State Electron.*, vol. 23, pp. 519–530, 1980.
35. M. V. Fischetti, "Monte Carlo Solution to the Problem of High-Field Electron Heating in SiO_2," *Physical Review Letters*, vol. 53, no. 3, p. 1755, 1984.
36. C. Jacoboni, A. Bertoni, P. Bordone, and R. Brunetti, "Wigner-function Formulation for Quantum Transport in Semiconductors: Theory and Monte Carlo Approach," *Mathematics and Computers in Simulations*, vol. 55, no. 1-3, pp. 67–78, 2001.
37. P. Bordone, A. Bertoni, R. Brunetti, and C. Jacoboni, "Monte Carlo simulation of quantum electron transport based on Wigner paths," *Mathematics and Computers in Simulation*, vol. 62, p. 307, 2003.
38. P. Lipavski, F. Khan, F. Abdolsalami, and J. Wilkins, "High-Field Transport in Semiconductors. I. Absence of the Intra-Collisional Field Effect," *Physical Review B*, vol. 43, no. 6, pp. 4885–4896, 1991.
39. T. Gurov, M. Nedjalkov, P. Whitlock, H. Kosina, and S. Selberherr, "Femtosecond relaxation of hot electrons by phonon emission in presence of electric field," *Physica B*, vol. 314, pp. 301–304, 2002.
40. M. Nedjalkov, D. Vasileska, E. Atanassov, and V. Palankovski, "Ultrafast Wigner Transport in Quantum Wires," *Journal of Computational Electronics*, vol. 6, pp. 235–238, 2007.
41. C. Ringhofer, M. Nedjalkov, H. Kosina, and S. Selberherr, "Semi-Classical Approximation of Electron-Phonon Scattering beyond Fermi's Golden Rule," *SIAM Journal of Applied Mathematics*, vol. 64, pp. 1933–1953, 2004.
42. M. Herbst, M. Glanemann, V. Axt, and T. Kuhn, "Electron-phonon quantum kinetics for spatially inhomogenenous excitations," *Phisical Review B*, vol. 67, pp. 195305–1–195305–18, 2003.
43. P. Bordone, A. Bertoni, R. Brunetti, and C. Jacoboni, "Monte Carlo Simulation of Quantum Electron Transport Based on Wigner Paths," *Mathematics and Computers in Simulation*, vol. 62, pp. 307–314, 2003.

44. R. Brunetti, C. Jacoboni, and F. Rossi, "Quantum theory of transient transport in semiconductors: A Monte Carlo approach," *Physical Review B*, vol. 39, pp. 10781–10790, May 1989.
45. B. K. Ridley, *Quantum processes in semiconductors*. Oxford University Press, fourth ed., 1999.
46. K.-Y. Kim and B. Lee, "On the high order numerical calculation schemes for the Wigner transport equation," *Solid-State Electronics*, vol. 43, pp. 2243–2245, 1999.
47. Y. Yamada, H. Tsuchiya, and M. Ogawa, "Quantum Transport Simulation of Silicon-Nanowire Transistors Based on Direct Solution Approach of the Wigner Transport Equation," *IEEE Trans. Electron Dev.*, vol. 56, pp. 1396–1401, 2009.
48. S. Barraud, "Phase-coherent quantum transport in silicon nanowires based on Wigner transport equation: Comparison with the nonequilibrium-Green-function formalism," *Journal of Applied Physics*, vol. 106, p. 063714, 2009.
49. H. Tsuchiya and U. Ravaioli, "Particle Monte Carlo Simulation of Quantum Phenomena in Semiconductor Devices," *J.Appl.Phys.*, vol. 89, pp. 4023–4029, April 2001.
50. R. Sala, S. Brouard, and G. Muga, "Wigner Trajectories and Liouville's theorem," *J. Chem. Phys.*, vol. 99, pp. 2708–2714, 1993.
51. P. Vitanov, M. Nedjalkov, C. Jacoboni, F. Rossi, and A. Abramo, "Unified Monte Carlo Approach to the Boltzmann and Wigner Equations," in *Advances in Parallel Algorithms* (Bl. Sendov and I. Dimov, eds.), pp. 117–128, IOS Press, 1994.
52. D. Ferry, R. Akis, and D. Vasileska, "Quantum Effect in MOSFETs: Use of an Effective Potential in 3D Monte Carlo Simulation of Ultra-Schort Channel Devices," *Int.Electron Devices Meeting*, pp. 287–290, 2000.
53. L. Shifren, R. Akis, and D. Ferry, "Correspondence Between Quantum and Classical Motion: Comparing Bohmian Mechanics with Smoothed Effective Potential Approach," *Phys.Lett.A*, vol. 274, pp. 75–83, 2000.
54. S. Ahmed, C. Ringhofer, and D. Vasileska, "An Effective Potential Aprroach to Modeling 25nm MOSFET Devices," *Journal of Computational Electronics*, vol. 2, pp. 113–117, 2003.
55. C. Ringhofer, C. Gardner, and D. Vasileska, "An Effective Potentials and Quantum Fluid Models: A Thermodynamic Approach," *Journal of High Speed Electronics and Systems*, vol. 13, pp. 771–801, 2003.
56. S. Haas, F. Rossi, and T. Kuhn, "Generalized Monte Carlo approach for the study of the coherent ultrafast carrier dynamics in photoexcited semiconductors," *Physical Review B*, vol. 53, no. 12, pp. 12855–12868, 1996.
57. M. Nedjalkov, I. Dimov, F. Rossi, and C. Jacoboni, "Convergency of the Monte Carlo Algorithm for the Wigner Quantum Transport Equation," *Journal of Mathematical and Computer Modelling*, vol. 23, no. 8/9, pp. 159–166, 1996.
58. K. L. Jensen and F. A. Buot, "The Methodology of Simulating Particle Trajectories Trough Tunneling Structures Using a Wigner Distribution Approach," *IEEE Trans.Electron Devices*, vol. 38, no. 10, pp. 2337–2347, 1991.
59. H. Tsuchiya and T. Miyoshi, "Simulation of Dynamic Particle Trajectories through Resonant-Tunneling Structures based upon Wigner Distribution Function," *Proc. 6th Int. Workshop on Computational Electronics IWCE6, Osaka*, pp. 156–159, 1998.
60. M. Pascoli, P. Bordone, R. Brunetti, and C. Jacoboni, "Wigner Paths for Electrons Interacting with Phonons," *Physical Review B*, vol. B 58, pp. 3503–3506, 1998.
61. V. Sverdlov, A. Gehring, H. Kosina, and S. Selberherr, "Quantum transport in ultra-scaled double-gate MOSFETs: A Wigner function-based Monte Carlo approach," *Solid-State Electronics*, vol. 49, pp. 1510–1515, 2005.
62. D. Querlioz, J. Saint-Martin, V. N. Do, A. Bournel, and P. Dollfus, "A Study of Quantum Transport in End-of-Roadmap DG-MOSFETs Using a Fully Self-Consistent Wigner Monte Carlo Approach," *IEEE Trans. Nanotechnology*, vol. 5, pp. 737–744, 2006.
63. D. Querlioz, J. Saint-Martin, V. N. Do, A. Bournel, and P. Dollfus, "Fully quantum self-consistent study of ultimate DG-MOSFETs including realistic scattering using a Wigner Monte-Carlo approach," *Int. Electron Device Meeting Tech. Dig. (IEDM)*, pp. 941–944, 2006.

64. L. Shifren and D. K. Ferry, "A Wigner function based ensemble Monte Carlo approach for accurate incorporation of quantum effects in device simulation," *Journal of Computational Electronics*, vol. 1, pp. 55–58, 2002.
65. D. Querlioz, P. Dollfus, V. N. Do, A. Bournel, and V. L. Nguyen, "An improved Wigner Monte-Carlo technique for the self-consistent simulation of RTDs," *Journal of Computational Electronics*, vol. 5, pp. 443–446, 2006.
66. D. Querlioz and P. Dollfus, *The Wigner Monte Carlo Method for Nanoelectronic Devices - A particle description of quantum transport and decoherence*. ISTE-Wiley, 2010.
67. M. Nedjalkov, H. Kosina, S. Selberherr, C. Ringhofer, and D. K. Ferry, "Unified particle approach to Wigner-Boltzmann transport in small semiconductor devices," *Physical Review B*, vol. 70, p. 115319, 2004.
68. A. Bertoni, P. Bordone, G. Ferrari, N. Giacobbi, and C. Jacoboni, "Proximity effect of the contacts on electron transport in mesoscopic devices," *Journal of Computational Electronics*, vol. 2, pp. 137–140, 2003.
69. C. Jacoboni and L. Reggiani, "The Monte Carlo Method for the Solution of Charge Transport in Semiconductors with Applications to Covalent Materials," *Rev.Mod.Phys.*, vol. 55, no. 3, pp. 645–705, 1983.
70. H. Kosina, "Wigner function approach to nano device simulation," *International Journal of Computational Science and Engineering*, vol. 2, no. 3/4, pp. 100 – 118, 2006.
71. S. Ermakow, *Die Monte-Carlo-Methode und verwandte Fragen*. München, Wien: R. Oldenburg Verlag, 1975.
72. J. Hammersley and D. Handscomb, *Monte Carlo Methods*. New York: John Wiley, 1964.
73. H. Kosina and M. Nedjalkov, *Handbook of Theoretical and Computational Nanotechnology*, vol. 10, ch. Wigner Function Based Device Modeling, pp. 731–763. Los Angeles: American Scientific Publishers, 2006.
74. M. Nedjalkov, R. Kosik, H. Kosina, and S. Selberherr, "Wigner Transport Through Tunneling Structures - Scattering Interpretation of the Potential Operator," in *Proc. Simulation of Semiconductor Processes and Devices*, (Kobe, Japan), pp. 187–190, Publication Office Business Center for Academic Societies Japan, 2002.
75. H. Kosina, M. Nedjalkov, and S. Selberherr, "A Monte Carlo Method Seamlessly Linking Classical and Quantum Transport Calculations," *Journal of Compuational Electronics*, vol. 2, no. 2-4, pp. 147–151, 2003.
76. H. Kosina, V. Sverdlov, and T. Grasser, "Wigner Monte Carlo Simulation: Particle Annihilation and Device Applications," in *Proc. Simulation of Semiconductor Processes and Devices*, (Monterey, CA, USA), pp. 357–360, Institute of Electrical and Electronics Engineers, Inc., Sept. 2006.
77. R. Tsu and L. Esaki, "Tunneling in a finite superlattice," *Appl. Phys. Lett.*, vol. 22, pp. 562–564, 1973.
78. L. L. Chang, L. Esaki, and R. Tsu, "Resonant tunneling in semiconductor double barriers," *Appl. Phys. Lett.*, vol. 24, pp. 593–595, 1974.
79. T. J. Shewchuk, P. C. Chapin, P. D. Coleman, W. Kopp, R. Fischer, and H. Morkoç, "Resonant Tunneling Oscillations in a GaAs-AlxGa1-xAs Heterostructure at Room-Temperature," *Appl. Phys. Lett.*, vol. 46, pp. 508–510, 1985.
80. H. Mizuta and T. Tanoue, *The physics and applications of resonant tunnelling diodes*. Cambridge University Press, 1995.
81. G. Iannaccone, G. Lombardi, M. Macucci, and B. Pellegrini, "Enhanced Shot Noise in Resonant Tunneling: Theory and Experiment," *Phys. Rev. Lett.*, vol. 80, pp. 1054–1057, 1998.
82. Y. M. Blanter and M. Büttiker, "Transition from sub-Poissonian to super-Poissonian shot noise in resonant quantum wells," *Phys. Rev. B*, vol. 59, pp. 10217–10226, 1999.
83. W. Song, E. E. Mendez, V. Kuznetsov, and B. Nielsen, "Shot noise in negative-differential-conductance devices," *Appl. Phys. Lett.*, vol. 82, pp. 1568–1570, 2003.
84. S. S. Safonov, A. K. Savchenko, D. A. Bagrets, O. N. Jouravlev, Y. V. Nazarov, E. H. Linfield, and D. A. Ritchie., "Transition from sub-Poissonian to super-Poissonian shot noise in resonant quantum wells," *Phys. Rev. Lett.*, vol. 91, p. 136801, 2003.

85. X. Oriols, A. Trois, and G. Blouin, "Self-consistent simulation of quantum shot noise in nanoscale electron devices," *Appl. Phys. Lett.*, vol. 85, pp. 3596–3598, 2004.
86. V. Y. Aleshkin, L. Reggiani, N. V. Alkeev, V. E. Lyubchenko, C. N. Ironside, J. M. L. Figueiredo, and C. R. Stanley, "Coherent approach to transport and noise in double-barrier resonant diodes," *Phys. Rev. B*, vol. 70, p. 115321, 2004.
87. V. N. Do, P. Dollfus, and V. L. Nguyen, "Transport and noise in resonant tunneling diode using self-consistent Green's function calculation," *J. Appl. Phys.*, vol. 100, p. 093705, 2006.
88. T. J. Park, Y. K. Lee, S. K. Kwon, J. H. Kwon, and J. Jang, "Resonant tunneling diode made of organic semiconductor superlattice," *Appl. Phys. Lett.*, vol. 89, p. 151114, 2006.
89. T. Kanazawa, R. Fujii, T. Wada, Y. Suzuki, M. Watanabe, and M. Asada, "Room temperature negative differential resistance of CdF2/CaF2 double-barrier resonant tunneling diode structures grown on Si(100) substrates," *Appl. Phys. Lett.*, vol. 90, p. 092101, 2007.
90. M. V. Petrychuk, A. E. Belyaev, A. M.Kurakin, S. V. Danylyuk, N. Klein, and S. A. Vitusevich, "Mechanisms of current formation in resonant tunneling AlN/GaN heterostructures," *Appl. Phys. Lett.*, vol. 91, p. 222112, 2007.
91. J.-P. Colinge, "Multiple-gate SOI MOSFETs," *Solid-State Electronics*, vol. 48, pp. 897–905, 2004.
92. J. Saint-Martin, A. Bournel, and P. Dollfus, "Comparison of multiple-gate MOSFET architectures using Monte Carlo simulation," *Solid-State Electronics*, vol. 50, pp. 94–101, 2006.
93. http://www.itrs.net/reports.html.
94. D. J. Frank, R. H. Dennard, E. Nowak, P. M. Solomon, Y. Taur, and H. S. P. Wong, "Device scaling limits of Si MOSFETs and their application dependencies," *Proc. IEEE*, vol. 89, pp. 259–288, 2001.
95. P. Dollfus, A. Bournel, S. Galdin, S. Barraud, and P. Hesto, "Effect of discrete impurities on electron transport in ultrashort MOSFET using 3-D MC simulation," *IEEE Trans. Electron Devices*, vol. 51, pp. 749–756, 2004.
96. T. Skotnicki, "Materials and device structures for sub-32 nm CMOS nodes," *Microelectronic Engineering*, vol. 84, pp. 1845–1852, 2007.
97. D. Reid, C. Millar, G. Roy, S. Roy, and A. Asenov, "Analysis of threshold voltage distribution due to random dopants: a 100 000-sample 3-D simulation study," *IEEE Trans. Electron Devices*, vol. 56, pp. 2255–2263, 2009.
98. J. Widiez, J. Lolivier, M. Vinet, T. Poiroux, B. Previtali, F. Daugé, M. Mouis, and S. Deleonibus, "Experimental evaluation of gate architecture influence on DG SOI MOSFETs performance," *IEEE Trans. Electron Devices*, vol. 52, pp. 1772–1779, 2005.
99. M. Vinet, T. Poiroux, J. Widiez, J. Lolivier, B. Previtali, C. Vizioz, B. Guillaumot, Y. L. Tiec, P. Besson, B. Biasse, F. Allain, M. Casse, D. Lafond, J.-M. Hartmann, Y. Morand, J. Chiaroni, and S. Deleonibus, "Bonded planar double-metal-gate NMOS transistors down to 10 nm," *IEEE Electron Device Lett.*, vol. 26, pp. 317–319, 2005.
100. J. Widiez, T. Poiroux, M. Vinet, M. Mouis, and S. Deleonibus, "Experimental comparison between Sub-0.1-μm ultrathin SOI single- and double-gate MOSFETs: Performance and Mobility," *IEEE Trans. Nanotechnol.*, vol. 5, pp. 643–648, 2006.
101. V. Barral, T. Poiroux, M. Vinet, J. Widiez, B. Previtali, P. Grosgeorges, G. L. Carval, S. Barraud, J.-L. Autran, D. Munteanu, and S. Deleonibus, "Experimental determination of the channel backscattering coefficient on 10-70 nm-metal-gate Double-Gate transistors," *Solid-State Electronics*, vol. 51, pp. 537–542, 2007.
102. J. Saint-Martin, A. Bournel, V. Aubry-Fortuna, F. Monsef, C. Chassat, and P. Dollfus, "Monte Carlo simulation of double gate MOSFET including multi sub-band description," *J. Computational Electronics*, vol. 5, pp. 439–442, 2006.
103. A. Bournel, V. Aubry-Fortuna, J. Saint-Martin, and P. Dollfus, "Device performance and optimization of decananometer long double gate MOSFET by Monte Carlo simulation," *Solid-State Electronics*, vol. 51, pp. 543–550, 2007.

104. M. Vinet, T. Poiroux, C. Licitra, J. Widiez, J. Bhandari, B. Previtali, C. Vizioz, D. Lafond, C. Arvet, P. Besson, L. Baud, Y. Morand, M. Rivoire, F. Nemouchi, V. Carron, and S. Deleonibus, "Self-aligned planar double-gate MOSFETs by bonding for 22-nm node, with metal gates, high-κ dielectrics, and metallic source/drain," *IEEE Electron Device Lett.*, vol. 30, pp. 748–750, 2009.
105. E. Joos, *Decoherence and the Appearance of a Classical World in Quantum Theory*. Springer-Verlag, 2003.
106. D. Querlioz, "Phnomnes quantiques et dcohrence dans les nano-dispositifs semiconducteurs : tude par une approche Wigner Monte Carlo," *PhD Dissertation, Univ. Paris-Sud, Orsay*, 2008.
107. M. V. Fischetti and S. E. Laux, "Monte Carlo study of electron transport in silicon inversion layers," *Phys. Rev. B*, vol. 48, pp. 2244–2274, 1993.
108. J. Saint-Martin, A. Bournel, F. Monsef, C. Chassat, and P. Dollfus, "Multi sub-band Monte Carlo simulation of an ultra-thin double gate MOSFET with 2D electron gas," *Semicond. Sci. Techn.*, vol. 21, pp. L29–L31, 2006.
109. L. Lucci, P. Palestri, D. Esseni, L. Bergagnini, and L. Selmi, "Multisubband Monte Carlo Study of Transport, Quantization, and Electron-Gas Degeneration in Ultrathin SOI n-MOSFETs," *IEEE Trans. Electron Devices*, vol. 54, pp. 1156–1164, 2007.
110. D. Querlioz, J. Saint-Martin, K. Huet, A. Bournel, V. Aubry-Fortuna, C. Chassat, S. Galdin-Retailleau, and P. Dollfus, "On the Ability of the Particle Monte Carlo Technique to Include Quantum Effects in Nano-MOSFET Simulation," *IEEE Trans. Electron Devices*, vol. 54, pp. 2232–2242, 2007.
111. F. Monsef, P. Dollfus, S. Galdin-Retailleau, H. J. Herzog, and T. Hackbarth, "Electron transport in Si/SiGe modulation-doped heterostructures using Monte Carlo simulation," *J. Appl. Phys.*, vol. 95, pp. 3587–3593, 2004.
112. S. M. Goodnick, D. K. Ferry, C. W. Wilmsen, Z. Liliental, D. Fathy, and O. L. Krivanek, "Surface roughness at the Si(100)-SiO2 interface," *Phys. Rev. B*, vol. 32, pp. 8171–8186, 1985.
113. H. Sakaki, T. Noda, K. Hirakawa, M. Tanaka, and T. Matsusue, "Interface roughness scattering in GaAs/AlAs quantum wells," *Appl. Phys. Lett.*, vol. 51, pp. 1934–1936, 1987.
114. D. Esseni, A. Abramo, L. Selmi, and E. Sangiorgi, "Physically based modeling of low field electron mobility in ultrathin single- and double-gate SOI n-MOSFETs," *IEEE Trans. Electron Devices*, vol. 50, pp. 2445–2455, 2003.
115. V. N. Do, "Modelling and simulation of quantum electronic transport in semiconductor nanometer devices," *PhD Dissertation, Univ. Paris-Sud, Orsay*, 2007.
116. J. Saint-Martin, A. Bournel, and P. Dollfus, "On the ballistic transport in nanometer-scaled DG MOSFETs," *IEEE Trans. Electron Devices*, vol. 51, pp. 1148–1155, 2004.

Chapter 6
Simulating Transport in Nanodevices Using the Usuki Method

Richard Akis, Matthew Gilbert, Gil Speyer, Aron Cummings, and David Ferry

Abstract To calculate the conductance of mesoscopic structures such as quantum wires and dots at low temperature and bias, one typically employs the Landauer–Büttiker formalism, which relates quantum mechanical transmission probability to conductance. In this chapter, we discuss a numerically stable method to solve this transmission problem, the Usuki method, which is closely related to both the scattering matrix approach and recursive Green's functions. It has a major advantage over the latter in that the electron density can be obtained far more efficiently. Various applications of this approach are presented: transport through open quantum dots, the study of spin filtering effects in quantum wire structures, computing the conductance of molecules and the application of the method to study MOSFETS. The extensions to the basic method required for each case are also discussed, the most extensive of which are required for the MOSFET problem, where inelastic scattering effects play a crucial role.

Keywords Nanostructures · Quantum dots · Quantum wires · Molecular electronics · MOSFETs · Spin-Hall Effect

1 Introduction

Going back to the early 1990s, the Nanostructures Research Group at Arizona State University has been carrying out fully quantum mechanical transport simulations of a variety of nanoscale devices. Our interest has been twofold. First, we have been working to achieve a fundamental physical understanding of the behavior exhibited by these structures. In particular, a fundamental issue in quantum mechanics concerns the manner in which the discrete level spectrum of an isolated system is modified when it is coupled to some external, macroscopic measuring environment. An ideal system for the study of this issue is provided by semiconductor

R. Akis (✉)
Department of Electrical, Computer, and Energy Engineering,
Arizona State University, Tempe, AZ, USA
e-mail: richard.akis@asu.edu

D. Vasileska and S.M. Goodnick (eds.), *Nano-Electronic Devices: Semiclassical and Quantum Transport Modeling*, DOI 10.1007/978-1-4419-8840-9_6,

quantum dots, which are quasi-zero dimensional semiconductor structures in which the flow of electrical current is confined on length scales comparable to the size of the electron itself [36]. The basic idea is that current flow between the macroscopic source and drain reservoirs is forced to occur via a central cavity whose size is small enough that the quantum energy level strongly modulates the current. In recent years, our group has been able to make a correspondence between the conductance fluctuations exhibited by open quantum dots and a process known as einselection [90], whereby certain quantum states survive the decoherence effects induced by the coupling to the external environment. Ironically, the quantum states that do survive are typically strongly influenced by the underlying classical dynamics of the system [4,5,12,13,15,17–20,29,33–35,90,91].

Secondly, we have been concerned with device applications. Recent advances in CMOS technology have reduced transistor gate lengths beyond the projections of the semiconductor roadmap [65]. Thus, new device paradigms that exploit, rather than are hindered by, quantum-mechanical phenomena are being proposed. Among the potential solutions to this problem are *quantum computing* [62] and *spintronics* [89]. For such applications, quantum dots are possible components, as are *quantum wires* [36] which are quasi-one-dimensional electron waveguides. Novel functionality can be expected by coupling such components together in such a way as to give rise to new behavior characteristic of the coupled quantum system alone. We have been exploring such applications through our simulation studies. Besides these new types of devices, we have also placed much effort into applying our quantum simulation techniques to more traditional devices such as Metal-Oxide-Semiconductor Field-Effect Transistors (MOSFETs). As they get smaller and smaller, quantum mechanical effects obviously become more significant, and one eventually expects a breakdown of the simple scaling behavior characterized by Moore's law [65]. Correspondingly, the traditional semi-classical tools of device simulation are fast becoming limited.

Our method of choice for carrying out the many quantum transport studies is one originally developed by Usuki and coworkers [82,83]. As described in the next section, it is a technique closely related to the cascading scattering matrix approach, as well as the Green's function approach which may be the most popular method for carrying out these kinds of calculations [7,28,51,56]. The Usuki technique however has a major advantage over the latter, as we will describe.

This chapter thus reviews and describes the Usuki method and how we have applied it to various types of devices. It is organized as follows. In Sect. 2, the basic Usuki technique is outlined and our application of the method to the study of low temperature and low bias transport in quantum dots is outlined. In Sect. 3, we discuss an advanced application of the technique, whereby the introduction of spin–orbit coupling (which requires an extension of the method) in a quantum wire system can be used to achieve spin-filtering effects. In Sect. 4, we apply the technique to the simulation of current flow through molecules. In Sect. 5, MOSFETs are the focus. Besides requiring that the simulations be made fully three dimensional, their room temperature operation required the most sophisticated extension to the method that we thus have made, that is, the explicit inclusion of inelastic scattering effects, which has been done in such a way that current conservation is maintained, thus avoiding a major problem encountered by previous approaches. Section 6 provides a brief summary.

2 The Basic Usuki Method and Its Application to the Study of Open Quantum Dots

2.1 A Prototypical Nanodevice: The Quantum Dot

In the vast majority of cases in which we have used the Usuki technique, it has been applied to simulating quantum dots realized using the split-gate technique [81]. According to this approach, metal gates with a fine-line pattern defined by electron beam lithography are first deposited on the surface of a GaAs/AlGaAs heterojunction. Figure 6.1a shows a Scanning Electron Microscope (SEM) image taken of such

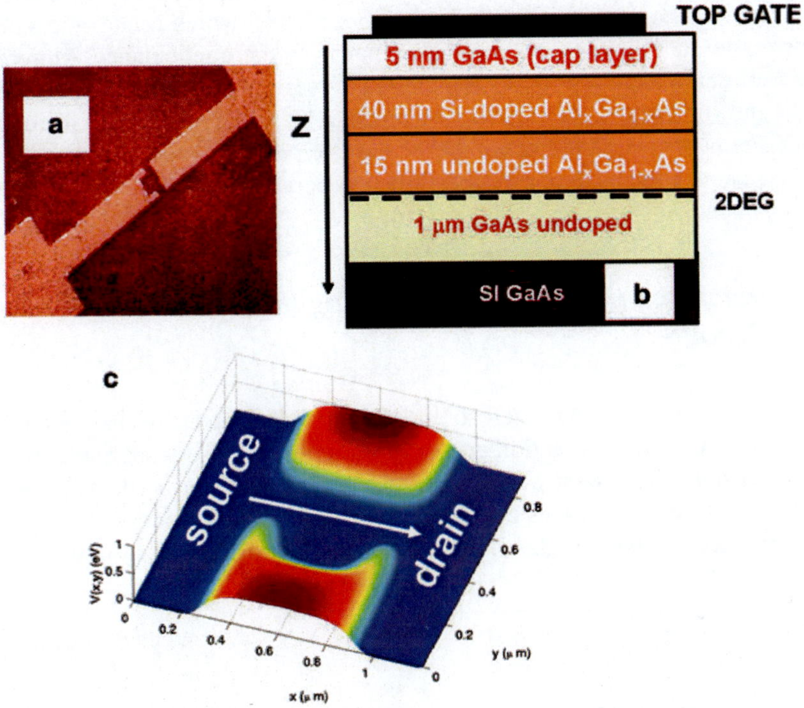

Fig. 6.1 **(a)** An SEM image of a split-gate used to create a quantum dot. **(b)** The heterostructure that is underneath. A 2DEG forms at the position indicated. **(c)** The confining potential, $V(x,y)$, seen by the electrons in the 2DEG

a structure [13]. The interior cavity here is 0.4 μm square. However, edge depletion around the gates makes the actual dot only about 0.3 μm square.

Figure 6.1b shows the heterostructure beneath the gates. Conducting electrons get trapped within a two-dimensional electron gas (2DEG) layer confined near the interface between GaAs and AlGaAs layers. Application of a suitable negative bias to the gates depletes the regions of 2DEG from directly underneath them, forming a dot whose lead openings are defined by means of quantum point contacts (QPCs) Fig. 6.1c displays the corresponding confining potential, theoretically calculated by solving Poisson's equation [13]. Conducting electrons move freely here in regions in which the potential is flat, and the "fingers" are the QPCs. In between them is the dot itself, which typically has a rounded potential, as shown, enclosed an area somewhat smaller than the lithographic dimensions of the gate cavity (0.3 by 0.3 μm in this case). The extent of this reduction in dimension can be determined experimentally by studying the transport at very high magnetic fields, where the edge states within the dot will exhibit Aharonov–Bohm oscillations [13, 15]. From the magnetic period of these oscillations, one can determine the dot area.

The situation we have been primarily interested in over the years is the *open* case, whereby there is one or more propagating mode allowed through the QPCs, such that the conductance of the dots is greater than $2e^2/h$ when computed using the Landauer–Büttiker formalism [21, 22, 36, 57, 58], which relates the sum of the transmission coefficients of the various modes to the conductance. Moreover, the bias between source and drain, which causes current flow in the direction of the arrow shown in Fig. 6.1c, is usually assumed to be vanishingly small in comparison to the Fermi energy. Another condition that is typically assumed is that the system is close enough to absolute zero that finite temperature effects can be neglected.

2.2 *Obtaining the Conductance Using the Usuki Simulation Technique*

Within the GaAs-AlGaAs heterostructure the conducting electrons are confined in a 2DEG, so that the z-direction generally need not be considered explicitly, in particular, if only the lowest quantum energy level associated with this direction is occupied. The electrons thus behave as free electrons with energy E, however they bear the effective mass of GaAs, $m^* = 0.067$ and obey the 2D Schrödinger equation

$$\frac{-\hbar^2}{2m^*}\left(\frac{d^2}{dx^2}+\frac{d^2}{dy^2}\right)\psi(x,y)+V(x,y)\psi(x,y)=E\psi(x,y). \tag{6.1}$$

For some simple problems, one can start directly with this equation, expressing the wave functions $\psi(x,y)$ in terms of analytical functions. However, for maximum flexibility, it is best to map the simulation domain onto a finite difference grid. Using a rectangular finite-difference lattice with lattice constant, a, position can thus be specified as $x = ia$ and $y = ja$, where i and j are integers. Keeping only the lowest

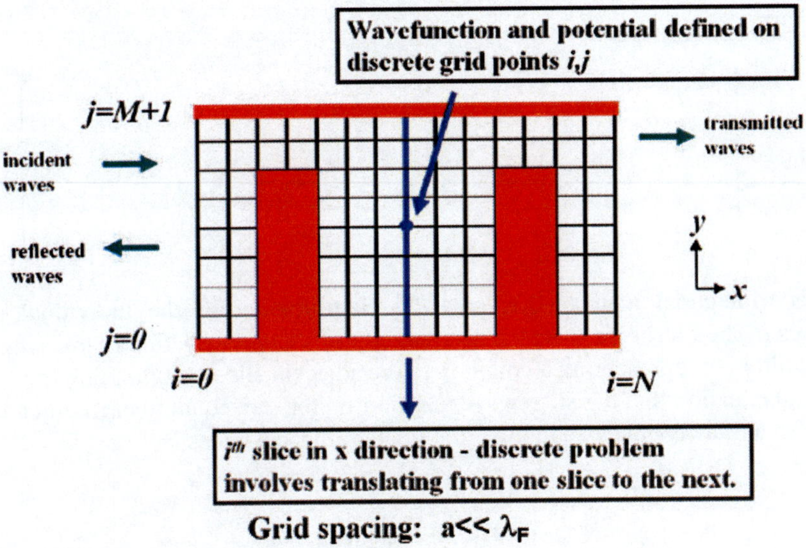

Fig. 6.2 Representation of the 2D transmission problem

order terms in the approximations of the second derivatives, the 2D Schrödinger equation becomes

$$-t(\psi_{i+1,j}+\psi_{i-1,j}+\psi_{i,j+1}+\psi_{i,j-1})+(V_{i,j}+4t)\,\psi_{i,j}=E\,\psi_{i,j} \tag{6.2}$$

where $\psi_{i,j}$ and $V_{i,j}$ represent, respectively the wavefunction and potential at site i,j, and $t=\hbar^2/(2m^*a^2)$, where $\hbar$ is the reduced Planck's constant. Since we are interested in current flow, the typical situation we consider is one in which the device is enclosed inside an ideal quantum wire, which extends outward to $\pm\infty$ along the x-axis. This is illustrated in Fig. 6.2, which is meant to represent a simplified version of the quantum dot confining potential shown in Fig. 6.1c. Portions of the picture that are shaded represent regions where $V_{i,j}$ is made to be a large number in comparison to E, and $V_{i,j}=0$ in the unshaded regions. More general cases where $V_{i,j}$ is an arbitrary varying function can be treated in trivial fashion, once the problem is set up. As indicated, the grid spacing should be much less than the Fermi wavelength, $\lambda_F=(h^2/2m^*E)^{1/2}$. Ideally, $a/\lambda_F\sim 0.1$ or smaller.

Along the top and bottom boundaries we use Dirichlet boundary conditions, so for a wire M lattice spacings high,

$$\psi_{i,j=0}=\psi_{i,j=M+1}=0. \tag{6.3}$$

Given this, the wavefunction along a particular slice i on the x-axis can be specified by a M-dimensional vector. Defining the diagonal matrix $\boldsymbol{t}=t\boldsymbol{I}$, (6.2) can be rewritten as a matrix equation relating these slice vectors:

$$\mathbf{H_{0i}}\,\vec{\psi}_i-\mathbf{t}\vec{\psi}_{i+1}-\mathbf{t}\vec{\psi}_{i-1}=E\mathbf{I}\vec{\psi}_i, \tag{6.4a}$$

where

$$\mathbf{H}_{0i} = \begin{bmatrix} (V_{i,M}+4t) & -t & 0 & \cdots & & \\ -t & (V_{i,M-1}4t) & -t & \cdots & & \\ & & & \ddots & & \\ & & \cdots & -t & (V_{i,2}+4t) & -t \\ & & \cdots & 0 & -t & (V_{i,1}+4t) \end{bmatrix}. \quad (6.4b)$$

This tridiagonal matrix represents the Hamiltonian for the individual isolated slices i. The t terms in (6.4a) can be thought of as a perturbation to this single slice Hamiltonian representing a coupling to the adjacent slices. Combining this with the trivial equation that the slice wave function vectors are equal to each other, one can derive a transfer-matrix equation that relates adjacent slices:

$$\begin{bmatrix} \vec{\psi}_i \\ \vec{\psi}_{i+1} \end{bmatrix} = \begin{bmatrix} 0 & \mathbf{I} \\ -\mathbf{I} & \left(\frac{\mathbf{H_{0i}}-E}{t}\right) \end{bmatrix} \begin{bmatrix} \vec{\psi}_{i-1} \\ \vec{\psi}_i \end{bmatrix} = \mathbf{T_i} \begin{bmatrix} \vec{\psi}_{i-1} \\ \vec{\psi}_i \end{bmatrix}. \quad (6.5)$$

Since the quantum wire acts as a waveguide, the actual current is carried by the propagating modes of the wire. Thus, we begin the calculation by turning to Bloch's theorem, and solving the eigenvalue problem for the transfer-matrix on the first slice:

$$\mathbf{T}_1 \begin{bmatrix} \vec{\psi}_1 \\ \vec{\psi}_0 \end{bmatrix} = \begin{bmatrix} \mathbf{T}_{11} & \mathbf{T}_{12} \\ \mathbf{T}_{12} & \mathbf{T}_{22} \end{bmatrix} \begin{bmatrix} \vec{\psi}_1 \\ \vec{\psi}_0 \end{bmatrix} = \lambda \begin{bmatrix} \vec{\psi}_1 \\ \vec{\psi}_0 \end{bmatrix}. \quad (6.6)$$

Since two adjacent slices are always considered in tandem, the eigenvectors of (6.6) have the form

$$\begin{bmatrix} \vec{u}_m(\pm) \\ \lambda_m(\pm)\, \vec{u}_m(\pm) \end{bmatrix}. \quad (6.7)$$

If there are q propagating wave modes ($|\lambda| = 1$) and $M - q$ evanescent modes ($|\lambda| \neq 1$), the corresponding eigenvalues can be expressed as

$$\begin{aligned} \lambda_m(\pm) &= e^{\pm i k_m{}^a}, m = 1, \ldots, q \\ \lambda_m(\pm) &= e^{\mp \kappa_m{}^a}, m = q+1, \ldots, M \end{aligned} \quad (6.8)$$

The $\pm$ symbol refers to the fact that the modes actually come in pairs, those that travel to the right (+) and those to the left (−). For the transmission problem, it is useful to collect these together in a $2\mathrm{M} \times 2\mathrm{M}$ matrix

$$\mathbf{T}_0 = \begin{bmatrix} \mathbf{U}_+ & \mathbf{U}_- \\ \lambda_+ \mathbf{U}_+ & \lambda_- \mathbf{U}_- \end{bmatrix}, \quad (6.9a)$$

where

$$\mathbf{U}_\pm = \begin{bmatrix} \vec{u}_1(\pm) & \ldots & \vec{u}_m(\pm) \end{bmatrix}, \quad (6.9b)$$

and

$$\lambda_{\pm} = diag\left[\lambda_1(\pm) \;\ldots\; \lambda_M(\pm)\right]. \tag{6.9c}$$

Multiplying by $\mathbf{T}_0$ converts the representation from the mode basis to the site basis, while multiplying by its inverse reverses this operation. To calculate the transmission through a device, one imposes the boundary conditions that the + modes are injected, each with unit amplitude, from the left side and there are no − modes coming from the right. For a structure N slices long, one must thus solve the transfer matrix problem:

$$\begin{bmatrix} \mathbf{t} \\ \mathbf{0} \end{bmatrix} = \mathbf{T}_0^{-1}\mathbf{T}_N\mathbf{T}_{N-1}\ldots\mathbf{T}_1\mathbf{T}_0 \begin{bmatrix} \mathbf{I} \\ \mathbf{r} \end{bmatrix}, \tag{6.10}$$

where $\mathbf{t}$ is a $2M$ by $2M$ matrix of transmission amplitudes of waves exiting from the right part of the structure, and $\mathbf{r}$ is the corresponding matrix of amplitudes of waves reflected back towards the left. The unit matrix, $\mathbf{I}$, and the zero matrix, $\mathbf{0}$, set the transport boundary conditions mentioned above. Given the matrix elements of $\mathbf{t}$, one can calculate the conductance, G, using the Landauer–Büttiker formula:

$$G = \frac{2e^2}{h}\sum_{m,n}\frac{v_n}{v_m}\left|t_{n,m}\right|^2, \tag{6.11}$$

where $t_{n,m}$ represents the transmission amplitude of mode n to mode m and the summation is only over propagating modes. Here v_n represents the velocity in the x-direction of nth mode, which can be obtained by taking the modal matrix elements of the probability current operator in the x-direction [9]. One finds that, up to a constant prefactor,

$$v_n = \sum_j 2t\sin(k_n a)\left|u_{n,j}\right|^2. \tag{6.12}$$

Unfortunately, (6.10) in its current form is made numerically unstable by the exponentially growing and decaying contributions of the evanescent modes that accumulate when the product of transfer matrices is taken. Usuki and colleagues [82,83] overcame this difficulty be rewriting the transfer matrix problem in terms of an iterative scheme. Rather than using the simple relationship given by (6.10), slices i and $i+1$ can be related by:

$$\begin{bmatrix} \mathbf{C}_1^{l+1} & \mathbf{C}_2^{l+1} \\ \mathbf{0} & \mathbf{I} \end{bmatrix} = \mathbf{T}_l \begin{bmatrix} \mathbf{C}_1^{l} & \mathbf{C}_2^{l} \\ \mathbf{0} & \mathbf{I} \end{bmatrix} \mathbf{P}_l, \tag{6.13a}$$

$$\mathbf{P}_l = \begin{bmatrix} \mathbf{I} & \mathbf{0} \\ \mathbf{P}_{l1} & \mathbf{P}_{l1} \end{bmatrix}, \tag{6.13b}$$

$$\mathbf{P}_{l1} = -\mathbf{P}_{l2}\mathbf{T}_{l21}\mathbf{C}_1^l, \tag{6.13c}$$

$$\mathbf{P}_{l2} = \left(\mathbf{T}_{l21}\mathbf{C}_2^l + \mathbf{T}_{l22}\right)^{-1}. \tag{6.13d}$$

The iteration is started by the condition $\mathbf{C}_1^0 = \mathbf{I}$ and $\mathbf{C}_2^0 = \mathbf{0}$. As shown schematically, in Fig. 6.2, this implies a situation in which the modes start off incident (**I**) from the left with unit amplitude and there are no waves coming in from the right. What results are reflected (**R**) and transmitted (**T**) waves. At the right end of the structure, the final transmission matrix **t** obeys the relationship:

$$\mathbf{t} = -(\mathbf{U}^+\lambda^+)^{-1}\left[\mathbf{C}_1^{N+1} - \mathbf{U}^+(\mathbf{U}^+\lambda^+)^{-1}\right]^{-1}. \tag{6.14}$$

The numerical stability of the Usuki et al. method in large part stems from the fact that the iteration implied by (6.13) involves products of matrices with *inverted* matrices. Taking such products tends to cancel out most of the troublesome exponential factors. In this regard, the Usuki method is closely related to the "cascading scattering matrix" method developed by Ko and Inkson [54]. Beyond the fact that their method involves a right to left recursion instead of a left to right one, the derived formulas have a very similar form.

2.3 *Reconstructing the Wave Functions and Computing the Probability Density*

Besides calculating the conductance, we can also obtain the electron density by reconstructing the electron wave functions using the $\mathbf{P}_{l1}$ and $\mathbf{P}_{l2}$ matrices. Usuki et al. [82,83] outlined a method for doing this starting from the left and working back to the end of the structure. Unfortunately, it entails performing a calculation similar to that for obtaining the conductance, but for every single slice. As a result, while the time it takes to calculate G goes as N, the time to reconstruct the wave function instead goes as $N!$, which makes it very time consuming. This poses a particular problem in cases where one wishes to study cases where self-consistently computed potentials are required, as they generally require that the density to be recalculated numerous times before convergence is achieved. We have found a simple way to make the reconstruction far more efficient. Instead of going from left to right, one starts at the end of the structure and works backword. Manipulating Usuki et al.'s equations, it can be shown [5] that for the final slice:

$$\psi_N = \mathbf{P}_{N2}. \tag{6.15}$$

Note here that ψ_N is a matrix, the columns of which represent the separate contributions of the individual modes to the total wave function on slice N. Going towards the left, one then does the iteration:

$$\psi_i = \mathbf{P}_{i1} + \mathbf{P}_{i2}\psi_{i+1}. \tag{6.16}$$

The **P**'s here are the same ones obtained during the conductance calculation and so are recalled from memory rather than being recalculated. The density at site i, j, given there are q propagating modes, becomes

$$n(x,y) = n(i,j) = \sum_{k=1}^{q} |\psi_{ijk}|^2. \tag{6.17}$$

Obtaining $n(x,y)$ in this modified way takes about the same amount of time as the original G calculation and can be *orders of magnitude faster* than the original technique described by Usuki et al. depending on the size of the structure.

2.4 Comparison with the Green's Function Method

Given that we have now described how to obtain the density, now is a logical point to compare this method with the Green's function approach [7,28,51,56].

As described in an appendix to their second paper [83], Usuki et al. recognized that their recursion technique, as described by (6.13), was closely related to a formulation of the recursive Green's functions method specifically described by Ando (6.5). Specifically, the C_1^{l+1} and C_2^{l+2} matrices can also be written in terms of a recursive Green's function. The Usuki recursion can be thought of as representing a form of the Dyson's equation, with the coupling **t** matrices representing a self-energy term.

That said, the process of obtaining the density in the case of Green's functions is significantly more cumbersome than the modified Usuki approach described above. In particular, one is required to do a complex energy contour integral to obtain the density matrix [51]

$$D = \frac{1}{2\pi i} \int_{-\infty}^{\infty} d\varepsilon G^{<}(\varepsilon), \tag{6.18}$$

where $G^{<}$ is the lesser Green's function. To obtain the density from D, one does a projection over the local wave function basis. When continuous functions, ϕ_μ, are used, the density has the form

$$n(r) = \sum_{\mu,\nu} \phi_\mu(r) \mathrm{Re}[D_{\mu,\nu}] \phi_\nu(r). \tag{6.19}$$

Importantly, to obtain n in this manner, the Green's function and associated interaction terms have to be calculated for *all* complex energy values along the chosen contour.

Our method avoids this contour integration; only one energy, E, ever need be used to obtain n. Moreover, even with a finite temperature Fermi–Dirac distribution, the summation over allowed energies is done *in the contact slice only*, rather than at each and every grid point. This follows as the wave function in the interior has a value relative to its value on the contact slice.

2.5 Modifications Required to Incorporate the Effects of a Magnetic Field

The most common complication to the basic problem that we typically have dealt with is the inclusion of a perpendicular magnetic field

$$\vec{B} = (0,0,B). \tag{6.20}$$

The most suitable choice for including this field in the problem is a vector potential in the Landau gauge

$$\vec{A} = (-By,0,0). \tag{6.21}$$

The standard procedure when performing calculations on a lattice is to include the field via Peierls phase factors, which are obtained by performing the path integral over the vector potential between adjacent lattice sites. The paths are assumed to follow straight lines, that is

$$for \quad x_i \leq x \leq x_{i+1} \quad y_j(x) = y_{i,j} + \frac{(y_{i+1,j} - y_{i,j})}{(x_{i+1} - x_i)}(x - x_i). \tag{6.22}$$

Thus, for the right and left directions on the lattice, the appropriate phase factors are

$$\theta_{R,i,j} = \frac{2\pi e}{h}\int \vec{A} \bullet \vec{dl} = \frac{2\pi e}{h}\int_{x_i}^{x_{i+1}} -By_j(x)dx = -\frac{\pi eB}{h}[(y_{i+1,j} + y_{i,j})(x_{i+1} - x_i)] \tag{6.23a}$$

and

$$\theta_{L,i,j} = \frac{2\pi e}{h}\int_{x_i}^{x_{i-1}} -By_j(x)dx = -\frac{\pi eB}{h}[(y_{i,j} + y_{i-1,j})(x_{i-1} - x_i)]. \tag{6.23b}$$

Since we chose the Landau gauge, the phase factors for the upper and lower directions are zero, $\theta_{U,i,j} = \theta_{D,i,j} = 0$. For the case of a uniform grid, which we have chosen to use here,

$$\theta_{L,i,j} = \theta_{R,i,j} = -2\pi eBa^2/h. \tag{6.24}$$

Given these factors, the 2D Schrödinger matrix equation, our (6.9a), must be modified

$$\mathbf{H_{0i}}\vec{\psi}_i - \tilde{\mathbf{t}}_{R,i}\,\vec{\psi}_{i+1} - \tilde{\mathbf{t}}_{L,i}\,\vec{\psi}_{i-1} = E\mathbf{I}\vec{\psi}_i, \tag{6.25}$$

where

$$\tilde{t}_{R,i,j} = e^{i\theta_{R,i,j}}t\,\delta_{i,j} \tag{6.26a}$$

and

$$\tilde{t}_{L,i,j} = e^{i\theta_{L,i,j}t}\,\delta_{i,j}. \tag{6.26b}$$

In addition, the eigenvalues of the wave modes also become shifted:

$$\lambda_m(\pm) = e^{\pm ik_m^{a+i\theta}{}_{R,0,j}},\ m = 1,\ldots,q$$

$$\lambda_m(\pm) = e^{\mp\kappa_m{}^{a+i\theta}{}_{R,0,j'}},\ m = q+1,\ldots,M \tag{6.27}$$

where j' is the index for which $y_{0,j'} = y_{max}/2$.

The expression used for determining the mode velocities now takes the form

$$v_n = \sum_j 2t_{R,0,j}\sin(k_n a + \theta_{R,0,j})\left|u_{n,j}\right|^2. \tag{6.28}$$

2.6 *Magnetotransport in a Quantum Dot*

We now present an illustrative example of the Usuki method, a square dot for which the confinement has been assumed to be hard wall in nature, similar to the schematic shown in Fig. 6.2. The dot in this case is a 0.3 μm square, with 0.04 μm QPC openings, which allow two propagating modes to enter and exit the dot for the given Fermi energy of 16 meV [3]. This value of energy was chosen in the cited work as it coincided well with the electron density found in experimental dot that was being compared with [11, 12], a value of $n_{2D} = 4\times10^{11}\,\text{cm}^{-2}$ (note that the relationship between energy and density is $E = \hbar^2\pi n_{2D}/2m^*$).

In Fig. 6.3a, the fluctuations in conductance, δG are plotted for the dot as a function of magnetic field. These are obtained from the raw conductance by doing a background subtraction done in such a way that the average value of δG approaches zero. There are a number of resonances evident. Importantly, rather than being aperiodic, there are sets of resonances that occur with virtually periodic spacing, with $\Delta B \sim 0.11$T. The Fourier transform of the fluctuations yields a peak that corresponds to this period. Such a peak was also found in the conductance fluctuations in the experimental dot that was being compared to [12].

The insets to Fig. 6.3a show the wave functions corresponding to the periodic resonances. In each case, they show the same recurring diamond pattern. As it turns out, these resonant states are all "scarred" [46] by the same classical orbit (i.e. a trajectory that retraces itself) shown in the inset of Fig. 6.3b. The fact that their amplitude is highly localized along this orbit makes them quite robust to the coupling of the dot to the external environment through the QPCs. In fact they are still present in the dot even if the QPCs are widened to almost half the size of the dot. Other resonant states that have considerable amplitude in the QPC regions, in particular,

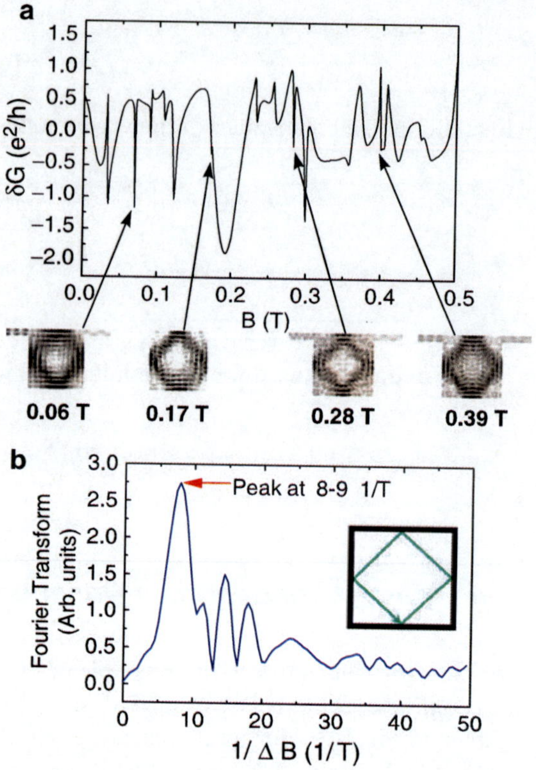

Fig. 6.3 **(a)** The conductance fluctuations vs. magnetic field for the 0.3 μm dot discussed in the text. Four resonances that appear in the curve are also indicated, with the wave function amplitude, $n(x,y)^{1/2}$, vs. x and y in each case plotted as insets. *Darker shading* corresponds to higher amplitude. **(b)** The Fourier transform of the conductance fluctuations. The *inset* shows a classical periodic orbit allowed in a square cavity

those that do not show a close association with a particular periodic orbit, do not survive the coupling to the environment. As our understanding of open quantum dots has progressed over the years, we come to understand what is being observed is a process known as *einselection* [18,33–35,90] and that the "diamond" is just one example of what is known as a *pointer state* [90].

A further illustration of the einselection process is shown in Fig. 6.4, which shows a comparison between the conductance for an open dot and the *energy spectrum* for a *closed* 0.3 μm dot as a function of energy and magnetic field (it should be noted that both are symmetric with field). Lighter shading corresponds to larger values of conductance, which takes on values ranging from $\sim 2e^2/h$ to $\sim 8e^2/h$ (the QPCs in this case has been made wider, 0.065 μm). Resonance and lines corresponding to the "diamond" state in the open dot are indicated by the arrows. With regards to the closed dot, rather than using a square, we have used a "T" shaped geometry that takes into account the perturbation effect produced by the QPCs. We have found that it generally necessary to do this, as the perturbed system provides a much more appropriate basis set for comparing open and closed systems. To obtain the closed dot spectrum, we assume Dirichlet boundary conditions ($\psi = 0$ along all boundaries). In this case, the eigenvalue equation for a closed dot that is N lattice spacings long along the x-direction and M spacings high becomes:

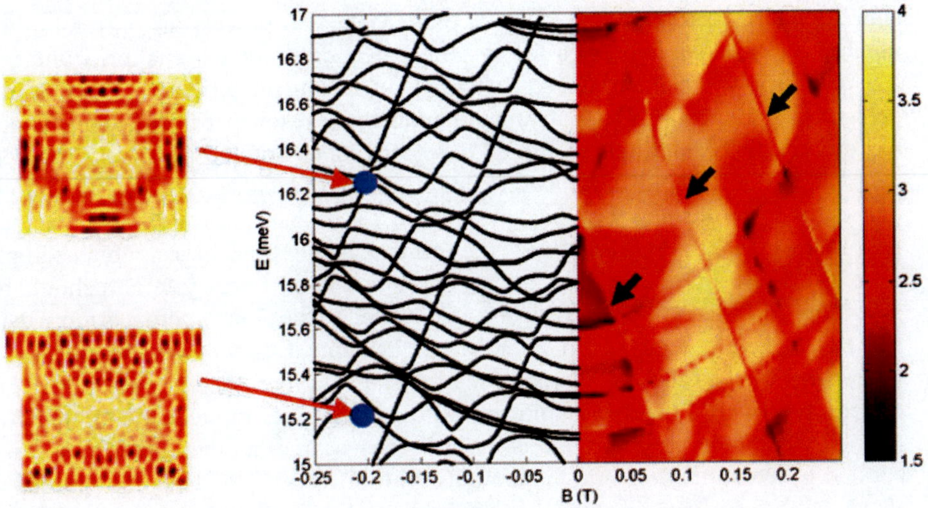

Fig. 6.4 On the *left side* of the picture is the energy spectrum as a function of magnetic field for a *closed square dot*, while the *right side* shows a shaded contour plot of the computed conductance for the *open dot*, with *lighter shading* corresponding to higher conductance, with the range in units of $2e^2/h$ indicated on the *right*. The *closed dot* eigenstates corresponding to the points indicated are shown in the *insets*. The arrows on the *right* indicate the lines of resonances on which the "diamond" resonance occurs in the *open dot*

$$\mathbf{H}_{dot}\,\vec{\psi} = \begin{bmatrix} \mathbf{H}_{01} & -\tilde{\mathbf{t}}_{R,1} & \mathbf{0} & \cdots & & & \\ -\tilde{\mathbf{t}}_{L,2} & \mathbf{H}_{02} & -\tilde{\mathbf{t}}_{R,2} & \cdots & & & \\ & & & \ddots & & & \\ & & & \cdots & -\tilde{\mathbf{t}}_{L,M-1} & \mathbf{H}_{0M-1} & -\tilde{\mathbf{t}}_{R,M-1} \\ & & & \cdots & \mathbf{0} & -\tilde{\mathbf{t}}_{L,M} & \mathbf{H}_{0M} \end{bmatrix}$$

$$\times \begin{bmatrix} \vec{\psi}_1 \\ \vec{\psi}_2 \\ \vdots \\ \vec{\psi}_{M-1} \\ \vec{\psi}_M \end{bmatrix} = E_n\,\vec{\psi}, \qquad (6.29)$$

Since $\mathbf{H}_{dot}$ is made up entirely of blocks that are either diagonal, tridiagonal, or zero (in the above equation, $\mathbf{0}$ denotes an M by M zero matrix) this is a sparse matrix problem which can be efficiently solved numerically by using employ Lanczos/Arnoldi factorization, such as the ARPACK fortran subroutines which are publicly available (www.caam.rice.edu/software/ARPACK/index.html).

Note that, while the left and right side of Fig. 6.4 are similar, there is not a one to one correspondence between the lines of eigenstates evident in the closed dot spectrum and resonance lines in the conductance, with the latter showing a significantly simplified structure. This is a result of the einselection process, as only a limited number of states survive when the system is opened to the external environment. The insets show two closed dot eigenstates. One, having the form of the "diamond", persists in the open system and causes conductance resonances. The other, which has considerable amplitude in the QPC regions, does not and so there is no corresponding resonance feature. In general, the states that do survive have amplitude concentrated in the interior of the dot. As such, it is probably not surprising that they tend to correspond to classical periodic orbits. Chaotic trajectories, which sample the entire phase space of the dot, are not supported in the open system. It should be noted that, if one continues to make the QPC wider, to the point where it is more than half the dot height, then the "diamond" resonance line itself will also disappear [5].

Besides the fact that the diamond scarred state survives the coupling to the external environment, another significant aspect is that it is replicated as a function of energy and field. A similar effect has also been seen as a function of the voltage applied to the split gates in quantum dots with different QPC lead configurations [5, 13, 15, 29]. Such replicated scars can be thought of as being part of a family of states in which there is a key state which is connected to a set of "offspring" states in the sense of a concept known as quantum Darwinism [16, 18, 69, 91]. One often thinks of a single scarred state as being unstable, but the proliferation of these states through quantum Darwinism can lead to significant robustness for the entire family of states [91].

Generally speaking, while experiments yield the kind of periodic conductance fluctuations that are seen in our simulations, the sharp conductance resonances are absent. Importantly, the calculations we have shown thus far have completely neglected thermal broadening and the phase breaking effects that occur as a result of inelastic scattering in the system. Needless to say, these are always present when experiments are performed. The easiest way to account for both thermal smearing and dephasing in our simulations is by introducing an effective temperature, T^*, such that $T^* > T$ [14]. At a given Fermi energy (E), the energy averaged magnetoconductance may then be computed by solving the following convolution integral numerically:

$$G_{av}(E) = \int G(E') \left[-\frac{df(E'-E)}{dE'} \right] dE'. \tag{6.30}$$

While (6.30) shows that G_{av} is determined by convolving the resistance with the derivative of the Fermi–Dirac distribution function (note: T^* is substituted for T) which in principle should to extend to $\pm\infty$, it is sufficient in practice to integrate over an energy window which is centered on the Fermi energy and is a few $k_B T^*$ wide. We found that the choice of $T^* = 0.5\,\text{K}$ yielded the best comparison with experiment for this size dot [14]. It should be noted that the actual temperature that the experiments was done at was ~30 mK. Figure 6.5 shows G_{av} as a function of E, and B. While the sharp resonances have been averaged out, the underlying periodic structure still remains. The inset shows that the "diamond" feature can also survive the effects of broadening. This image was obtained by doing the same convolution integral, but with $n(x,y)$ in the integrand.

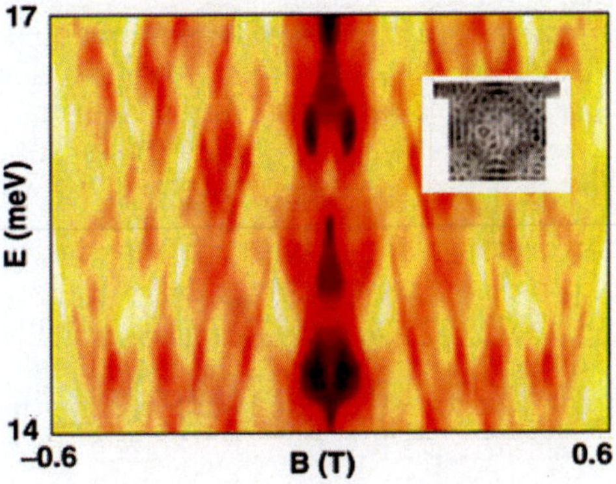

Fig. 6.5 The thermally averaged conductance ($T^* = 0.5\,\mathrm{K}$) for the same dot, with the *inset* showing a thermally averaged version of the "diamond" resonance

We conclude this section by noting that theoretical and experimental evidence for einselection in open quantum dots has been obtained by probing these systems in different ways. Analogous results have been obtained when fluctuations as a function of gate voltage were studied [13]. It is also evident in linear arrays of coupled quantum dots [18]. Most recently, scanning gate microscopy measurements have yielded results that indicated the presence of "scarred" states in rectangular dots somewhat larger ($\sim 1\,\mu\mathrm{m}$) than the example shown here [19,20]. Einselection also is evident in dots composed of graphene [47,48].

3 Introducing Spin-Dependent Effects into the Usuki Method

3.1 Spin–Orbit Coupling

As mentioned in the introduction, we have also been interested in potential device applications for nanostructures. Spintronics [89] is a field of electronics that, instead of utilizing an electron's charge, utilizes its spin to perform computational tasks. One major hurdle with this approach is that the spin of an electron is a degenerate quantity in many situations, and therefore some source of perturbation is needed to lift this degeneracy. One method of doing this involves the use of ferromagnetic contacts to inject spin-polarized carriers into semiconductors. Another method entails the application of external magnetic fields to manipulate the spin densities in quantum structures. While both of these lines of attack hold promise, there are also some problems. Historically, the spin injection efficiency of ferromagnetic contacts has been rather poor, and externally-applied magnetic fields are difficult to control to the precise degree needed to develop complicated circuits (although recent work in magnetic domain walls may soon change this). Furthermore, both of these solutions are difficult to integrate into today's circuit fabrication technology. For this reason, many groups, including ourselves, have turned their attention toward an all-electrical means of spin manipulation–spin–orbit coupling [86].

Spin–orbit coupling is a quasi-relativistic effect where an electron moving in an external electric field "feels" an effective magnetic field in its rest frame. In a vacuum, the spin–orbit term is not very strong, but, in a semiconductor, charge carriers are subjected to large local electric fields caused by the Coulomb interaction with the cores of the atoms in the crystal. The result is that in many semiconductors, electrons and holes can experience relatively large spin–orbit coupling. In the Bloch representation, the wave functions of electrons and holes are characterized by two parts, a lattice-periodic part and a slowly varying envelope function. The lattice-periodic part couples to the fields from the atomic cores of the crystal, while the envelope function couples to the macroscopic fields. Given this, it turns out that there are two primary sources of spin–orbit coupling in semiconductors, bulk-inversion asymmetry (BIA), and structural-inversion asymmetry (SIA). BIA arises from the strong microscopic fields and the effect arises in materials whose unit cell lacks inversion symmetry [31], while SIA results from a combination of the microscopic and macroscopic electric fields and arises when the lattice itself lacks a global inversion symmetry such as occurs in a triangular quantum well [86]. While the strength of BIA-induced spin–orbit coupling is more or less constant (it is dependent on the material and the fixed geometry of the quantum well), it has been shown that the SIA-induced spin–orbit coupling can be tuned by applying an external electric field. Usually, the SIA term is described by the Rashba Hamiltonian [23]

$$H_{SO} = \boldsymbol{\alpha} \cdot (\boldsymbol{\sigma} \times \mathbf{k}). \tag{6.31}$$

In this expression, $\boldsymbol{\alpha}$ is proportional to the electric field, $\boldsymbol{\sigma}$ represents the appropriate Pauli spin matrix, and $\mathbf{k}$ is the electron wave vector. If we assume a 2D electron gas in the x-y plane and an electric field applied along the z-axis, we can substitute the operator form of the wave vector. Then, the spin–orbit Hamiltonian becomes

$$H_{SO} = i\alpha_z \left(\begin{bmatrix} 0 & -i \\ i & 0 \end{bmatrix} \cdot \frac{\partial}{\partial x} - \begin{bmatrix} 0 & 1 \\ 1 & 0 \end{bmatrix} \cdot \frac{\partial}{\partial y} \right). \tag{6.32}$$

The total Hamiltonian of the system can be divided into two terms, $H = H_0 + H_{SO}$, where H_0 is the part of the Hamiltonian without spin, i.e. the left side of (6.1). With the spin–orbit coupling included, the Schrödinger equation must written with the wave function now split into spin-up and spin-down components:

$$\hat{\psi} = \begin{bmatrix} \psi^{\uparrow} \\ \psi^{\downarrow} \end{bmatrix}. \tag{6.33}$$

The equation that must be solved then becomes

$$(H_0 - EI)\hat{\psi} - \alpha_z \begin{bmatrix} 0 & \left(-\frac{\partial}{\partial x} + i\frac{\partial}{\partial y}\right) \\ \left(\frac{\partial}{\partial x} + i\frac{\partial}{\partial y}\right) & 0 \end{bmatrix} \hat{\psi} = 0. \tag{6.34}$$

To study Rashba spin–orbit coupling in nanostructures using the Usuki method, we must first rewrite this equation in discretized form. Assuming that we are in a system where wave propagation is along the y direction, and confinement is along x, we write [26]

$$\begin{bmatrix} (H_{0j}-EI) & T_{SO}^{\uparrow} \\ T_{SO}^{\downarrow} & (H_{0j}-EI) \end{bmatrix} \hat{\psi}_j - \begin{bmatrix} t & 0 \\ 0 & t \end{bmatrix} (\hat{\psi}_{j+1}+\hat{\psi}_{j-1}) - \begin{bmatrix} 0 & it_{SO} \\ it_{SO} & 0 \end{bmatrix} (\hat{\psi}_{j+1}-\hat{\psi}_{j-1}) = 0 \tag{6.35a}$$

where $t_{SO} = \alpha_z/2a$ and

$$T_{SO}^{\uparrow,\downarrow} = \begin{bmatrix} 0 & \pm t_{SO} & 0 & \dots & 0 \\ \mp t_{SO} & 0 & \pm t_{SO} & \dots & 0 \\ 0 & \mp t_{SO} & 0 & \dots & \dots \\ \dots & \dots & \dots & \dots & \pm t_{SO} \\ 0 & 0 & \dots & \mp t_{SO} & 0 \end{bmatrix}. \tag{6.35b}$$

Grouping adjacent slice vectors together as was done in the previous section, one can rewrite the above as the following (6.24):

$$\begin{bmatrix} \hat{\Psi}_j \\ \hat{\Psi}_{j+1} \end{bmatrix} = \begin{bmatrix} 0 & I \\ K^{-1}Q & K^{-1}H_j' \end{bmatrix} \begin{bmatrix} \hat{\Psi}_{j-1} \\ \hat{\Psi}_j \end{bmatrix} \tag{6.36a}$$

where

$$K = \begin{bmatrix} t & it_{SO}I \\ it_{SO}I & t \end{bmatrix},$$
$$H_j' = \begin{bmatrix} (H_{0j}-EI) & T_{SO}^{\uparrow} \\ T_{SO}^{\downarrow} & (H_{0j}-EI) \end{bmatrix},$$
$$Q = \begin{bmatrix} -t & it_{SO}I \\ it_{SO}I & -t \end{bmatrix}. \tag{6.36b}$$

The Usuki recursion scheme can then be applied to (6.36). To modify the above set of equations to account for the presence of a perpendicular magnetic field, Peierls phase factors are introduced the manner shown in Sect. 2.5, though in this case, the appropriate choice of gauge is $(0, Bx, 0)$. Importantly, the presence of the field also introduces an additional correction

$$-\alpha_z eBx\sigma_x/\hbar \tag{6.37}$$

that must be added to the $\boldsymbol{H}_0$ portion of the Hamiltonian (e here is the charge of the electron).

To be completely general, one should also add to $\boldsymbol{H}_0$ a spin dependent term that accounts for Zeeman splitting, $\pm g^* \mu_B B$, where g^* is the effective g-factor for the material and μ_B is the Bohr magneton. This is typically neglected as the effect is comparatively small (thus its omission in the quantum dot calculations). However, for the simulations shown in the next section it is included, in particular for the case where a magnetic field in the plane of the 2DEG has been introduced, directed along the x-direction. The effect of the field under those circumstances comes only through the Zeeman term and does not introduce any Peierls phase factors.

3.2 The Spin Hall Effect in a Double Y Branch Device

In this section, we discuss an application of the spin–orbit formalism that we have now introduced into the Usuki technique. One of the more remarkable features of Rashba spin–orbit coupling is that it gives rise to an intrinsic spin Hall effect where a longitudinal charge current is accompanied by a transverse spin current, polarized normal to the plane of the 2DEG [76]. In finite systems such as quantum wires, the transverse spin current leads to an accumulation of oppositely polarized spins on *opposite* sides of the wire. This has led to proposals of Y-shaped branching structures [25, 27, 88] as a means of generating spin-polarized currents in mesoscopic systems, a first stage towards possible spintronics applications. Most experimental efforts to measure the spin Hall effect in semiconductor systems have focused on optical techniques. The primary reason for this is because the transverse spin current is not accompanied by a transverse charge current, and there is no known way to directly measure a "spin voltage." However, an indirect method of measuring the spin Hall effect in a mesoscopic system was been proposed by Hankiewicz et al. [43] and Cummings et al. [27], and experimental measurements have been carried out by Jacob et al. [49]. The device in question utilizes a double Y-branch quantum wire structure, in conjunction with the spin Hall effect, to generate and detect spin-polarized currents in InAs quantum wells in a purely electrical measurement.

The simulated version of this device is shown in Fig. 6.6, which depicts a double Y-branch structure, where the branch points are labeled X and Y. The typical input quantum wire port is marked A, while B, C, and D typically represent output ports. The device utilizes sets of side gates to create QPC contrictions in each wire segment at the indicated locations. In this device, an unpolarized electronic current is injected at port A. Due to the spin Hall effect, spin up ($+z$ polarization) and spin down ($-z$ polarization) move to opposite sides of wire 2, and at junction X spin up and spin down electrons are separated into wires 1 and 3. Therefore, junction X acts like a spin filter. Because the electrons entering wire 3 have a finite spin polarization along the z-axis, they will undergo spin precession and undergo a process known as *jitter* or its German translation *zitterbewegung* as they move toward junction Y. What this means is that, as the electron moves down the y-axis, it will actually wobble back and forth in the plane of the 2DEG and perpendicular to its direction of travel [74]. The period of oscillation of this phenomenon is equal to the spin precession length,

$$\pi\hbar^2/m^*\alpha_z.$$

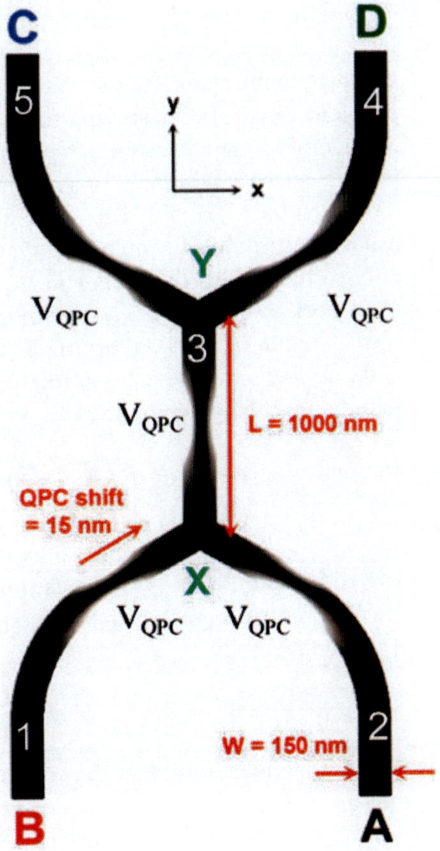

Fig. 6.6 The double Y branch quantum wire structure, consisting of five quantum wire segments used for the simulations of the device measured by Jacob et al. [49]. Positions in the device where potential constrictions are located are marked V_{QPC}, while X and Y mark the two branching points. The *black regions* correspond to points where the potential $V(x, y)$ is zero, the white regions where it is above the Fermi energy, and the *grey regions* where it falls in between

Upon reaching junction Y, the positions of the electrons are likely to be off center due to this jitter, resulting in an imbalance in the output currents at terminals C and D. Thus, in the absence of any structural asymmetry, imbalanced output currents will be an indicator of spin polarization resulting from the spin filter at junction X.

As the 2DEG lies in an InAs quantum well, the appropriate effective mass is $m^* = 0.023$. The density of the 2DEG in the experimental device was measured to be $n_{2D} = 5.3 \times 10^{11}\text{cm}^{-2}$, and we have adjusted our Fermi energy accordingly. The Rashba coefficient was measured to be $\alpha_z = 20\,\text{meV}\cdot\text{nm}$. In the experiments [49], a number of devices were fabricated with different wire widths, W, and filter spacings, L. As indicated in the figure, we have chosen to use the $W = 150\,\text{nm}, L = 1{,}000\,\text{nm}$ for our simulations. Measurements indicated that there was a small shift in the positions of the QPCs in the direction indicated in the figure, thus in the simulations a QPC shift of 15 nm has been applied. One minor difference between the theoretical and experimental configuration is that, in the simulated structure, the input and output wires are curved to align with the y-axis, as shown, while the experimental device is a true double Y, rather than this double "tuning fork", which is simpler to model.

One important complication with this device is that there are four ports here, instead of just the two considered by the basic Usuki formalism. However, one can get away with using the two port formalism by using a simple trick. To begin the problem, one starts by solving for the propagating modes that occur at the A–B end of the device. One then sorts these modes according to whether they are localized in the A branch or B branch. In a similar manner, one also categorizes the propagating modes on the C–D end. The conductance at the C and D ports is obtained from the usual transmission formula, (6.11), except the summation over m,n is now limited only to the modes that occur in the appropriate ports. To obtain the conductance from port A to port B, G_{AB}, is slightly more complicated. Because they are being considered in tandem on the "incident" end of the device, what is actually required is the *reflection* matrix, $\boldsymbol{r}$. As it happens, Usuki et al. also constructed a recursion scheme for this, given by [82]

$$\begin{bmatrix} \mathbf{I} & \mathbf{0} \\ \mathbf{D}_2^{l+1} & \mathbf{D}_2^{l+1} \end{bmatrix} = \begin{bmatrix} \mathbf{I} & \mathbf{0} \\ \mathbf{D}_2^{l} & \mathbf{D}_2^{l} \end{bmatrix} \mathbf{P}_l. \tag{6.38}$$

The reflection matrix iteration is started by the condition $\mathbf{D}_1^0 = 0$ and $\mathbf{D}_2^0 = \mathbf{I}$, and the final reflection matrix is obtained as $\boldsymbol{r} = \mathbf{D}_1^{\mathrm{N}+1}$. The conductance from port A to B is then given by

$$G_{AB} = \frac{2e^2}{h} \sum_{m,n} \frac{v_n}{v_m} \left| r_{n,m} \right|^2, \tag{6.39}$$

where it is understood that the summation over m is for modes in port A only, and the summation over n is for modes in port B.

Away from the QPCs, the quantum wires are modeled as square well waveguides. At each QPC, the potential is modeled as

$$V_{QPC} = \frac{m^* \omega^2}{2 \cdot \cosh((y-y_0)/l_{QPC}) \cdot \cosh((x-x_0)/l_{QPC})}, \tag{6.40}$$

where $\hbar\omega$ is a harmonic oscillator energy that characterizes the strength of the QPC constriction, (x_0, y_0) is the center of a given QPC, and l_{QPC} defines the length of the QPC along the wire. The QPC length, l_{QPC}, is 100 nm. The total size of the device in Fig. 6.6 is 1,900 by 3,500 nm, with a grid spacing of 5 nm along each axis. Finally, superimposed over the potential landscape shown in Fig. 6.5 is a random disorder potential with a Lorentzian energy distribution. This was included to match the total conductance of the experimental device when no voltage is applied to the QPCs and port A is used as the input, that is, $G_{total} \approx 4G_0$, where $G_0 = e^2/h$. The disorder strength necessary to achieve this is $\Gamma = 1$ meV.

Figures 6.7–6.10 show the results of the simulations. Due to the disorder potential and the fact that these simulations are run at zero temperature, the sweeps of conductance with respect to the QPC strength are extremely noisy. Thus, the curves have been smoothed using a binomial smoothing algorithm over 12 points in order to highlight the trend of the data.

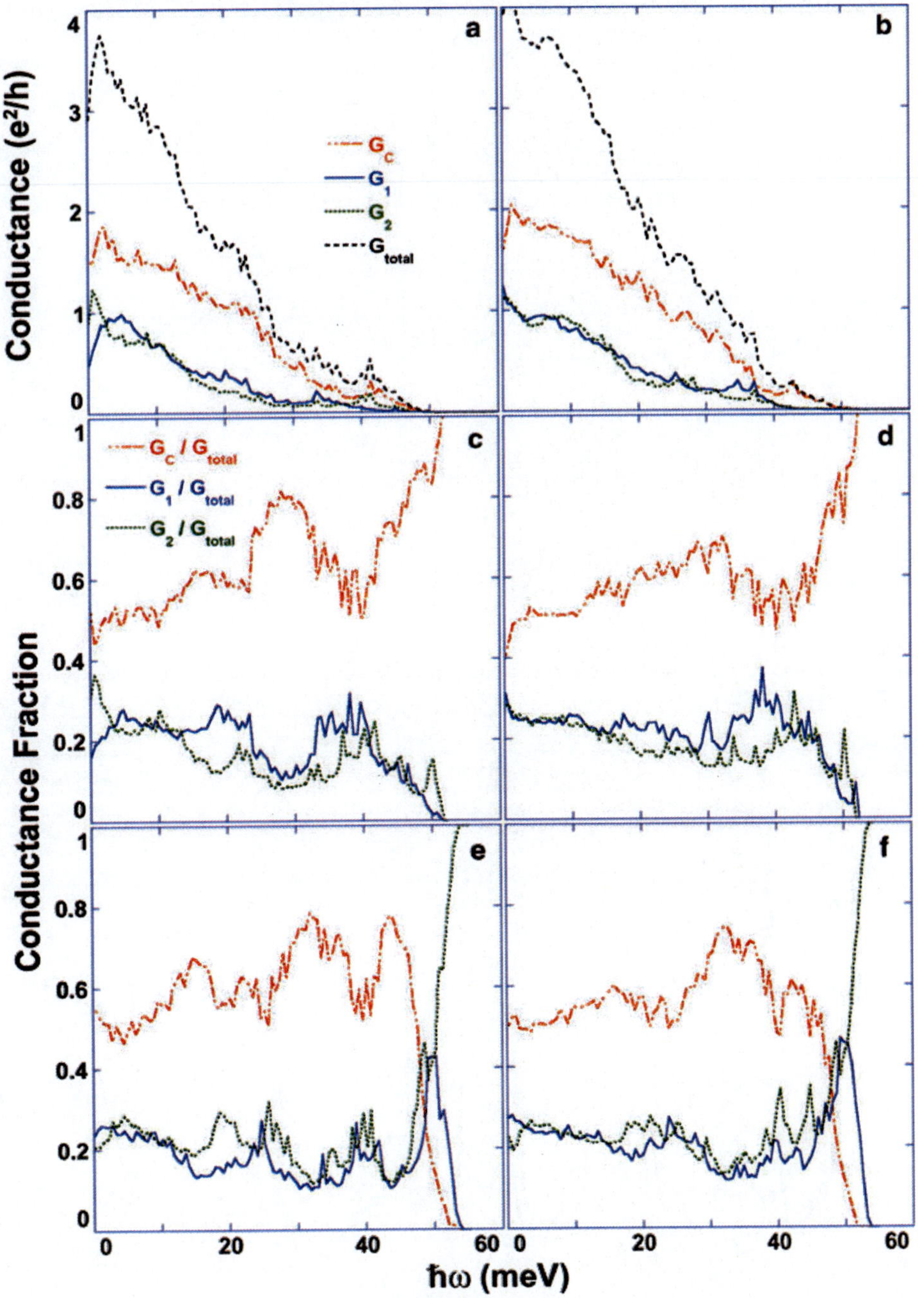

Fig. 6.7 Conductance through the device vs. QPC strength, $\hbar\omega$. In the *left column* of subplots no spin–orbit coupling is present, while in the *right column* $\alpha_z = 20\,\text{meV}\cdot\text{nm}$. The *top row* shows the absolute value of the conductance, assuming branch A is the input, the *middle row* shows the relative conductance of each branch, assuming branch A is the input, and the *bottom row* shows the relative conductance of each branch, assuming branch B is the input

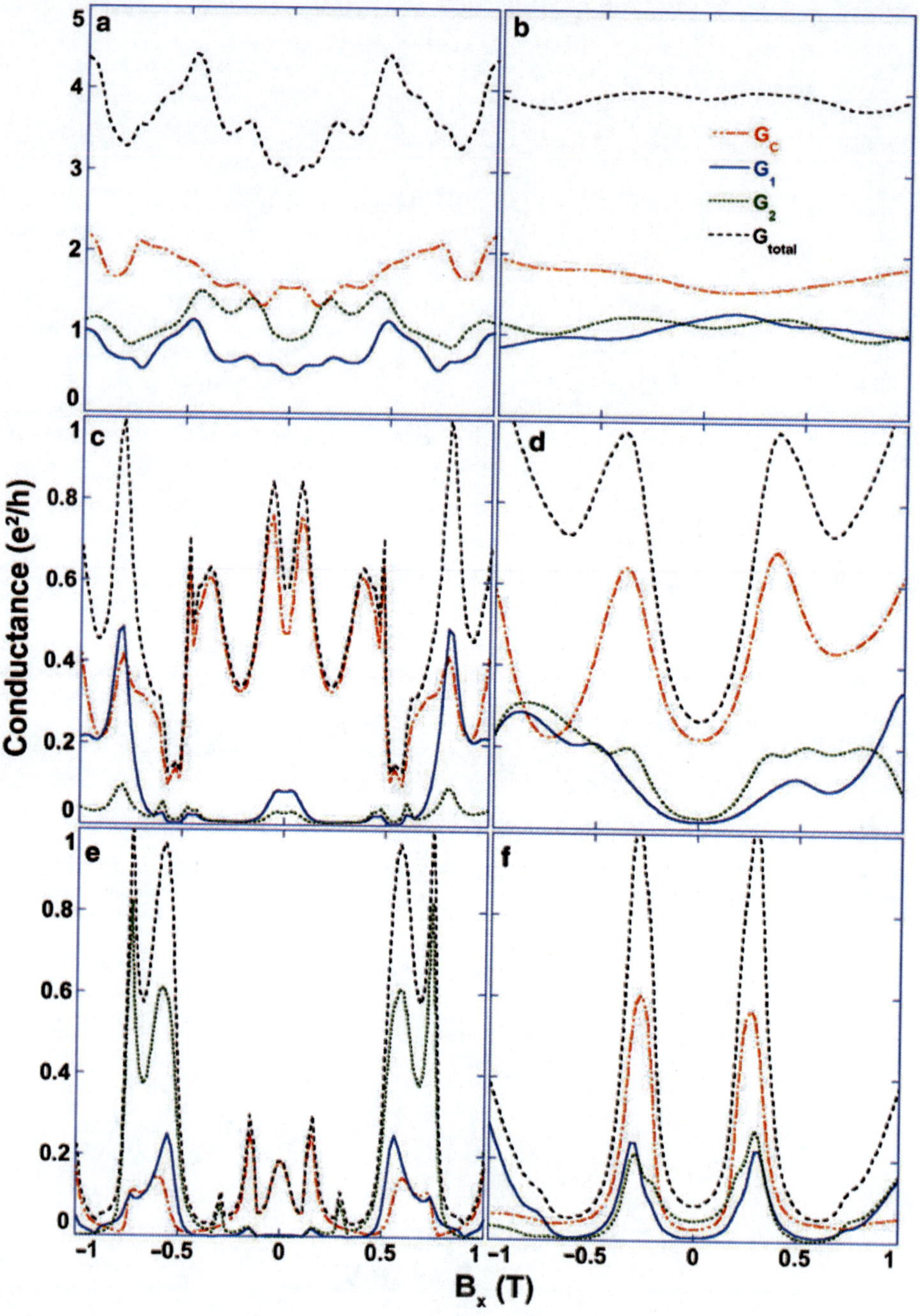

Fig. 6.8 Conductance through the device vs. the in-plane magnetic field, B_x. In the *left column* of subplots no spin–orbit coupling is present, while in the *right column* $\alpha_z = 20\,\text{meV}\cdot\text{nm}$. The *top, middle*, and *bottom rows* show the absolute value of the conductance, assuming branch A is the input, for QPC strengths of 0, 35, and 45 meV, respectively

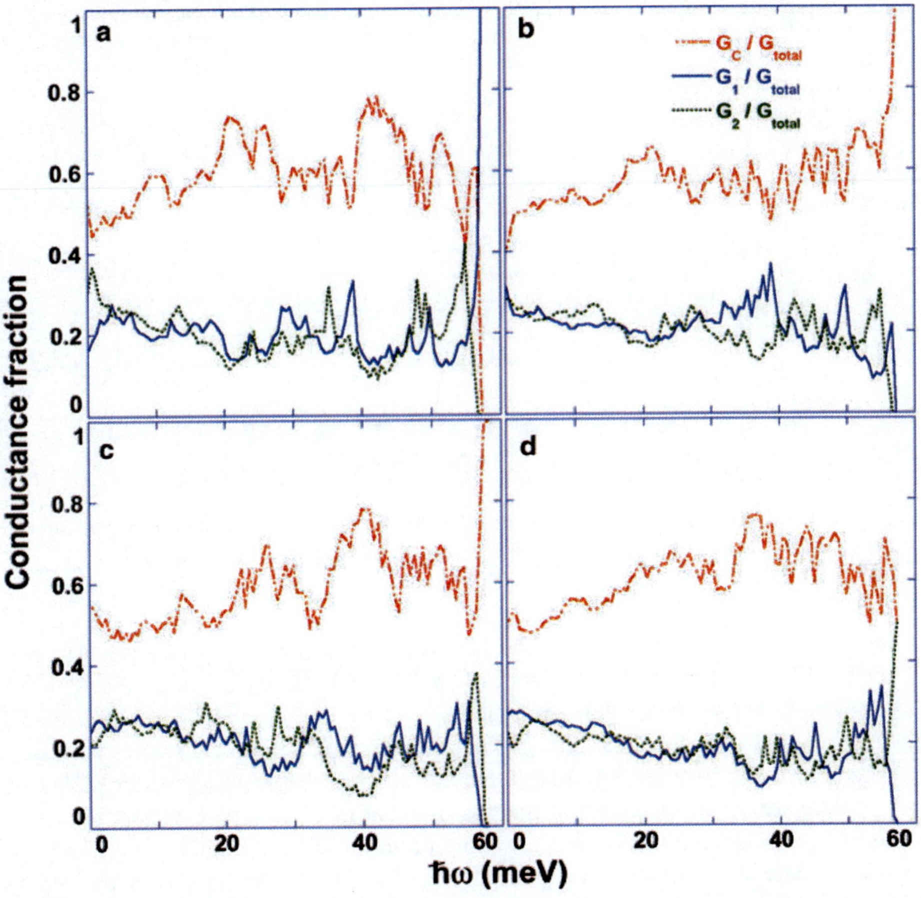

Fig. 6.9 Conductance through the device vs. QPC strength with a zero QPC offset. In the *left column* no spin–orbit coupling is present, while in the *right column* $\alpha_z = 20\,\text{meV}\cdot\text{nm}$. In the *top row* branch A is the input, and in the *bottom row* branch B is the input

Figure 6.7 shows the conductance through the device as a function of the QPC strength. In the left column of subplots no spin–orbit coupling is present, while in the right column $\alpha_z = 20\,\text{meV}\cdot\text{nm}$. The top row shows the absolute value of the conductance of each output branch, as well as the total conductance, assuming branch A is the input. The middle row shows the conductance of each branch relative to the total conductance, assuming branch A is the input, and the bottom row shows the conductance of each branch relative to the total conductance, assuming that branch B is the input. Adopting the nomenclature used by Jacob et al. [49], the label G_C refers to conductance from the input to the nearest output, passing only through the first filter, i.e. either G_{AB} or $G_{\text{BA}} \cdot G_1$ refers to conductance from the input to the output on the opposite side of the device, i.e. either G_{AC} or $G_{\text{BD}} \cdot G_2$ refers to conductance of the output on the second spin filter that is on the same side of the device as the input, i.e. either G_{AD} or $G_{\text{BC}} \cdot G_{total}$ is the total conductance from the

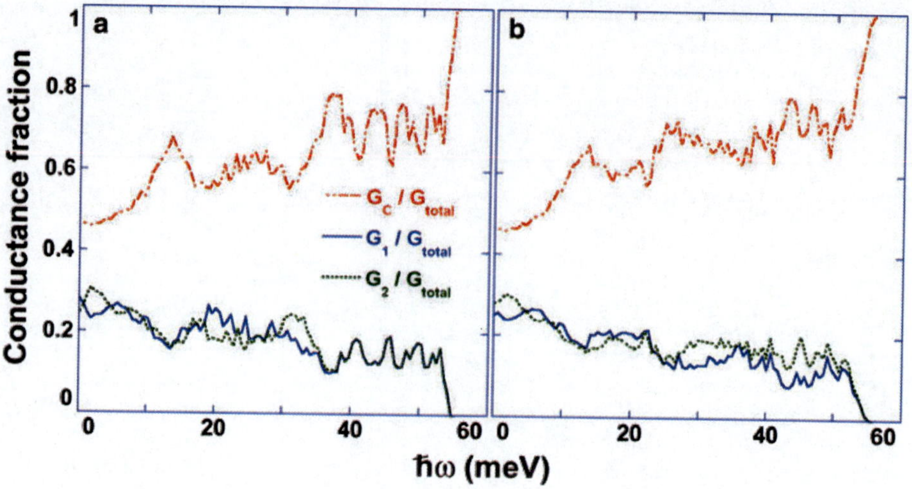

Fig. 6.10 Conductance through the simulated device vs. QPC strength, with zero QPC offset and no disorder. In **(a)** no spin–orbit coupling is present, while in part **(b)** $\alpha_z = 20\,\mathrm{meV \cdot nm}$

input to the other three outputs, and is given by $G_{total} = G_C + G_1 + G_2$. As can be seen in Figs. 6.7a, b, the conductance of each branch decreases with increasing QPC strength, as the wires are being pinched off. The magnitude of G_C is larger than G_1 or G_2, because it represents the lowest-resistance path. One also notes that, in the presence of the disorder potential, there are no discernable conductance plateaus.

There are several interesting features present in Figs. 6.6c–f. First, either G_1 or G_2 is primarily dominant over the whole range of QPC strength, with periodic regions where they become approximately equal. At 35–40 meV, near the point of pinchoff, the imbalance between G_1 and G_2 increases. Finally, G_1 is larger when A is the input, and G_2 is larger when B is the input. These are all features present in the experimental observations, when the equivalent experimental parameter, gate voltage, is varied.

It is also interesting to compare the plots with and without spin–orbit coupling. In comparing Fig. 6.7c–d, one can see a region between 30 and 40 meV where the dominance of G_1 over G_2 appears to be stronger in Fig. 6.7d, for which $\alpha_z = 20\,\mathrm{meV \cdot nm}$. However, between 15 and 25 meV, the dominance of G_1 over G_2 appears to be much stronger in Fig. 6.7c, where there is no spin–orbit coupling present. A similar situation appears between Figs. 6.7e, f. These discrepancies illustrate the fact that spin–orbit coupling alone is not sufficient to account for the imbalances between G_1 and G_2. In addition to the spin Hall effect, the disorder potential and the offset of the QPCs are also likely to be contributing to the conductance imbalance. This situation can be at least partly elucidated with the application of an in-plane magnetic field, B_x.

Figure 6.8 shows the absolute value of the conductance through the device as a function of an in-plane field, B_x. As in the previous figure, no spin–orbit coupling is present in the left column of subplots, while in the right column $\alpha_z = 20\,\mathrm{meV \cdot nm}$.

The top, middle, and bottoms rows show magnetic field sweeps for QPC strengths of 0, 35, and 45 meV, respectively. The labeling of the conductance curves is the same as in the previous figure, and port A is assumed to be the input in all six cases.

As one might expect, the conductance plots here are symmetric about $B_x = 0$ when no spin–orbit coupling is present, and asymmetric when the Rashba spin–orbit coupling strength is nonzero. Of greater interest in this figure is the degree of asymmetry of the plots in the second column, or the lack thereof. When the QPC strength is zero, as in Fig. 6.8b, there is quite noticeable asymmetry in G_1 and G_2. However, with finite QPC strength, as in Fig. 6.8d, f the asymmetry almost vanishes, even though $\alpha_z = 20\,\text{meV}\cdot\text{nm}$ in all three cases. This result suggests two conclusions. The first is that the asymmetry of Fig. 6.8b shows that the disorder potential does not play a significant role in the conductance imbalance when the QPCs are not active. However, this is for a large Fermi energy compared to the strength of the disorder potential, so disorder could still play a significant role near pinchoff. The second conclusion is that the symmetry of Fig. 6.8d, f shows that either the disorder potential or the offset QPCs (or both) play a more significant role than the spin Hall effect near pinch off.

To determine the roles of the QPC offset and the disorder potential, the relative conductance of the device is again plotted as a function of QPC strength in Fig. 6.9, but this time with the QPC offset equal to zero. Figure 6.9 differs from Fig. 6.7 in two key aspects. The first is that, over the whole range of QPC strength, one output port does not appear to be favored over the other. This is in contrast to Fig. 6.7, where G_1 was dominant for essentially the whole range. The second is that the difference between G_1 and G_2 in Fig. 6.7 appears to be much smaller on average than what was seen in Fig. 6.6. Note as well that the regions of QPC strength in Fig. 6.9b where G_1 is larger than G_2 correspond to regions in (d) where G_2 is larger than G_1. These results indicate that the QPC offset is the primary reason for the favoring of one output port over another in Fig. 6.7. It also suggests that the disorder potential, while not as strong an effect as the QPC offset, does contribute to the opposite behavior of inputs A and B. To verify this behavior, the conductance is plotted as a function of QPC strength, without disorder or a QPC offset, in Fig. 6.10.

Figure 6.10a shows the case without spin–orbit coupling, while (b) shows the case where $\alpha_z = 20\,\text{meV}\cdot\text{nm}$. In this figure, three important features stand out. The first is that the conductance imbalance between G_1 and G_2 above 40 meV is entirely due to the spin Hall effect, since the other sources of asymmetry have been removed. The second is that the results are identical whether one uses input A or input B, due to the perfect symmetry of the structure. The third is that while the conductance imbalance due to the spin Hall effect is present, it is not any larger than the imbalance induced by either the disorder potential or the QPC offset.

With the set of simulations presented above, it is possible to draw some general conclusions about the experimental results. Given the qualitative agreement between the theory and experiment, it appears that most of the conductance asymmetry seen in the experimental measurements is due to the offset of the QPCs. The opposite behavior for different inputs is also due primarily to the QPC offset, and secondarily to whatever underlying disorder potential may exist in the structure. However, while overwhelmed by these effects, the role of spin–orbit coupling is not negligible, and can be teased out with in-plane magnetic field sweeps. Furthermore, the degree of asymmetry in these sweeps can be used to determine just how much of a role the spin Hall effect is playing at any particular value of QPC strength.

4 Molecular Electronics Applications

4.1 Conduction Through Molecules

When Mark Reed's experimental group at Yale University published the current–voltage characteristic across a single molecule using their mechanical break junction technique, an entirely new domain of electronic devices became available for research [73]. Shortly after this, other experimental groups were able to perform similar measurements using a variety of other techniques. Unfortunately, the analysis of the electronic characteristics of isolated individual molecules connected to metallic contacts encountered major difficulties, with the initial calculations yielding conductance values varying over several orders of magnitude. The details of contact geometry and adsorption chemistry in the myriad of variations were discovered to play a pivotal role in theoretical analysis [30]. All the contact specific minutiae could be lumped and parameterized to match experimental measurements, but this ad hoc fitting gave little insight into the nature of molecular conduction. On the other hand, from a first-principles standpoint, the number of contact configurations and variations multiplied, never seeming to cover all the possibilities, and never satisfactorily producing the quantitative end result.

Our simulation work in this area took a heuristic approach to the problem of molecular conduction [60,77,78,80]. Instead of targeting a numerical range for the conductance, the nature of the problem itself was examined. What barriers to an accurate solution actually exist, and if they cannot be surmounted, can any meaningful information be gleaned? Besides the direct transport calculations, for which of course used the Usuki method, as we shall describe, the problem was also looked at using information gleaned from other approaches, such as examining the complex bandstructure associated with molecules [60, 77, 80], which gives insight into tunneling behavior, and localized orbitals and density functional theory to study the electronic structure [60, 78–80]. The aim of using several theoretical methods in conjunction with each other was to provide a spectrum of theoretical conclusions by which some insight into experimentally observed phenomena might be understood.

With regards to the transport problem, the Usuki technique allows the conductance to be calculated quickly, and therefore comparisons can be drawn between slightly differing systems to infer insights into experimental phenomena. One application for which it was particularly useful, was to examine what occurs when a molecule is gradually being stretched [78, 80].

4.2 Extending the Usuki Method to Deal with the Molecular Problem

A molecular conductance calculation can be started by using a tight binding/Linear Combination of Atomic Orbitals (LCAO) model. Electron motion can then be characterized as occurring via hopping between atomic orbitals as they move from atom to atom through the molecular system. Using tight-binding, a Hamiltonian in its matrix representation features site energies for each orbital j on atom i, with

$$\langle j,i|H|j,i\rangle = \varepsilon_{j,i} \tag{6.41}$$

as its diagonal terms. Off-diagonal terms, which correspond to the hopping energy for an electron in orbital j of atom i to orbital l on atom k, are

$$\langle j,i|H|l,k\rangle = t_{l,k,j,i}. \tag{6.42}$$

When the orbital basis is chosen to model the system in question, semi-empirical parameters or simple π-orbital terms can then be employed in transmission calculations. In that regard, it is no coincidence that the parameter t was used in the descretized Schrodinger equation shown earlier. The discretization process maps the equation onto a simple s-orbital tight binding model. The site energies above are equivalent to the $V_{i,j}$ terms in the discrete case used in the previous sections.

Given the Hamiltonian matrix elements, the molecular and contact atomic site energies can be mapped onto our usual Usuki discrete lattice. This scheme is shown pictorially in Fig. 6.1 for the specific case of two ring oligoaniline molecule that is connected to two gold contacts. The rings themselves are composed of carbon atoms, while nitrogen atoms connect the rings to each other and the gold contacts. Because this system is actually three dimensional, the slices represent a collapsed space where the terms of the Hamiltonian preserve the spatial relationships. In general, each atom may have multiple orbitals with distinct coupling information to neighboring atomic orbitals as well as to the other orbitals on the same atom. For example, hydrogens have one orbital; carbons, nitrogens and sulfurs have four; and golds have nine orbitals. As indicated in the figure, instead of just one hopping parameter, t, there are multiple ones that need to be included, depending on which types of atoms are bonded together, as well as the nature of the individual bonds.

For the periodic gold slabs shown in Fig. 6.11, each atom-orbital is assigned a row in the slice matrix H_{0i}, with nonzero off-diagonal terms representing the hopping terms to other atoms in the slab. Edge atoms have these hopping terms to their neighbors periodically linked across the cell (in contrast, the previous examples discussed in this chapter imposed Dirichlet boundary conditions). As it happens, the spread of the metallic wavefunction leaves few nonzero matrix elements in the matrix. The molecule is collapsed into a reduced number of slices as shown (in a simpler one ring case, only one slice would be necessary). Importantly, the molecular wave functions are much more localized, so the slice matrices for them is sparser than that for the gold slab slices. The coupling matrices between the slices, $H_{i,i-1}$

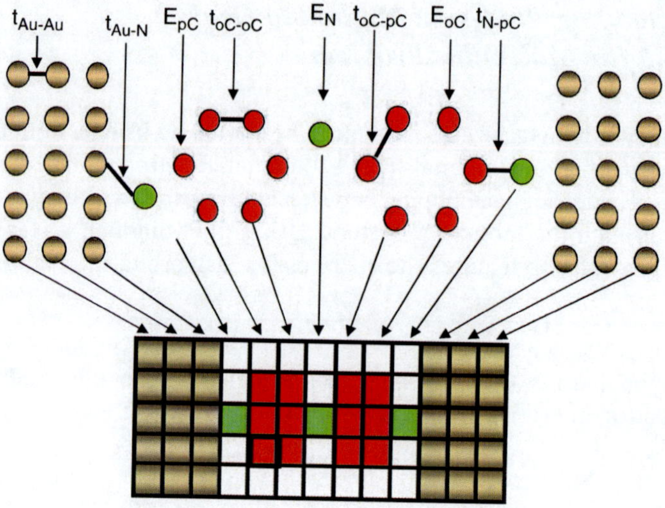

Fig. 6.11 Lattice site assignment scheme for the example of a two ring polyaniline molecule. Some of the relevant tight binding parameters are as indicated

and $H_{i,i+1}$, comprise the hopping terms between adjacent atoms. Since the formalism as derived cannot treat next nearest slice coupling, this information cannot be included in the calculation. This includes any tunneling information from left to right contact. As such, the conductance calculations represent the transmission exclusively through the molecule.

The vector representing the wave function has a length M equal to the number of atoms in the slab, usually in some sequential order, multiplied by the number of orbitals per atom. In the simplest case, where all the atoms were hydrogen, then the problem would be essentially identical to the finite difference formalism. The general inter-slice transfer matrix can be written [80]

$$T_i = \begin{bmatrix} 0 & I \\ -H_{i,i+1}^{-1} H_{i,i-1} & H_{i,i+1}^{-1}(E_F - H_{0i}) \end{bmatrix}, \tag{6.43}$$

which has a more general form than (6.5), but can just as easily be placed into the Usuki iteration scheme. As mentioned earlier, the advantage of the Usuki technique over a Green's function method is that the transmission matrix and wave function for all points can be derived simultaneously, and this directly yields the density at each point, while the Green's function method requires integrals over the density of states. In the molecular problem, since the potentials must all be calculated self-consistently, this speed advantage becomes very significant. The self-consistency employs a Poisson solver which calculates the change in electronic potential and adds these corrective terms back into the Hamiltonian matrix.

$$\nabla^2 \delta V(r) = \delta\rho(r). \tag{6.44}$$

During the course of the calculation, one must do a self consistent loop, that is calculate the transmission, solve Poisson's equation to obtain a new potential, recalculate the transmission, and repeat until convergence is achieved. Solving Poisson's equation is complicated by the fact that the coefficients obtained from the transmission calculation must be projected onto wave functions for each atomic orbital centered at the accurate positions in three dimensional space to completely reconstruct the density profile. The Poisson solver that we employed is the symmetric successive over-relaxation preconditioned bi-conjugate gradient stabilized algorithm (SSOR-BiCGSTAB) originally developed by van der Vorst [84]. To comply with the non-orthogonal mesh given by the unit cell form the energy spectrum code, the solver works over a 15 diagonal matrix. It should be added that due to the low applied bias used in the experiments which were analyzed in most of our simulation work, the self-consistent potential introduced fairly small corrections.

4.3 Application: Transmission Through Polymers

As mentioned, a tight binding model for molecular conduction can be easily treated using the Usuki method. One major difficulty lies, not in the method itself, but in the difficulty in specifying the contact parameters, specifically the Fermi level and the molecule-contact coupling. In Fig. 6.11, a representation of a two ring polyanline molecule was depicted. Polyaniline was one of the first conducting polymers discovered [64]. Transport in polyanilines has generated much interest due to the abrupt switching in electrical conductivity as the ambient acid-base chemistry is changed, with the base being the insulating form. Using the tight-binding parameters from the work of Vignolo et al. [85], we studied the transmission through various oligoanilines in order to gain some insight into some experiments by He et al. [45], which used an electromigration nanojunction technique to place the molecules between the gold contacts. Using the tight binding parameters in question for the various polyaniline models, we were able to reproduce the experimentally measured bandstructures to considerable accuracy [77, 80].

In Fig. 6.12, we show transmission plots that we obtained for the two, three and four ring oligoanilines of the fully benzenoid variant, called leucoemeraldine base. The region of energy where each curve shows a large decrease in transmission can be associated with a band gap. In that regard, the transmission decreases exponentially as the ring number is increased (note transmission here is on logarithmic scale) with the decay constant being consistent with the complex wave number one obtains in gap regions from a complex band structure calculation. Exactly the same sort of behavior is seen in linear arrays of quantum dots [17]. The semiempirical parameters that were used were specifically chosen to fit this gap to experimental values as accurately as possible. It should be noted that our transmission calculation was calibrated to the experiments by varying the hopping energy from the gold contact to the Nitrogen (labeled t_{Au-N} in Fig. 6.11). While changing this parameter does not alter the position in energy of the gap or the resonance structure to a significant

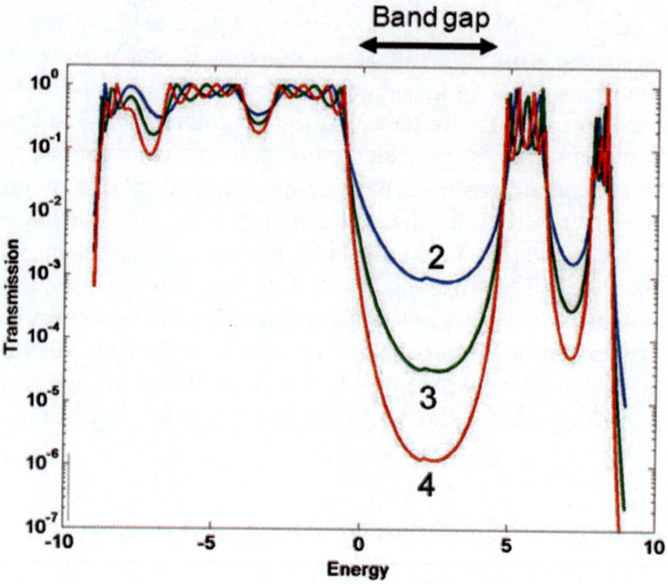

Fig. 6.12 Transmission through 2, 3, and 4 membered leucoemeraldine base oligoanilines

degree, what it does do is shift the transmission curves up or down, and, in turn, the current, which is the experimentally measured quantity. If the Fermi level happens to be in the region around a gap, it is evident how small shifts in its placement could yield widely differing conductance results. For this reason, it is desirable to use the most sophisticated bandstructure calculations, which accurately include the extended states in the contacts and their hybridization with the molecular levels to obtain truly trustworthy results.

5 Applying the Usuki Method to the Study of MOSFETS

5.1 Motivation for Using the Usuki Method

The MOSFET has been the workhorse of the semiconductor industry for many years, with progress in size and speed following Moore's law [65]. However, as devices get smaller and smaller, quantum mechanical effects become more significant, and one eventually expects a breakdown of the simple scaling behavior. Correspondingly, the traditional semi-classical tools of device simulation are fast becoming limited. There have been efforts to expand such methods as Monte Carlo and drift-diffusion to incorporate quantum effects via an effective quantum potential [32], which provides a computationally efficient way to do so. This effort has been most notable in Monte Carlo where the effective potential has found some success in predicting some of the quantum phenomena arising in next generation devices, such as charge setback from the gate. Unfortunately, simple quantum corrective tools such as the effective potential cannot account for quantum phenomena such as tunneling.

Different fully quantum mechanical methods used to model MOSFETS include simple analytical models [8, 67], Green's function approaches [28, 56], coupled Schrodinger approaches [59,71] and Pauli master equation approaches [38]. In each of these methods, the length and the depth are typically modeled rigorously, while the third dimension is usually included through the assumption that there is no interesting physics to capture in this dimension. Therefore, the third dimension is usually treated using a basis expansion which is then included in the Hamiltonian, or the simplifying assumption that only one subband in the orthogonal direction is occupied, therefore making higher-dimensional transport considerations unnecessary. In general, this is certainly not a valid assumption. In the source of the device, the modes that are excited are three dimensional (3D) in nature. These modes are then propagated from the 3D section of the source to the channel. The excitation of different modes changes as one approaches the drain, due to the large source-drain bias. Moreover, as the doping and the Fermi level in short channel MOSFETs increases, we can no longer assume that there is only one occupied subband even at the source.

In the applications of the Usuki method discussed in the preceding sections, the systems under study were generally at low temperature, meaning that inelastic scattering is almost completely suppressed. At room temperature, the dissipation produced by these mechanisms cannot be ignored, nor can they be dealt with using the low temperature broadening model used earlier.

Unfortunately, quantum simulators typically encounter great difficulty in properly accounting for dissipation. Statistical approaches introduce random phase fluctuations into the simulations [10, 70], however, a large sample space is required over which to average, and this entails a great many runs to have any valuable results. Another method is to add an imaginary term to the Hamiltonian which represents the phase breaking time of the electron in the system under consideration [68]. Unfortunately, the imaginary term is typically constant throughout the device, and therefore fails to consider the inhomogeneous density in the out of equilibrium system. Moreover, this approach does not conserve current. Dissipation may also be included through the use of Büttiker probes [21]. While this approach is an improvement over the use of a phase-breaking related term, in that it is current conserving, it suffers from the fact that an additional loop must be included to ensure that the probes do not change the number of electrons in the system, nor does it account for the spatial inhomogeneity of the density and the scattering. Moreover, a fitting parameter must be used to calibrate the probes to the proper low field mobility. A relaxation time approximation has also been used in approaches utilizing either the density matrix [53] or the Wigner function [42].

These difficulties led us to attempt to overcome them by using an extended version of the Usuki technique, which is not only fully 3D, but, more importantly, accounts for dissipation in a manner which preserves current conservation and does not require fitting parameters [6, 34, 35, 41]. We have found that we can account for dissipation in a device by introducing a real space self-energy term, and we can accurately model such effects as electron–phonon scattering.

5.2 Extending the Usuki Method to Deal with a Three Dimensional Silicon Device

In the case of silicon, there are six equivalent ellipsoids that make up the conduction band. The 3D Schrödinger equation for the wave function contribution from valley i is given by:

$$\frac{-\hbar^2}{2}\left(\frac{1}{m^*_{x,i}}\frac{d^2}{dx^2}+\frac{1}{m^*_{y,i}}\frac{d^2}{dy^2}+\frac{1}{m^*_{z,i}}\frac{d^2}{dz^2}\right)\Psi^{(i)}+V(x,y,z)\Psi^{(i)}=E\Psi^{(i)}. \tag{6.45}$$

Here, it is assumed that the effective masses are constant, in order to simplify the equations (to generalize this to nonparabolic bands, the reciprocal mass would enter between the partial derivatives). The values of the effective masses that enter into (6.1) depend on how one chooses to orient the device with respect to the crystal axes, which is why the valley index is included in them in (6.1).

For the 3D finite difference grid, we again assume a uniform spacing a, with $x = sa$, $y = la$ and $z = \eta a$, where s, k and η are integers. The 3D finite difference Hamiltonian becomes [34, 35]

$$\begin{aligned}&-t_x^{(i)}({\psi^{(i)}}_{s+1,l,\eta}+{\psi^{(i)}}_{s-1,k,\eta})-t_y^{(i)}({\psi^{(i)}}_{s,l+1,\eta}+{\psi^{(i)}}_{s,k-1,\eta})\\&-t_z^{(i)}({\psi^{(i)}}_{s,l,\eta+1}+{\psi^{(i)}}_{s,l,\eta-1})+(V_{s,l,\eta}+2t_z^{(i)}+2t_z^{(i)}+2t_z^{(i)}){\psi^{(i)}}_{s,l,\eta}=E{\psi^{(i)}}_{s,l,\eta}\end{aligned} \tag{6.46}$$

with hopping energies given by

$$t_x^{(i)}=\frac{\hbar^2}{2m_{x,i}{}^*a^2},\quad t_y^{(i)}=\frac{\hbar^2}{2m_{y,i}{}^*a^2},\quad t_z^{(i)}=\frac{\hbar^2}{2m_{z,i}{}^*a^2}. \tag{6.47}$$

Given the tight-binding form, an artificial band structure is created. The band along each direction has a cosinusoidal variation with momentum eigenvalue, with the total width of this band being

$$W=2t_z^{(i)}+2t_z^{(i)}+2t_z^{(i)}. \tag{6.48}$$

As in the 2D case, the discrete Schrödinger equation (6.46) can be used to obtain transfer matrices that allows one to translate across the structure. However, since the devices now being considered are fully three dimensional, there are two dimensions corresponding to the transverse direction instead of just one. Each transverse plane contains $N_y \times N_z$ grid points. We re-order the coefficients into a $N_yN_z \times 1$ first-rank tensor (i.e. a vector), so that the propagation is handled by a simpler matrix multiplication. Since the smallest dimension in our calculations is generally in the z direction, we use N_z for the expansion, and write the vector wave function as

$$\Psi^{(i)} = \begin{bmatrix} \psi^{(i)}_{1,Ny} \\ \psi^{(i)}_{2,Ny} \\ \dots \\ \psi^{(i)}_{Nz,Ny} \end{bmatrix}. \tag{6.49}$$

Now, (6.46) can be rewritten as a matrix equation

$$H^{(i)}\Psi^{(i)}(s) - T_x^{(i)}\Psi^{(i)}(s-1) - T_x^{(i)}\Psi^{(i)}(s+1) = EI\Psi^{(i)}(s). \tag{6.50}$$

Here, I is the unit matrix, E is the energy to be found from the eigenvalue equation, while

$$H(i) = \begin{bmatrix} H_0^{(i)}(\mathbf{r}) & \tilde{t}_z^{(i)} & \dots & 0 \\ \tilde{t}_z^{(i)} & H_0^{(i)}(\mathbf{r}) & \dots & \dots \\ \dots & \dots & \dots & \tilde{t}_z^{(i)} \\ 0 & \dots & \tilde{t}_z^{(i)} & H_0^{(i)}(\mathbf{r}) \end{bmatrix}, \tag{6.51}$$

is a Hamiltonian corresponding to an individual slice, and

$$T_x^{(i)} = \begin{bmatrix} \tilde{t}_x^{(i)} & 0 & \dots & 0 \\ 0 & \tilde{t}_x^{(i)} & \dots & 0 \\ \dots & \dots & \dots & \dots \\ 0 & 0 & \dots & \tilde{t}_x^{(i)} \end{bmatrix}. \tag{6.52}$$

represents the inter-slice coupling. The dimension of these two super-matrices is $N_z \times N_z$, while the basic H_0 terms of (6.51) have dimension of $N_y \times N_y$, so that the total dimension of the above two matrices is $N_yN_z \times N_yN_z$. In general, if we take k and j as indices along y, and η and ν as indices along z, then

$$(\tilde{t}_z^{(i)})_{\eta\nu} = \tilde{t}_z^{(i)}\delta_{\eta\nu}, \quad (\tilde{t}_y^{(i)})_{kj} = t_y^{(i)}\delta_{kj}, \quad (\tilde{t}_x^{(i)})ss' = t_z^{(i)}\delta ss', \tag{6.53}$$

and

$$H_0^{(i)}(\mathbf{r}) = \begin{bmatrix} V(s,1,\eta+W) & 0 & \dots & 0 \\ t_y^{(i)} & \tilde{t}_x^{(i)} & \dots & 0 \\ \dots & \dots & \dots & \dots \\ 0 & 0 & \dots & \tilde{t}_x^{(i)} \end{bmatrix}. \tag{6.54}$$

Using the above, one can construct a transfer matrix equation that relates adjacent pairs of slices:

$$T_s = \begin{bmatrix} 0 & -I \\ -I & (H^{(i)} - E)(T_x^{(i)})^{-1} \end{bmatrix}. \tag{6.55}$$

At this point, one can apply the Usuki recursion technique as described earlier to compute the transmission, summing over the net contributions from all the valleys.

5.3 *Introduction of Separable Scattering Mechanisms*

As it turns out, by their very form, it is quite simple to modify the Usuki recursion formulas by the addition of an on-site self-energy Σ that provides a correction due to scattering to the local potential. It generally has both real and imaginary parts, with the latter representing the dissipative interactions. In semiconductors, the scattering is weak, and is traditionally treated by first-order, time-dependent perturbation theory, which yields the common Fermi golden rule for scattering rates. With such weak scattering, the real part of the self-energy can generally be ignored for the phonon interactions, and the part that arises from the carrier–carrier interactions is incorporated into the solutions of Poisson's equation.

In the many-body formulations of the self-energy, it is expressed as a two-site function [37]:

$$\Sigma(\mathbf{r}_1, \mathbf{r}_2). \tag{6.56}$$

Since we are using transverse modes in the quantum wire, this may be rewritten as

$$\Sigma(i, j; i', j', x_1, x_2). \tag{6.57}$$

Here, the scattering accounts for transitions from transverse mode i, j at position x_1 to i', j' at position x_2. Generally, one then makes a center-of-mass transformation [50]

$$X = \frac{x_1 + x_2}{2}, \quad \xi = x_1 - x_2, \tag{6.58}$$

and then Fourier transforms on the difference variable to give

$$\Sigma(i, j; i', j', X, k_x) = \frac{1}{2\pi} \int d\xi \, e^{i\xi k_x} \Sigma(i, j; i', j', X, \xi). \tag{6.59}$$

The center-of-mass position X remains in the problem as the mode structure may change as one moves along the channel. At this point, the left-hand side of (6.59) is the self-energy computed by the normal scattering rates, such as is done in quantum wells and quantum wires [36, 55]. As mentioned above, since scattering in semiconductors is relatively weak, it is sufficient to compute these using Fermi's golden rule, which is an evaluation of the bare self-energy in (6.59), rather than incorporating more elaborate many-body effects. Since the Usuki recursion is in the site representation, we have to reverse the Fourier transform in (6.59) to get the x-axis variation, and do a mode-to-site unitary transformation to get the self-energy in the form necessary for the recursion. We thus proceed by using the Fermi golden rule expression for each scattering process of interest and generating a real space self-energy from it. The imaginary part of the self-energy is related to the scattering rate via

$$\mathrm{Im}\{\Sigma(i,j;i',j',X,k_x)\} = \hbar \left(\frac{1}{\tau}\right)_{i,j}^{i',j'}. \tag{6.60}$$

It is the latter scattering rate which we calculate, which is a function of the x-directed momentum (which is related, in turn, to the energy of the carrier) in a cross-section of the device, which can be thought of locally as a quantum wire. This scattering rate must be converted to the site representation with a unitary transformation. At site s,l,η, the correction due to scattering that gets added to the local potential $V_{s,l,\eta}$ is given by

$$\Gamma(s,l,\eta) = \mathrm{Im}\{\Sigma\} = U_s^+ \left(\frac{\hbar}{\tau}\right)_{i,j}^{i',j'} U_s, \tag{6.61}$$

where U_s is the matrix of modes obtained for the sth slice along the device in the x-direction and U_s^+ is its conjugate.

The Fermi golden rule scattering rate for acoustic phonons is treated in nearly all textbooks (see, for example, Lundstrom [62]), and the only modification is to account for the transverse modes of the quantum wire. Rather than repeat the microscopic details of such a calculation, we begin with the general form

$$\left(\frac{1}{\tau}\right)_{i,j}^{i',j'} = \frac{2\pi}{\hbar}\frac{D_{ac}^2 k_B T}{2\rho v_s^2} I(i,j,i',j') \sum_{\vec{k}'} \delta\left(E_{k'} - E_k\right), \tag{6.62}$$

where D_{ac} is the acoustic deformation potential, ρ is the density, v_s is the velocity of sound, and $I(i,j,i',j')$ is an intermodal overlap integral. Here, the acoustic phonon is treated, as is normal, as quasi-elastic in that the energy transferred to the acoustic mode is considerably smaller than the carrier energy, and the delta function in (6.62) serves to conserve the energy in the process.

In the above equations, E_k and $E_{k'}$ are the energies corresponding to the initial and final energy states assuming parabolic subbands. This may be visualized using Fig. 6.13, which illustrates a simple two subband model to define the initial and final energies. We define $E_{0,ij}$ as the energy value corresponding to $k_x = 0$ in the initial subband, while $E_{0,i'j'}$ corresponds to the value of the energy in the final subband with a $k_x' = 0$ value. With these definitions,

$$E_k = E_{0,ij} + \frac{\hbar^2 k_1^2}{2m_x^*}, \tag{6.63}$$

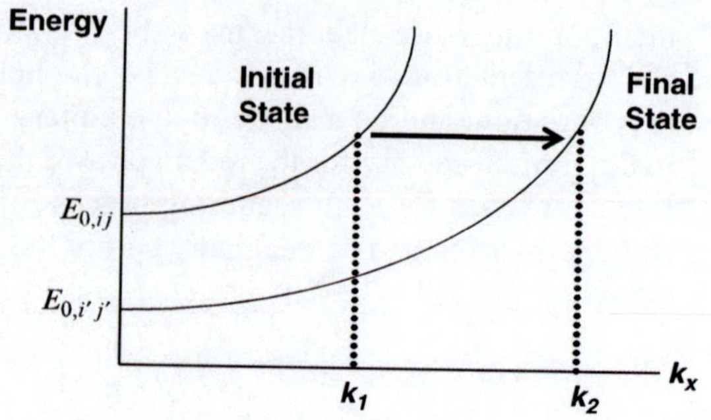

Fig. 6.13 Schematic of the parabolic bandstructure used in the formulation of the scattering method

and

$$E_{k'} = E_{0,i'j'} + \frac{\hbar^2 k_2^2}{2m_x^*}. \tag{6.64}$$

From this, the difference between the initial and final energies becomes

$$E_k - E_{k'} = E_{0,ij} - E_{0,i'j'} + \frac{\hbar^2}{2m_x^*}(k_1^2 - k_2^2). \tag{6.65}$$

To solve for k_2 in terms of k_1 and the difference between the initial and final energies, we define

$$\Delta_{ij}^{i'j'} = E_{0,ij} - E_{0,i'j'}, \tag{6.66}$$

and we thus write

$$k_2^2 = k_1^2 + \frac{2m_x^*}{\hbar^2}\Delta_{ij}^{i'j'}. \tag{6.67}$$

Following the usual procedure in computing scattering rates, the summation over final k' states is replaced with an integration, as

$$\sum_{k'} \to \frac{L}{2\pi}\int_{-\infty}^{\infty} dk', \tag{6.68}$$

where L is length along x. We now combine (6.68) with (6.62) to obtain

$$\left(\frac{1}{\tau_{ac}}\right)_{i,j}^{i',j'} = \frac{\pi D_{ac}^2 k_B T}{\hbar \rho v_s^2} I(i,j,i',j') \frac{L}{2\pi} \frac{1}{\left|\frac{\partial E_{k'}}{\partial k'}\right|}, \tag{6.69}$$

where the last term is evaluated using (6.66). The scattering rate is then

$$\left(\frac{1}{\tau_{ac}}\right)_{i,j}^{i',j'} = \frac{m_x^* D_{ac}^2 k_B T}{2\hbar^3 \rho v_s^2} \frac{LI(i,j,i',j')}{\sqrt{k^2 + \frac{2m_x^*\Delta_{ij}^{i'j'}}{\hbar^2}}} \theta\left(k^2 + \frac{2m_x^*\Delta_{ij}^{i'j'}}{\hbar^2}\right), \tag{6.70}$$

where θ is the Heaviside step function ($\theta(x) = 1$ for $x > 0$, and 0 for $x < 0$).

Now, for our real space quantum transport approach, we need to reverse the Fourier transform in (6.59). That is, we use the inverse transform to real space from momentum space and obtain the final form for the acoustic deformation potential scattering rate. The Fourier integral is

$$\left(\frac{1}{\tau}\right)_{i,j}^{i',j'}(x-x') = \frac{m_x^* D_{ac}^2 k_B T}{2\hbar^3 \rho v_s^2}(LI(i,j,i',j'))\frac{1}{\sqrt{2\pi}}\int_{\beta}^{\infty}\frac{e^{ik(x-x')}}{\sqrt{k^2+\frac{2m_x^*\Delta_{ij}^{i'j'}}{\hbar^2}}}dk,$$

$$\beta = \sqrt{-\frac{2m_x^*\Delta_{ij}^{i'j'}}{\hbar^2}}. \tag{6.71}$$

The lower limit in the integration results in zero if $\Delta_{ij}^{i'j'} > 0$. From Fig. 6.13, it can be seen that scattering cannot occur from the lower subband to the upper subband unless there is a minimum momentum (or energy), and this accounts for the non-zero lower limit in the integration for such situations. Assuming $\Delta_{ij}^{i'j'} \leq 0$, the integration can then be carried out easily to yield (the other cases are also easily done)

$$\left(\frac{1}{\tau}\right)_{i,j}^{i',j'}(x-x') = \frac{m_x^* D_{ac}^2 k_B T}{2\hbar^3 \rho v_s^2}(LI(i,j,i',j'))\frac{1}{\sqrt{2\pi}}\cdot$$
$$\left\{\frac{\pi}{2} - \beta Si[-i\beta(x-x')]\cosh[\beta(x-x')] - \beta Ci[-i\beta(x-x')]\sinh[\beta(x-x')]\right\}. \tag{6.72}$$

The term in curly brackets is sharply peaked around $x = x'$, which implies the scattering is local with regard to the individual slices in the recursion. There is coupling between the modes within a slice, but this local (to the slice) behavior is just the normal assumption in quasi-classical cases, where the scattering is assumed to be local in space [100]. Yet we need to know the total scattering rate within the slice, so this is achieved by integrating over x' in order to find the resultant scattering weight

$$\left(\frac{1}{\tau_{ac}}\right)_{i,j}^{i',j'} = \frac{m_x^* D_{ac}^2 k_B T}{4\hbar^3 \rho v_s^2}(LI(i,j,i',j'))\sqrt{\frac{\pi}{2}}. \tag{6.73}$$

For its use in the Usuki recursion, this scattering rate must then be converted to the site representation with the unitary transformation given by (6.61).

In a similar manner, we have also derived expressions for the intervalley scattering in silicon due to f and g processes, which is carried out by high energy optical modes. The reader is referred elsewhere [41] for the full details.

Given the transmission as a function of E, drain bias, V_{ds} and gate voltage, V_g, obtained using the Usuki method with scattering, one can calculate the current flowing through the through the device for a given source-drain bias, V_{sd} as follows [36]:

$$I_{ds} = \frac{2e}{h}\int (f(E) - f(E - eV_{ds}))T(E, V_{ds}, V_g)dE. \tag{6.74}$$

5.4 *Application of the Method: Determining the Ballistic to Diffusive Crossover in a SOI MOSFET*

In this section, we show an application of the methodology described above. Experiments have shown it possible to fabricate MOS transistors in a Silicon-On-Insulator (SOI) environment with channel widths as small as 2 nm [24, 52, 66]. The semiconductor industry believes that the sizes of what effectively are quantum wires, along with the improved scalability associated with SOI technologies, would be ideal as next generation transistors and interconnects. Given the typically small film thickness associated with these devices (∼6 nm), it is clear that the transport in such devices will be one dimensional (1D) in nature. With this in mind, we considered an SOI MOSFET [6, 41], with its channel aligned along the (110) direction. Our goal here was to determine the point where this MOSFET exhibits a ballistic to diffusive crossover as a function of channel length. It has been suggested that the transport in small transistors is ballistic, and that once a carrier enters the channel it will continue to the drain, with no chance to scatter back to the source [61]. However, it has been shown that scattering within the channel will cause second-order effects which do affect the terminal characteristics of the transistor [75], and thus far, the search for ballistic behavior has not been so successful.

Figure 6.14 shows an overview cutaway cross-section of the device under consideration in the *x*-*y* plane, showing source, channel and drain, and dopant positions

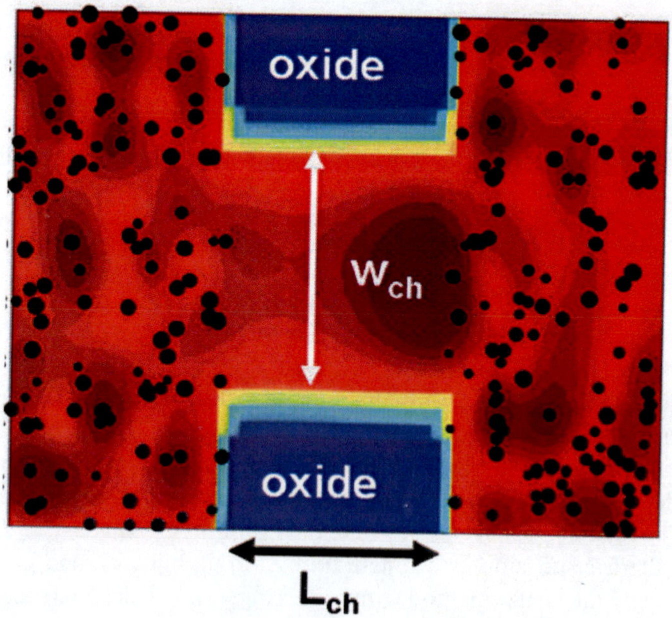

Fig. 6.14 Cutaway overview of the SOI MOSFET device, showing dopant atoms in the source and drain. The interior shading indicates the electron density that was obtained during this particular simulation. For clarity, a relatively long channel length was used to generate this picture

indicated. Oxide barriers, as shown are placed on either side of the channel to simulate the appearance of a hard wall boundary that would be present in an actual experimental system. The thickness of the silicon layer in the z-direction is 6.51 nm. The source and drain of the device are 36.93 nm wide and 27.15 nm in length. The source and drain of the device are discretely doped n-type with a doping concentration of $1 \times 10^{20}\,\mathrm{cm}^{-3}$, while the channel is undoped. The quantum wire that forms the channel of the device has metal gates on three sides to form a trigate-type transistor. The gate oxide thickness (SiO_2) on this device was 1 nm. It should be noted that the effect of the presence of dopants in our approach is included as a local correction to the potential (generally appearing as a local potential well). No scattering rate calculation is performed or required for them using our method.

As would be the case in a real device, the dopants placed in random locations, according to the prescription outlined by Wong and Taur [87]. The same doping profile is used for each of the cases we examined, and the source and drain regions were left untouched at the channel length and channel width were varied. In every case, the gate voltage was set at Vg = 0.6 V, while the source-drain bias was Vsd = 0.01 V.

When the transport is ballistic, the resistance of the channel will be determined by the inverse of the Landauer conductance, as

$$R_{ballistic} = \left[\frac{2e^2}{h}N\right]^{-1}, \tag{6.75}$$

where N is the number of transverse modes propagating through the wire. Importantly, there is no dependence upon the length of the wire in the ballistic limit. On the other hand, in the diffusive case, when the resistance is determined by the mobility (μ) and carrier density(n), then the appropriate expression is

$$R_{diffusive} = \left[\frac{1}{\sigma}\right]\frac{L_{ch}}{A} = \frac{1}{ne\mu}\frac{L_{ch}}{A}, \tag{6.76}$$

where L_{ch} is the channel length and A is the "cross-sectional area" of the inversion layer. We use the area here, rather than just the width of the two-dimensional layer, as we are dealing with a three-dimensional wire with full quantization in the transverse direction. In this limit, the resistance increases linearly with the channel length. It is the value of L_{ch} where there is a change from (6.75) to (6.76) that we were interested in determining.

In Fig. 6.15, the computed resistance for a device with a channel width of 6.5 nm is shown. It should be noted that rather than the Landauer formula, the resistance was obtained from the I–V characteristic in this case. Phonon scattering due to acoustic and f and g processes is included. Note that at 300 K, the resistance increases almost linearly with channel length, in line with the diffusive prediction. It should be noted that different random impurity configurations will yield different curves, but with the same general trend. In contrast, at a temperature of 100 K, the electron–phonon scattering is largely suppressed, and in this case resistance quickly rises as function of channel length and then essentially saturates at $L_{ch} \sim 2$ nm. This is indicative of a

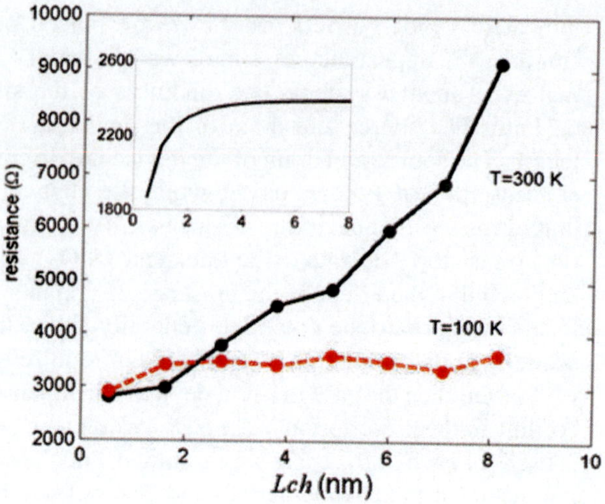

Fig. 6.15 Resistance vs. L_{ch} for a device with channel width $w_{ch} = 6.5\,\mathrm{nm}$ at the indicated temperatures. *Inset* result of a calculation for a perfect, ballistic wire without scattering

ballistic to diffusive crossover at that channel length. Note as well that the resistance is actually lower for the $T = 300\,\mathrm{K}$ case for very short channels lengths. To understand what is special about $L_{ch} \sim 2\,\mathrm{nm}$, a calculation for the ideal, ballistic case was also done, as shown in the inset. Here, a *perfect* wire is used without any scattering at all from phonons or impurities. Finite temperature is accounted for using the simple thermal broadening used earlier with $T^* = 300\,\mathrm{K}$. As is evident, $L_{ch} \sim 2\,\mathrm{nm}$ is the approximate length for which the resistance saturates in the ballistic case. Beyond this length, the wire acts as an ideal quantum point contact for which the transmission is quantized and proportional to the number of propagating modes in the wire. As this result clearly indicates, one cannot get away with doing a simple, nondissipative calculation to capture the physics of what happens in such devices at room temperature, a point we made at the beginning of this section.

6 Summary

In this chapter, we have outlined the Usuki method for calculating electronic transport in nanostructures and have provided a few illustrative examples of its application. Many other cases that we have studied over the years could have been discussed, such as its use in interpreting the results of scanning gate microscopy experiments in quantum dots, its application to the problem of understanding of the ∼0.7 conductance plateau phenomena observed in quantum point contacts [2], and its use in simulating quantum wave processing [1] and qubit applications [39,40,44], amongst others. Beyond the advantages over other quantum based methods that we have already outlined in the preceding discussion, another is its simplicity. Regardless of what computer language that it is written in, the core piece of any code that implements it is only a few dozen lines long, at most. While our most computationally intensive applications of the method have generally been written in fortran, some students have coded it instead in MATLAB and obtained publishable results with it [72]. Recently, we have also implemented a parallelized version of our fortran code. We have achieved quite good speedup while at the same time we are now able to look at problems that are significantly larger than what has ever been attempted before. Problems like the 3D MOSFET, which previously could take several days of computer time, are now far more tractable.

Acknowledgements We would like to thank the financial support from the Office of Naval Research, the Department of Energy and Intel Corporation. The experiments of Prof. Jonathan Bird and colleagues at ASU and at the University of Buffalo were the inspiration for much of the simulation work we have done over the years. The group of Prof. Yuichi Ochiai at Chiba University provided similar inspiration. At ASU, we have also had fruitful collaborations with the groups of Prof. Dragica Vasileska, Prof. Ying-Chen Lai, Prof. Otto Sankey, and Prof. Stephen Goodnick. The team of Roland Brunner and his adviser Prof. Friedemar Kuchar at the University of Leoben helped illuminate the correspondence between our quantum simulations and the classical behavior in quantum dots. Our thanks also go out to Jan Jacob and his advisors Prof. Meier and Prof. Matsuyama at the University of Hamburg for the collaborative work that they initiated on the spin Hall effect.

References

1. Akis R., Ferry D.K Quantum waveguide array generator for performing Fourier transforms: Alternate route to quantum computing. Appl. Phys. Lett. **79**, 2823–2825 (2001).
2. Akis R., Ferry D.K.: Simulations of Spin Filtering Effects in a Quantum Point Contact. J. Phys. Condensed Matter (2008). doi: 10.1088/0953-8984/20/16/164201
3. Akis R., Bird J.P., Ferry D.K. Magnetotransport fluctuations in regular semiconductor ballistic quantum dots. Phys. Rev. B **54**, 17705–17715 (1996).
4. Akis R., Bird J.P., Ferry D.K.: The persistence of eigenstates in open quantum dots. Appl. Phys. Lett. (2002). doi: 10.1063/1.1490404
5. Akis R., Bird J.P., Vasileska D., Ferry D.K., deMoura A.P.S., Lai Y.-C.: On the Influence of Resonant States on Ballistic Transport in Open Quantum Dots: Spectroscopy and Tunneling in the Presence of Multiple Conducting Channels, In: Bird J.P. (ed.) Electron Transport in Quantum Dots, pp. 209–276. Kluwer Academic Publishers, Boston (2003)
6. Akis R., Gilbert M., Ferry D.K.: Fully quantum mechanical simulations of gated silicon quantum wire structures: investigating the effects of changing wire cross-section on transport. J. Phys. Conf. Series **36**, 87–90 (2006).
7. Ando T.: Quantum point contacts in magnetic fields. Phys. Rev. B **44**, 8017–8027 (1991).
8. Assad F., Ren Z., Vasileska D., Datta S., M. Lundstrom: On the performance limits for Si MOSFETs: a theoretical study. IEEE Trans. Elec. Dev. **47**, 232–240 (2000)
9. Baranger H.U., Stone A. D.: Electrical linear-response theory in an arbitrary magnetic field: A new Fermi-surface formation. Phys. Rev. B **40**, 8169–8193 (1989).
10. Benisty H.: Reduced electron-phonon relaxation rates in quantum-box systems: Theoretical analysis. Phys. Rev. B **51**, 13281–13292 (1995).
11. Bird J. P., Olatona D. M., Newbury R., Taylor R. P., Ishibashi K., Stopa M., Aoyagi Y., Sugano T., Ochiai Y.: Lead-induced transition to chaos in ballistic mesoscopic billiards. Phys. Rev. B **52**, R14336–R14339 (1995).
12. Bird J.P., Ferry D.K. R., Ishibashi K., Aoyagi Y., Sugano T., Ochiai Y.: Periodic conductance fluctuations and stable orbits in mesoscopic semiconductor billiards. Europhys. Lett. **35**, 529–534 (1996).

13. Bird J.P., Akis R., Ferry D.K. Vasileska D., Cooper J., Aoyagi Y., Sugano T.: Lead-orientation-dependent wave function scarring in open quantum dots. Phys. Rev. Lett. **82**, 4691–4694 (1999a).
14. Bird J.P., Akis R., Ferry D.K.: Magnetoprobing of the discrete level spectrum of open quantum dots. Phys. Rev. B **60**, 13676–13681 (1999b).
15. Bird J.P., Akis R., Ferry D.K., de Moura A.P.S., Lai Y.-C., Indlekofer K.M.: Interference and interactions in open quantum dots. Rep. Prog. Phys. **66**, 583–632 (2003).
16. Blume-Kohout R., Zurek W.H.: Quantum Darwinism in Quantum Brownian Motion. Phys. Rev. Lett. (2008). doi: 10.1103/PhysRevLett.101.240405
17. Brunner R., Kuchar F., Meisels R., Akis R., Ferry D.K., Bird J.P.,: Draining of the Sea of Chaos: Role of Resonant Transmission and Reflection in an Array of Billiards. Phys. Rev. Lett. (2007). doi: 10.1103/PhysRevLett.98.204101
18. Brunner R., Akis R., Ferry D.K., Kuchar F., Meisels R.: Coupling-induced bipartite pointer states in arrays of electron billiards: Quantum Darwinism in action?. Phys. Rev. Lett. (2008). doi: 10.1103/PhysRevLett.101.024102
19. Burke, A. M., Akis, R., Day T. E., Speyer G., Ferry D.K., Bennett B.R.: Imaging scarred states in quantum dots. J. Phys. Condensed Matter (2009). doi: 10.1088/0953–8984/21/21/212201
20. Burke, A. M., Akis, R., Day T. E., Speyer G., Ferry D.K., Bennett B.R.: Periodic Scarred States in Open Quantum Dots as Evidence of Quantum Darwinism. Phys. Rev. Lett. (2010). doi: 10.1103/PhysRevLett.104.176801
21. Büttiker M.: Role of quantum coherence in series resistors. Phys. Rev. B **33**, 3020–3026 (1986).
22. Büttiker M., Imry Y., Landauer R., Pinhas S.: Generalized many-channel conductance formula with application to small rings. Phys. Rev. B **31**, 6207–6215 (1985).
23. Bychkov Y.A., Rashba E.I.: Oscillatory effects and the magnetic susceptibility of carriers in inversion layers. J. Phys. C. 17, 6039–6045 (1984).
24. Cui Y., Lieber C. M.: Functional nanoscale electronic devices assembled using silicon nanowire building blocks. Science **291**, 851–853 (2001).
25. Cummings A.W., Akis R., Ferry D.K.: Electron spin filter based on Rashba spin-orbit coupling. Appl. Phys. Lett. (2006). doi: 10.1063/1.2364859
26. Cummings A.W., Akis R., Ferry D.K.: The Rashba Effect and Non-Abelian Phase in Quantum Wire Devices. J. Comp. Elect. **6**, 101–104 (2007).
27. Cummings A.W., Akis R., Ferry D.K., Jacob J., Matsuyama T., Merkt U., Meier G.: Cascade of Y-shaped spin filters in InGaAs/InAs/InGaAs quantum wells. J. Appl. Phys. (2008). doi: 10.1063/1.2980328
28. Datta S.: Nanoscale device modeling: the Green's function method. Superlattices and Microstructures **28**, 253–278 (2000).
29. de Moura A.P.S., Lai Y.-C., Akis R., Bird J.P., Ferry D.K.: Tunneling and Nonhyperbolicity in Quantum Dots. Phys. Rev. Lett. (2002). doi: 10.1103/PhysRevLett.88.236804
30. Di Ventra M., Pantelides S. T., Lang, N. D.: First-Principles Calculation of Transport Properties of a Molecular Device. Phys. Rev. Lett. **84**, 979–982 (2000).
31. Dresselhaus G.: Spin-Orbit Coupling Effects in Zinc Blende Structures. Phys. Rev., **100**, 580–586, (1955).
32. Ferry D. K.: Effective potentials and the onset of quantization in ultrasmall MOSFETs. Superlatt. Microstruct. **28**, 419–423 (2000).
33. Ferry D.K., Akis R., Bird J.P.: Einselection in action: Decoherence and pointer states in open quantum dots. Phys. Rev. Lett. (2004). doi: 10.1103/PhysRevLett.93.026803
34. Ferry D.K., Akis R., Bird J.P.: Einselection and the quantum to classical transition in quantum dots. J. Phys. Condensed Matter (2005a). doi: 10.1088/0953–8984/17/13/001
35. Ferry D.K., Akis R., Gilbert M.J., Ramey S.M.: Physics of Silicon Nanodevices. In: Oda S., Ferry D.K. (eds.) Silicon Nanoelectronics, pp. 200–210. Taylor & Francis, Boca Raton (2005b)
36. Ferry D.K., Goodnick S.M., Bird J.P.: Transport in Nanostructures, Second Edition Cambridge, Cambridge (2009)
37. Fetter A. L., Walecka J. D.: Quantum Theory of Many-Particle Systems. McGraw-Hill, New York (1971)

38. Fischetti M.V.: Theory of electron transport in small semiconductor devices using the Pauli master equation. J. Appl. Phys., **83**, 270–291 (1988).
39. Gilbert M.J., Akis R., Ferry D.K: Magnetically and electrically tunable semiconductor quantum waveguide inverter. Appl. Phys. Lett. (2002). doi: 10.1063/1.1525073
40. Gilbert M.J., Akis R., Ferry D.K.: Dual computational basis qubit in semiconductor heterostructures. J. Appl. Phys. (2003). doi: 10.1063/1.1599633
41. Gilbert M.J., Akis R., Ferry D.K.: Phonon-assisted ballistic to diffusive crossover in silicon nanowire transistors. J. Appl. Phys. (2005). doi: 10.1063/1.2120890
42. Grubin H.L., Kreskovsky J.P., Govindan T.R., Ferry D.K.: Uses of the quantum potential in modeling hot-carrier semiconductor devices. Semicond. Sci. Technol. **9**, 855–858 (1994)
43. Hankiewicz E. M., Molenkamp L. W., Jungwirth T., and Sinova J.: Manifestation of the spin Hall effect through charge-transport in the mesoscopic regime. Phys. Rev. B, vol. 70, p., Dec. 2004. doi: 10.1103/PhysRevB.70.241301
44. Harris J., Akis R., Ferry D.K.: Magnetically switched quantum waveguide qubit", Appl. Phys. Lett. **79**, 2214–2215 (2001).
45. He H., Zhu J., Tao N. J., Nagahara L. A., Amlani I., Tsui R.: A Conducting Polymer Nanojunction Switch. J. Am. Chem. Soc., **123**, 7730–7731 (2001).
46. Heller E. J.:. Bound-State Eigenfunctions of Classically Chaotic Hamiltonian Systems: Scars of Periodic Orbits. Phys. Rev. Lett. 53, 1515–1518 (1984).
47. Huang L., Lai Y.-C., Ferry D. K., Goodnick S. M., Akis R.: Transmission and scarring in graphene quantum dots. Phys. Condensed Matter (2009). doi: 10.1088/0953–8984/21/34/344203
48. Huang L., Lai Y.-C., Ferry D. K., Goodnick S. M., Akis R.: Relativistic quantum scars. Phys.Rev. Lett. (2010). doi: 10.1103/PhysRevLett.103.054101
49. Jacob J., Meier G., Peters S., Matsuyama T., Merkt U., Cummings A.W., Akis R., Ferry D.K.: Generation of highly spin-polarized currents in cascaded In As spin filters. J. Appl. Phys. (2009). doi: 10.1063/1.3124359
50. Kadanoff L. P., Baym G.: Quantum Statistical Mechanics. Benjamin/Cummings, Reading (1962)
51. Ke S.-H., Baranger H.U., Yang W.: Electron transport through molecules: Self-consistent and non-self-consistent approaches. Phys. Rev B (2004). doi: 10.1103/PhysRevB.70.085410
52. Kedzierski J., Bokor J., Anderson E.: Novel method for silicon quantum wire transistor fabrication. J. Vac. Sci. Tech. B **17**, 3244–3247 (1999).
53. Kluksdahl N.C., Kriman A.M., Ferry D.K., and Ringhofer C.: Self-consistent study of the resonant tunneling diode. Phys. Rev. B, **39**, 7720–7735 (1989).
54. Ko D.Y.K., Inkson J.C.: Matrix method for tunneling in heterostructures: Resonant tunneling in multilayer systems. Phys. Rev. B, **38**, 9945–9951 (1988).
55. Kotlyar R., Obradovic B., Matagne P., Stettler M., Giles M.D.: Assessment of room-temperature phonon-limited mobility in gated silicon nanowires. Appl. Phys. Lett. **84**, 5270–5272 (2004).
56. Lake R., Klimeck G., Bowen R.C., Jovanovic D: Single and multiband modeling of quantum electron transport through layered semiconductor devices. J. App. Phys. (1997). doi: 10.1063/1.365394
57. Landauer R.: Spatial variation of currents and fields due to localized scatterers in metallic conduction. IBM J. Res. Develop. 1, 223–231 (1957).
58. Landauer R.: Electrical resistance of disordered one-dimensional lattices. Phil. Mag. **21**, 863–867 (1970).
59. Laux S.E., Kumar A., Fischetti M.V.: Ballistic FET modeling using QDAME: quantum device analysis by modal expansion. IEEE Trans. Nano. **1**, 255–259 (2002).
60. Lee M.H., Speyer G., Sankey O.F.: Theory of electron transport through single molecules of polyaniline. J. Phys. Condensed Matter (2007). doi: 10.1088/0953–8984/19/21/215204
61. Lundstrom M.: Elementary scattering theory of the Si MOSFET. IEEE Elect. Dev. Lett. **18**, 361–363 (1997).
62. Lundstrom M.: Fundamentals of Carrier Transport. Cambridge, Cambridge (2000).

63. Marinescu D.C., Marinescu G.M.: Approaching Quantum Computation. Pearson Prentice Hall, Upper Saddle River (2005).
64. MacDiarmid A.G., Chiang J.-C., Richter A.F., Epstein A.J.: Polyaniline: A New Concept in Conducting Polymers. Synth. Met., **18**, 285 (1987)
65. Moore G.E.: Cramming more components onto integrated circuits. Electronics **38** (1965).
66. Namatsu H., Kurihara K., Nagase M., Makino T.: Fabrication of 2 nm wide silicon quantum wires through a combination of a partially-shifted resist pattern and orientation-dependent etching. Appl. Phys. Lett. **70**, 619–621 (1997).
67. Natori K.: Ballistic metal-oxide semiconductor field effect transistor. J. Appl. Phys. **76**, 4879–4890 (1994).
68. Neofotistos G., Lake R., Datta S.: Inelastic-scattering effects on single-barrier tunneling. Phys. Rev. B **43**, 2442–2445 (1991).
69. Ollivier H., Poulin D., Zurek W.H.: Objective Properties from Subjective Quantum States: Environment as a Witness. Phys. Rev. Lett. (2004). doi: 10.1103/PhysRevLett.93.220401
70. Pala M.G., Iannaccone G.: Effect of dephasing on the current statistics of mesoscopic devices. Phys. Rev. Lett. **93**, 256803 (2004).
71. Pikus F.G., Likharev K.K.: Nanoscale field effect transistors: an ultimate size analysis. Appl. Phys. Lett. **71**, 3661–3663 (1997).
72. Ramamoorthy A., Akis, R., Bird J.P: Influence of Realistic Potential Profile of Coupled Electron Waveguide on Electron Switching Characteristics. IEEE Trans. Nanotechnology (2006). doi: 10.1109/TNANO.2006.883478
73. Reed M. A., Zhou C., Muller C. J., Burgin T. P., Tour J. M.: Conductance of a Molecular Junction. Science, **278**, 252–254 (1997).
74. Schliemann J., Loss D., Westervelt R.M: Zitterbewegung of ElectronicWave Packets in III-V Zinc-Blende Semiconductor Quantum Wells. Phys. Rev. Lett. (2005). doi: 10.1103/PhysRevLett.94.206801
75. Svishenko A. and Anantram M.P.: Role of scattering in nanotransistors. IEEE Trans. Elect. Dev. **50**, 1459–1466 (2003).
76. Sinova J., Culcer D., Niu Q., Sinitsyn N. A., Jungwirth T., MacDonald A.H: Universal Intrinsic Spin Hall Effect. Phys. Rev. Lett. (2004). doi: 10.1103/PhysRevLett.92.126603
77. Speyer G., Akis R., Ferry D.K.: Rapid molecular conductance calculations using transfer matrix method. Physica E, **19**, 145–148 (2003).
78. Speyer G., Akis R., Ferry D.K.: Conductance investigations of stretched molecules. IEEE Trans. Nanotechnology (2005). doi: 10.1109/TNANO.2005.851287
79. Speyer G., Akis R., Ferry D.K.: Using local orbitals in DFT to examine oligothiophene conductance anomalies. J. Phys. Conf. Ser. (2006). doi: 10.1088/1742–6596/38/1/007
80. Speyer G., Akis R., Ferry D.K.: Complexities of the Molecular Conductance Problem. In: Lyshevski S.E. (ed.) Nano and Molecular Electronics Handbook, pp 21–1–68. CRC Press, Boca Raton (2007)
81. Thornton T.J., Pepper M., Ahmed H., Andrews D., Davies G.J.: One-Dimensional Conduction in the 2D Electron Gas of a GaAs-AlGaAs Heterojunction. Phys. Rev. Lett. **56**, 1198–1201 (1986).
82. Usuki T., Saito M., Takatsu M., Kiehl R.A., Yokoyama N.: Numerical analysis of electron wave detection by a wedge shaped point contact. Phys. Rev. B **520**, 7615–7625 (1994).
83. Usuki T., Saito M., Takatsu M., Kiehl R.A., Yokoyama N.: Numerical analysis of ballistic electron transport in magnetic-fields by using a quantum point contact and a quantum wire. Phys. Rev. B **52**, 8244–8255 (1995).
84. van der Vorst H. A.: Bi-CGSTAB: A fast and smoothly converging variant of Bi-CG for the solution of non-symmetric linear systems. J. SIAM J. Sci. Stat. Comp. **13**, 631–644 (1992).
85. Vignolo P., Farchioni R., Grosso G.: Tight-Binding Effective Hamiltonians for the Electronic States of Polyaniline Chains. Phys. Stat. Sol. B **223**, 853–866 (2001).
86. Winkler R.: Spin–Orbit Coupling Effects in Two-Dimensional Electron and Hole Systems. Springer, Berlin (2003).
87. Wong H.S., Taur Y.: Three-dimensional "atomistic" simulation of discrete random dopant distribution effects in sub-0.1 μm MOSFETs. IEDM Tech. Dig., 705–708 (1993).

88. Yamamoto M., Ohtsuki T., Kramer B: Spin polarization in a T-shaped conductor induced by strong Rashba spin-orbit coupling. Phys. Rev. B (2005). doi: 10.1103/PhysRevB.72.115321
89. Zutic I., Fabian J., Das Sarma S.: Spintronics: Fundamentals and applications. Rev. Mod. Phys. **76**, 323–410 (2004).
90. Zurek W.H.: Decoherence, einselection, and the quantum origins of the classical. Rev. Mod. Phys. **75**, 715–775 (2003).
91. Zurek W.H.: Quantum Darwinism. Nature Physics (2009). doi: 10.1038/nphys1202

Chapter 7
Quantum Atomistic Simulations of Nanoelectronic Devices Using QuADS

Shaikh Ahmed, Krishnakumari Yalavarthi, Vamsi Gaddipati, Abdussamad Muntahi, Sasi Sundaresan, Shareef Mohammed, Sharnali Islam, Ramya Hindupur, Ky Merrill, Dylan John, and Joshua Ogden

Abstract As semiconductor devices shrink into the nanoscale regime and new classes of nanodevices emerge, device performance is increasingly being dominated by the *granularity* in the underlying material and the *quantum mechanical* effects in the electronic states. At nanoscale, modeling and simulation approaches based on a *continuum* representation of the underlying material typically used by device engineers become invalid. On the other side, various ab initio materials science methods offer intellectual appeal, but can only model very small systems having $\sim$100 atoms. The variety of geometries, materials, and doping configurations in semiconductor devices at the nanoscale suggests that a *general* nanoelectronic modeling tool is needed. This paper describes our on-going efforts to develop a multiscale Quantum Atomistic Device Simulator (QuADS) to address these needs. QuADS bridges the gap (and crosses the intellectual boundary) between continuum and ab initio modeling paradigms and enable the quantum-corrected atomistic numerical modeling of non-equilibrium charge and phonon transport phenomena in realistically-sized systems containing more than 100 million atoms! QuADS is primarily being built upon extended versions of three modules: (a) Open source LAMMPS molecular dynamics code for geometry construction and modeling structural relaxations. To enhance accuracy, ab initio ABINIT tool is used for parameterization of force and polarization coefficients and model bandstructure calculations; (b) Open source NEMO 3-D tool, which employs a variety of tight-binding models (s, sp^3s^*, $sp^3d^5s^*$), for the calculation of excitonic and phonon spectra and optical transition rates; and (c) A quantum-corrected (benchmarked against the non-equilibrium Green function formalism) 3-D Monte Carlo electron–phonon transport kernel. Using QuADS, nanoelectronic device designers will be able to address many challenging issues including crystal atomicity, defects, interfaces and surfaces, strain relaxation, piezoelectric and pyroelectric polarization, quantum confinement, highly-interacting and dissipative current and phonon paths, and performance in harsh environments – all

S. Ahmed (✉)
Department of Electrical and Computer Engineering, Southern Illinois University at Carbondale, 1230 Lincoln Drive, Carbondale, IL 62901, USA
e-mail: ahmed@siu.edu

D. Vasileska and S.M. Goodnick (eds.), *Nano-Electronic Devices: Semiclassical and Quantum Transport Modeling*, DOI 10.1007/978-1-4419-8840-9_7,

on an equal footing. With the multi-million atom handling capability, the simulator creates new engineering routes for optimizing the efficiency and reliability of nanoelectronic and optoelectronic devices that were previously infeasible. Successful applications of QuADS are demonstrated by three examples: (1) Effects of internal fields in InN/GaN quantum dots; (2) Importance of second order polarization in InAs/GaAs quantum dots; and (3) Modeling unintentional single charge effects in silicon nanowire FETs. QuADS uses several novel, memory-miserly, parallel and fast algorithms, and incorporates state-of-the-art fault-tolerant software design approaches, which enables the simulator to assess the reliability of available *petaflop* computing platforms (TeraGrid, NCCS, NICS). A web-based online interactive version for educational purposes will soon be available on http://www.nanoHUB.org.

Keywords Semiconductor device simulation · Quantum effects · Quantum dots · Solid-state lighting · Nanowire · Tight-binding · Monte Carlo simulation · Effective potential · High-performance scientific computing · QuADS

1 Introduction

1.1 Progress in Nanoelectronics

Since the invention of the point-contact bipolar transistor in 1947, advanced fabrication technologies, introduction of new materials with unique properties, and broadened understanding of the underlying physical processes have resulted in tremendous growth in the number and variety of semiconductor devices and literally changed the world. To date, there are about 60 major devices, with over 100 device variations related to them [1]. The most important factor driving the continuous device improvement has been the semiconductor industry's relentless effort to reduce the cost *per function* (historically, ∼25–29% per year) on a chip. This is done by *device scaling*, which involves reducing the transistor size while keeping the electric field constant from one generation to the next. Device scaling has paved the way for a continuous and systematic increase in transistor density in a chip and improvements in system performance (described by Moore's Law [2]) for the past 40 years. For example, regarding conventional silicon MOSFETs, the most critical device for today's advanced integrated circuits, the device size is scaled in all dimensions, resulting in smaller oxide thickness, junction depth, channel length, channel width, and isolation spacing. However, recent studies by many researchers around the globe reveal the fact that, as the silicon industry moves into the 45 nm node regime and beyond, two of the most important challenges facing us are the growing dissipation of *standby power* and the increasing *variability and mismatch* in device characteristics.

The Semiconductor Industry Association (SIA) forecasts [3] that the current rate of transistor performance improvement can be sustained for another 10–15 years, but only through the development and introduction of new materials and transistor structures. One-dimensional nanomaterials such as semiconductor nanowires (NWs) can function both as nanoscale transistors and interconnects and play a key role in future semiconductor industry. Given the central role of silicon in the semiconductor industry, silicon nanowires (SiNWs) represent a particularly attractive class of building blocks for nanoelectronics [4–6] because their diameter and electronic properties can be controlled during synthesis in a predictable manner. It is expected that using these new technologies intrinsic device speed may exceed 1 THz and integration densities will be more than 1 billion transistors/cm^2 [6].

At the same time, rapid progress in nanofabrication technologies has led to the emergence of *new classes* of nanoscale devices that are expected to bring about revolutionary changes in electronic, photonic, computation, information processing, biotechnology, and medical industries. For example, semiconducting quantum dots (QDs) grown by self-assembly are of particular importance in optoelectronics [7,8], since they can be used as detectors of infrared radiation, optical memories, and in laser applications. The strongly peaked energy dependence of density of states and the strong overlap of spatially confined electron and hole wavefunctions provide ultra-low laser threshold current densities, high temperature stability of the threshold current, and high material and differential quantum gain/yield. Strong oscillator strength and non-linearity in the optical properties have also been observed [9, 10]. Self-assembled quantum dots also have potential for applications in quantum cryptography as single photon sources and quantum computation [11–13]. In electronic applications QDs have been used to operate like a single-electron transistor and demonstrate a pronounced Coulomb blockade effect.

1.2 Need for Simulations

As semiconductor devices shrink into the nanoscale regime, there arise problems related to not only the understanding of the device operation but the complicated manufacturing processes also. This fact signifies that the traditional *trial-and-error approach* of device optimization by actually making the devices will no more be feasible since it is both time-consuming and too expensive. Since computers are considerably cheaper resources, *simulation* is becoming an indispensable tool for device engineers working in semiconductor industries today. On the other hand, for researchers, besides offering the possibility to test hypothetical devices which have not (or could not have) yet been manufactured, device simulation offers unique insight into device behavior by allowing the observation of internal phenomena that cannot be measured. Thus, a critical facet of the nanoscale device development is the creation of modeling and simulation tools that can quantitatively explain or even predict experiments. In particular it would be very desirable to explore the design space before, or in conjunction with, the (typically time consuming and expensive) experiments. A *general tool* that is applicable over a large set of materials and geometries is highly desirable. But the tool development itself is not enough. The tool needs to be deployed to the user community so it can be made more reliable, flexible, and accurate.

1.3 Nanoelectronic Device Modeling Challenges

It is well-known that performance and efficiency of novel nanoscale devices are determined by an intricate interplay of electronic and phonon bandstructure effects, dynamics of charge and phonon transport phenomena, and various (internal and external) electromagnetic fields used to drive the device. Therefore, any efforts of modeling nanodevices must involve a multiscale-multiphysics problem and tackle a large number of identified hurdles of scientific uncertainty. A list of these crucial issues is delineated in the following: (1) The lack of spatial symmetry in the overall geometry of novel nanoscale devices requires explicit three-dimensional (3-D) representation and simulation on an *atomic* lattice; (2) Assembly of lattice-mismatched semiconductors in many of these devices leads to a strong inhomogeneous, non-linear, and *long-range strain fields* (for example, in InAs/GaAs superlattices the range is typically ~20 nm [14–16]), which strongly modifies the energy spectrum and the bandstructure parameters (density-of-states, effective mass, mass anisotropy, bandgap, and deformation potential); (3) A variety of III–IV materials are piezoelectric. Any spatial non-symmetric distortion in nanostructures made of these materials will create piezoelectric fields. In contrast to devices in most other material systems, such as the well-known InAs/GaAs system, the *piezoelectric* polarization effects play a dominant role in wurtzite crystal structure based devices for two reasons [17–20] – First, in wurtzite semiconductors, biaxial strain in the basal plane [(0001) plane] causes a piezoelectric field parallel to the C axis ([0001] axis). Since most wurtzite heterostructures are grown on the (0001) plane, the resulting biaxial strain is usually large. Second, due to the high ionicity of the underlying bonds, the piezoelectric constants of these hybrid materials are significantly larger than those of most other semiconductor materials. Piezoelectric field modifies the potential landscape, lower the crystal symmetry, lead to a global shift and a strong band-mixing in the energy spectrum, and hence must be taken into account; (4) Additionally, *spontaneous/pyroelectric* polarization occurs in wurtzite crystals. In many of these systems, the built-in potential resulting from the spontaneous polarization has been found to be of the same order of magnitude as the one resulting from piezoelectric effects [20]; (5) At nanoscale structural and surface relaxations, alloy disorder, formation of defects and amorphous interlayer, and atom clustering are all important. These phenomena are usually temperature sensitive [21, 22] and molecular dynamics simulations are often required. (6) For materials having wide bandgap and large exciton binding energy proper treatment of many-body excitonic states becomes crucial in device modeling; (7) Also, in nanostructures, various properties such as mobility, exciton energy, and radiative lifetime are expected to be strongly affected by *quantum confinement*. Strong quantum confinement results in significant modification of optical phonon (lattice vibration) modes in comparison with bulk phonons and demand special careful treatment; (8) Modeling *transport phenomena* must expose an intricate interplay of classical electrostatics, quantum tunneling, dynamic charge screening, and scattering enhanced carrier recombination [23] in the current and heat paths.

1.4 *Size of the Computational Domain*

What all these novel *nanostructured* devices have in common is that they exploit physical processes at nanometer scale where the atoms in the *active region* are on the order of 10,000 to more than 100 million! For example, self-assembled QDs, with an average height of 1–5 nm, are typically of size (base length/diameter) 5–50 nm and consist of 5,000–2,000,000 atoms. Arrays of quantum-mechanically coupled (stacked) quantum dots that are used as optically active regions in high-efficiency room-temperature lasers consist of 3–7 QDs with typical lateral extension of 10–50 nm and dot height of 1–3 nm. Such dots contain 5–50 million atoms in total, where atomistic details of surrounding material matrix (substrate and cap layers) and interfaces are extremely important. While system sizes of tens of millions of atoms appear at first sight huge and wasteful, in [24] we have demonstrated that the underlying physical problems *require* such large scale analysis. The thickness of an isolating buffer around the active quantum dot region does influence the energy of the confined states, and the buffer size must be chosen adequately large.

1.5 *State-of-the-Art Modeling Approaches*

What the above discussion suggests is that the design of reliable nanostructures must consider simulation domains containing millions of atoms, which, in other words, demands the solution of quantum mechanics in systems having more than 10^7 degrees of freedom! Also, at the atomic scale of novel nanostructured semiconductors the distinction between new device and new material is blurred and device physics and material science meet. However, contemporary material and device modeling efforts are disjoint, divided mainly between device engineers using commercial simulators with jellium/continuum models, and material scientists using ab initio approaches that can handle only $\sim$100 atoms. A pen-picture of the limitations and potentialities of the available approaches of semiconductor device modeling is delineated below:

1.5.1 Commercial Simulators

The existing commercial device modeling tools (e.g. SILVACO [25], APSYS [26], Synopsys [27]) cannot fully predict device behavior at the *nanometer* scale, where the *granular* representation of the underlying material, the effects of internal fields and quantum mechanical size-quantization in the electronic states, and the highly-interacting *transport* paths in the device operation are all important. Commercial tool vendors painstakingly calibrate to suites of experimental data and patch atomistic corrections into their codes but the resulting models are *unphysical* and *non-predictive*. To give an example, recently the APSYS simulator was used to explore a number of possible design approaches for optimizing the internal quantum efficiency (IQE) of InGaN light-emitters [28]. The authors of this January 2010 IEEE article concluded stating, "substantial work has still to be done to reach a stage where truly fitting-parameter-free modeling can be achieved since several significant physical parameters of the III–nitride material system are not well established yet".

1.5.2 Research Tools

As of now, to the best of the authors' knowledge, there is no large-scale fully *atomistic* simulator available for modeling nanoscale devices that comprehensively treats structural relaxation, bandstructure, and transport calculations all on an equal footing. Research tools that are available today usually isolate device phenomena at different levels of hierarchy, which makes these tools unsuitable for an engineer to use in practical device design and optimization. A description is as follows:

Electronic Structure: Theoretical knowledge of the electronic structure of semiconductor devices is the ***first and essential step*** toward the interpretation and the understanding of the experimental data and reliable device design. It is clear that, at nanoscale, modeling approaches based on a ***continuum*** representation (such as effective mass [29], and $k \bullet p$ [20]) are invalid. Continuum models assume the symmetry of the nanostructure to be that of its overall geometric shape. For example, in quantum dot simulations using continuum models, dome-shaped dots are assumed to have continuous cylindrical symmetry $C_{\infty v}$, whereas pyramidal dots are assumed to have C_{4v} symmetry. In a recent effort on modeling $In_{1-x}Ga_xN$ quantum dots using $k \bullet p$ approach [20], it was found that the envelope S function reproduces the symmetry of the confining potential, the excited P and D states are energetically degenerate and optically isotropic – a group of observations that clearly suppresses the true fundamental atomistic symmetry of the underlying crystal and thus *overestimates* the quantum efficiency of the light emitters in these quantum dots. On the other side, various ab initio atomistic materials science methods (fundamental many-electron correlated methods based on perturbation theory, quantum Monte Carlo method, or GW approach) offer intellectual appeal, but can only predict masses and bandgaps for very small systems (around 100 atoms). Thus, for quantum dot simulations, the simulation domain requiring multimillion atoms prevent the use of ab initio methods. Empirical methods (***Pseudopotentials*** [30] and *Tight Binding* [31–33]), which eliminate enough unnecessary details of core electrons, but are finely tuned to describe the atomistically dependent behavior of valence and conduction electrons, are attractive in realistically-sized nanodevice (containing millions of atoms) simulations. Tight-binding is a local basis representation, which naturally deals with finite device sizes, alloy-disorder and hetero-interfaces and it results in very sparse matrices. The requirements of storage and processor communication are therefore minimal compared to pseudopotential implementations and perform extremely well on inexpensive Linux clusters. A comparative pen-picture of the pros and cons of different large-scale bandstructure solver is depicted in Table 7.1.

Table 7.1 Comparison of bandstructure methods

Band model	Realistic size	Random alloy	Interface roughness	Internal fields	Non-parabolic dispersion
EM	√	X	X	√	x
k•p	√	X	X	√	x
EPM	x	√	√	√	√
TB	√	√	√	√	√

EM effective mass, *EPM* empirical pseudopotential method, *TB* tight-binding formalism

Charge Transport: Almost all ab initio and quantum chemistry codes treat *closed* systems close to or at equilibrium. Therefore, full ab initio methods are usually not used to simulate current flow. On the other hand, full *atomistic* quantum treatment of modeling charge transport in practical *open* systems demands the solution of *non-equilibrium* statistical mechanics in multimillion-atom systems in excess of 10^7 complex degrees of freedom! Hence, approximate methods that resolve the physics of the valence electrons with stable bonds and required phonon modes and capture the essential size-quantization effects are appropriate for such open devices/systems. For quite some time, quantum mechanical size-quantization effects in nanostructures [34–36] have been analyzed using density matrices, Wigner functions [37], Feynman path integrals [38], and non-equilibrium Green's functions (NEGF) [39–43] with varying success. In contrast to, for example, the Wigner function approach (which is Markovian in time), the Green's functions method allows one to consider simultaneously correlations in space and time or space and energy, both of which are expected to be important in nanoscale devices. Today (although the full-band NEGF transport formalism has been well-established) accurate and reliable atomistic 3-D modeling of *scattering-dominated* diffusive/dissipative transport in practical nanoscale devices using the NEGF approach is prohibitively expensive. For example, to simulate a silicon nanowire containing 30,000 atoms using the NEGF method, the memory requirement becomes $\sim$100 GB needing $60,000+$ hours of computational time!

Phonon Transport: The current state-of-the-art approaches of modeling thermal transport are as follow [44]: (a) The Fourier heat conduction theory in conjunction with the interface thermal resistance, or the *Kapitza* resistance, which is applicable when the phonon mean free path (MFP) is shorter than the characteristic length of the nanodevice such as the particle diameter and/or interparticle separation distance; (b) Another approach in the investigation of the thermal conductivity is through the calculation of the phonon dispersion in periodic structures [45]; (c) Due to the short wavelength of the dominant phonon heat carriers, the phonon scattering at interfaces is often diffuse. For practical nanoscale devices, where the diffuse interface scattering not only reduces the phonon mean free path but also destroys the coherence of phonons, the classical size effect models such as the phonon Boltzmann transport equation (BTE) can be applicable.

From the foregoing discussion, one can infer that: (1) novel nanoscale devices are unique examples of systems where different branches of physics (molecular dynamics, quantum electronic structure, charge and phonon transport, statistical physics and thermodynamics, classical electrostatics, and optics) meet together spanning across different spatial and time scales. Nanodevice modeling, therefore, is a *multiscale and multiphysics* problem; (2) To be relevant to device designers, structures to be modeled must have realistic extent (millions of atoms) and represent bandgaps and masses extremely well; and (3) Since the continuum methods are clearly incapable of capturing essential physics at nanoscale and the best available ab initio materials science models (although offer greater accuracy) can scale to systems with only ~100 atoms, one must consider *empirical* approaches for modeling realistically-extended nanostructures. Also, the variety of geometries, materials, and doping configurations in semiconductor devices at the nanoscale suggests that a *general* nanoelectronic modeling tool is needed. This paper describes our on-going efforts to develop a multiscale Quantum Atomistic Device Simulator (QuADS) to address these needs.

2 Our Simulator: Quantum Atomistic Device Simulator

QuADS essentially bridges the gap (and crosses the intellectual boundary) between continuum and ab initio modeling paradigms and enable the quantum-corrected atomistic numerical modeling of non-equilibrium charge and phonon transport phenomena in realistically-sized systems containing more than 100 million atoms! The simulation strategy, which is divided into different computational phases, spanning from the molecular structure of the constituting elements, to the electron and phonon band structure, transport and optical coupling, is depicted in Fig. 7.1. The Figure also shows the various length and time scales and the associated observables and how one passes between them, and the codes used (and to be used) along with their interdependencies. A short description of the core packages used in QuADS is as follows.

2.1 *NEMO 3-D*

For computing *electronic structure* (energy eigenvalues, wavefunctions, $E-k$ for nanowires), we have used the open source NEMO 3-D tool. Detail description of this package can be found in [24, 32, 33, 46]. NEMO 3-D bridges the gap between the *large* size (millions of atom) classical semiconductor device models and the *molecular* level (few atoms) modeling. NEMO 3-D currently enables the computation of electronic structure using a variety of *tight-binding models* $(s, sp^3s^*, sp^3d^5s^*)$ that are optimized with a genetic algorithm tool. Whereas, for the calculation of *atomistic* (non-linear) strain relaxation, NEMO 3-D currently employs the atomistic valence-force field (VFF) with strain-dependent Keating potentials [47]. From the single-particle eigenstates various physical properties can be calculated in NEMO 3-D such as optical matrix elements, Coulomb and exchange matrix elements,

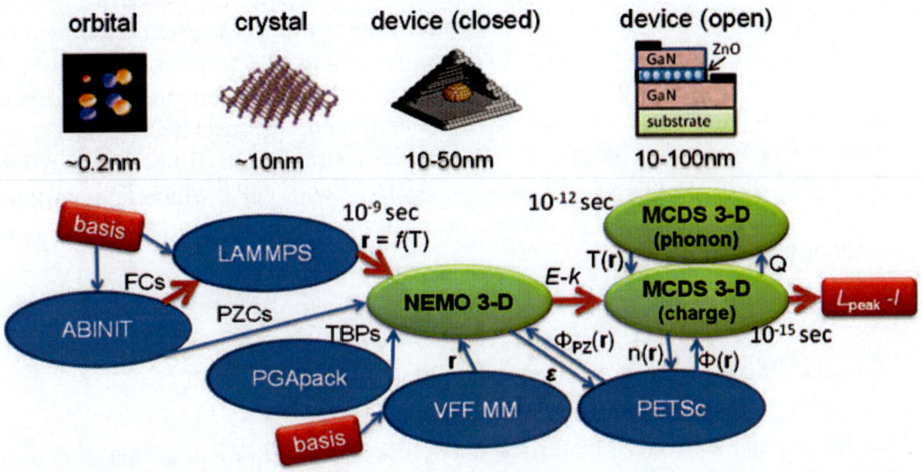

Fig. 7.1 QuADS simulation platform. Shown are the domains with length scales, program flow (major up-stream flow in *bold arrow*), the data interfaces, codes used, the associated observables, how one passes between phases, and their interdependencies, and relevant time scales. basis: atomic basis to define a crystal, ABINIT: ab initio module, FCs: force constants, PZCs: polarization coefficients, TBPs: tight-binding parameters, LAMMPS: massively parallel molecular dynamics code, **r**: atom positions, $f(\mathrm{T})$: as a function of temperature, VFF MM: valence force-field molecular mechanics, NEMO 3-D: nanoelectronics modeling tool for bandstructure calculations, *E-k*: energy bandstructures (effective masses, bandgaps, density-of-states), PETSc: parallel linear algebra solver, T(**r**): temperature distribution, Q: heat, $\Phi(\mathbf{r})$: potential distribution, n(**r**): charge density distribution, $L_{\mathrm{peak}} - I$: peak intensity vs. current characteristic

approximate single cell bandstructures from supercell bandstructure. Effects of interaction with external electromagnetic fields are also included. This versatile software currently allows the calculation of *single-particle* electronic states and optical response of various semiconductor structures including bulk materials, quantum dots, impurities, quantum wires, quantum wells and nanocrystals. NEMO 3-D includes spin in its fundamental atomistic tight binding representation. The complexity and generality of physical models in NEMO 3-D can place high demands on computational resources. For example, in the 20-band electronic calculation the discrete Hamiltonian matrix is of order 20 times the number of atoms. Thus, in a computation with 20 million atoms, the matrix is of order 400 million. Computations of that size can be handled because of the parallelized design of the package. NEMO 3-D is implemented in ANSI C, C++ with MPI used for message-passing, which ensures its portability to all major high-performance computing platforms, and allows for an efficient use of distributed memory and parallel execution mechanisms. New features of NEMO 3-D include 3-D domain decomposition parallelism. The algorithms/solvers available in NEMO 3-D include the PARPACK library, a custom implementation of the Lanczos method, Block Lanczos method, the spectrum folding method and the Tracemin method. The NEMO 3-D package is maintained primarily by the Network for Computational Nanotechnology (NCN) at Purdue University West Lafayette under the supervision of Professor Gerhard Klimeck. Recent benchmarks show [48] that 3-D domain decomposition scheme can be utilized exceeding 32,000 cores on realistic electronic structures comprised of one billion atoms. We are not aware of any other semiconductor device simulation code that can simulate such large number of atoms and hence incorporated NEMO 3-D as the *primary computing engine* of QuADS. For simulations from first principles, the group's primary choice has been the ABINIT code [49], which is mainly used for parameterizations of force fields, internal polarization, and model bandstructure calculation.

2.2 LAMMPS

Lattice mismatch between materials in a hybrid structure sometimes leads to plastic relaxation even for a thin (few monolayers) active layer [50, 51]. In these cases, the hybrid nanostructure cannot be considered pseudomorphic and accurate computation of the relaxed atom positions beg for a detailed molecular dynamics analysis of the problem. For modeling *plasticity* in strain relaxation and, thereby, complement the VFF Keating model, and the calculations of temperature-dependent *structural relaxations*, and phonon modes (thermal conductivity), we use the massively parallel classical molecular dynamics package LAMMPS [52]. LAMMPS can model systems in liquid, solid, or gaseous states with only a few particles up to *millions or billions*. It can model atomic, polymeric, biological, metallic, granular, and coarse-grained systems using a variety of force fields and boundary conditions. The open source LAMMPS code is freely-available from Sandia National Lab, and is distributed under the terms of the GNU public use license. LAMMPS is designed to be easy to modify or extend with new capabilities, such as new force fields, atom types, boundary conditions, or diagnostics. LAMMPS is also designed to allow it to be coupled to other codes. For example, a quantum mechanics code might compute forces on a subset of atoms and pass those forces to LAMMPS. Recently, in collaboration with Professor Mesfin Tsige at the U of Akron Polymer Science department, LAMMPS has been ported to our in-house NSF funded cluster *Maxwell*, and used in the simulations of time-evolution of the molecular layering in a thin $C_{16}F_{34}$ film on an α-quartz substrate. It was observed that the extent of layering oscillates in time with an amplitude and period that depend strongly on temperature. The scaling of LAMMPS on ORNL's Jaguar HPC cluster is shown in Fig. 7.2 for up to 120,000 processors and 32,000 atoms/processor, maximum number of atoms simulated being $\sim$3.84 billions!

2.3 MCDS 3-D

In QuADS, to model various transport phenomena, an in-house 3-D atomistic particle-based Monte Carlo device simulator (MCDS 3-D) has been incorporated.

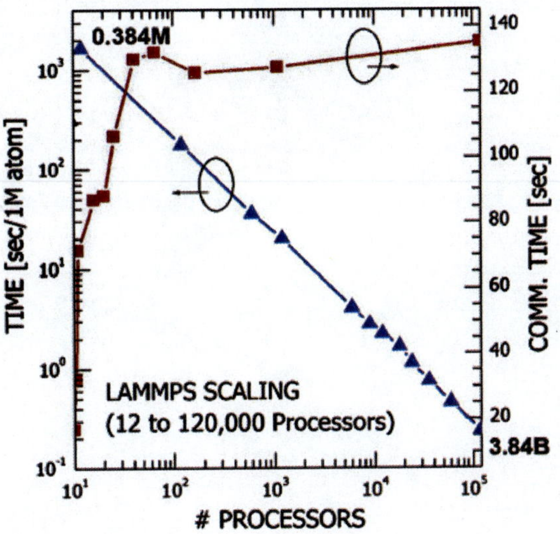

Fig. 7.2 LAMMPS scaling on Cray XT5 HPC machine at ORNL

Quantum mechanical size-quantization effects have been accounted for via a *parameter-free* effective potential scheme [53]. The approach is based on a perturbation theory around thermodynamic equilibrium and derived from the idea that the semiclassical Boltzmann equation with the quantum corrected potential and the Wigner equation should possess the same steady state. It leads to an effective potential/field, which takes into account the discontinuity at the Si/SiO_2 barrier interface due to the difference in the semiconductor and the oxide affinities. The effective potential possesses no fitting parameters, as the size of the electron (wavepacket) is determined from its energy. The resultant quantum potential is, in general, two-degrees smoother than the original Coulomb and barrier potentials of the device, i.e. possesses two more classical derivatives, which essentially eliminates the problem of the statistical noise. The calculated quantum barrier field (QBF) for low-energy (left column) and high-energy (right column) electrons are shown in Fig. 7.3 with the following salient features: (1) QBF decays almost exponentially with distance from the Si/SiO_2 interface proper; (2) QBF increases with increasing the wavevector of the carriers along the normal (crystal growth) direction; (3) The contour plots clearly reveal the fact that the electrons with lower momentum *feel* the quantum field far from the interface proper, whereas can easily approach the interface as their momentum increases; and (4) A similar trend is also observed with the variation in electron energy. Electrons with higher energy can reach the vicinity on the interface, thus, behaving as classical point-like particles. The Incomplete Lower-Upper (ILU) decomposition method has been employed for the solution of the 3-D Poisson equation. To treat full Coulomb (electron–ion and electron–electron) interactions properly, the simulator implements two real-space molecular dynamics (MD) schemes: the particle-particle-particle-mesh (P^3M) method and the corrected Coulomb approach. The effective force on an electron is computed as a combination of the short-range molecular dynamics force and the long-range Poisson force.

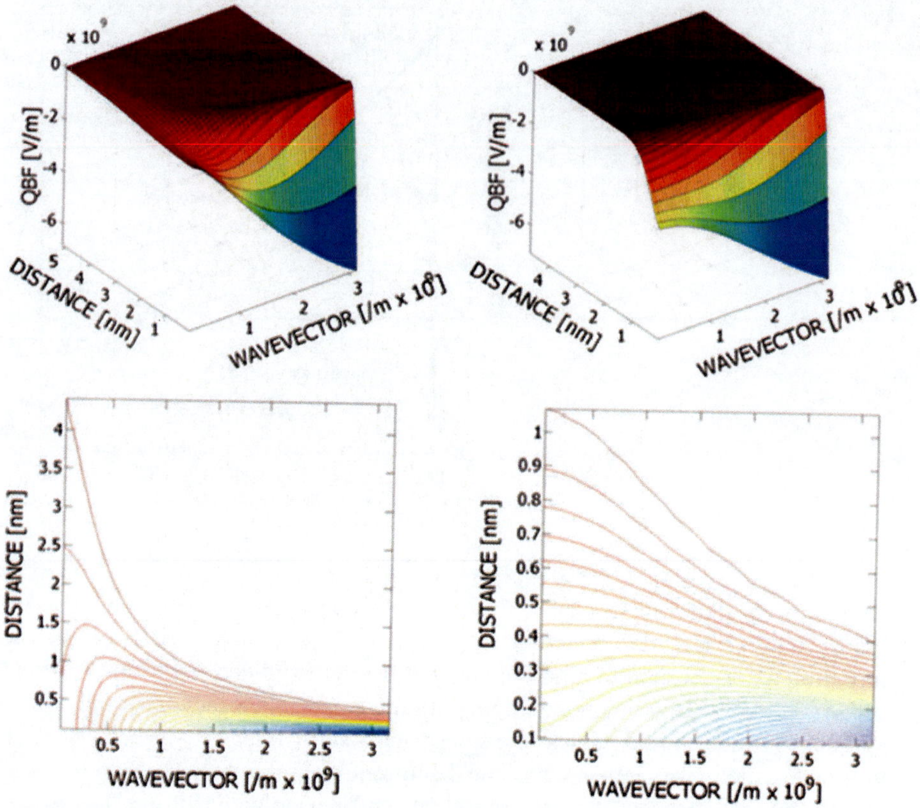

Fig. 7.3 Variation of the quantum barrier field (QBF) as a function of distance from the Si/SiO_2 interface (in a MOSFET) and wavevector k_y along the depth (*left panel*: low energy electrons, *right panel*: high energy electrons)

The implementation details of these models and methodologies have been discussed in [54]. Regarding the Monte Carlo transport kernel, both the intravalley (for example acoustic phonons) and intervalley scattering mechanism (g- and f-phonon processes) have been included. Also, necessary *event-biasing algorithms* [55] are used in the simulator that enhance the carrier statistics and result in a faster convergence of the channel current. Enhancement algorithms in the Monte Carlo simulations are especially useful when the device behavior is governed by *rare events* (for example subthreshold condition, tunneling, etc.) in the carrier transport process.

It is worth mentioning that, to properly treat the charge transport and the heating effects without any approximations, one in principle has to solve the *coupled* electron/hole-optical phonons–acoustic phonons–heat bath problem, where each sub-process has to be addressed in a somewhat individual manner and included in the global picture via a self-consistent loop [56–61]. The system is nonlinear, as the probabilities depend on the product of the electron and phonon distribution functions, and poses a multi-scale problem since the sub-processes involve different time scales: the velocity of the phonons is two orders of magnitude lower than the velocity of the electrons. Accordingly, the heat transfer by the lattice is much slower process than the charge transfer. In collaborative effort with Professor Dragica Vasileska at Arizona State University, we are planning to solve self-consistently the Boltzmann transport equation (BTE) for the carriers using a particle based Monte Carlo simulator MCDS 3-D (thus taking into account hot carrier and other non-stationary effects such as velocity overshoot) with the microscopic BTE for the phonons. In this formalism, phonons will be treated on an equal basis with electrons as superparticles with group velocity dependent on the type and mode of phonons (e.g. longitudinal, acoustic, etc.). The outputs of the thermal simulations are the temperature distribution $T(\mathbf{r})$, which will be used as an input to charge transport kernel of the integrated simulator.

2.4 Graphical User Interface

For effective and interactive 3-D visualizations, QuADS is combined with a graphical user interface (GUI) based on *Rappture* [62] toolkit developed (and freely available) by Network for Computational Nanotechnology at Purdue University. Two approaches can be followed: (1) The legacy application is not modified at all and a *wrapper script* translates Rappture I/O to the legacy code; and (2) Rappture is integrated into the source code to handle all I/O. Figure 7.4 shows the rappturization approach and the essential steps involved therein. The first step is to declare the parameters associated with one's tool by describing Rappture objects in the

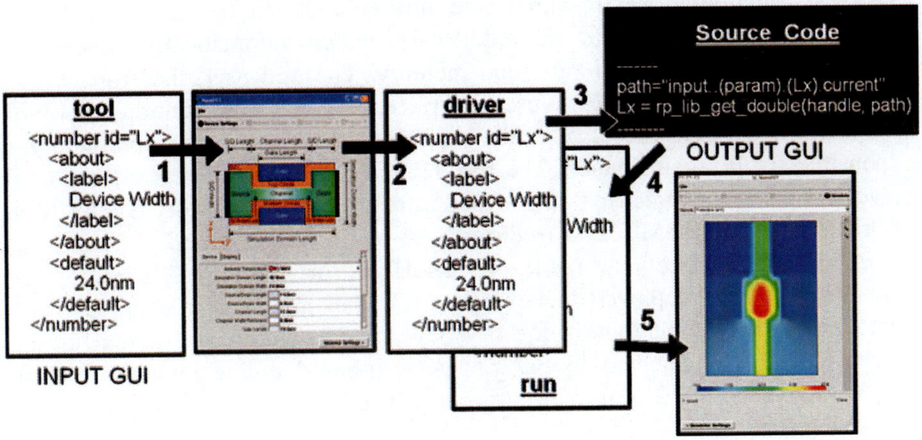

Fig. 7.4 *Rappture*: Revolutionizing tool development

Extensible Markup Language (XML). Rappture reads the XML description for a tool and generates the GUI automatically. The second step is that the user interacts with the GUI, entering values, and eventually presses the Simulate button. At that point, Rappture substitutes the current value for each input parameter into the XML description, and launches the simulator with this XML description as the driver file. In the third step, using parser calls within the source code, the simulator gets access to these input values. Rappture has parser bindings for a variety of programming languages, including C/C++, Fortran, Python, and MATLAB. And finally, the simulator reads the inputs, computes the outputs, and sends the results through run file back to the GUI for the user to explore.

2.5 Deployment Plan

QuADS will eventually be an *open source* under GNU General Public (GPL) license and be deployed online on www.nanoHUB.org. The nanoHUB is a multi-university (led by Purdue University), NSF-funded initiative and offers a set of *free* cyber services including *interactive* online simulation, tutorials, seminars, and online courses packaged using *e*-learning standards. We consider nanoHUB as an excellent forum to host QuADS and related modules/tutorials because over the past few years it has gained significant visibility, momentum, and credibility in the nanotechnology community of engineers, scientists, educators, and most importantly with students. The nanoHUB's well-established processes provide simple procedures for developing, deploying, revising, and (importantly) evaluating nanotechnology related educational content. Access to these tools is granted to users via the web browsers, without the necessity of any local installation by the remote users. The definition of specific sample layout and parameters is done using a dedicated GUI in the remote desktop (VNC) technology. The necessary computational resources are further assigned to the simulation dynamically by the web-enabled middleware, which automatically allocates the necessary amount of CPU time and memory. The end user, therefore, has access not only to the code, a user interface, and the computational resources necessary to run it but also to the scientific and engineering community responsible for its maintenance. The nanoHUB is currently considered one of the leaders in science gateways and cyber infrastructure. A prototype 2-D Monte Carlo simulator QuaMC 2-D (reduced web-based online *interactive* version of QuADS 3-D) is freely available on nanoHUB for educational purposes [63]. QuaMC 2-D was deployed in February 2007 and has been used by 371 users who ran 6,420 simulations in the last 3 years. The number of global users of QuaMC 2-D are depicted in Fig. 7.5. A web-based online interactive version of 3-D QuADS for educational purposes will soon be available on http://www.nanoHUB.org.

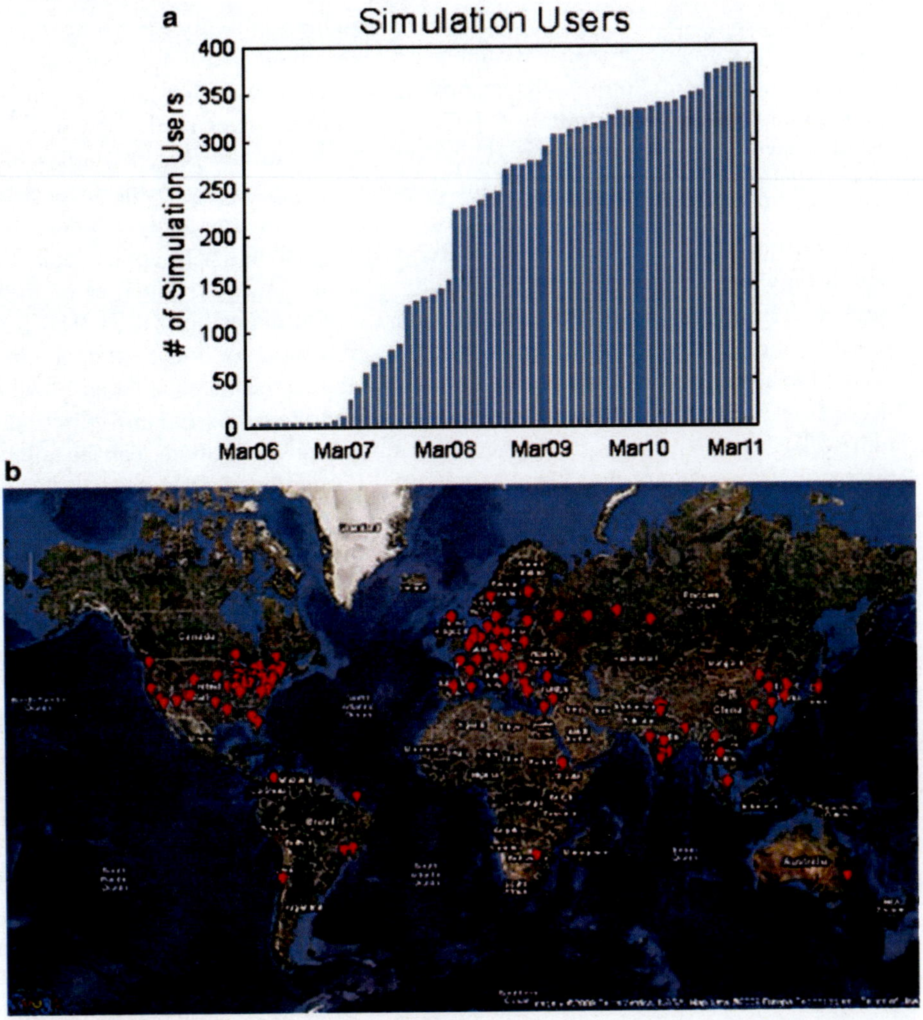

Fig. 7.5 (**a**) Number of annual users who have run at least one simulation using QuaMC 2-D simulator. (**b**) World map of QuaMC 2-D users

3 Recent Simulations Using QUADS

3.1 Effects of Internal Fields in InN/GaN Quantum Dots

3.1.1 Objective

In the last decade, GaN and its related alloys especially InGaN have been the subject of intense experimental and theoretical research due mainly to their wide range of emission frequencies, high stability against defects, and potential for applications in various optoelectronic, solid-state lighting, and high-mobility electronic devices [64, 65]. Since the heteroepitaxy of InN on GaN involves a lattice mismatch up to ~11%, a form of Stranski–Krastanov mode can be used for growing InN on GaN by molecular beam epitaxy (MBE). This finding gives rise to the possibility of growing InN quantum dots (QDs) on GaN substrates. Recent studies have shown that the strain between InN and GaN can be relieved by misfit dislocations at the hetero-interface after the deposition of the first few InN bilayers and before the formation of InN islands [66–68]. Relaxed InN islands with controllable size and density can be formed by changing the growth parameters (such as temperature) in either MBE or metalorganic chemical vapour deposition (MOCVD) [69]. Relaxation of elastic strain at free surfaces in semiconductor dots (and nanowires) allows the accommodation of a broader range of lattice mismatch and band-lineups in coherent nanostructures than is possible in conventional bulk and thin-film (quantum well) heterostructures, and, therefore, threading dislocations can be all but nonexistent in quantum dots (and nanowires). Furthermore, QDs used in the active region of optical devices provide better electron confinement (due to strongly peaked energy dependence of density of states) and thus a higher temperature stability of the threshold current and the luminescence than quantum wells. It is also clear that high-quality bulk GaN is an ideal substrate material for nitride nanostructures. Pure GaN crystal is five times more thermally conductive than sapphire, and optically transparent at visible and near-UV wavelengths [70]. Very recently, July 2010 issue of the IEEE Spectrum covers the story of a small polish company establishing a huge technical edge by manufacturing nearly perfect 2-in. crystal of GaN [71].

Knowledge of the electronic bandstructure of nanostructures is the first and an essential step towards the understanding of the optical performance (luminescence) and reliable device design. Hexagonal group-III nitride 2-D quantum well (QW) heterostructures have experimentally been shown to demonstrate *polarized transitions* in quantized electron and hole states and non-degeneracy in the first excited state in various spectroscopic analyses [72, 73]. These observations suggest the existence of certain *symmetry lowering mechanisms* (structural and electrostatic fields) in these low dimensional nanostructures. While 0-D QDs promise better performance, only very few and recent experimental results exist concerning the photoluminescence (PL) and electroluminescence (EL) of nitride QDs in the visible spectral region [68, 74, 75], and experiments revealing *polarization anisotropy* in InN QDs are rare. Similar to the 2-D QW structures, the optical properties of the QDs are expected, to a large extent, to be determined by an intricate interplay between the structural and the electronic properties, and (since not yet been fully assessed experimentally) demand detailed theoretical investigations.

In this section, using mainly the NEMO 3-D tool in QuADS, we study the electronic bandstructure of wurtzite InN/GaN quantum dots having *three* different geometries, namely, box, dome, and pyramid. The main objectives are twofold – (1) To explore the nature and quantify the role of crystal atomicity, strain fields, piezoelectric, and pyroelectric potentials in determining the energy spectrum and the wavefunctions, and (2) To address a group of related phenomena including shift in the energy state, symmetry-lowering and non-degeneracy in the first excited state,

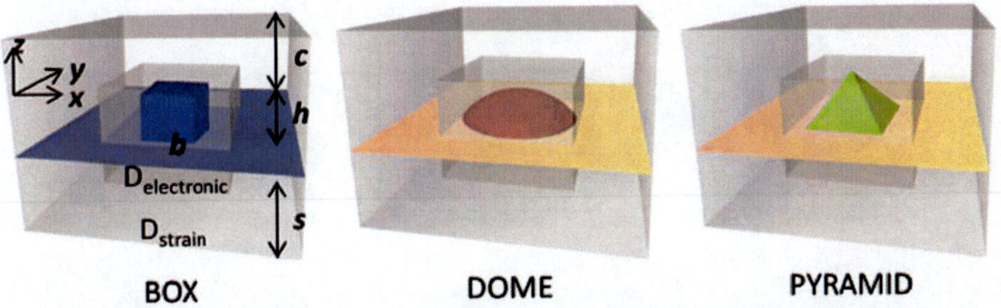

Fig. 7.6 Simulated InN/GaN quantum dots on a thin InN wetting layer. Two major computational domains are also shown. D_{elec}: central smaller domain for electronic structure (quantum) calculation, and D_{strain}: outer domain for strain calculation. In the figure: s is the substrate height, c is the cap layer thickness, h is the dot height, and d is the dot diameter/base length as appropriate

strong band-mixing in the overall conduction band electronic states, and strongly suppressed and optically anisotropic interband transitions. We have also demonstrated the importance of full 3-D atomistic material representation and the need for using *realistically-extended* substrate and cap layers (multimillion atom modeling) in the study of electronic structure of these reduced-dimensional QDs.

3.1.2 Simulation Results

Figure 7.6 shows the simulated quantum dots with box, dome, and pyramid geometries. The InN QDs grown in the [0001] direction and embedded in a GaN substrate used in this study have (unless otherwise stated) diameter/base length, $d \sim 10.1$ nm and height, $h \sim 5.6$ nm, and are positioned on an InN wetting layer of one atomic-layer thickness. The simulation of strain is carried out in the large computational box, while the electronic structure computation is restricted to the smaller inner domain. All the strain simulations fix the atom positions on the bottom plane to the GaN lattice constant, assume periodic boundary conditions in the lateral dimensions, and open boundary conditions on the top surface. The strain parameters used in this work were validated through the calculation of Poisson ratio of the bulk materials. The inner electronic box assumes a closed boundary condition with passivated dangling bonds.

Figure 7.7 shows the topmost valence (HOMO) and first four conduction band wavefunctions (projected on the X-Y plane) for the quantum dots *without* strain relaxation. Here, both the InN dot and the GaN barrier assume the lattice positions of perfect wurtzite GaN. The topmost valence (HOMO) state in all three quantum dots has orbital S-character retaining the geometric symmetry of the dots. The first electronic state is S-like, while the next three states are P-like and the *split* (non-degeneracy) in these levels originate from the crystal fields alone. Note that the magnitude of the split (defined as $\Delta P = E_{010} - E_{100}$) in the P level is largest in a box ($\sim$45.294 meV) and minimum in a dome ($\sim$1.476 meV). Also, the anisotropy in the P states assumes different *orientations* – for box and pyramid dots, the first P state is oriented along the [010] direction and the second along [100] direction, while the

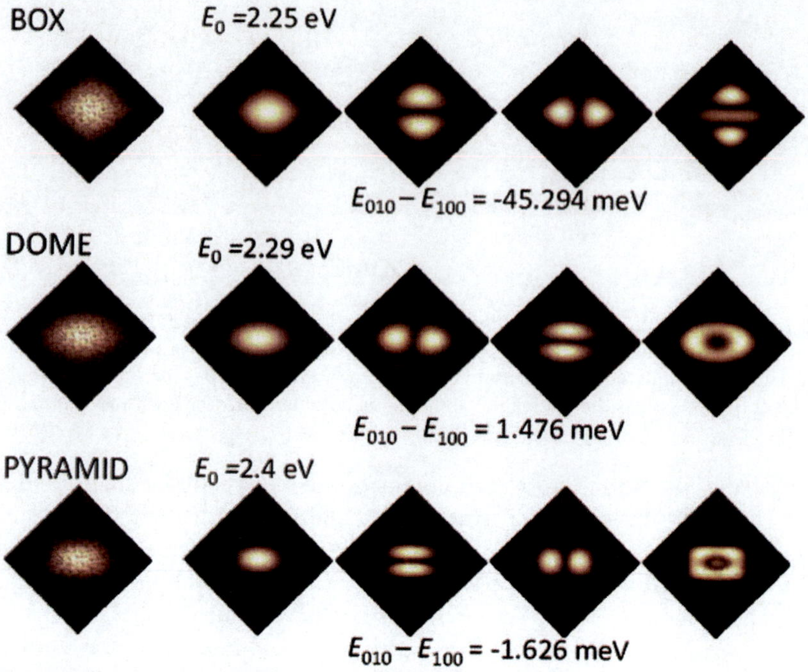

Fig. 7.7 Topmost valence (HOMO) and first four conduction band wavefunctions due to fundamental crystal and interfacial symmetry. Noticeable are the split and the anisotropy in the P level. Number of atoms simulated: 1.78 million (strain domain), 0.8 million (electronic domain)

converse occurs in a dome. It is clear that the fundamental crystal atomicity and the interfaces (between the dot material InN and the barrier material GaN) lower the geometric shape symmetry *even in the absence of strain relaxation*. Therefore, the interface plane creates a short-range interfacial potential and cannot be treated as a reflection plane.

Next, we introduce atomistic strain relaxation in our calculations using the VFF method with the Keating potential. In this approach, the total elastic energy of the sample is computed as a sum of bond-stretching and bond-bending contributions from each atom. The equilibrium atomic positions are found by minimizing the total elastic energy of the system [32]. However, piezoelectricity is neglected in this step. The total elastic energy in the VFF approach has only one global minimum, and its functional form in atomic coordinates is quartic. The conjugate gradient minimization algorithm in this case is well-behaved and stable.

Strain modifies the effective confinement volume in the device, distorts the atomic bonds in length and angles, and hence modulates the confined states. From our calculations, atomistic strain was found to be anisotropic and long-ranged penetrating deep (~20 nm) into both the substrate and the cap layers stressing the need for using *realistically-extended* substrate and cap layers (multimillion-atom modeling) in the study of electronic structure of these reduced-dimensional QDs. Figure 7.8 shows the wavefunction distributions for the topmost valence band and

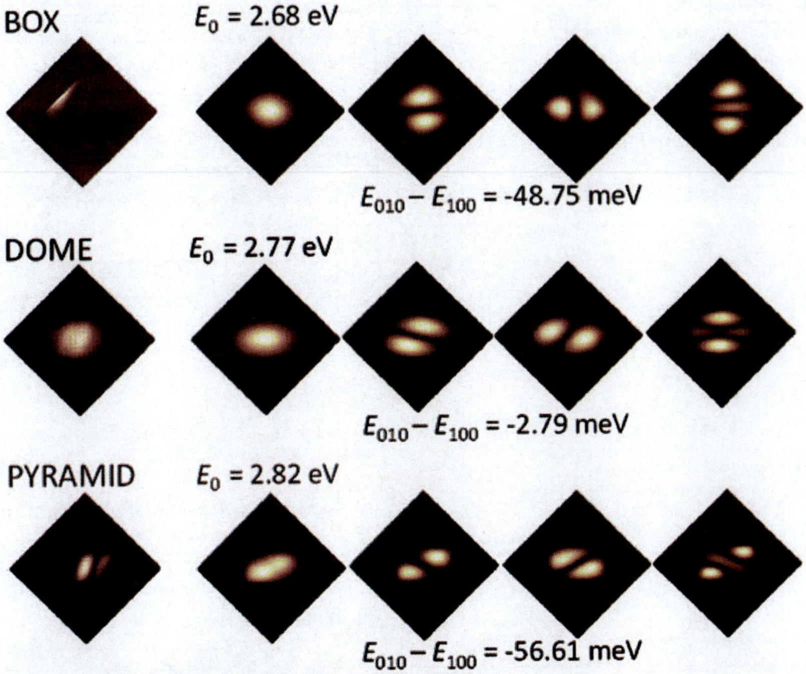

Fig. 7.8 Topmost valence (HOMO) and first four conduction band wavefunctions due to combined effects of atomicity and strain relaxation. Noticeable are the strongly displaced HOMO (hole) wavefunctions, deformed LUMO states, and split and the anisotropy in the P level

first 4 (four) conduction band electronic states in a 2-D projection. Noticeable are the deformed LUMO (electronic) states, and the pronounced optical anisotropy and non-degeneracy in the P levels. Strain introduces *uniform* orientational pressure (adds negative potential) in all three quantum dots with ΔP to be largest in a pyramid and minimum in a dome. Also, strain relaxation causes *blue* shift in the conduction band electronic states and results in strongly displaced HOMO (hole) wavefunctions. These observations will have significant implications on the optical polarization and performance of devices based on these nanostructures.

In pseudomorphically grown heterostructures, the presence of non-zero atomistic stress tensors results in a deformation in the crystal lattice and leads to a combination of piezoelectric and pyroelectric field, which has been incorporated in the Hamiltonian as an external potential (within a non-selfconsistent approximation). The resulting potential distributions along the growth direction are shown in Fig. 7.9 for all three quantum dots. One can see that the potential (both the piezoelectric and pyroelectric), in accordance with the dot volume, has the largest *magnitude* in a box, and is minimum in a pyramid. The pyroelectric potential is significantly larger ($\sim$5 times) than the piezoelectric counterpart and tends to oppose the latter. This also suggests that for an appropriate choice of alloy composition and quantum dot size/geometry, spontaneous and piezoelectric fields may be caused to cancel out!

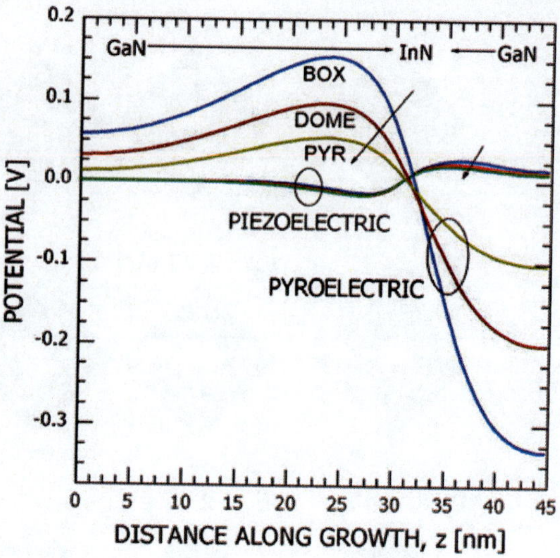

Fig. 7.9 Induced piezoelectric and pyroelectric potential distributions along the *z* (growth) direction. Note the spread/penetration in the surrounding material matrix

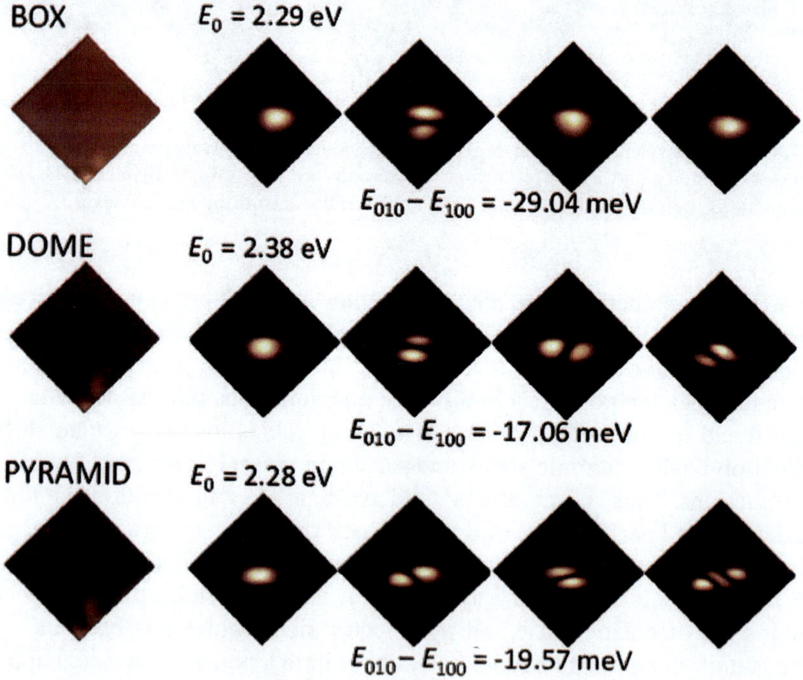

Fig. 7.10 Topmost valence band (HOMO) and first four conduction band wavefunctions including interfacial effects, strain, and piezoelectricity

Figure 7.10 shows the topmost valence band and first 4 (four) conduction band wavefunctions for all three quantum dots including the strain relaxation and the piezoelectric potential. The piezoelectric potential introduces a global *red* shift

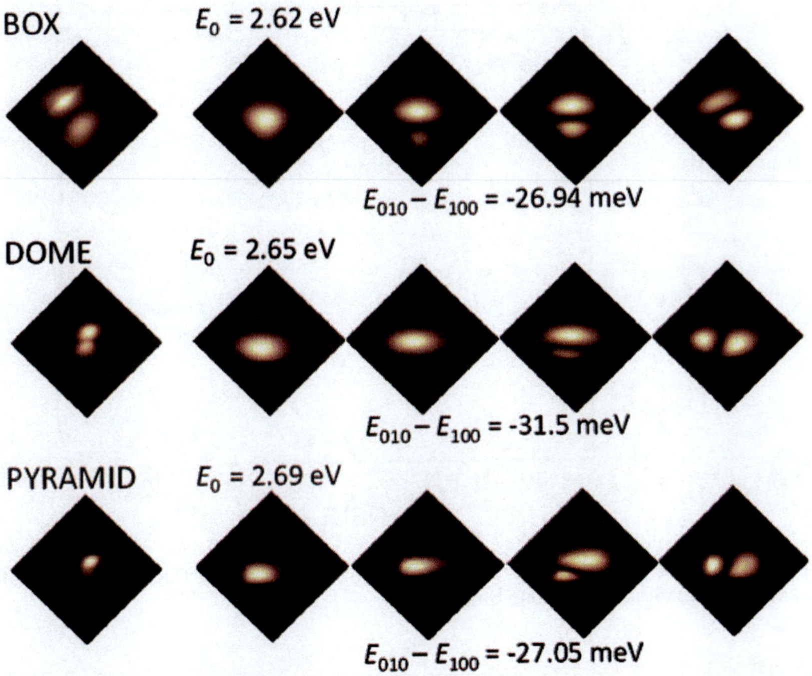

Fig. 7.11 Topmost valence band (HOMO) and first four conduction band wavefunctions including all four (4) competing internal fields originating from interfacial effects, strain, piezoelectricity and pyroelectricity

in the energy spectrum and *opposes* the strain induced field (without any significant modifications in the wavefunction orientations) in the box and pyramid dots. Figure 7.11 shows the same wavefunctions for all three quantum dots including the combined effects of *all four types* of internal fields, namely, interface, strain, piezoelectricity, and pyroelectricity in the calculations. The pyroelectric potential was found to be large enough to create band mixing and strong wavefunction anisotropy in the conduction band energy landscape.

Figure 7.12 shows the interband optical transition rates between ground hole (HOMO) and ground electronic states (LUMO) in all three quantum dots revealing significant suppression and strong polarization anisotropy (peak occurring at angles greater than zero) due to spatial irregularity (displacement) in the wavefunctions originating from the combined effects of all four internal fields. The true atomistic symmetry of the quantum dots, thus, influences the electronic bandstructure and in general the strengths of the optical transitions differ for different geometry. The transition rates were found to be inversely proportional to volume of QD with values maximum in the pyramid and minimum for the box structures.

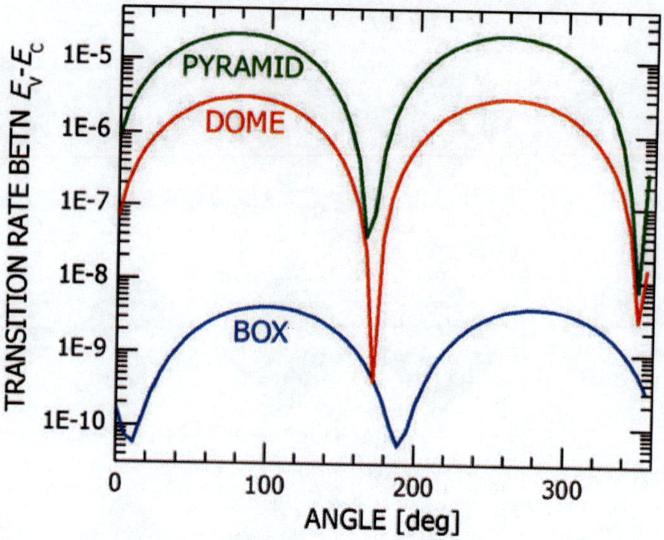

Fig. 7.12 Interband optical transition rates between ground hole (HOMO) and ground electronic states (LUMO) in all three quantum dots

3.2 Importance of Second Order Polarization in InAs/GaAs Quantum Dots

3.2.1 Objective

In the case of the InAs/GaAs quantum dots, the lattice mismatch is around 7% and leads to a strong long-range strain field within the extended neighborhood of each quantum dot. Strain can be atomistically inhomogeneous, involving not only biaxial components but also non-negligible shear components. Therefore, any spatial non-symmetric distortion in quantum dots (and other nanostructures) made of these materials will create piezoelectric fields, which will modify the electrostatic potential landscape. It is well known that the piezoelectric polarization is generally a *non-linear* function of strain, the non-linearity becoming important for large epitaxial strains. Recently, it has been shown [76] that the piezoelectric polarization in strained InAs/GaAs systems has strong contributions from second-order effects that have so far been neglected. In this calculation, the piezoelectric tensor is given by [77]: $\tilde{e}_{\mu,j} = \tilde{e}^0_{\mu,j} + \sum_k \tilde{B}_{\mu,j,k}\eta_k$, where $\eta_j (j = 1,6)$ is used to denote strain in the Voigt notation. Here $\tilde{e}^0_{\mu,j}$ is the reduced proper piezoelectric tensor of the *unstrained* material, while $\tilde{B}_{\mu,j,k}$ is a fifth rank tensor with Cartesian coordinates, and μ is the strain index in Voigt notation j,k and represents the first-order change of the reduced piezoelectric tensor with strain. In [76], Bester et al. have found that, for [111] -oriented $In_xGa_{1-x}As$ quantum wells, the linear and the quadratic piezoelectric coefficients have the opposite effect on the field, and for large strains (large

In concentration) the quadratic terms even dominate! Thus, the piezoelectric field turns out to be a rare example of a physical quantity for which the first-order and second-order contributions are of comparable magnitude.

In this section, we study the electronic properties of Zincblende InAs/GaAs quantum dots having three different geometries, namely, box, dome, and pyramid (as depicted in Fig. 7.6. In particular, for piezoelectricity, for the first time within the framework of $sp^3d^5s^*$ tight-binding theory, *four* different recently-proposed polarization models (linear and non-linear) have been considered in this study. In contrast to recent studies of similar quantum dots, our calculations yield a *non-vanishing* net piezoelectric contribution to the built-in electrostatic field.

3.2.2 Simulation Results

For the calculations of the piezoelectric polarization in InAs/GaAs QDs, we have considered *four* different models and followed the recipe in [76]: (1) Linear approximation using *experimental* (bulk) values for polarization constants ($-0.045\,\mathrm{C/m^2}$ for InAs, and $-0.16\,\mathrm{C/m^2}$ for GaAs); (2) Linear (first-order) approximation using microscopically-determined values for polarization constants ($-0.115\,\mathrm{C/m^2}$ for InAs, and $-0.230\,\mathrm{C/m^2}$ for GaAs); (3) Second-order (quadratic) polarization using microscopically-determined values for polarization constants $\beta_{114} = -0.531$, $\beta_{124} = -4.076, \beta_{156} = -0.120$ for InAs, and $\beta_{114} = -0.439, \beta_{124} = -3.765, \beta_{156} = -0.492$ for GaAs); and (4) A combination of the first and the second order effects using the above mentioned microscopically-determined values for polarization constants.

The piezoelectric potential along the growth (z) direction using the four different models are shown in Fig. 7.13. From this figure, one can extract at least three important features: (1) Piezoelectric potential has its largest *magnitude* in a pyramidal dot with the peak being located near the pyramid tip, and the minimum in a box; (2) The spread of the potential is largest in a box and minimum in a pyramid; and (3) *Within the quantum dot region*, the second-order effect has comparable/similar magnitude as the first-order contribution, and, indeed, the two terms oppose each other. However, noticeable is the fact that the first-order contribution, as compared to the quadratic term, penetrates *deeper* inside the surrounding material matrix. This particular effect, we believe, in contrast to the findings in [77], results in a non-vanishing and reasonably large *net* (1st + 2nd) piezoelectric potential within the region of interest. The fact that the 1st order and the second order terms oppose each other is also noticeable in Fig. 7.14, which depicts the surface plots of the piezoelectric potential distribution in the X–Y plane. Note that the 1st order term has somewhat larger magnitude and spread than the quadratic term. Also, associated with both these two terms, noticeable is the asymmetry and inequivalence (in terms of potential magnitude and distribution) along the [110] and the $[1\bar{1}0]$ directions.

Figure 7.15 shows the first 4 (four) conduction band wavefunctions for all three quantum dots including *both* the strain and the piezoelectric fields (fourth model) in the calculations. The piezoelectric potential introduces a global shift in the energy

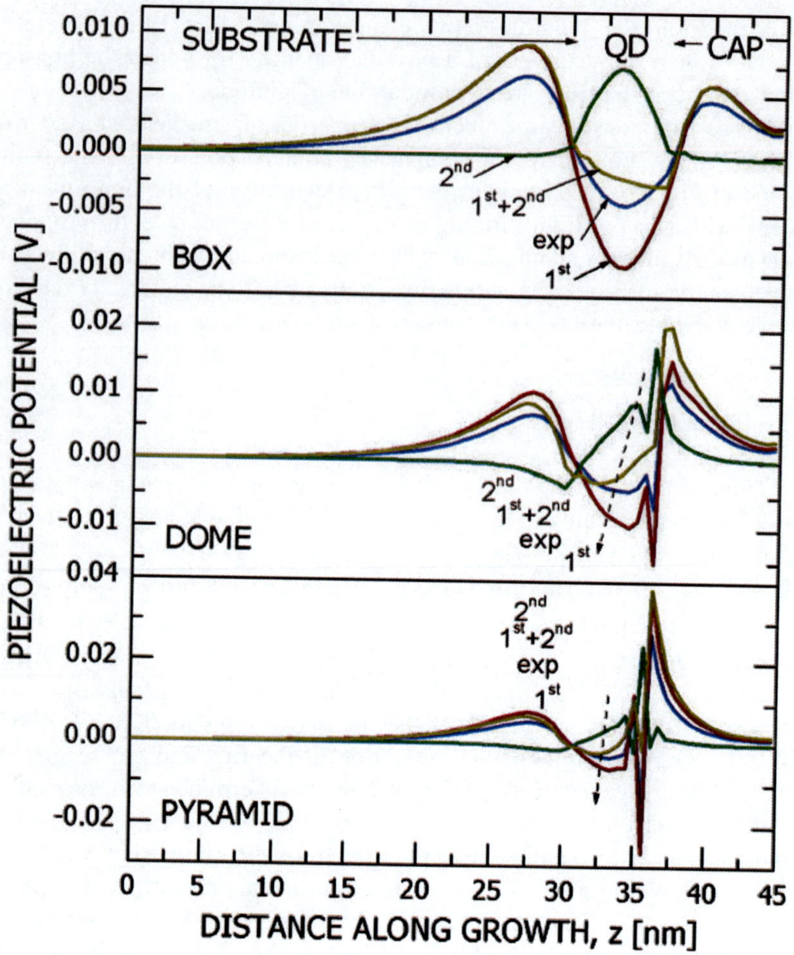

Fig. 7.13 Induced piezoelectric potential along the z (growth) direction in all three quantum dots. Four different models for the polarization constants have been used in the calculations: (1) linear and experimentally measured, (2) linear through ab initio calculations, (3) quadratic through ab initio calculations, and (4) combination of first and second order components. Also in this figure, note the varying spread/penetration of the potential in the surrounding material matrix as a function of dot shape

spectrum and generally opposes the strain induced field. In box and dome shaped dots, the net piezoelectric potential is found to be strong enough to fully offset the combined effects of interface and strain fields and, thereby, *flip* the optical polarization anisotropy. Also shown in this figure are the splits in the P levels (ΔP) for all three quantum dots. In order to fully assess the piezoelectric effects, we have prepared Table 7.2 that quantifies the *individual* net contributions from crystal atomicity and interfaces, strain, and the various components of piezoelectric fields in the spilt of the P level. The net piezoelectric contribution is found to be largest in a box and

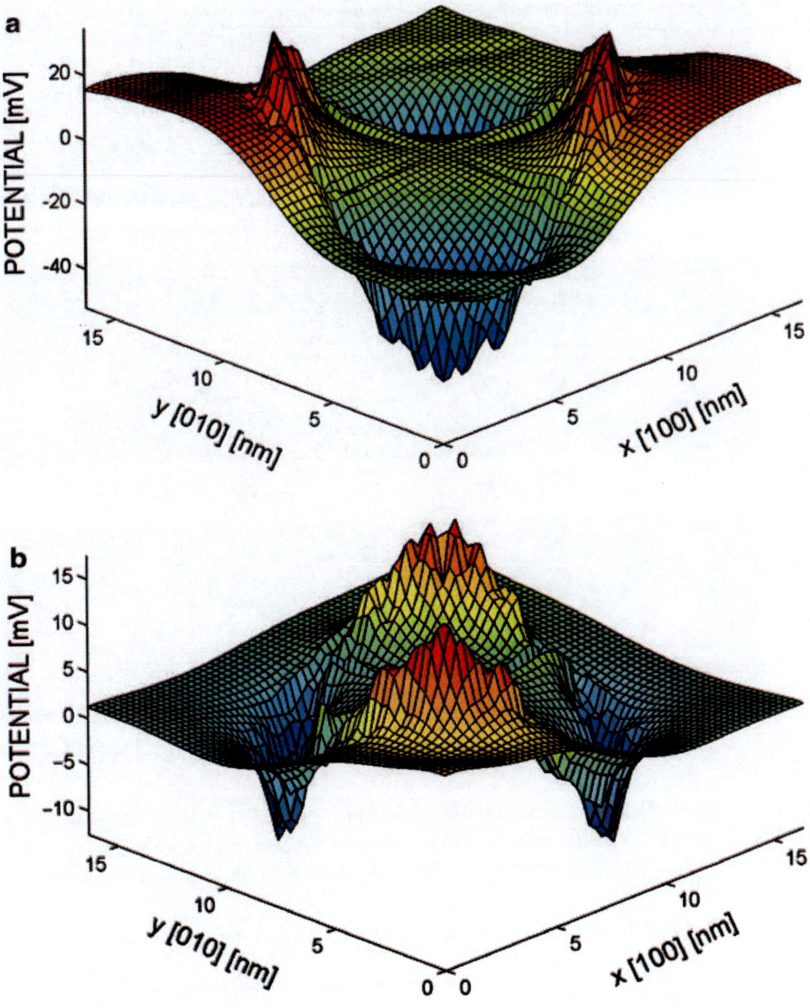

Fig. 7.14 Linear and quadratic contributions of the piezoelectric potential in the *X–Y* plane halfway through the dot height. Note the magnitude, orientation, and anisotropy in the induced potential

minimum in a pyramid, which clearly establishes a direct correspondence between the piezoelectric potential and the volume of the quantum dot under study.

Figure 7.16 shows the influence of the three major types of internal fields on the single-particle conduction band ground states (*S*-orbital) in the quantum dots. Due to size-quantization, it is found that the ground energy increases as the volume of the quantum dot decreases. Noticeable is that the strain relaxation introduces pronounced *blue* shifts in the conduction band ground state; whereas the piezoelectricity causes a much smaller red shift therein.

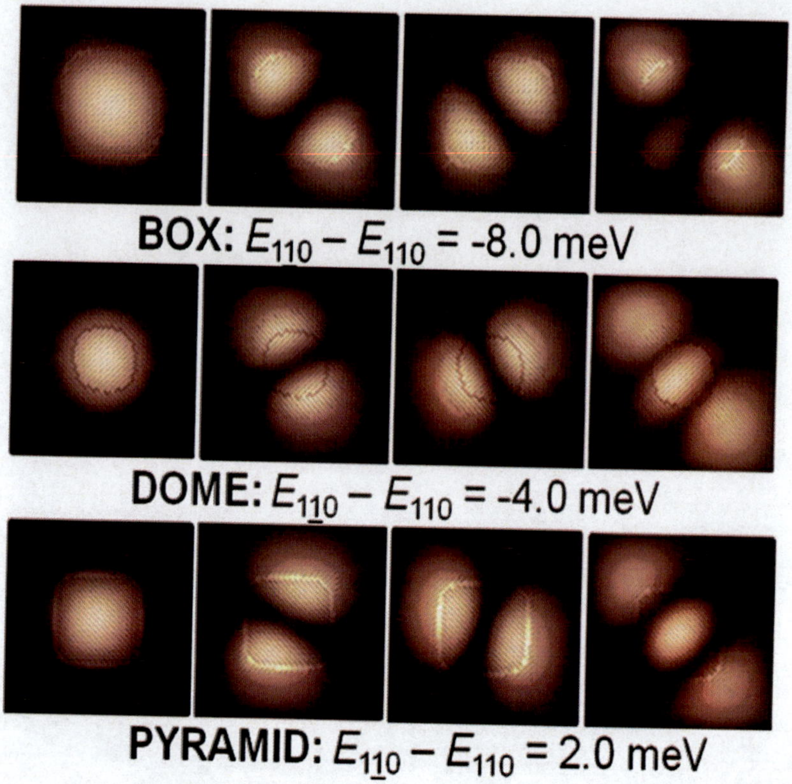

Fig. 7.15 First 4 electronic wavefunctions and split in the P levels in all three quantum dots including atomicity/interfacial effects, strain, and piezoelectricity. Note the varying piezoelectric contributions, which can be attributed mainly to the volume of the quantum dot under study

Table 7.2 Net contribution (in meV) of various effects in P-splitting

Effect	Box	Dome	Pyramid
Atomicity/interface	1.2	−7.4	3.56
Strain relaxation	1.4	13.3	3.24
Interface + strain	2.6	5.9	6.8
PZ (1st order)	−12.6	−16.9	−10.8
PZ (2nd order)	0.4	5.1	5.2
PZ (1st + 2nd order)	−10.6	−9.9	−4.8

3.3 Modeling Unintentional Single Charge Effects in Silicon Nanowire FETs

3.3.1 Objective

In recent years, the study of discrete charge induced random telegraphic signal (RTS) in emerging devices (FinFETs, nanowires, carbon nanotube FETs) has

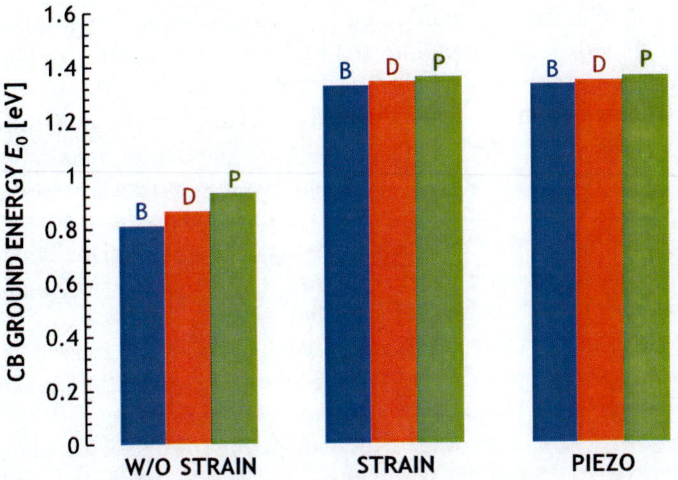

Fig. 7.16 Conduction band ground states in box (B), dome (D), and pyramid (P) shaped quantum dots including interface effects (w/out strain), strain, and piezoelectricity

attracted much attention. RTS results from the capture and emission of charged particles from defect states/traps and leads to modulation in carrier density (electrostatics) and mobility (dynamics), dominates the low frequency noise (LFN) performance around and below threshold, and significantly changes the ON-state current in nanoscale FETs. At nanoscale, the effect of even a single unintentional impurity or a defect located in the channel region of the transistor can have deleterious effects on the overall performance. Experimental data for gate-all-around silicon nanowire FET having gate width of ~100 nm and length of ~200 nm shows the dependence of RTS on the gate voltage and that the on current fluctuation can be as high as 25% [78]. Fast switching time constants observed in these studies suggest that interface states/bulk traps rather than oxide traps are responsible for the RTS [79]. In this section, we investigate the effects of a *single* channel charge on the performance characteristics of *n*-channel silicon nanowire (NW) FETs. It is shown that the percentage change in the ON-current depends on an intricate interplay of device size, geometry, channel orientation, gate bias, and energetics and spatial location of the charge.

3.3.2 Simulation Results

Gate-all-around rectangular silicon nanowires (SiNWs) with three different channel orientations, namely [100], [110] and [111], are considered in this work. The *n*-type channel is 18 nm long and undoped. The channel cross sectional area varies from 2×2 to $10 \times 10\,\text{nm}^2$. The oxide thickness is 2 nm and the gate is assumed to be a metal gate with workfunction equal to the semiconductor workfunction. Regarding the Monte Carlo transport kernel, intravalley scattering is limited to acoustic phonons. For the intervalley scattering, both g- and f- phonon processes have been included. At present, impact ionization and surface-roughness scattering are not included in the model. They are omitted, as they tend to mask the role of the space-quantization effects on the overall device performance. Impact ionization is neglected, as, for the drain biases used in the simulation (maximum 0.8 V), electron energy is typically insufficient to create electron–hole pairs. Also, band-to-band tunneling and generation and recombination mechanisms have not been included in these simulations. For the Ohmic contacts, the charge-neutral method has been used. Reflecting boundary conditions are employed at the artificial boundaries. A quasi-static assumption has been made for the holes.

Nanowire bandstructure parameters (bandgap, effective masses, and density of states) have been computed using the NEMO 3-D package in QuADS. The calculated bandgap (Fig. 7.17a) and Gamma-valley effective masses (Fig. 7.17b) of the silicon nanowires are found to be widely deviating from their *bulk* counterparts and show strong dependence on NW dimensions. These (confinement) effective masses were implemented in the MCDS 3-D transport kernel in QuAD and used in the single-band (Γ-valley for [100] and [110], and X-valley for [111]) transport studies of silicon nanowires.

The effective potential method used in this work was validated using a fully quantum mechanical simulator, *nanoWIRE*, freely available on www.nanoHUB.org [80]. The *nanoWIRE* tool simulates the quantum mechanical size quantization in the inversion layer and *phase coherent* and *ballistic* transport properties in three-dimensional FET devices. The overall simulation framework consists of the mode-space effective mass non-equilibrium Green function (NEGF) equations solved self-consistently with Poisson's equation. The (ballistic) drain current vs. gate over-drive characteristic for a [100] $2 \times 2\,\text{nm}^2$ silicon nanowire was computed using both QuADS and the *nanoWIRE* tool and shows reasonable agreement (Fig. 7.18a). The electron distribution in the nanowire X–Z plane (Fig. 7.18b) depicts a charge set-back from the interface proper by almost 1.5 nm leading to a pronounced quantum capacitance.

To study the effect of unintentional single charge effects, a single negative charge (or trapped electron) was placed deterministically in three different locations (source-end, channel-center, drain-end) within the channel region of silicon nanowires having [100], [110], and [111] crystal directions and varying cross-sections. For each configuration, the ON-current was calculated for an applied bias set of $V_G = V_D = 0.8\,\text{V}$. A single negative charge in the channel region changes both the electrostatics and the carrier dynamics of the transistor under study through modifying (raising) the conduction band locally and reducing the carrier velocity (kinetic energy), respectively. Changes in the kinetic energy profile, in turn, affect the thermodynamics of the system leading to lesser inelastic carrier scattering. Figure 7.19a shows the 2-D conduction band profile of a [100] $3 \times 3\,\text{nm}^2$ silicon nanowire having a charge at the source end. Noticeable in this figure is the potential barrier, which sits on top of the conduction band and extends fairly into the whole channel cross-section blocking a significant portion of source electrons flowing into the channel region. The device operation is affected by this localized barrier from both

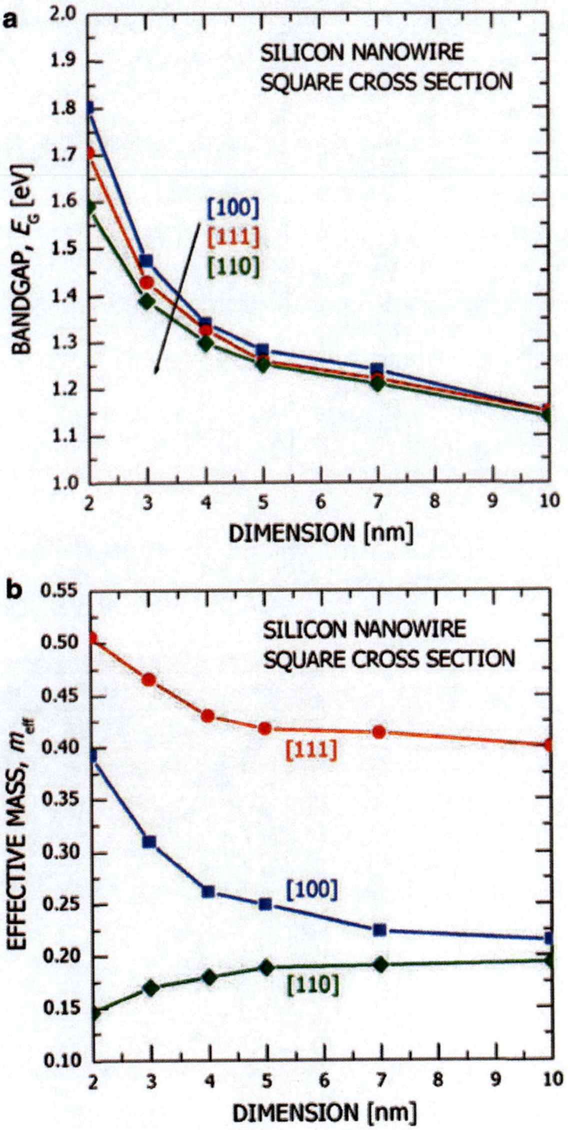

Fig. 7.17 (**a**) Energy bandgap and (**b**) Gamma-valley effective mass, as a function of the dimension of silicon nanowires

electrostatics (effective increase in channel doping and the threshold voltage) and dynamics (transport) points of view. The transport is affected through modulation of carrier velocity and energy (both potential and kinetic) characteristics as shown in Fig. 7.19b where the dip is due to the presence of a single impurity in the source-end of the channel region. The influence of a negative channel charge on the ON-state current and the percentage change in the ON-current for [100], [110], and [111] silicon nanowires with varying cross-sectional area is depicted in Fig. 7.20. Negative charge at the source-end affects the drain current most, while when located at the drain-end causes least shift in the device drive characteristics. In consistent with

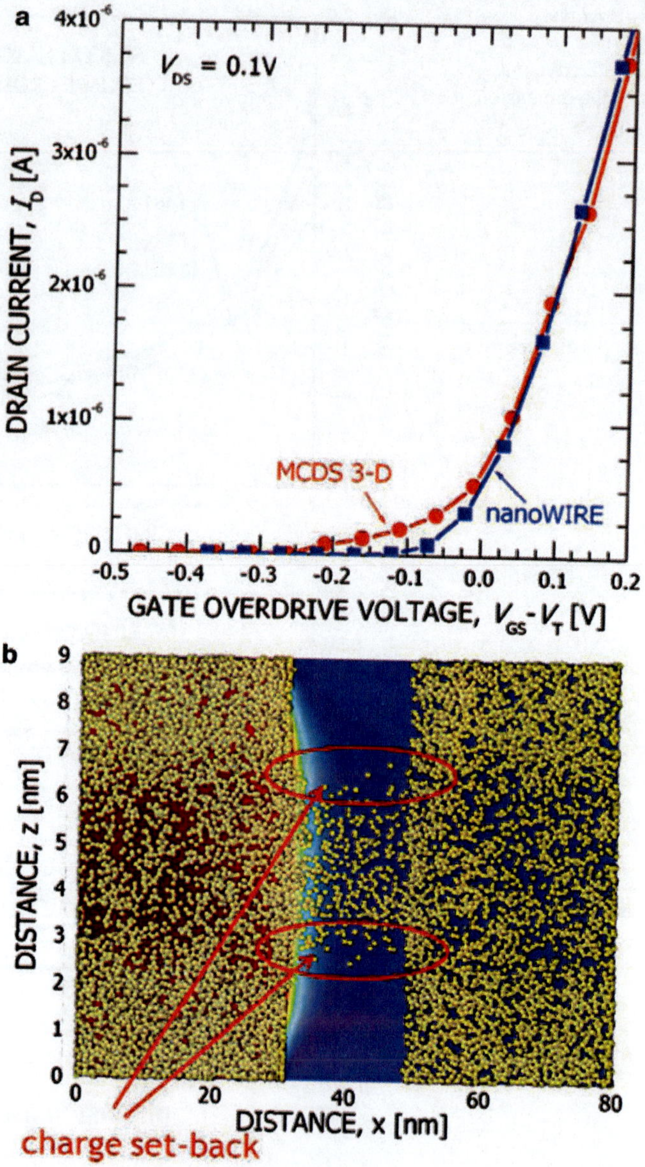

Fig. 7.18 (**a**) Comparison of QuADS and *nanoWIRE* simulations. (**b**) Electron (*dots*) distribution in the nanowire active region

the effective masses (Fig. 7.17b), silicon nanowire with [110] crystal orientation was found to deliver the maximum ON-current, while [111] delivering the least. Also, fluctuation in the ON-current is minimum in nanowires with [111] channel orientation and decreases in all three channel orientations as the cross-sectional area (size) of the nanowire increases.

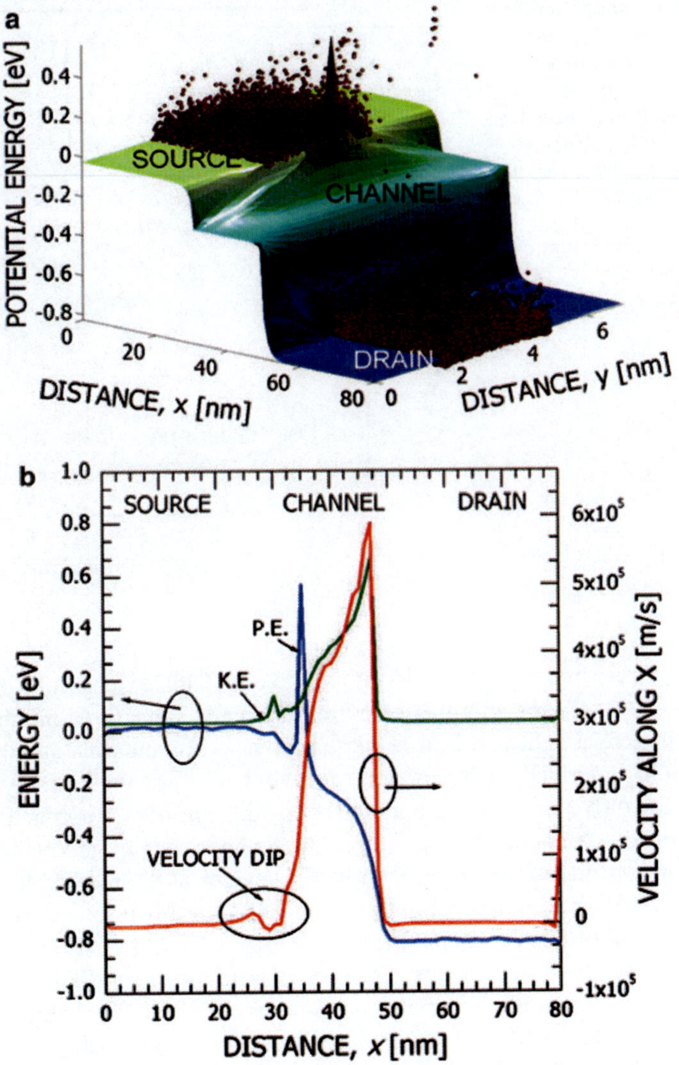

Fig. 7.19 (**a**) 2-D potential plot in the X–Z plane of the nanowire transistor showing the single charge induced barrier and the electrons. (**b**) Velocity and energy plots for $V_G = V_D = 0.8\,V$ when a single impurity is present at the source-end of the channel

4 Conclusion

Nanoscale devices are unique examples of systems where different branches of physics (molecular dynamics, quantum electronic structure, charge and phonon transport, statistical physics and thermodynamics, classical electrostatics, and optics) meet together spanning across different spatial and time scales. Nanodevice modeling, therefore, is a *multiscale and multiphysics* problem. To be relevant to device

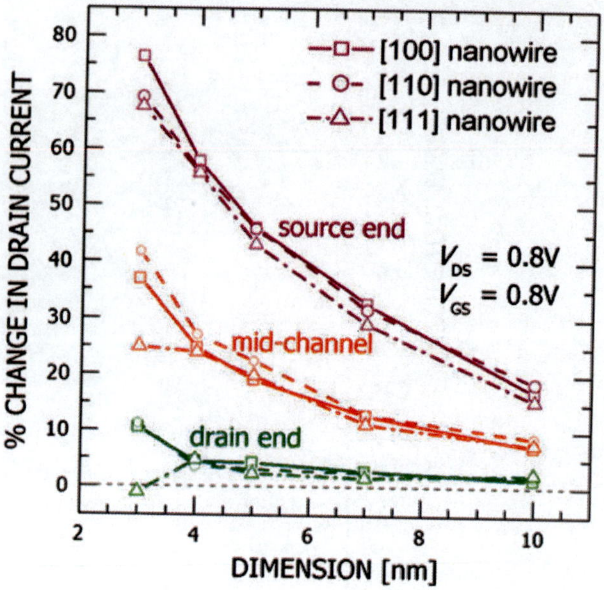

Fig. 7.20 Comparison of the ON-state current (I_D) and percentage change in the ON-current due to the presence of a single negative charge in [100], [110], and [111] silicon nanowires. $V_G = V_D = 0.8\,V$

designers, structures to be modeled must have realistic extent (millions of atoms) and represent bandgaps and masses extremely well. Since the continuum methods are clearly incapable of capturing essential physics at nanoscale and the best available ab initio materials science models (although offer greater accuracy) can scale to systems with only ~100 atoms, one must consider *empirical* approaches for modeling realistically-extended nanostructures. Also, the variety of geometries, materials, and doping configurations in semiconductor devices at the nanoscale suggests that a *general* nanoelectronic modeling tool is needed. In this paper we have described our on-going efforts to develop a multiscale Quantum Atomistic Device Simulator (QuADS) to address these needs. QuADS is primarily being built upon extended versions of three modules, namely, open source LAMMPS molecular dynamics code, the open source NEMO 3-D bandstructure tool, and the in-house MCDS 3-D transport kernel.

QuADS demonstrates the capability of modeling a large variety of relevant, realistically sized nanoelectronic devices. Atomistic simulations using the NEMO 3-D package in QuADS have been carried out to study the influence of internal electrostatic fields in Wurtzite InN/GaN quantum dots having three different geometries, namely, box, dome, and pyramid, all having a diameter/base length of 10.1 nm and a height of 5.6 nm. Atomistic strain and the resulting piezoelectric and pyroelectric potentials are found to be long-ranged and penetrating deep (~20 nm) inside both the substrate and the cap layers. This stresses the need for using *realistically-extended* substrate and cap layers containing at least three million atoms in the theoretical study of electronic structure of these reduced-dimensional QDs. In contrast to the interfacial symmetry, strain is found to have a general (uniform) tendency to orient the electronic wavefunctions along the [010] direction and further lowers the symmetry of the system under study. The induced piezoelectric and pyroelectric potentials are significantly large (tens of meV in some cases), opposing, and anisotropic in the QD planes. All four types of internal fields introduce a global shift and a band mixing in the energy spectrum, and lead to significant suppression and strong polarization anisotropy in the interband optical transitions. In case of Zinblende InAs/GaAs QDs, for the first time, 4 (four) different models for polarization have been implemented within the atomistic tight-binding description. In contrast to some recent small-scale numerical experiments, our calculations yield a non-vanishing and reasonably large *net* piezoelectric potential, which can be attributed to the fact that the potential from the linear term, as compared to the quadratic counterpart, penetrates *deeper* into the surrounding material matrix. Also, motivated by a number of recent experiments, we have numerically investigated the influence of single negative channel charges (trapped electrons) on the performance characteristics of gate-all-around [100], [110], and [111] silicon nanowire FETs having varying cross-sectional areas. It has been demonstrated that the device parameters related to both the electrostatics (density) and the carrier transport (velocity, mobility, energy) are modified due to the presence of a single channel charge. Simulation results indicate that unintentional channel charges located/induced at the source end of the device are most critical to the ON-current fluctuations. In $3 \times 3\,\text{nm}^2$ silicon nanowires, the maximum ON-current fluctuations were as high as $\sim$70%. This work also suggests that design optimization for RTS may involve the use of nanowires with different crystal orientations. All these QuADS calculations underline the importance to represent explicitly the atomistically resolved physical system containing millions of atoms with a physics based local orbital representation. The complexity of the system demands the use of well qualified, tuned, optimized algorithms and modern HPC platforms. The full version of QuADS will soon be available for device engineers, material scientists, educators, and students through the nanoHUB, powered by the NSF Teragrid. Tool documentation, tutorials, and case studies will be posted on nanoHUB as supplemental material.

Acknowledgment This work is supported by the ORAU/ORNL High-Performance Computing Grant 2009. Computational resources supported by the National Science Foundation under Grant No. 0855221 and the Rosen Center for Advanced Computing (RCAC) at Purdue University are also acknowledged. The development of the NEMO 3-D tool involved a large number of individuals at JPL and Purdue University, whose work has been cited. Shaikh Ahmed would like to thank Gerhard Klimeck at Purdue University and Dragica Vasileska at Arizona State University for many useful discussions.

References

1. S. M. Sze and G. May, *Fundamentals of Semiconductor Fabrication*, John Wiley and Sons Inc., 2003.
2. G. Moore, "*Progress in digital integrated electronics,*" *IEDM Tech. Digest*, pp. 11–13, 1975.
3. Semiconductor Industry Association (SIA) *International Technology Roadmap for Semiconductors 2009* (http://www.itrs.net/Links/2009ITRS/Home2009.htm).

4. Y. Wu et al. "Controlled growth and structures of molecular-scale silicon nanowires," *Nano Lett.*, vol. 4, pp. 433–436, 2004.
5. Y. Cui, X. Duan, J. Hu, and C. M. Lieber, "Doping and Electrical Transport in Silicon Nanowires," *J. Phys. Chem. B*, vol. 104, 5213, 2000.
6. Y. Cui, Y. Zhong, Z. Wang, D. Wang, C. M. Lieber, "High performance silicon nanowire field effect transistors," *Nano Lett.*, vol. 3, pp. 149–152, 2003.
7. P. Michler, A. Kiraz, C. Becher, W. V. Schoenfeld, P. M. Petroff, Lidong Zhang, E. Hu, A. Imamoglu, "A Quantum Dot Single-Photon Turnstile Device", *Science*, vol. 290, pp. 2282–2285, 2000.
8. Y. Arakawa, H. Sasaki, "Multidimensional quantum well laser and temperature dependence of its threshold current" *Appl. Phys. Lett.*, vol. 40, pp. 939, 1982.
9. E. Moreau, I. Robert, L. Manin, V. Thierry-Mieg, J. Gérard, I. Abram, "Quantum Cascade of Photons in Semiconductor Quantum Dots", *Phys. Rev. Lett.*, vol. 87, pp. 183601, 2001.
10. M. Maximov, Y. Shernyakov, A. Tsatsul'nikov, A. Lunev, A. Sakharov, V. Ustinov, A. Egorov, A. Zhukov, A. Kovsch, P. Kop'ev, L. Asryan, A. Alferov, N. Ledentsov, D. Bimberg, A. Kosogov, P. Werner, "High-power continuous-wave operation of a InGaAs/AlGaAs quantum dot laser", *J. Appl. Phys.*, vol. 83, pp. 5561, 1998.
11. B. Kane, "A Silicon-based Nuclear Spin Quantum Computer", *Nature*, vol. 393, pp. 133, 1998.
12. D. Loss, DP. DiVincenzo, "Quantum computation with quantum dots", *Phys. Rev. A*, vol. 57, pp. 120, 1998.
13. M. Friesen, P. Rugheimer, D. Savage, M. Lagally, D. van der Weide, R. Joynt, M. Eriksson, "Practical design and simulation of silicon-based quantum-dot qubits", *Phys. Rev. B*, vol. 67, 121301, 2003.
14. S. Ahmed, M. Usman, C. Heitzinger, R. Rahman, A. Schliwa, and G. Klimeck, "Atomistic Simulation of Non-Degeneracy and Optical Polarization Anisotropy in Zincblende Quantum Dots," *The 2nd Annual IEEE International Conference on Nano/Micro Engineered and Molecular Systems* (IEEE-NEMS), Jan 2007, Bangkok, Thailand.
15. A. J. Williamson, L. W. Wang, and Alex Zunger, "Theoretical interpretation of the experimental electronic structure of lens-shaped self-assembled InAs/GaAs quantum dots," *Phys. Rev. B*, vol. 62, 12963 – 12977, 2000.
16. Olga L. Lazarenkova, Paul von Allmen, Fabiano Oyafuso, Seungwon Lee, and Gerhard Klimeck, "Effect of anharmonicity of the strain energy on band offsets in semiconductor nanostructures", *Appl. Phys. Lett.* vol. 85, 4193, 2004.
17. Fabio Bernardini and Vincenzo Fiorentinia, "First-principles calculation of the piezoelectric tensor d of III–V nitrides," *Appl Phys. Lett.*, vol. 80, 22, pp. 4145–47, June 2002.
18. N. Baer, S. Schulz, S. Schumacher, P. Gartner, G. Czycholl, and F. Jahnke, "Optical properties of self-organized wurtzite InN/GaN quantum dots: A combined atomistic tight-binding and full configuration interaction calculation," *Appl Phys. Lett.*, vol. 87, 231114, 2005.
19. T. Saito, Y. Arakawa, "Electronic structure ofpiezoelectric In0:2Ga0:8N quantum dots in GaN calculated using a tight-binding method," *Physica E*, vol. 15, 169–181, 2002.
20. Momme Winkelnkemper, Andrei Schliwa, and Dieter Bimberg, "Interrelation of structural and electronic properties in $In_xGa_{1-x}N/GaN$ quantum dots using an eight-band $k{\bullet}p$ model," *Phys. Rev. B*, vol. 74, 155322, 2006.
21. G. Binnig, H. Rohrer, Ch. Gerber, and E. Weibel. *Phys. Rev. Lett.*, 50, 120–126, 1983.
22. Karl D Brommer, M Needels, B.E. Larson, and J.D. Joannopoulous., *Phys. Rev. Lett.*, vol. 68, 1355, 1992.
23. I.D. Parker, "Carrier tunneling and device characteristics in polymer light-emitting diodes," *Journal of Applied Physics*, 75, 3, 1656–1666, 1994.
24. Shaikh Ahmed, Neerav Kharche, Rajib Rahman, Muhammad Usman, Sunhee Lee, Hoon Ryu, Hansang Bae, Steve Clark, Benjamin Haley, Maxim Naumov, Faisal Saied, Marek Korkusinski, Rick Kennel, Michael Mclennan, Timothy B. Boykin, and Gerhard Klimeck, "Multimillion Atom Simulations with NEMO 3-D," In Meyers, Robert (Ed.) *Encyclopedia of Complexity and Systems Science*, 6, 5745–5783. Springer New York 2009.
25. http://www.silvaco.com/

26. APSYS User's Manual 2005, http://www.crosslight.com
27. http://www.synopsys.com/home.aspx
28. Simone Chiaria, Enrico Furno, Michele Goano, and Enrico Bellotti, "Design Criteria for Near-Ultraviolet GaN-Based Light-Emitting Dioded", *special issue of IEEE Transactions on Electron Devices on LEDs*, vol. 57, 1, pp. 60–70, January 2010.
29. C. Pryor, J. Kim, L.W. Wang, A. J. Williamson, and A. Zunger, "Comparison of two methods for describing the strain profiles in quantum dots", *J. Apl. Phys.*, vol 83, 2548, 1998.
30. Gabriel Bester and Alex Zunger, Cylindrically shaped zinc-blende semiconductor quantum dots do not have cylindrical symmetry: Atomistic symmetry, atomic relaxation, and piezoelectric effects, Phys. Rev. B 71 (2005) 045318.
31. J. M. Jancu, F. Bassani, F. Della Sala, R. Scholz, Transferable tight-binding parametrization for the group-III nitrides, Appl. Phys. Lett. 81 (2002) 4838.
32. G. Klimeck, S. Ahmed, N. Kharche, H. Bae, S. Clark, B. Haley, S. Lee, M. Naumov, H. Ryu, F. Saied, M. Prada, M. Korkusinski, and T. B. Boykin, Atomistic simulation of realistically-sized nanodevices using NEMO 3-D, *IEEE Trans. on Elect. Dev.* 54 (2007) 2079–2099.
33. S. Ahmed, S. Islam, and S. Mohammed, Electronic Structure of InN/GaN Quantum Dots: Multimillion Atom Tight-Binding Simulations, *IEEE Trans. on Elect. Dev.* 57 (2010) 164–173.
34. S. Datta, *Electronic Transport in Mesoscopic Systems*, Cambridge Studies in Semiconductor Physics and Microelectronic Engineering, 1995.
35. D. K. Ferry and S. M. Goodnick, *Transport in Nanostructures*, Cambridge University Press, 1997.
36. S. Datta, *Quantum Transport: Atom to Transistor*, Cambridge University Press, 2005.
37. E. Wigner, "On the quantum correction for thermodynamic equilibrium," *Phys. Rev.*, vol. 40, pp. 749–759, 1932.
38. P. Feynman and H. Kleinert, "Effective classical partition functions," *Phys. Rev. A*, vol. 34, pp. 5080–5084, 1986.
39. R. Lake, G. Klimeck, R.C. Bowen, and D. Jovanovic, *J. Appl. Phys.*, vol. 81, 7845, 1997.
40. A. Buin, A. Verma, A. Svizhenko and M. P. Anantram, "Enhancement of hole mobility in [110] Silicon Nanowires," *Nano Lett.*, vol. 8, p. 760—765, 2008.
41. Neophytos Neophytou, Shaikh Ahmed, Gerhard Klimeck, "Influence of vacancies on metallic nanotube transport performance", *Appl. Phys. Lett.*, vol. 90, 182119, 2007.
42. I. Knezevic, "Decoherence due to contacts in ballistic nanostructures," *Physical Review B*, vol. 77, 125301, 2008.
43. A. Svizhenko, M. P. Anantram, T. R. Govindan, B. Biegel and R. Venugopal, "Two Dimensional Quantum Mechanical Modeling of Nanotransistors," *J. Appl. Phys.*, vol. 91, p. 2343, 2002.
44. Ming-Shan Jeng, Ronggui Yang, David Song, Gang Chen, "Modeling the Thermal Conductivity and Phonon Transport in Nanoparticle Composites Using Monte Carlo Simulation," *Journal of Heat Transfer*, vol. 130, 2008.
45. D. Donadio, G. Galli, "Atomistic simulations of heat transport in silicon nanowires," *Phys. Rev. Lett.* 102, 195901, 13 May 2009.
46. G. Klimeck, F. Oyafuso, T. Boykin, R. Bowen, and P. von Allmen, "Development of a Nanoelectronic 3-D (NEMO 3-D) Simulator for Multimillion Atom Simulations and Its Application to Alloyed Quantum Dots," *Computer Modeling in Engineering and Science*, 3, pp. 601, 2002.
47. P. Keating, "Effect of Invariance Requirements on the Elastic Strain Energy of Crystals with Application to the Diamond Structure", *Phys. Rev.*, vol. 145, 1966.
48. Benjamin P. Haley, Sunhee Lee, Mathieu Luisier, Hoon Ryu, Faisal Saied, Steve Clark, Hansang Bae, and Gerhard Klimeck, "Advancing nanoelectronic device modeling through peta-scale computing and deployment on nanoHUB," *Journal of Physics: Conference Series*, vol. 180, 012075, 2009. Also, http://cobweb.ecn.purdue.edu/~gekco/nemo3D/index.html
49. http://www.abinit.org/
50. E. Bellet-Amalric, C. Adelmann, E. Sarigiannidou, J. L. Rouvière, G. Feuillet, E. Monroy, and B. Daudin., "Plastic strain relaxation of nitride heterostructures," *J. Appl. Phys.*, vol. 95, 1127, 2004.

51. J. G. Lozano, A. M. Sánchez, R. García, D. González, M. Herrera, N. D. Browning, S. Ruffenach, and O. Briot, "Configuration of the misfit dislocation networks in uncapped and capped InN quantum dots," *Appl. Phys. Lett.*, vol. 91, 071915, 2007.
52. http://lammps.sandia.gov/
53. S. Ahmed, C. Ringhofer, D. Vasileska, "Parameter-Free Effective Potential Method for Use in Particle-Based Device Simulations," *IEEE Trans. Nanotech.*, vol. 4, pp. 465–471, July 2005.
54. D. Vasileska and S. Ahmed, "Narrow-Width SOI Devices: The Role of Quantum Mechanical Size Quantization Effect and the Unintentional Doping on the Device Operation," *IEEE Trans. Elect. Dev.*, vol. 52, pp. 227–236, 2005.
55. M. Nedjalkov, S. Ahmed, and D. Vasileska, "A self-consistent event biasing scheme for statistical enhancement," *J. Comp. Elect.*, vol. 3, pp. 305–309, 2004.
56. P. Lugli, P. Bordone, L. Reggiani, M. Rieger, P. Kocevar, and S. M. Goodnick, "Monte Carlo Studies of Nonequilibrium Phonon Effects in Polar Semiconductors and Quantum Wells," *Phys. Rev. B*, vol. 39, pp. 7852—7875, 1989.
57. C. Jacoboni and L. Reggiani, 'The Monte Carlo Method for the Solution of Charge Transport in Semiconductors with Applications to Covalent Materials," *Rev. Modern Phys.*, vol. 55, pp. 645–705, 1983.
58. M. Fischetti, and S. Laux, "Monte Carlo study of electron transport in silicon inversion layers," *Phys. Rev. B*, vol. 48, pp. 2244–2274, 1993.
59. M. Lundstrom, *Fundamentals of Carrier Transport*, Cambridge University Press, 2000.
60. K. Tomizawa, *Numerical Simulation of Submicron Semiconductor Devices*, Artech House, Boston, 1993.
61. J. Bude, "Scattering mechanisms for semiconductor transport calculations," *Monte Carlo Device Simulation: Full Band and Beyond*, Kluwer Academic Publishers, pp. 27–66, 1991.
62. https://developer.nanohub.org/projects/rappture/
63. https://nanohub.org/resources/1092
64. FA Ponce and DP Bour, "Nitride-based semiconductors for blue and green light-emitting devices," *Nature*, 386, 351–359, 1997.
65. H. Morkoç, and S. N. Mohammad, "High-luminosity blue and blue-green gallium nitride light-emitting diodes," *Science*, vol. 267, pp. 51–55, 1995.
66. S. Ruffenach, B. Maleyre, O. Briot, B. Gil, "Growth of InN quantum dots by MOVPE," *physica status solidi (c)*, vol. 2, 826–832, 2005.
67. W. Ke, C. Fu, C. Chen, L. Lee, C. Ku, W. Chou, W.-H Chang, M. Lee, W. Chen, and W. Lin, "Photoluminescence properties of self-assembled InN dots embedded in GaN grown by metal organic vapor phase epitaxy," *Appl. Phys. Lett.*, vol. 88, 191913, 2006.
68. J. Kalden, C. Tessarek, K. Sebald, S. Figge, C. Kruse, D. Hommel, and J. Gutowski, "Electroluminescence from a single InGaN quantum dot in the green spectral region up to 150 K," *Nanotechnology*, vol. 21, 015204, 2010.
69. H. Wang, D. Jiang, J. Zhu, D. Zhao, Z. Liu, Y. Wang, S. Zhang, and Yang, H, "Kinetically controlled InN nucleation on GaN templates by metalorganic chemical vapour deposition," *J. Phys. D*, vol. 42, 145410, 2009.
70. X. A. Cao and S. D. Arthur, "High-power and reliable operation of vertical light-emitting diodes on bulk GaN," *Appl. Phys. Lett.*, vol. 85, 3971, 2004.
71. R. Stevenson, "The world's best gallium nitride," *IEEE Spectrum*, vol. 47, 40–45, 2010.
72. J. Bhattacharyya, S. Ghosh, M. R. Gokhale, B. M. Arora, H. Lu, and W. J. Schaff, "Polarized photoluminescence and absorption in A-plane InN films," *Appl. Phys. Lett.*, vol. 89, 151910, 2006.
73. P. Waltereit, O. Brandt, A. Trampert, H. T. Grahn, J. Menniger, M. Ramsteiner, M. Reiche, and K. H. Ploog, "Nitride semiconductors free of electrostatic fields for efficient white light-emitting diodes," *Nature*, vol. 406, pp. 865–868, 2000.
74. A. Jarjour, R. Taylor, R. Oliver, M. Kappers, C. Humphreys, and A. Tahraoui, "Electrically driven single InGaN/GaN quantum dot emission," *Appl. Phys. Lett.*, vol. 93, 233103, 2008.
75. M. Senes, K. Smith, T. Smeeton, S. Hooper, and J. Heffernan, "Strong carrier confinement in InGaN/GaN quantum dots grown by molecular beam epitaxy," *Phys. Rev. B*, vol. 75, 045314, 2007.

76. Gabriel Bester, Xifan Wu, David Vanderbilt, and Alex Zunger, "Importance of second-order piezoelectric effects in zincblende semiconductors," *Phys. Rev. Lett.*, vol. 96, 187602, 2006.
77. Gabriel Bester, Alex Zunger, Xifan Wu, and David Vanderbilt, "Effects of linear and nonlinear piezoelectricity on the electronic properties of InAs/GaAs quantum dots," *Phys. Rev. B*, vol. 74, 081305, 2006.
78. C. Wei, Y. Jiang, Y. Z. Xiong, X. Zhou, N. Singh, S. C. Rustagi, G. Q. Lo, and D. Lee Kwong, "Impact of Gate Electrodes on 1/f Noise of Gate-All-Around Silicon Nanowire Transistors," *IEEE Elect. Dev. Lett.*, vol. 30, No. 10, October 2009.
79. Z. Jing, R. Wang, R. Huang, Y. Tian, L. Zhang, D. W. Kim, D. Park, and Y. Wang, "Investigation of low-frequency noise in silicon nanowire MOSFETs," *IEEE Elect. Dev. Lett.*, vol. 30, no. 1, pp. 57–60, Jan. 2009.
80. *nanowire* simulator at http://nanohub.org/tools/nanowire/. Accessed on March 21, 2010.